The Mycota

Edited by
K. Esser

The Mycota

I *Growth, Differentiation and Sexuality*
1st edition ed. by J.G.H. Wessels and F. Meinhardt
2nd edition ed. by U. Kües and R. Fischer

II *Genetics and Biotechnology*
Ed. by U. Kück

III *Biochemistry and Molecular Biology*
Ed. by R. Brambl and G. Marzluf

IV *Environmental and Microbial Relationships*
Ed. by D. Wicklow and B. Söderström

V *Plant Relationships*
Ed. by G. Carroll and P. Tudzynski

VI *Human and Animal Relationships*
Ed. by D.H. Howard and J.D. Miller

VII *Systematics and Evolution*
Ed. by D.J. McLaughlin, E.G. McLaughlin, and P.A. Lemke†

VIII *Biology of the Fungal Cell*
Ed. by R.J. Howard and N.A.R. Gow

IX *Fungal Associations*
Ed. by B. Hock

X *Industrial Applications*
Ed. by H.D. Osiewacz

XI *Agricultural Applications*
Ed. by F. Kempken

XII *Human Fungal Pathogens*
Ed. by J.E. Domer and G.S. Kobayashi

XIII *Fungal Genomics*
Ed. by A.J.P. Brown

The Mycota

A Comprehensive Treatise
on Fungi as Experimental Systems
for Basic and Applied Research

Edited by K. Esser

Fungal Genomics

Volume Editor:
A.J.P. Brown

With 37 Figures, 6 in Color, and 16 Tables

Springer

Series Editor

Professor Dr. Dr. h.c. mult. Karl Esser
Allgemeine Botanik
Ruhr-Universität
44780 Bochum, Germany

Tel.: +49 (234)32-22211
Fax.: +49 (234)32-14211
e-mail: Karl.Esser@rub.de

Volume Editor

Professor Dr. Alistair J.P. Brown
Aberdeen Fungal Group
School of Medical Sciences
Institute of Medical Sciences
University of Aberdeen
Foresterhill
Aberdeen AB25 2ZD, UK

Tel.: +44 (1224) 555883
Fax: +44 (1224) 555844
e-mail: al.brown@abdn.ac.uk

Library of Congress Control Number: 2005928810

ISBN-10 3-540-25594-X Springer Berlin Heidelberg New York
ISBN-13 978-3-540-25594-9 Springer Berlin Heidelberg New York

Springer is a part of Springer Science+Business Media
springeronline.com

Printed in Germany

Editor: Dr. Dieter Czeschlik, Heidelberg, Germany
Desk editor: Dr. Andrea Schlitzberger, Heidelberg, Germany
Cover design: Erich Kirchner, Heidelberg, Germany
Production: Katja Röser, Leipzig, Germany
Typesetting: LE-TEX Jelonek, Schmidt & Vöckler GbR, Leipzig, Germany

Printed on acid-free paper 31/3152 5 4 3 2 1 0

Karl Esser

(born 1924) is retired Professor of General Botany and Director of the Botanical Garden at the Ruhr-Universität Bochum (Germany). His scientific work focused on basic research in classical and molecular genetics in relation to practical application. His studies were carried out mostly on fungi. Together with his collaborators he was the first to detect plasmids in higher fungi. This has led to the integration of fungal genetics in biotechnology. His scientific work was distinguished by many national and international honors, especially three honorary doctoral degrees.

Alistair J. P. Brown

(born 1955) is Professor in Molecular and Cell Biology at the University of Aberdeen (UK). His postdoctoral work at MIT was followed by faculty positions in the Genetics Department at Glasgow University (UK), and then in the School of Medical Sciences at Aberdeen University. His work has focussed on yeast gene regulation and the control of *Candida albicans* virulence, morphogenesis and stress responses. His group routinely combines genomic technologies with molecular and cellular approaches. Along with his colleagues Neil Gow and Frank Odds, he leads the *Aberdeen Fungal Group*. He also leads Proteomics and Transcript Profiling facilities at Aberdeen University.

Series Preface

Mycology, the study of fungi, originated as a subdiscipline of botany and was a descriptive discipline, largely neglected as an experimental science until the early years of this century. A seminal paper by Blakeslee in 1904 provided evidence for selfincompatibility, termed "heterothallism", and stimulated interest in studies related to the control of sexual reproduction in fungi by mating-type specificities. Soon to follow was the demonstration that sexually reproducing fungi exhibit Mendelian inheritance and that it was possible to conduct formal genetic analysis with fungi. The names Burgeff, Kniep and Lindegren are all associated with this early period of fungal genetics research.

These studies and the discovery of penicillin by Fleming, who shared a Nobel Prize in 1945, provided further impetus for experimental research with fungi. Thus began a period of interest in mutation induction and analysis of mutants for biochemical traits. Such fundamental research, conducted largely with *Neurospora crassa*, led to the one gene: one enzyme hypothesis and to a second Nobel Prize for fungal research awarded to Beadle and Tatum in 1958. Fundamental research in biochemical genetics was extended to other fungi, especially to *Saccharomyces cerevisiae*, and by the mid-1960s fungal systems were much favored for studies in eukaryotic molecular biology and were soon able to compete with bacterial systems in the molecular arena.

The experimental achievements in research on the genetics and molecular biology of fungi have benefited more generally studies in the related fields of fungal biochemistry, plant pathology, medical mycology, and systematics. Today, there is much interest in the genetic manipulation of fungi for applied research. This current interest in biotechnical genetics has been augmented by the development of DNAmediated transformation systems in fungi and by an understanding of gene expression and regulation at the molecular level. Applied research initiatives involving fungi extend broadly to areas of interest not only to industry but to agricultural and environmental sciences as well.

It is this burgeoning interest in fungi as experimental systems for applied as well as basic research that has prompted publication of this series of books under the title *The Mycota*. This title knowingly relegates fungi into a separate realm, distinct from that of either plants, animals, or protozoa. For consistency throughout this Series of Volumes the names adopted for major groups of fungi (representative genera in parentheses) are as follows:

Pseudomycota

Division: Oomycota (*Achlya*, *Phytophthora*, *Pythium*)
Division: Hyphochytriomycota

Eumycota

Division: Chytridiomycota (*Allomyces*)
Division: Zygomycota (*Mucor*, *Phycomyces*, *Blakeslea*)
Division: Dikaryomycota

Subdivision:	Ascomycotina
Class:	Saccharomycetes (*Saccharomyces, Schizosaccharomyces*)
Class:	Ascomycetes (*Neurospora, Podospora, Aspergillus*)
Subdivision:	Basidiomycotina
Class:	Heterobasidiomycetes (*Ustilago, Tremella*)
Class:	Homobasidiomycetes (*Schizophyllum, Coprinus*)

We have made the decision to exclude from *The Mycota* the slime molds which, although they have traditional and strong ties to mycology, truly represent nonfungal forms insofar as they ingest nutrients by phagocytosis, lack a cell wall during the assimilative phase, and clearly show affinities with certain protozoan taxa.

The Series throughout will address three basic questions: what are the fungi, what do they do, and what is their relevance to human affairs? Such a focused and comprehensive treatment of the fungi is long overdue in the opinion of the editors.

A volume devoted to systematics would ordinarily have been the first to appear in this Series. However, the scope of such a volume, coupled with the need to give serious and sustained consideration to any reclassification of major fungal groups, has delayed early publication. We wish, however, to provide a preamble on the nature of fungi, to acquaint readers who are unfamiliar with fungi with certain characteristics that are representative of these organisms and which make them attractive subjects for experimentation.

The fungi represent a heterogeneous assemblage of eukaryotic microorganisms. Fungal metabolism is characteristically heterotrophic or assimilative for organic carbon and some nonelemental source of nitrogen. Fungal cells characteristically imbibe or absorb, rather than ingest, nutrients and they have rigid cell walls. The vast majority of fungi are haploid organisms reproducing either sexually or asexually through spores. The spore forms and details on their method of production have been used to delineate most fungal taxa. Although there is a multitude of spore forms, fungal spores are basically only of two types: (i) asexual spores are formed following mitosis (mitospores) and culminate vegetative growth, and (ii) sexual spores are formed following meiosis (meiospores) and are borne in or upon specialized generative structures, the latter frequently clustered in a fruit body. The vegetative forms of fungi are either unicellular, yeasts are an example, or hyphal; the latter may be branched to form an extensive mycelium.

Regardless of these details, it is the accessibility of spores, especially the direct recovery of meiospores coupled with extended vegetative haploidy, that have made fungi especially attractive as objects for experimental research.

The ability of fungi, especially the saprobic fungi, to absorb and grow on rather simple and defined substrates and to convert these substances, not only into essential metabolites but into important secondary metabolites, is also noteworthy. The metabolic capacities of fungi have attracted much interest in natural products chemistry and in the production of antibiotics and other bioactive compounds. Fungi, especially yeasts, are important in fermentation processes. Other fungi are important in the production of enzymes, citric acid and other organic compounds as well as in the fermentation of foods.

Fungi have invaded every conceivable ecological niche. Saprobic forms abound, especially in the decay of organic debris. Pathogenic forms exist with both plant and animal hosts. Fungi even grow on other fungi. They are found in aquatic as well as soil environments, and their spores may pollute the air. Some are edible; others are poisonous. Many are variously associated with plants as copartners in the formation of lichens and mycorrhizae, as symbiotic endophytes or as overt pathogens. Association with animal systems varies; examples include the predaceous fungi that trap nematodes, the microfungi that grow in the anaerobic environment of the rumen, the many insect-associated fungi and the medically important pathogens afflicting humans. Yes, fungi are ubiquitous and important.

There are many fungi, conservative estimates are in the order of 100,000 species, and there are many ways to study them, from descriptive accounts of organisms found in nature to laboratory experimentation at the cellular and molecular level. All such studies expand our knowledge of fungi and of fungal processes and improve our ability to utilize and to control fungi for the benefit of humankind.

We have invited leading research specialists in the field of mycology to contribute to this Series. We are especially indebted and grateful for the initiative and leadership shown by the Volume Editors in selecting topics and assembling the experts. We have all been a bit ambitious in producing these Volumes on a timely basis and therein lies the possibility of mistakes and oversights in this first edition. We encourage the readership to draw our attention to any error, omission or inconsistency in this Series in order that improvements can be made in any subsequent edition.

Finally, we wish to acknowledge the willingness of Springer-Verlag to host this project, which is envisioned to require more than 5 years of effort and the publication of at least nine Volumes.

Bochum, Germany
Auburn, AL, USA
April 1994

Karl Esser
Paul A. Lemke
Series Editors

Addendum to the Series Preface

In early 1989, encouraged by Dieter Czeschlik, Springer-Verlag, Paul A. Lemke and I began to plan *The Mycota*. The first volume was released in 1994, 12 volumes followed in the subsequent years. Unfortunately, after a long and serious illness, Paul A. Lemke died in November 1995. Thus, it was my responsibility to proceed with the continuation of this series, which was supported by Joan W. Bennett for Volumes X–XII.

The series was evidently accepted by the scientific community, because first volumes are out of print. Therefore, Springer-Verlag has decided to publish some of the completely revised and updated new editions of Volumes I, II, III, IV, VI, and VIII. I am glad that most of the volume editors and authors have agreed to join our project again. I would like to take this opportunity to thank Dieter Czeschlik, his colleague, Andrea Schlitzberger, and Springer-Verlag for their help in realizing this enterprise and for their excellent cooperation for many years.

Bochum, Germany
July 2005

Karl Esser

Volume Preface

The Fungal Genomics era is gathering pace, and this is an enormously exciting time for mycologists. Having trained during an era when whole PhD theses were devoted to the sequencing of a single gene, it is incredibly exciting to run research programmes in the current era, when a fungal genome sequence can be generated in a matter of weeks. Genome sequences for a whole range of fungal species have been made public in the last few years, including those for *Aspergillus fumigatus*, *Candida glabrata*, *Cryptococcus neoformans*, *Debaryomyces hansenii*, *Kluyveromyces lactis*, *Magnaporthe grisea*, *Neurospora crassa* and *Yarrowia lipolytica*. Furthermore, many more genome sequences (for other *Candida*, *Coccidioides*, *Histoplasma*, *Mycosphaerella*, *Pneumocystis* and *Saccharomyces* species, for example) will be released in the near future and, in all probability, the generation of additional fungal genome sequences will continue for some time at a rate of about one every 60 days. These genome sequences are providing a wealth of invaluable data about the evolution, life cycles, cell biology, and virulence of fungi.

Genomic technologies have developed to differing extents, depending upon the fungal species one is interested in. The genome sequences for some fungal species are only just becoming available. For these species, genomic analyses are in the very early stages of development. The initial process of annotating genes within a genome sequence is incomplete for many fungi, and the generation of microarrays for transcript profiling analysis is still some way off. In contrast, a decade has passed since the *Saccharomyces cerevisiae* genome sequence was first made public. This was one of the first genome sequences available for any cellular organism and, thanks to the sophisticated molecular toolbox available for bakers' yeast, this model organism is now leading the way in many areas of functional genomics and systems biology. (Nearly) complete sets of knockout, conditional and epitope tagged mutants are available for the genome-wide analyses of gene function in yeast. Researchers are spoilt for choice with respect to the type and format of yeast microarray which is available – oligonucleotide or gene arrays for transcript profiling, or intergenic arrays for genome-wide chromatin immunoprecipitation studies. Protein microarrays are now available for *S. cerevisiae*, and high-throughput yeast proteomics is now being exploited by numerous research groups. Massively parallel analyses of protein–protein interactions in yeast were published several years ago. Metabolomics is already being applied to system-wide analyses of yeast physiology. Therefore, for many research groups, the problem has shifted from the generation of data to the accurate interpretation and incisive exploitation of massive datasets. Thankfully, increasingly sophisticated software tools are being developed for the integration of genetic, transcriptomic, proteomic, interactomic and metabolomic datasets. As a result, genomics is now revealing important global perspectives of fungal cell biology which were not obvious using standard reductionist approaches. There were two main challenges in editing this volume on Fungal Genomics. First, it was impossible to provide comprehensive coverage of this immense field. Second, the speed of development of Fungal Genomics precluded the publication of a text which is bang up to date. New developments do not stop for editors! Therefore, this volume does not attempt to cover every topic or to include every last-minute development. (I apologise to those whose favourite topic is not covered.) Rather, my aims have been twofold. In the 13 chapters of this 13th volume of The Mycota, my first aim has been to illustrate the current impact and potential future impact of genomics in different fungal species – in those where genomics is mature, and others where genomics is relatively immature. Second, I wanted to show how genomics is being applied to a diverse range of interesting questions in mycological research. The chapters illustrate fundamental principles of fungal genomics which are universally applicable. Mycota XIII is divided into three sections, the first of which addresses *fungal systems biology and evolution*. The volume starts with a chapter on *Saccharomyces cerevisiae*, as befits its status as the pre-eminent model organism in the genomics era. Also, this chapter addresses the metabolomics and systems biology of bakers' yeast, thereby immediately reminding the reader that there is much more to genomics than genome sequencing and transcript profiling. The other chapters in this section illustrate the huge impact of genome sequencing upon our understanding of fungal evolution, and the exploitation of novel bioinformatic approaches in the identification of genes associated with fungal virulence.

The second section, on *fungal rhythms and responses*, addresses some of the most topical issues in fungal biology – circadian rhythms, programmed cell death, stress responses, secretion and polarised growth, and cellular morphogenesis. These chapters cover a range of fungal species, including filamentous fungi and yeasts. As well as describing the massive impact of fungal genomics in these species, they highlight areas where genomics has the potential to significantly accelerate our research efforts.

The last section focuses on *fungal pathogenesis*, an area of great medical significance. Chapters in this section discuss how both proteomic and transcriptomic approaches have provided, and are helping to provide, important new insights into the pathobiology of some of the major fungal pathogens of humans.

I am very grateful to the authors for their outstanding contributions to Mycota XIII. All are internationally renowned in their chosen fields, and all are extremely busy people. I appreciate their willingness to commit valuable time to this rewarding and interesting project.

Aberdeen, UK
July 2005

ALISTAIR J.P. BROWN
Volume Editor

Contents

List of Contributors

SARAH AHMAD School of Biological and Chemical Sciences, University of Exeter, Washington Singer Laboratories, Perry Road, Exeter EX4 4QG, UK

DAVID B. ARCHER (e-mail: david.archer@nottingham.ac.uk, Tel.: +44-115-9513313) School of Biology, University of Nottingham, University Park, Nottingham NG7 2RD, UK

OLIVER BADER Robert Koch Institut, NG4, Nordufer 20, 13353 Berlin, Germany

JÜRG BÄHLER (e-mail: jurg@sanger.ac.uk, Tel.: +44-1223-494861) The Wellcome Trust Sanger Institute, Hinxton, Cambridge CB10 1SA, UK

MADHUMITA C. BAROOAH School of Biological and Chemical Sciences, University of Exeter, Washington Singer Laboratories, Perry Road, Exeter EX4 4QG, UK

ALISTAIR J.P. BROWN (e-mail: al.brown@abdn.ac.uk, Tel.: +44-1224-555883, Fax: +44-1224-555844) Aberdeen Fungal Group, School of Medical Sciences, Institute of Medical Sciences, University of Aberdeen, Foresterhill, Aberdeen AB25 2ZD, UK

JUAN I. CASTRILLO (e-mail: juan.i.castrillo@man.ac.uk) Faculty of Life Sciences, Michael Smith Building, University of Manchester, Oxford Road, Manchester M13 9PT, UK

JAY C. DUNLAP Departments of Genetics and Biochemistry, 704 Remsen, Dartmouth Medical School, Hanover, NH 03755, USA

CHANTAL FRADIN Robert Koch Institut, NG4, Nordufer 20, 13353 Berlin, Germany

KEN HAYNES (e-mail: k.haynes@imperial.ac.uk, Tel.: +44-20-83831245, Fax: +44-20-83833394) Department of Infectious Diseases, Imperial College London, Du Cane Road, London W12 0NN, UK

BERNHARD HUBE (e-mail: HubeB@rki.de, Tel.: +49-1888-7542116, Fax: +49-30-45472328) Robert Koch Institut, NG4, Nordufer 20, 13353 Berlin, Germany

JENNIFER K. LODGE (e-mail: lodgejk@slu.edu, Tel.: +1-314-5778143, Fax: +1-314-5778156) Edward A. Doisy Department of Biochemistry and Molecular Biology, Saint Louis University School of Medicine, 1402 S. Grand Blvd., St. Louis, MS 63104, USA

JENNIFER J. LOROS (e-mail: jennifer.loros@dartmouth.edu, Tel.: +1-603-6501154, Fax: +1-603-6501128) Departments of Biochemistry and Genetics, 704 Remsen, Dartmouth Medical School, Hanover, NH 03755, USA

LOUIS J. MONTCALM Department of Genetics, Smurfit Institute, University of Dublin, Trinity College, Dublin 2, Ireland

CAROL A. MUNRO (e-mail: c.a.munro@abdn.ac.uk, Tel.: +44-1224-555882, Fax: +44-1224-555844) Aberdeen Fungal Group, School of Medical Sciences, Institute of Medical Sciences, Foresterhill, University of Aberdeen, Aberdeen AB25 2ZD, UK

ANDRÉ NANTEL (e-mail: andre@bri.nrc.ca) Biotechnology Research Institute, National Research Council of Canada, 6100 Royalmount Avenue, Montreal, Quebec H4P 2R2, Canada

REX T. NELSON Edward A. Doisy Department of Biochemistry and Molecular Biology, Saint Louis University School of Medicine, 1402 S. Grand Blvd., St. Louis, MS 63104, USA; present address: G329 Agronomy Hall, Iowa State University, Ames, IA 50011, USA

STEPHEN G. OLIVER (e-mail: steve.oliver@manchester.ac.uk, Tel.: +44-161-2751578; Fax: +44-161-2755082) Faculty of Life Sciences, Michael Smith Building, University of Manchester, Oxford Road, Manchester M13 9PT, UK

MARK RAMSDALE (e-mail: m.ramsdale@abdn.ac.uk, Tel.: +44-1224-555882, Fax: +44-1224-555844) Aberdeen Fungal Group, School of Medical Sciences, Institute of Medical Sciences, University of Aberdeen, Foresterhill, Aberdeen AB25 2ZD, UK

ANITA SIL (e-mail: sil@cgl.ucsf.edu, Tel.: +1-415-5021805, Fax: +1-415-476820) Department of Microbiology and Immunology, University of California, San Francisco, 513 Parnassus, P.O. Box 0414, San Francisco, CA 94143-0414, USA

DARREN M. SOANES School of Biological and Chemical Sciences, University of Exeter, Washington Singer Laboratories, Perry Road, Exeter EX4 4QG, UK

NICHOLAS J. TALBOT (e-mail: n.j.talbot@exeter.ac.uk, Tel.: +44-1392-264673) School of Biological and Chemical Sciences, University of Exeter, Washington Singer Laboratories, Perry Road, Exeter EX4 4QG, UK

GEOFFREY TURNER (e-mail: g.turner@shef.ac.uk, Tel.: +44-114-2226211) Department of Molecular Biology and Biotechnology, University of Sheffield, Firth Court, Western bank, Sheffield S10 2TN, UK

MALCOLM WHITEWAY (e-mail: Malcolm.Whiteway@cnrc-nrc.gc.ca, Tel.: +1-514-4966146, Fax: +1-514-4966213) Biotechnology Research Institute, National Research Council of Canada, 6100 Royalmount Avenue, Montreal, Quebec H4P 2R2, Canada

BRIAN T. WILHELM The Wellcome Trust Sanger Institute, Hinxton, Cambridge CB10 1SA, UK

KEN WOLFE (e-mail: khwolfe@tcd.ie, Tel.: +353-1-6081253, Fax: +353-1-6798588) Department of Genetics, Smurfit Institute, University of Dublin, Trinity College, Dublin 2, Ireland

Biocemistry and Molecular Genetics

1 Metabolomics and Systems Biology in *Saccharomyces cerevisiae*

J.I. Castrillo[1], S.G. Oliver[1]

CONTENTS

I. Introduction

The genomic revolution has been characterized by the generation of huge amounts of information in the form of genome sequences, and of their direct application in comprehensive molecular studies and comparative genomics strategies (e.g. Goffeau et al. 1996, 1997; The International Human Genome Mapping Consortium 2001a, 2001b; von Mering et al. 2002; Gavin and Superti-Furga 2003). In the post-genomic era, the explosion of new technologies, whose exploitation is becoming progressively more refined, is facilitating the study of biological systems on a genome-wide scale. These studies are being performed at different functional genomic levels, including the genome, transcriptome, proteome and metabolome, and in a progressively more integrative way (Oliver 1997, 2002; Oliver et al. 1998; ter Linde et al. 1999; Delneri et al. 2001; Castrillo and Oliver 2004 and references therein).

At the same time that these techniques are being applied and refined, the complexity of biological systems is being rediscovered. Living things are made up of thousands of components (genes, transcripts, proteins and metabolites), which are subject to modification by post-transcriptional and post-translational mechanisms. These components participate in anabolic, catabolic and regulatory networks in response to environmental and developmental signals, many of which are still to be elucidated (Castrillo and Oliver 2004 and references therein). In this context, the utilization of well-defined model systems under controlled conditions is of central importance in the drive towards an integrative systems biology perspective of the cell.

The purpose of this chapter is to present an up-to-date view of the functional genomics of *Saccharomyces cerevisiae* and its growing potential as a model system for systems biology studies. The importance of including metabolomics as part of this integrative approach to the study of the eukaryotic cell as a biological system is emphasised.

II. *Saccharomyces cerevisiae*: A Model Eukaryote and a Reference System in Biology

Saccharomyces cerevisiae is a species of budding yeast, a group of unicellular fungi belonging to the phylum *Ascomycetes*. *S. cerevisiae* is being used as a model eukaryote in biology because the basic mechanisms of DNA replication, chromosomal recombination, cell division, gene expression, and metabolism are generally conserved between yeast and higher eukaryotes (i.e. mammals; Rose and Harrison 1987–1995; Sherman 1998, 2002; Castrillo and Oliver 2004).

[1] Faculty of Life Sciences, Michael Smith Building, The University of Manchester, Oxford Road, Manchester M13 9PT, UK

The Mycota XIII
Fungal Genomics
Alistair J.P. Brown (Ed.)

Among the properties which make *S. cerevisiae* a particularly suitable organism for biological studies are its Generally Regarded As Safe (GRAS) status, rapid growth, well-dispersed cells, simple methods of cultivation under controlled conditions, ease of replica plating and mutant isolation. Moreover, yeasts represent a well-defined genetic system with facile techniques of genetic manipulation (Brown and Tuite 1998; Sherman 1998, 2002; Castrillo and Oliver 2004).

The favourable characteristics of this yeast, together with its easy accessibility due to its economic importance in beer- and bread-making, make it a cheap source for biochemical studies. Thus, many metabolites and enzymes of central metabolic processes (i.e. Embden-Meyerhoff pathway, tricarboxylic acid cycle, TCA), and the first complete map of central metabolic pathways were first unveiled in *S. cerevisiae*, which is used as a touchstone model for the study of the eukaryotic cell (Lehninger 1975; Rose and Harrison 1987–1995; Fell 1997; Alberts et al. 2002; Castrillo and Oliver 2004 and references therein). As a result of all this, a wide knowledge of the genetics, biochemistry and physiology of this yeast is presently available (Rose and Harrison 1987–1995; Brown and Tuite 1998; Sherman 1998, 2002; Burke et al. 2000; Guthrie and Fink 2002a, 2002b, 2004).

At the same time as being considered a platform for basic studies, *S. cerevisiae* has attracted considerable interest from the early days of microbiology as a 'biological system', capable of performing specific biotransformations of interest to the fermentation industry (Pasteur 1857; Rose and Harrison 1993; Olson and Nielsen 2000; Schwartz 2001; Ton and Rao 2004). As a consequence of this, *S. cerevisiae* cultures have been subjected to a number of modelling strategies directed towards the representation of different metabolic and cell biological processes. These were simple models at first, with limited information on the metabolism and internal regulatory mechanisms, i.e. basic unstructured models, in which cell growth was modelled as an autocatalytic process, and the specific rates of individual reactions were based on kinetic models (e.g. Monod; cf. Bailey and Ollis 1986; Sinclair and Cantero 1990). As a more complete knowledge of the metabolic pathways and control mechanisms became available, more structured models and comprehensive modelling strategies (i.e. metabolic steady-state flux models including information on yeast central metabolic pathways and cybernetic models) could be applied (Bailey and Ollis 1986; Castrillo and Ugalde 1994 and references therein; Cortassa and Aon 1994; Giuseppin and van Riel 2000; Lei et al. 2001). At this point, attempts to increase the flux through specific pathways by metabolic engineering techniques (Bailey 1991; Stephanopoulos and Vallino 1991) met with only limited success, revealing our lack of real understanding of the dynamics of metabolism in *S. cerevisiae*. Such an outcome was anticipated (at least, by some) as a direct consequence of the application of the Metabolic Control Analysis (MCA) theory, a fundamental framework in quantitative modelling and metabolic control (Kacser and Burns 1973; Kacser 1995; Fell 1997; see also last section).

If *S. cerevisiae* is to fulfil its potential as a model eukaryote, and as a reference 'biological system' at the cellular level, the progressive incorporation of functional genomic information at the different levels (i.e. genome, transcriptome, proteome and metabolome) into mathematical models representative of the global behaviour of the cell will be required (Kitano 2002; Ideker 2004a). *S. cerevisiae* was the first eukaryotic organism for which the complete genome was sequenced (Goffeau et al. 1996, 1997). Hence, it is at the forefront of the post-genomic era (Castrillo and Oliver 2004; see next section).

III. Functional Genomics of *S. cerevisiae*: State of the Art

A. Functional Genomics: Levels of Regulation

A schematic representation of the eukaryotic cell as a global system, with the different levels of functional genomics (genome, transcriptome, proteome and metabolome; Oliver 1997, 2002; Oliver et al. 1998; Delneri et al. 2001), their localization and interactions, main regulatory circuits and relationships is presented in Fig. 1.1. Our current view of the eukaryotic cell is that of a system characterized by a coordinate integration of the different functional genomic levels and individual networks, in direct relation with the environment. This system is intrinsically complex and involves the integration of regulatory mechanisms at the genomic, transcriptional, post-transcriptional, post-translational and metabolic

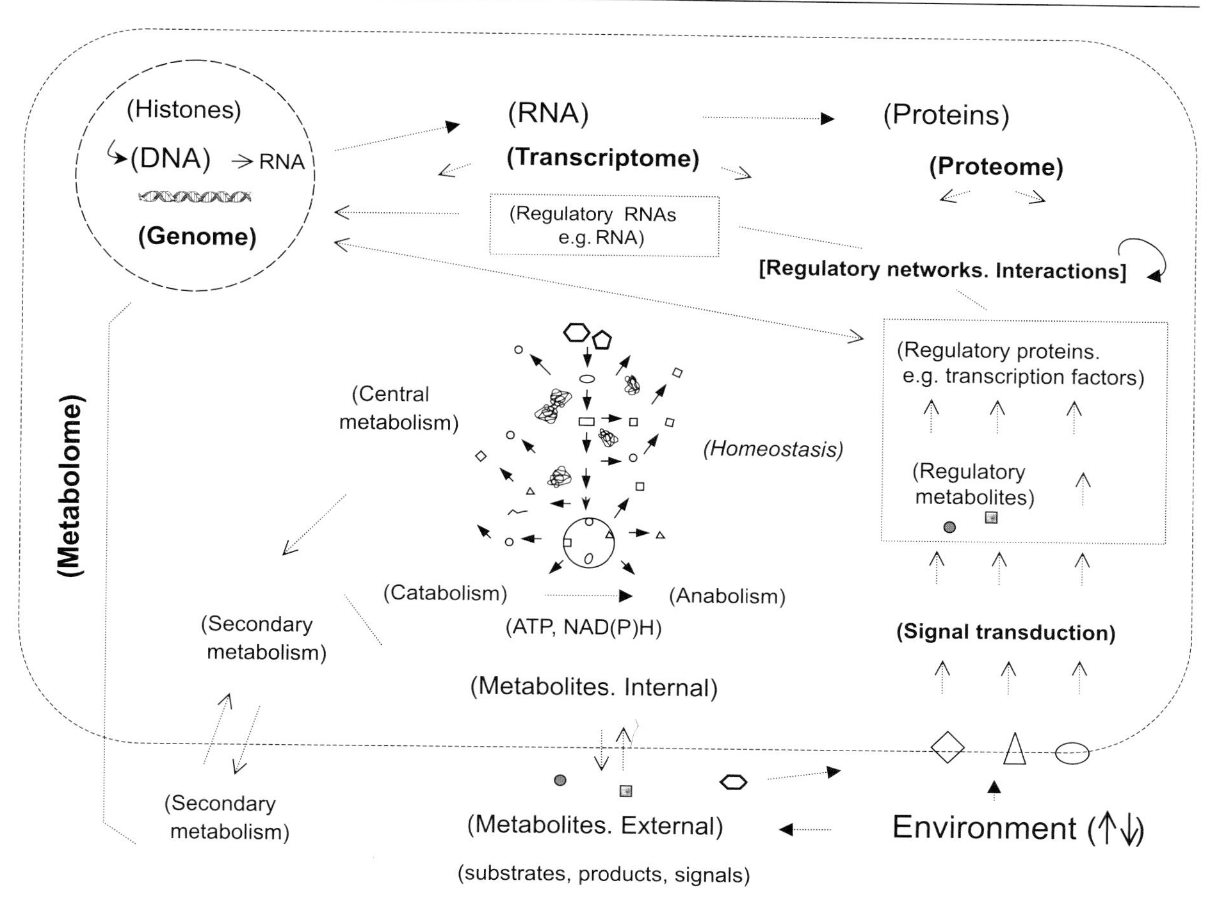

Fig. 1.1. Eukaryotic cell. Functional genomic levels and regulatory mechanisms

levels (Fafournoux et al. 2000; Muratani and Tansey 2003; Verger et al. 2003; Castrillo and Oliver 2004; Choudhuri 2004). Among the most relevant mechanisms are epigenetic mechanisms (e.g. DNA methylation, histone modifications), mRNA splicing and small regulatory RNAs, protein methylation, glycosylation, ubiquitination and sumoylation, protein–protein and protein–metabolite interactions, and participation of metabolites together with transcription factors and regulatory proteins in signal transduction pathways (Day and Tuite 1998; Castrillo and Oliver 2004 and references therein; Choudhuri 2004).

B. *S. cerevisiae* Functional Genomics: State of the Art

S. cerevisiae (laboratory strain S288C; *MATαSUC2 mal mel gal2 CUP1 flo1 flo8-1 hap1*; Mortimer and Johnston 1986; Gaisne et al. 1999; Sherman 2002) was the first eukaryotic organism for which the whole genome sequence was completed and made publicly available (Goffeau et al. 1996, 1997). This sequence has been further certified as the best annotated eukaryotic genome (Goffeau 2000). The collection of complete *S. cerevisiae* chromosome sequences and annotations is available at Goffeau et al. (1997, http://www.nature.com/genomics/papers/s_cerevisiae.html), and can also be obtained at the National Centre for Biotechnological Information (NCBI; http://www.ncbi.nlm.nih.gov/mapview/map_search. cgi? taxid=4932). In this case, the nucleotide and protein sequences are provided by the *Saccharomyces* Genome Database (SGD; http://www.yeastgenome.org) and are revised as SGD is updated.

The *S. cerevisiae* genome contains about 12 Mb of DNA distributed between 16 chromosomes which contain a total of about 6000 genes, with a relatively low proportion containing introns

Table. 1.1. Characteristics of *Saccharomyces cerevisiae* (Goffeau et al. 1996; Sherman 1998, 2002; National Centre for Biotechnological Information, NCBI, http://www.ncbi.nlm.nih.gov/mapview/map_search.cgi?taxid=4932)

S. cerevisiae	Characteristics	
Lineage	*Eukaryota; Fungi; Ascomycota; Saccharomycotina; Saccharomycetes; Saccharomycetales; Saccharomycetaceae; Saccharomyces*	
Nuclear genome	16 chromosomes (12,052 Kb) 2 μm circle plasmid (6.3 Kb)	
Mitochondrial genome	Mitochondrial DNA (850 Kb)	
ORFs	6183 ORFs encoding for 5773 proteins	
tRNA genes	262 tRNA genes	
Introns	3.8%	
Intracellular dsRNA viruses	0.1% of total nucleic acid content	
Cells	*Haploid*	*Diploid*
Volume per cell (μm^3)	70	120
Global content	1 mg wet weight or 0.25 mg dry weight contains ~ 0.28 μg of DNA 20 μg of total RNA ~ 1 μg of mRNA 0.10 mg total protein	1 mg wet weight or 0.25 mg dry weight contains ~ 0.42 μg of DNA 24 μg of total RNA ~ 1.2 μg of mRNA 0.10 mg total protein

(~ 4%; Goffeau et al. 1996, 1997; Brown and Tuite 1998; Sherman 1998, 2002). From the whole genome, a total of 5257 protein-coding genes have been annotated (i.e. reported to code for a specific protein with a gene ontology category curated with experimental evidence) as of the end of October 2004 (Yeast Proteome Database, http://proteome.incyte.com; Costanzo et al. 2001; Csank et al. 2002). The main characteristics of *S. cerevisiae* are summarized in Table 1.1.

Once the genome sequence is known, post-genomic studies entail, first, the design and implementation of advanced high-throughput methods and genomic strategies to extract the maximum information at the different functional genomic levels and, second, an efficient analysis of the huge amount of data generated, in order to extract valid conclusions and new knowledge (e.g. mechanisms of regulation and their integration). The most advanced functional genomics methods and strategies investigated in *S. cerevisiae* have been reviewed by Castrillo and Oliver (2004). A summary of the main databases and resources for analysis of yeast genomic data is presented in Table 1.2. The most relevant strategies are:

1. At the genome level, application of molecular genetic techniques on a large scale for the generation of comprehensive collections of yeast mutants. Thus, for example, single deletion mutants, double mutants and the 'TRIPLES' collection of mutants, obtained by random transposon insertion, are used for global analyses, functional profiling and gene characterization (Ross-Macdonald et al. 1999; Winzeler et al. 1999; Giaever et al. 2002; Scherens and Goffeau 2004 and references therein). To complement yeast knockout collections, the construction of a collection of yeast strains in which a tetracycline-responsive promoter is inserted upstream of individual essential genes will allow us to explore the function of essential genes via conditional and titratable promoter alleles (Eisenstein 2004; Mnaimneh et al. 2004).
2. At the gene expression (transcriptome) level, microarrays have been widely used for the global analysis of yeast gene expression patterns (Lashkari et al. 1997; Wodicka et al. 1997; Spellman et al. 1998), with more recent studies revealing the importance of a careful experimental design, controlled conditions and good strategies for data processing and statistical analysis (ter Linde et al. 1999; Hayes et al. 2002; Boer et al. 2003; Tilstone 2003). In addition, new approaches to analyse not only relative changes in gene expression (mRNA) levels but also net transcription rates on a genomic scale are progressively being incorporated (Iyer and Struhl 1996; Hirayoshi and Lis 1999; Garcia-Martinez et al. 2004).
3. At the proteome level: the first whole-proteome microarray was developed for yeast, and advanced strategies for the preparation of such

Table. 1.2. Main resources, tools and databases for individual and integrative analysis of yeast genomics data

Databases/resources	Reference
Genome databases	
S. cerevisiae Genome Database (SGD)	http://www.yeastgenome.org
National Centre for Biotechnological Information	http://www.ncbi.nlm.nih.gov/mapview/map_search.cgi?taxid=4932
Munich Information centre for Protein Sequences (MIPS)	http://mips.gsf.de/genre/proj/yeast/index.jsp
S. cerevisiae mutants collection	http://www.uni-frankfurt.de/fb15/mikro/euroscarf/complete.html
S. cerevisiae resource center	http://depts.washington.edu/~yeastrc
Gene expression (transcriptome) resources	
Microarray standards (MIAME)	http://www.mged.org/miame
ArrayExpress	http://www.ebi.ac.uk/arrayexpress
Stanford Microarray Database	http://genome-www5.stanford.edu
Promoter, regulatory sequences and transcription factor databases	
S. cerevisiae promoter database	http://cgsigma.cshl.org/jian/
Yeast Transcription Factors and related components (YTF)	http://biochemie.web.med.uni-muenchen.de/YTFD/
TRANSFAC	http://www.biobase.de/pages/products/transfac.html
Proteome resources	
Proteomics Standards Initiative (PSI)	http://psidev.sourceforge.net
Proteomics platform (PEDRo)	http://pedro.man.ac.uk/
Protein–protein interactions	(von Mering et al. 2002)
Yeast proteins localization	http://yeastgfp.ucsf.edu/
Yeast protein microarrays	(Zhu and Snyder 2003)
Metabolic pathways databases	
KEGG	http://www.genome.ad.jp/kegg/
BioCyC	http://biocyc.org
PathDB	http://www.ncgr.org/pathdb/
RIKEN	http://genome.gsc.riken.go.jp/DNA-Book/metabolome.shtml
Biomolecular interactions, BIND	http://www.blueprint.org/bind/bind.php
Tools for analysis and management of global genomic information	
Gene Ontology tools	http://www.geneontology.org/GO.tools.shtml
- GoMiner	http://discover.nci.nih.gov/gominer/
- GenMAPP	http://www.genmapp.org/
Genome Information Management System (GIMS)	http://www.cs.man.ac.uk/img/gims/ (Cornell et al. 2003)
Systems biology resources	
Yeast Systems Biology Network	http://www.ysbn.org
Systems Biology DataBase (SBDB)	http://www.sysbioldb.org

arrays have been developed (Zhu et al. 2001, 2003; Zhu and Snyder 2003). In addition, studies on sub-cellular localization of yeast proteins on a proteome-wide scale, phosphoproteome studies, protein–protein interaction maps and studies on protein turnover have all been undertaken (Ficarro et al. 2002; Gavin et al. 2002; Ho et al. 2002; von Mering et al. 2002; Pratt et al. 2002; Ghaemmaghami et al. 2003; Huh et al. 2003; Wohlschlegel and Yates 2003).

4. At the metabolome level, new methods for the analysis of yeast metabolites, strategies to ascribe function to unknown genes, and the classification of yeast mutants using metabolic fingerprinting and footprinting have been developed (Gonzalez et al. 1997; Raamsdonk et al. 2001; Allen et al. 2003; Castrillo et al. 2003).
5. At the bioinformatics level, new machine-learning methods for the analysis of transcriptome, proteome and metabolome data, and for the study of yeast regulatory networks have been derived. Relevant web resources, databases and methods for the global analysis of yeast genomic data, data repositories and data warehouses for storage and rapid access to yeast genomic raw data have all been generated. These include the *Saccharomyces* Genome Database (SGD, http://www.yeastgenome.org), microarray databases and repositories (e.g. Stanford Microarray Database and Array-

Express, http://genome-www5.Stanford.edu; http://www.ebi.ac.uk/arrayexpress), the Yeast Protein Database (YPD, http://proteome.incyte.com), the Proteomics Experiment Data Repository (PEDRo, http://pedro.man.ac.uk), metabolic pathways and metabolic databases (KEGG, http://www.genome.ad.jp/kegg/), global management systems of genomic information (GIMS, Cornell et al. 2003, http://www.cs.man.ac.uk/img/gims/) and gene ontology tools (e.g. GoMiner, MAPPFinder and GeneMAPP, http://www.geneontology.org/GO.tools.html).

6. Integrated functional analysis strategies incorporate growth competition experiments and high-throughput methods to quantify the results of the growth competition, for the elucidation of gene function (Baganz et al. 1998; Giaever et al. 2002; Merritt and Edwards 2004).

In summary, the favourable characteristics of *S. cerevisiae*, together with the advanced functional genomics resources, strategies and methods already available, make it an optimal reference organism for studies of integrative systems biology at the cellular level. In these studies, a number of important factors such as the role of metabolomics and new regulatory mechanisms should not be overlooked. In order to pursue comprehensive holistic approaches to unveil the dynamics of gene/protein networks and to integrate this information into progressively more realistic genome-wide models, particular care has to be taken to incorporate information and mechanisms at all of these different functional levels.

IV. Metabolomics in Comprehensive Post-Genomic Studies: Towards Systems Biology Using *S. cerevisiae* as a Model

A. Metabolomic Studies: Metabolic Networks and Participation of Metabolites in Regulation

One of the most relevant breakthroughs in functional genomics is the implementation of progressively more refined high-throughput techniques for genome-wide studies of the cell. In the post-genomic era, primary efforts are being invested mainly on analytical strategies at the genome, gene expression and proteome levels, with metabolomic studies starting later and receiving less attention. This situation is rapidly being corrected, and the latest studies and strategies are rapidly re-establishing the relevance of metabolites and metabolomics in integrative studies towards a system-level understanding of the cell (Teusink et al. 1998; Fiehn 2001; Raamsdonk et al. 2001; De la Fuente et al. 2002; Adams 2003; Allen et al. 2003; Harrigan and Goodacre 2003; Weckwerth 2003; Goodacre et al. 2004).

Metabolomics can be defined as the comprehensive analysis of the complete pool of cellular metabolites (the 'metabolome') closely interacting with the other functional genomic levels (Fig. 1.1). The relevance of metabolomics in the post-genomic era can be exemplified by (1) the scope and importance of metabolic network studies, and (2) the rediscovered role of metabolites in regulation at the different genomic levels.

The central metabolic pathways are the biological networks which have been most subject to comprehensive analyses. The basic information is presently accessible in metabolic pathways databases (e.g. BioCyc, http://biocyc.org; KEGG, http://www.genome.ad.jp/kegg/; RIKEN, http://genome.gsc.riken.go.jp/DNA-Book/metabolome.shtml; Metabolic Vision metabolic pathway database, http://www.ariadnegenomics.com/products/meta.html).

These metabolic networks have been the first for which conceptual framework theories have been defined. Some of these have focused on the description of the networks as flux maps such as Metabolic Flux Analysis (MFA; Varma and Palsson 1994). In this approach, the dynamic system is turned into a steady-state model in which no kinetic information is needed. MFA has the objective of, for example, quantifying all intracellular fluxes for exploitation in metabolic engineering strategies. This flux balance approach has some limitations, however, since not all reactions involving major co-factors, such as ATP, NADH and NADPH, are exactly known. Thus, in MFA, most recent studies have focused on the incorporation of new analytical strategies such as isotopomer labelling (e.g. 13C labelling applied to MFA; Wiechert 2001).

As soon as better theoretical frameworks become available, new strategies can be applied. In spite of this, the continuing limited success of attempts at metabolic engineering has pointed to the existence of overlooked concepts, and the need for truly comprehensive conceptual approaches in order to understand the dynamics and control of metabolic networks. Among these conceptual approaches, the most significant is Metabolic Control Analysis (MCA) theory (Kacser and Burns 1973; Kacser 1995; Fell 1997). The three main principles of MCA theory can be summarized as follows.

1. In a metabolic pathway, control is shared – that is, the control of flux through the pathway is 'distributed' between the different enzymatic steps and is not exerted by a single enzyme catalysing the so-called rate-limiting step.
2. The control exerted by each enzymatic step is measured by the 'flux control coefficient', C_e^J, which can be defined as the relative change in flux (J) caused by a specific modulation of the activity of an enzyme (e) at steady state.
3. For an unbranched pathway, the Summation Theorem states that the sum of the flux control coefficients of all steps is equal to unity.

The fundamental difference between the rate-limiting step concept and the MCA theory is the notion of distributed control in which each enzymatic step contributes to the global control of the metabolic flux. From this perspective, attempts to increase the flux through a central metabolic pathway will require the manipulation of several enzymatic steps in order to achieve a measurable effect on the global flux. Central metabolic pathways are tightly regulated and resist manipulation, whereas strategies involving the manipulation of peripheral metabolic pathways have been reported to be more successful (Brown 1997; Fell 1998; Stephanopoulos 1999). In modern MCA theory, other aspects (such as metabolic compartmentation) are also being considered, and new advances and conceptual principles are being incorporated (Peletier et al. 2003). For a good review on the basic principles of metabolic control analysis theory, the reader can refer to Fell (1997, 1998) and references therein. Additional strategies for modelling metabolic networks and approaches such as top-down analysis are being investigated and incorporated into the different levels of analysis (Brown et al. 1990; Quant 1993; Krauss and Quant 1996). Finally, among the latest studies on biochemical pathways are network constraint-based models which rely on the definition of minimal functional units of metabolism (elementary flux modes or extreme pathways; Klamt and Stelling 2003; Papin et al. 2004).

Apart from these metabolic network studies, a relevant aspect in metabolomics is the participation of metabolites in regulation. In the post-genomic era, considerable efforts are focused on the role of regulatory mechanisms (at the epigenetic, genetic, transcriptional and post-transcriptional levels) in the global behaviour of the cell. With this perspective, metabolites could be regarded as inert, with minor participation in regulation. However, the complete pool of internal and external cellular metabolites plays a remarkable role in regulation and control. Firstly, at the level of intermediary metabolism, metabolites such as fructose-1,6-diphosphate, ATP, ADP and citrate exert rapid short-term regulation upon central metabolic fluxes (e.g. by the rapid activation or inhibition of enzymes by reversible covalent modification or by allosteric effects; Monod et al. 1963; Fell 1997; Plaxton 2004). Secondly, the nature and levels of external metabolites (substrates, products and other external compounds) constitute essential environmental signals detected by the cell, usually via ligand-membrane receptor interactions. These signals are transduced into the cell via specific signal transduction pathways with the major participation of regulatory proteins (e.g. transcription factors; Hancock 1997; Sprague et al. 2004). However, an increasing number of studies are reporting an important role for internal metabolites (e.g. phosphate, cAMP, inositol phosphates, phosphatidic acid, amino acids) as regulatory molecules participating in signal transduction and complex regulatory mechanisms (Zaragoza et al. 1999; Hansen and Johannesen 2000; Muller et al. 2003; Sellick and Reece 2003; Auesukaree et al. 2004; Loewen et al. 2004; Sprague et al. 2004). Moreover, recent studies have reported new mechanisms by which metabolites may control gene expression (e.g. by direct interaction with mRNA-riboswitches, without the participation of proteins), or which can lead to post-translational histone modifications (Cech 2004; Dong and Xu 2004).

The advanced studies on metabolic networks and the sometimes overlooked roles of metabolites in regulation constitute new challenges. More importantly, they confirm the importance of metabolomics, together with other genome-wide high-throughput strategies (e.g. genome, transcriptome and proteome studies) in the generation of a complete, integrative description of the cell as a global system (Fig. 1.1), which is a prerequisite for comprehensive systems biology studies.

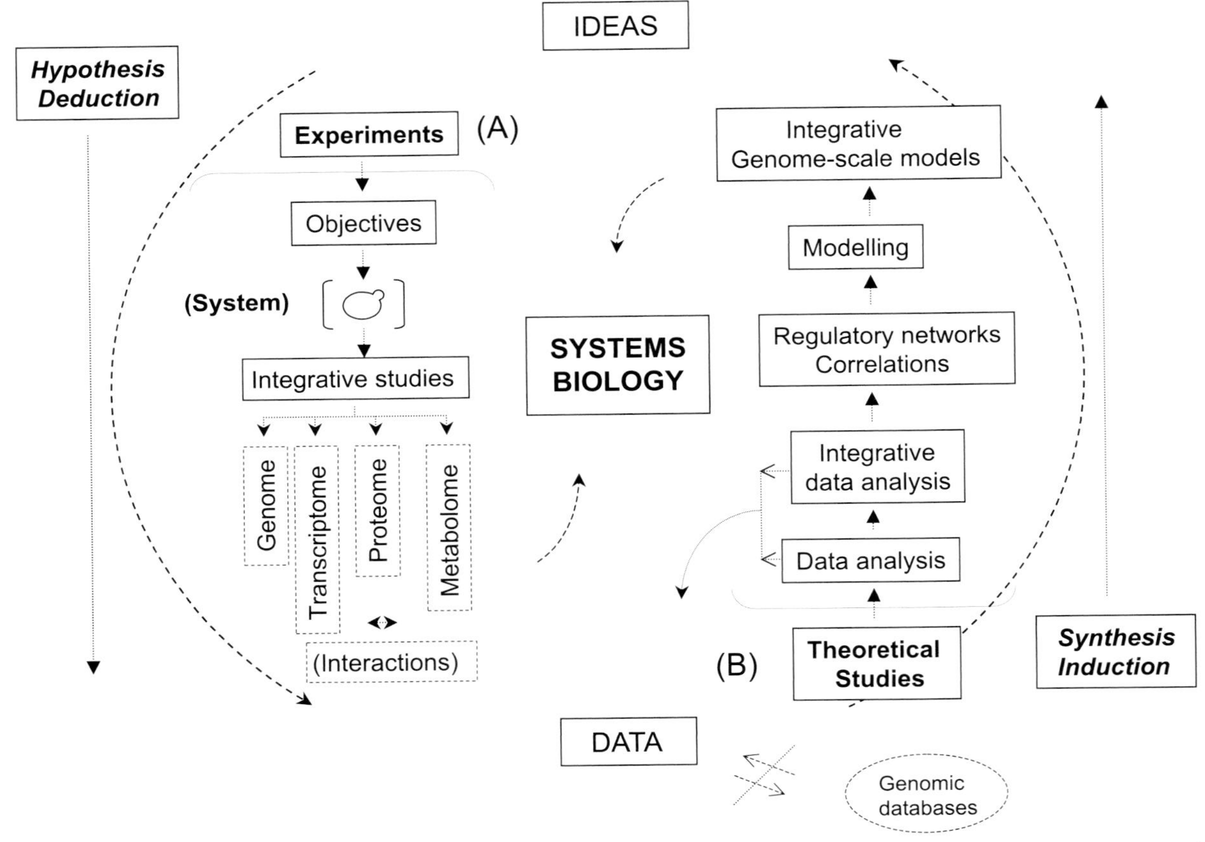

Fig. 1.2. Systems biology. Integration of global experimental (**A**) and theoretical studies (**B**) in the iterative cycle of knowledge (Kell and Oliver 2004)

B. Integrative Studies for Global Description of Cellular Complexity: Cellular Networks and Systems Biology

Genes, RNAs, proteins and metabolites exert their role in a 'cell system' rich in complexity, with the participation of mechanisms of regulation at the genomic, transcriptional, post-transcriptional and metabolic level (Fig. 1.1; Day and Tuite 1998; Castrillo and Oliver 2004; Choudhuri 2004). This fact has been demonstrated by different authors in several studies in which no (simple) direct correlation between levels of gene expression (mRNA), protein content and/or metabolic fluxes could be established (Gygi et al. 1999; Fell 2001; Ideker et al. 2001b; ter Kuile and Westerhoff 2001; Yoon and Lee 2002; Bro et al. 2003; Glanemann et al. 2003; Lee et al. 2003; Mehra et al. 2003; Yoon et al. 2003; Daran-Lapujade et al. 2004). All these results point to the existence of specific mechanisms of regulation at the different 'omic levels, and suggest that many of the circuits in the cellular networks are still to be elucidated. These cellular networks, their underlying mechanisms, interrelationships and flexibility in adapting to environmental changes constitute the main central processes controlling the global behaviour of the cell as a 'biological system'. Their elucidation constitutes one of the most daunting challenges in post-genomic biology. As a result, comprehensive integrative studies are increasingly being undertaken to advance the understanding of the logics of the responsible regulatory modules (Fell 2001; Oliver et al. 2002; Phelps et al. 2002; Nurse 2003; Castrillo and Oliver 2004; Stelling 2004).

Systems biology aims at understanding all of the genotype–phenotype relationships brought about by these cellular networks, as well as the principles and mechanisms governing the behaviour of biological systems (Kitano 2002; Nurse 2003; Stelling 2004). The main objective of systems biology is the construction of mathematical models by which to interrogate and iteratively refine our knowledge of the system (e.g. the eukaryotic cell; Kitano 2002; Nurse 2003; Ideker 2004a; Kafatos and Eisner 2004; Stelling 2004). Systems biology addresses this task by combining integrative experimental and theoretical approaches (Stelling 2004). This outlook is illustrated in Fig. 1.2, in the context of the iterative cycle of knowledge (Kell and Oliver 2004). First, high-throughput integrative experimental studies at several functional genomic levels (e.g. gene expression together with proteome and/or metabolome studies; Gygi et al. 1999; Bro et al. 2003; Lee et al. 2003; Urbanczyk-Wochniak et al. 2003) provide system-level sets of data (Fig. 1.2A). From this point, comprehensive theoretical studies (bioinformatics, 'in silico' methods), integrative data analysis and/or modelling approaches (e.g. Kell and King 2000; Mendes 2002; Yao 2002; Stelling 2004 and references therein) allow the generation of new knowledge at a system level and make predictions capable of being tested in further experimental studies (Fig. 1.2B). Hence, systems biology operates in two main ways: (1) by deduction, by perturbing the biological system in specifically designed experiments, monitoring the responses, integrating the data and formulating mathematical models able to reproduce this behaviour; and (2) in an inductive way, by perturbing the model, making predictions as to what should happen in response to these perturbations and testing them through laboratory experiments (Ideker et al. 2001a; Sage 2004).

The ultimate challenge in systems biology would be to integrate all genomic information for a biological system (e.g. mammalian cell line or tissue) under different conditions, to unveil all internal regulatory networks responsible for its behaviour, and to construct a global dynamic model which reliably reproduces this behaviour. However, the limitations of existing high-throughput techniques, the difficulty of obtaining all information at the different 'omic levels, the intrinsic complexity of the biological systems (with many mechanistic uncertainties), and the predictable limitations at the computational level are limiting the scope of a functional genomics approach to systems biology, primarily to studies of single-celled organisms and sub-cellular systems (Stelling 2004 and references therein). In this context, simple reference organisms will constitute the primary subjects of comprehensive analyses and, at this point, the optimal characteristics of *S. cerevisiae* as a touchstone model in the post-genomic era (Castrillo and Oliver 2004) are positioning it at the leading edge of advanced systems biology studies.

C. *S. cerevisiae* as a Reference Model in Systems Biology: Advanced Studies, Biological Networks and Genome-Scale Models, and Applications

Systems biology entails analysis and comparison of results from specifically designed integrative experiments and global theoretical studies (Fig. 1.2). This goal is still difficult to achieve. This is because the ability to combine two or more functional genomic strategies (e.g. microarrays and metabolome studies) in a single experiment in conjunction with comprehensive 'in silico' approaches is limited to relatively few research groups. As a result, only a few true systems biology studies, combining experimental work with a strong theoretical basis, have been published. These have concentrated on well-defined unicellular organisms (e.g. bacteria, yeast) and sub-cellular systems (Ideker et al. 2001b; Covert et al. 2004; Ozbudak et al. 2004).

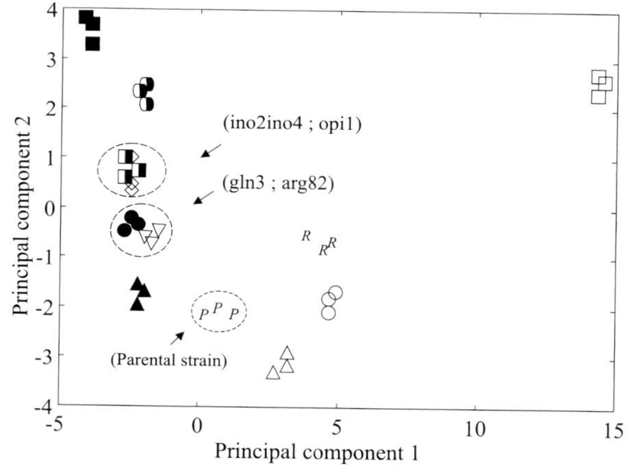

Fig. 1.3. Principal components analysis (PCA) of electrospray mass spectrometry (ES-MS) analysis of external metabolites corresponding to *S. cerevisiae* strains growing in exponential phase (OD_{600}=1.0). Reference: external medium (*R*). Parental strain: BY4709 (ATCC 200872) (*P*). Transactivator mutants: (1) carbon metabolism: BY4709*cat8*-[Delta] (△) and BY4709*hap4*-[Delta] (▲); (2) nitrogen assimilation: BY4709*nil1*-[Delta] (○), BY4709*gln3*-[Delta] (●), BY4709*arg82*-[Delta] (▽) and BY4709*nil1*-[Delta], *gln3*-[Delta] (◑); (3) lipid metabolism: BY4709*opi1*-[Delta] (◇), BY4709*ino2*-[Delta] (□), BY4709*ino4*-[Delta] (■) and BY4709*ino2*-[Delta], *ino4*-[Delta] (◨)

Reductions in both the costs and operational difficulty of functional genomic technologies, together with an increasing trend to perform such research in a collaborative manner, is advancing the systems biology agenda. Thus, systems biology is a rapidly emerging field in which novel 'in silico' approaches are being used to implement network- and genome-scale modelling strategies which integrate data from the different 'omic levels. However, the afore-mentioned difficulties mean that the majority of these in silico studies rely on data extracted from the literature and public genomic databases. A word of caution is therefore appropriate – particular care should be taken, since the validity of these studies will be critically dependent on the selection of primary datasets (often from different laboratories) which have been obtained under essentially equivalent conditions (von Mering et al. 2002; Castrillo and Oliver 2004). This emphasises the need for data standards (e.g. MIAME, http://www.mged.org/Workgroups/MIAME/miame.html; PEDRo, http://pedro.man.ac.uk; ArtMet; http://www.armet.org) to ensure there is sufficient metadata (i.e. well-defined formal descriptions of the experimental conditions and analytical procedures) to allow such judgements to be made.

As far as the implementation of new genome-wide techniques is concerned, the vast majority of the most advanced high-throughput genomic technologies are being validated in *S. cerevisiae*, at the genome, transcriptome, proteome, metabolome and interaction levels. This makes yeast the most advanced biological system for the application of global experimental strategies, combining studies at different 'omic levels (Raamsdonk et al. 2001; Gavin et al. 2002; Giaever et al. 2002; Hayes et al. 2002; Ho et al. 2002; Allen et al. 2003; Zhu et al. 2003; Scherens and Goffeau 2004). As an example of this, the principal components analysis of the metabolome profiles of selected *S. cerevisiae* transactivator mutants, corresponding to different nutrient assimilation pathways, is presented in Fig. 1.3. This metabolome analysis, combined with the use of selected mutants, shows that metabolomic studies are capable of discriminating between different transcription factor mutants. Moreover, when a number of transactivator mutants tended to cluster together, these corresponded to mutants of one and the same nutrient assimilation pathway. These results point to the possibility of using metabolomics to investigate new transcription factors, or families of transcription factors, and they illustrate the contribution that metabolomics can make to functional genomic studies.

In the field of theoretical systems biology, recent years have witnessed the appearance of advanced bioinformatics strategies, integrative data analysis, and network studies towards the implementation of genome-scale models, with *S. cerevisiae* (in the majority of cases) being the selected model. Thus, in the development of comprehensive data analysis tools, efforts have been directed mainly towards the development of new classification algorithms, clustering, and machine learning methods (unsupervised or supervised) to unveil characteristic patterns and their interrelationships (Kell and King 2000; McInerney 2002; Mendes 2002; Kapetanovic et al. 2004; Wei et al. 2004). The natural expansion of these studies is leading to the advent of a new conceptual framework in post-genomics science – network biology – dedicated to the study of the topology, modular components, and the governing principles of cellular networks (Barabási and Oltvai 2004; Papin and Palsson 2004; Stelling 2004). Such work has concentrated on metabolic, protein–protein interaction, signalling, and regulatory networks, with (again) *S. cerevisiae* as the reference model (Harbison et al. 2004; Ideker 2004b; Ihmels et al. 2004; Yeang et al. 2004). For example, studies on the dynamics of the transcriptional regulatory network of *S. cerevisiae* have been reported recently (Luscombe et al. 2004). These reveal large topological changes depending on environmental conditions, with transcription factors altering their interactions in response to stimuli, a few of them serving as permanent hubs but most acting transiently in certain conditions only.

The final objective in systems biology is to incorporate the experimental data into mathematical models which are descriptive of the cell system, and which are capable of predicting its behaviour. Genome-scale models of metabolic or other networks may be made at different levels of granularity, depending on the qualitative and quantitative characteristics of the analytical technologies employed. The most relevant approaches to the mathematical modelling of cellular networks (Stelling 2004) are as follows.

1. Interaction-based: these result in static models for the study of individual networks (e.g. gene expression; protein–protein interaction networks).

2. Constraint-based: network representations based on constraints include the stoichiometry and reversibility of metabolic reactions, and allow the calculation of fluxes and steady-state flux distributions.
3. Mechanistic representations: these complex dynamic models include stoichiometry and kinetic parameters but are more representative of the real system, and allow us to formulate precise, experimentally testable hypotheses.

Among the most remarkable examples of reconstructed models reported are an expanded model of *Escherichia coli* K-12 (Reed et al. 2003), the first genome-scale reconstructions and models of *S. cerevisiae* (Förster et al. 2003; US Patent 2003228567, US Patent Office 2003), and the construction and validation of a compartmentalized genome-scale model of *S. cerevisiae* (Duarte et al. 2004).

Given its favourable characteristics as a touchstone model in post-genomic studies (Castrillo and Oliver 2004) and the success of the network studies reported above, it is clear that *S. cerevisiae* is emerging as the best model eukaryote for systems biology studies at the single-cell level. Progressive advances in high-throughput methods, improved experimental strategies and more refined models leading to the elucidation of the components and patterns of regulatory networks are sure to result in direct applications in disciplines such as metabolic engineering (e.g. optimisation of bioconversion pathways; Stephanopoulos and Gill 2001; Gill and Dodge 2004; WO Patent 0107567, World Intellectual Property Organisation 2001a), biomedical and environmental sciences, and the pharmaceutical industry (e.g. biomarkers discovery, new tools for the identification of targets, and development of new drugs and therapeutic strategies; Fernie et al. 2004; Ilyin et al. 2004; Wang et al. 2004; WO Patent 0178652, World Intellectual Property Organisation 2001b). In order to realise the full potential of yeast as a reference model in systems biology, the high level of international collaborations between yeast research groups seen in the genome sequencing (Goffeau et al. 1996, 1997; http://www.yeastgenome.org) and functional genomics projects (Oliver 1997, 2002; Winzeler et al. 1999; Giaever et al. 2002) must be continued and even enhanced. To this end, SysBiolDB has been established to facilitate the deposition and exchange of information by systems biology researchers worldwide (http://www.sysbioldb.org), while the recently created Yeast System Biology Network (http://www.ysbn.org) will promote collaborations between yeast systems biologists. If such joint initiatives prosper, and succeed in leading the yeast research community in the right direction, we can anticipate a bright future for *S. cerevisiae* as a reference model in systems biology.

V. Conclusions

New advanced studies in the post-genomic era are unveiling the real picture of the cell as a system rich in complexity, with participation of mechanisms at the epigenomic, genomic, transcriptomic, proteomic and metabolomic levels in network interactions which require extensive and integrative investigation. These mechanisms configure the internal cellular networks responsible for the global behaviour of the cell, which include complex regulatory circuits and signal transduction pathways allowing the cell to respond in an appropriate manner to changes in its environment. Many of these regulatory mechanisms remain to be elucidated, and we urge that the role of the metabolome should not be overlooked in studies of regulatory networks in the post-genomics era.

In order to reach the goal of understanding the eukaryotic cell in a holistic and integrated manner, the utilization of good biological models, which are well characterized at the genetic and all functional genomics levels, and which can serve as a reference platform towards the study of more complex systems (e.g. mammalian cells), is essential. We submit that the biological characteristics of *S. cerevisiae*, together with its advanced position at the forefront of post-genomics studies, will make it the reference model for systems biology studies of the eukaryotic cell.

Acknowledgements. This work was supported by an EC contract to S.G. Oliver within the frame of the Garnish network of FP5 and the BBSRC's Investigating Gene Function Initiative within COGEME (Consortium for the Functional Genomics of Microbial Eukaryotes; http://www.cogeme.man.ac.uk). We thank Dr. M. Bolotin-Fukuhara (IGM, Université Paris XI, Orsay Cedex, France) for providing yeast strains.

References

Adams A (2003) Metabolomics: small-molecule omics. The Scientist 17:38–40

Alberts B, Johnson A, Lewis J, Raff M, Roberts K, Walter P (2002) Molecular biology of the cell, 4th edn. Garland Science, Taylor and Francis Group, New York

Allen J, Davey HM, Broadhurst D, Heald JK, Rowland JJ, Oliver SG, Kell DB (2003) High-throughput classification of yeast mutants using metabolic footprinting. Nat Biotechnol 21:692–696

Auesukaree C, Homma T, Tochio H, Shirakawa M, Kaneko Y, Harashima S (2004) Intracellular phosphate serves as a signal for the regulation of the PHO pathway in Saccharomyces cerevisiae. J Biol Chem 279:17289–17294

Baganz F, Hayes A, Farquhar R, Butler PR, Gardner DCJ, Oliver SG (1998) Quantitative analysis of yeast gene function using competition experiments in continuous culture. Yeast 14:1417–1427

Bailey JE (1991) Toward a science of metabolic engineering. Science 252:1668–1675

Bailey JE, Ollis DF (1986) Biochemical engineering fundamentals, 2nd edn. McGraw Hill, New York

Barabási AL, Oltvai ZN (2004) Network biology: understanding the cell's functional organization. Nat Rev Genet 5:101–113

Boer VM, de Winde JH, Pronk JT, Piper MDW (2003) The genome-wide transcriptional responses of Saccharomyces cerevisiae grown on glucose in aerobic chemostat cultures limited for carbon, nitrogen, phosphorus, or sulfur. J Biol Chem 278:3265–3274

Bro C, Regenberg B, Lagniel G, Labarre J, Montero-Lomeli M, Nielsen J (2003) Transcriptional, proteomic, and metabolic responses to lithium in galactose-grown yeast cells. J Biol Chem 278:32141–323149

Brown AJP (1997) Control of metabolic flux in yeasts and fungi. Trends Biotechnol 15:445–447

Brown AJP, Tuite MF (1998) Yeast gene analysis. Academic Press, San Diego, Methods in Microbiology 26

Brown GC, Hafner RP, Brand MD (1990) A 'top-down' approach to the determination of control coefficients in metabolic control theory. Eur J Biochem 188:321–325

Burke D, Dawson D, Stearns T (2000) Methods in yeast genetics, 2000 edn. A Cold Spring Harbor Laboratory Course Manual. Cold Spring Harbor Laboratory Press, New York

Castrillo JI, Oliver SG (2004) Yeast as a touchstone in post-genomic research. Strategies for integrative analysis in functional genomics. J Biochem Mol Biol 37:93–106

Castrillo JI, Ugalde UO (1994) A general model of yeast energy metabolism in aerobic chemostat culture. Yeast 10:185–197

Castrillo JI, Hayes A, Mohammed S, Gaskell SJ, Oliver SG (2003) An optimised protocol for metabolome analysis in yeast using direct infusion electrospray mass spectrometry. Phytochemistry 62:929–937

Cech TR (2004) RNA finds a simpler way. Nature 428:263–264

Choudhuri S (2004) The nature of gene regulation. Int Arch Biosci 2004:1001–1015

Cornell M, Paton NW, Hedeler C, Kirby P, Delneri D, Hayes A, Oliver SG (2003) GIMS: an integrated data storage and analysis environment for genomic and functional data. Yeast 20:1291–1306

Cortassa S, Aon MA (1994) Metabolic control analysis of glycolysis and branching to ethanol production in chemostat cultures of *Saccharomyces cerevisiae* under carbon, nitrogen, or phosphate limitations. Enzyme Microb Technol 16:761–770

Costanzo MC, Crawford ME, Hirschman JE, Kranz JE, Olsen P, Robertson LS, Skrzypek MS, Braun BR, Hopkins KL, Kondu P et al. (2001) YPD, PombePD and WormPD: model organism volumes of the BioKnowledge library, an integrated resource for protein information. Nucleic Acids Res 29:75–79

Covert MW, Knight EM, Reed JL, Herrgard MJ, Palsson BØ (2004) Integrating high-throughput and computational data elucidates bacterial networks. Nature 429:92–96

Csank C, Costanzo MC, Hirschman J, Hodges P, Kranz JE, Mangan M, O'Neill K, Robertson LS, Skrzypek MS, Brooks J et al. (2002) Three yeast proteome databases: YPD, PombePD, and CalPD (MycoPathPD). Methods Enzymol 350:347–373

Daran-Lapujade P, Jansen ML, Daran JM, van Gulik W, de Winde JH, Pronk JT (2004) Role of transcriptional regulation in controlling fluxes in central carbon metabolism of *Saccharomyces cerevisiae*. A chemostat culture study. J Biol Chem 279:9125–9138

Day DA, Tuite MF (1998) Post-transcriptional gene regulatory mechanisms in eukaryotes: an overview. J Endocrinol 157:361–371

De la Fuente A, Snoep JL, Westerhoff HV, Mendes P (2002) Metabolic control in integrated biochemical systems. Eur J Biochem 269:4399–4408

Delneri D, Brancia FL, Oliver SG (2001) Towards a truly integrative biology through the functional genomics of yeast. Curr Opin Biotechnol 12:87–91

Dong L, Xu CW (2004) Carbohydrates induce mono-ubiquitination of H2B in yeast. J Biol Chem 279:1577–1580

Duarte NC, Herrgard MJ, Palsson BØ (2004) Reconstruction and validation of *Saccharomyces cerevisiae* iND750, a fully compartmentalized genome-scale metabolic model. Genome Res 14:1298–1309

Eisenstein M (2004) Getting down to the bare essentials. Nat Methods 20 July 2004. DOI 101038/nmteh030

Fafournoux P, Bruhat A, Jousse C (2000) Amino acid regulation of gene expression. Biochem J 351:1–12

Fell DA (1997) Understanding the control of metabolism. Portland Press, London

Fell DA (1998) Increasing the flux in metabolic pathways: a metabolic control analysis perspective. Biotechnol Bioeng 58:121–124

Fell DA (2001) Beyond genomics. Trends Genet 17:680–682

Fernie AR, Trethewey RN, Krotzky AJ, Willmitzer L (2004) Metabolite profiling: from diagnostics to systems biology. Nat Rev Mol Cell Biol 5:763–769

Ficarro SB, McCleland ML, Stukenberg PT, Burke DJ, Ross MM, Shabanowitz J, Hunt DF, White FM (2002) Phosphoproteome analysis by mass spectrometry and its application to *Saccharomyces cerevisiae*. Nat Biotechnol 20:301–305

Fiehn O (2001) Combining genomics, metabolome analysis and biochemical modelling to understand metabolic networks. Comp Funct Genomics 2:155–168

Förster J, Famili I, Fu P, Palsson BØ, Nielsen J (2003) Genome-scale reconstruction of the *Saccharomyces cerevisiae* metabolic network. Genome Res 13:244–253

Gaisne M, Bécam AM, Verdière J, Herbert CJ (1999) A 'natural' mutation in *Saccharomyces cerevisiae* strains de-

rived from S288c affects the complex regulatory gene HAP1 (CYP1). Curr Genet 36:195–200
Garcia-Martinez J, Aranda A, Perez-Ortin JE (2004) Genomic run-on evaluates transcription rates for all yeast genes and identifies gene regulatory mechanisms. Mol Cell 15:303–313
Gavin AC, Superti-Furga G (2003) Protein complexes and proteome organization from yeast to man. Curr Opin Chem Biol 7:21–27
Gavin AC, Bosche M, Krause R, Grandi P, Marzioch M, Bauer A, Schultz J, Rick JM, Michon AM, Cruciat CM et al. (2002) Functional organization of the yeast proteome by systematic analysis of protein complexes. Nature 415:141–147
Ghaemmaghami S, Huh WK, Bower K, Howson RW, Belle A, Dephoure N, O'Shea EK, Weissman JS (2003) Global analysis of protein expression in yeast. Nature 425:737–741
Giaever G, Chu AM, Ni L, Connelly C, Riles L, Veronneau S, Dow S, Lucau-Danila A, Anderson K, Andre B et al. (2002) Functional profiling of the *Saccharomyces cerevisiae* genome. Nature 418:387–391
Gill RT, Dodge T (2004) Special issue on inverse metabolic engineering. Metab Eng 6:175–176
Giuseppin ML, van Riel NA (2000) Metabolic modelling of *Saccharomyces cerevisiae* using the optimal control of homeostasis: a cybernetic model definition. Metab Eng 2:14–33
Glanemann C, Loos A, Gorret N, Willis LB, O'Brien XM, Lessard PA, Sinskey AJ (2003) Disparity between changes in mRNA abundance and enzyme activity in *Corynebacterium glutamicum* and implications for DNA microarray analysis. Appl Microbiol Biotechnol 61:61–68
Goffeau A (2000) Four years of post-genomic life with 6000 yeast genes. FEBS Lett 480:37–41
Goffeau A, Barrell BG, Bussey H, Davis RW, Dujon B, Feldmann H, Galibert F, Hoheisel JD, Jacq C, Johnston M et al. (1996) Life with 6000 genes. Science 274:546–567
Goffeau A, Aert R, Agostini-Carbone ML, Ahmed A, Aigle M, Alberghina L, Albermann K, Albers M, Aldea M, Alexandraki D et al (1997) The yeast genome directory. Nature 387 Suppl no 6632 (http://www.nature.com/genomics/papers/s_cerevisiae.html)
Gonzalez B, François J, Renaud M (1997) A rapid and reliable method for metabolite extraction in yeast using boiling buffered ethanol. Yeast 13:1347–1355
Goodacre R, Vaidyanathan S, Dunn WB, Harrigan GG, Kell DB (2004) Metabolomics by numbers: acquiring and understanding global metabolite data. Trends Biotechnol 22:245–252
Guthrie C, Fink GR (2002a) Guide to yeast genetics and molecular and cell biology. Part B. Academic Press, Elsevier Science, San Diego, Methods in Enzymology vol 350
Guthrie C, Fink GR (2002b) Guide to yeast genetics and molecular and cell biology. Part C. Academic Press, Elsevier Science, San Diego, Methods in Enzymology vol 351
Guthrie C, Fink GR (2004) Guide to yeast genetics and molecular biology. Part A. Academic Press, Elsevier Science, San Diego, Methods in Enzymology vol 194
Gygi SP, Rochon Y, Franza BR, Aebersold R (1999) Correlation between protein and mRNA abundance in yeast. Mol Cell Biol 19:1720–1730
Hancock JT (1997) Cell signalling. Prentice Hall, Harlow
Hansen J, Johannesen PF (2000) Cysteine is essential for transcriptional regulation of the sulfur assimilation genes in *Saccharomyces cerevisiae*. Mol Gen Genet 263:535–542
Harbison CT, Gordon DB, Lee TI, Rinaldi NJ, Macisaac KD, Danford TW, Hannett NM, Tagne JB, Reynolds DB, Yoo J et al. (2004) Transcriptional regulatory code of a eukaryotic genome. Nature 431:99–104
Harrigan GG, Goodacre R (2003) Metabolic profiling: its role in biomarker discovery and gene function analysis. Kluwer, Boston
Hayes A, Zhang N, Wu J, Butler PR, Hauser NC, Hoheisel JD, Lim F, Sharrocks AD, Oliver SG (2002) Hybridization array technology coupled with chemostat culture: tools to interrogate gene expression in *Saccharomyces cerevisiae*. Methods 26:281–290
Hirayoshi K, Lis JT (1999) Nuclear run-on assays: assessing transcription by measuring density of engaged RNA polymerases. Methods Enzymol 304:351–362
Ho Y, Gruhler A, Heilbut A, Bader GD, Moore L, Adams SL, Millar A, Taylor P, Bennett K, Boutilier K et al. (2002) Systematic identification of protein complexes in *Saccharomyces cerevisiae* by mass spectrometry. Nature 415:180–183
Huh WK, Falvo JV, Gerke LC, Carroll AS, Howson RW, Weissman JS, O'Shea EK (2003) Global analysis of protein localization in budding yeast. Nature 425:686–691
Ideker T (2004a) Systems biology 101 – what you need to know. Nat Biotechnol 22:473–475
Ideker T (2004b) A systems approach to discovering signalling and regulatory pathways – or, how to digest large interaction networks into relevant pieces. Adv Exp Med Biol 547:21–30
Ideker T, Galitski T, Hood L (2001a) A new approach to decoding life: systems biology. Annu Rev Genomics Hum Genet 2:343–372
Ideker T, Thorsson V, Ranish JA, Christmas R, Buhler J, Eng JK, Bumgarner R, Goodlett DR, Aebersold R, Hood L (2001b) Integrated genomic and proteomic analyses of a systematically perturbed metabolic network. Science 292:929–934
Ihmels J, Levy R, Barkai N (2004) Principles of transcriptional control in the metabolic network of *Saccharomyces cerevisiae*. Nat Biotechnol 22:86–92
Ilyin SE, Belkowski SM, Plata-Salaman CR (2004) Biomarker discovery and validation: technologies and integrative approaches. Trends Biotechnol 22:411–416
Iyer V, Struhl K (1996) Absolute mRNA levels and transcriptional initiation rates in *Saccharomyces cerevisiae*. Proc Natl Acad Sci USA 93:5208–5212
Kacser H (1995) Recent developments beyond metabolic control analysis. Biochem Soc Trans 23:387–391
Kacser H, Burns JA (1973) The control of flux. Symp Soc Exp Biol 27:65–104
Kafatos FC, Eisner T (2004) Unification in the century of biology. Science 303:1257
Kapetanovic IM, Rosenfeld S, Izmirlian G (2004) Overview of commonly used bioinformatics methods and their applications. Ann N Y Acad Sci 1020:10–21

Kell DB, King RD (2000) On the optimization of classes for the assignment of unidentified reading frames in functional genomics programmes: the need for machine learning. Trends Biotechnol 18:93–98
Kell DB, Oliver SG (2004) Here is the evidence, now what is the hypothesis? The complementary roles of inductive and hypothesis-driven science in the post-genomic era. Bioessays 26:99–105
Kitano H (2002) Systems biology: a brief overview. Science 295:1662–1664
Klamt S, Stelling J (2003) Two approaches for metabolic pathway analysis? Trends Biotechnol 21:64–69
Krauss S, Quant PA (1996) Regulation and control in complex, dynamic metabolic systems: experimental application of the top-down approaches of metabolic control analysis to fatty acid oxidation and ketogenesis. J Theor Biol 182:381–388
Lashkari DA, DeRisi JL, McCusker JH, Namath AF, Gentile C, Hwang SY, Brown PO, Davis RW (1997) Yeast microarrays for genome wide parallel genetic and gene expression analysis. Proc Natl Acad Sci USA 94:13057–13062
Lee PS, Shaw LB, Choe LH, Mehra A, Hatzimanikatis V, Lee KH (2003) Insights into the relation between mRNA and protein expression patterns: II. Experimental observations in *Escherichia coli*. Biotechnol Bioeng 84:834–841
Lehninger AL (1975) Biochemistry, 2nd edn. Worth Publishers, New York
Lei F, Rotboll M, Jorgensen SB (2001) A biochemically structured model for *Saccharomyces cerevisiae*. J Biotechnol 88:205–221
Loewen CJ, Gaspar ML, Jesch SA, Delon C, Ktistakis NT, Henry SA, Levine TP (2004) Phospholipid metabolism regulated by a transcription factor sensing phosphatidic acid. Science 304:1644–1647
Luscombe NM, Babu MM, Yu H, Snyder M, Teichmann SA, Gerstein M (2004) Genomic analysis of regulatory network dynamics reveals large topological changes. Nature 431:308–312
McInerney JO (2002) Bioinformatics in a post-genomics world – the need for an inclusive approach. Pharmacogenomics J 2:207–208
Mehra A, Lee KH, Hatzimanikatis V (2003) Insights into the relation between mRNA and protein expression patterns. I. Theoretical considerations. Biotechnol Bioeng 84:822–833
Mendes P (2002) Emerging bioinformatics for the metabolome. Brief Bioinformatics 3:134–145
Merritt J, Edwards JS (2004) Assaying gene function by growth competition experiment. Metab Eng 6:212–219
Mnaimneh S, Davierwala AP, Haynes J, Moffat J, Peng WT, Zhang W, Yang X, Pootoolal J, Chua G, Lopez A et al. (2004) Exploration of essential gene functions via titratable promoter alleles. Cell 118:31–44
Monod J, Changeux J-P, Jacob F (1963) Allosteric proteins and cellular control systems. J Mol Biol 6:306–329
Mortimer RK, Johnston JR (1986) Genealogy of principal strains of the yeast genetic stock center. Genetics 113:35–43
Muller D, Exler S, Aguilera-Vazquez L, Guerrero-Martin E, Reuss M (2003) Cyclic AMP mediates the cell cycle dynamics of energy metabolism in *Saccharomyces cerevisiae*. Yeast 20:351–367
Muratani M, Tansey WP (2003) How the ubiquitin-proteasome system controls transcription. Nat Rev Mol Cell Biol 4:192–201
Nurse P (2003) Systems biology: understanding cells. Nature 424:883
Oliver SG (1997) Yeast as a navigational aid in genome analysis. Microbiology 143:1483–1487
Oliver SG (2002) Functional genomics: lessons from yeast. Philos Trans R Soc B 357:17–23
Oliver SG, Winson MK, Kell DB, Baganz F (1998) Systematic functional analysis of the yeast genome. Trends Biotechnol 16:373–378
Oliver DJ, Nikolau B, Wurtele ES (2002) Functional genomics: high-throughput mRNA, protein, and metabolite analyses. Metab Eng 4:98–106
Olson OS, Nielsen J (2000) Metabolic engineering of *Saccharomyces cerevisiae*. Microbiol Mol Biol Rev 64:34–50
Ozbudak EM, Thattai M, Lim HN, Shraiman BI, van Oudenaarden A (2004) Multistability in the lactose utilization network of *Escherichia coli*. Nature 427:737–740
Papin JA, Palsson BØ (2004) Topological analysis of mass-balanced signaling networks: a framework to obtain network properties including crosstalk. J Theor Biol 227:283–297
Papin JA, Stelling J, Price ND, Klamt S, Schuster S, Palsson BØ (2004) Comparison of network-based pathway analysis methods. Trends Biotechnol 22:400–405
Pasteur L (1857) Mémoire sur la fermentation appelée lactique. In: Mémoires Société Sciences Agriculture Arts Lille, séance 3 août 1857, 2ème Série, vol V, pp 13–26
Peletier MA, Westerhoff HV, Kholodenko BN (2003) Control of spatially heterogeneous and time-varying cellular reaction networks: a new summation law. J Theor Biol 225:477–487
Phelps TJ, Palumbo AV, Beliaev AS (2002) Metabolomics and microarrays for improved understanding of phenotypic characteristics controlled by both genomics and environmental constraints. Curr Opin Biotechnol 13:20–24
Plaxton WC (2004) Principles of metabolic control. In: Storey KB (ed) Functional metabolism of cells: control, regulation, and adaptation. Wiley, New York, pp 1–23
Pratt JM, Petty J, Riba-Garcia I, Robertson DHL, Gaskell SJ, Oliver SG, Beynon RJ (2002) Dynamics of protein turnover, a missing dimension in proteomics. Mol Cell Proteomics 1:579–591
Quant PA (1993) Experimental application of top-down control analysis to metabolic systems. Trends Biochem Sci 18:26–30
Raamsdonk LM, Teusink B, Broadhurst D, Zhang N, Hayes A, Walsh MC, Berden JA, Brindle KM, Kell DB, Rowland JJ et al. (2001) A functional genomics strategy that uses metabolome data to reveal the phenotype of silent mutations. Nat Biotechnol 19:45–50
Reed JL, Vo TD, Schilling CH, Palsson BØ (2003) An expanded genome-scale model of *Escherichia coli* K-12 (iJR904 GSM/GPR). Genome Biol 4:R54
Rose AH, Harrison JS (1987–1995) The yeasts, vols 1–6. Academic Press, London
Rose AH, Harrison JS (1993) The yeasts, vol 5. Academic Press, London
Ross-Macdonald P, Coelho PS, Roemer T, Agarwal S, Kumar A, Jansen R, Cheung KH, Sheehan A, Symoniatis D, Umansky L et al. (1999) Large-scale analysis

of the yeast genome by transposon tagging and gene disruption. Nature 402:413–418
Sage L (2004) Genome-scale model predicts gene regulation. The Scientist 18:42
Scherens B, Goffeau A (2004) The uses of genome-wide yeast mutant collections. Genome Biol 5:229 (http://genomebiologycom/2004/5/7/229)
Schwartz M (2001) The life and works of Louis Pasteur. J Appl Microbiol 91:597–601
Sellick CA, Reece RJ (2003) Modulation of transcription factor function by an amino acid: activation of Put3p by proline. EMBO J 22:5147–5153
Sherman F (1998) An introduction to the genetics and molecular biology of the yeast *Saccharomyces cerevisiae* (http://dbburmcrochester.edu/labs/sherman_f/yeast/)
Sherman F (2002) Getting started with yeast. Methods Enzymol 350:3–41 (updated in http://dbburmcrochester.edu/labs/sherman_f/startedyeast.pdf)
Sinclair CG, Cantero D (1990) Fermentation modelling. In: McNeil B, Harvey LM (eds) Fermentation. A practical approach. IRL Press, Oxford University Press, Oxford, pp 65–112
Spellman PT, Sherlock G, Zhang MQ, Iyer VR, Anders K, Eisen MB, Brown PO, Botstein D, Futcher B (1998) Comprehensive identification of cell cycle-regulated genes of the yeast *Saccharomyces cerevisiae* by microarray hybridization. Mol Biol Cell 9:3273–3297
Sprague GF Jr, Cullen PJ, Goehring AS (2004) Yeast signal transduction: regulation and interface with cell biology. In: Opresko LK, Gephart JM, Mann MB (eds) Advances in experimental medicine and biology. Kluwer/Plenum, New York, Advances in Systems Biology vol 547, pp 91–105
Stelling J (2004) Mathematical models in microbial systems biology. Curr Opin Microbiol 7:513–518
Stephanopoulos G (1999) Metabolic fluxes and metabolic engineering. Metab Eng 1:1–11
Stephanopoulos G, Gill RT (2001) After a decade of progress, an expanded role for metabolic engineering. Adv Biochem Eng Biotechnol 73:1–8
Stephanopoulos G, Vallino JJ (1991) Network rigidity and metabolic engineering in metabolite overproduction. Science 252:1675–1681
ter Kuile BH, Westerhoff HV (2001) Transcriptome meets metabolome: hierarchical and metabolic regulation of the glycolytic pathway. FEBS Lett 500:169–171
ter Linde JJ, Liang H, Davis RW, Steensma HY, van Dijken JP, Pronk JT (1999) Genome-wide transcriptional analysis of aerobic and anaerobic chemostat cultures of *Saccharomyces cerevisiae*. J Bacteriol 181:7409–7413
Teusink B, Baganz F, Westerhoff HV, Oliver SG (1998) Metabolic control analysis as a tool in the elucidation of the function of novel genes. In: Brown AJ, Tuite MF (eds) Methods in Microbiology, vol 26. Academic Press, London, pp 297–336
The International Human Genome Mapping Consortium (2001a) Initial sequencing and analysis of the human genome. Nature 409:860–921
The International Human Genome Mapping Consortium (2001b) A physical map of the human genome. Nature 409:934–941
Tilstone C (2003) Vital statistics. Nature 424:610–613
Ton VK, Rao R (2004) Functional expression of heterologous proteins in yeast: insights into Ca^{2+} signaling and Ca^{2+}-transporting ATPases. Am J Physiol Cell Physiol 287:580–589
Urbanczyk-Wochniak E, Luedemann A, Kopka J, Selbig J, Roessner-Tunali U, Willmitzer L, Fernie AR (2003) Parallel analysis of transcript and metabolic profiles: a new approach in systems biology. EMBO Rep 4:989–993
US Patent Office (2003) US Patent 2003228567. Compositions and methods for modelling *Saccharomyces cerevisiae* metabolism. US Patent Office, Alexandria, VA
Varma A, Palsson BØ (1994) Metabolic flux balancing: Basic concepts, scientific and practical use. Bio/Technology 12:994–998
Verger A, Perdomo J, Crossley M (2003) Modification with SUMO. A role in transcriptional regulation. EMBO Rep 4:137–142
von Mering C, Krause R, Snel B, Cornell M, Oliver SG, Fields S, Bork P (2002) Comparative assessment of large-scale data sets of protein–protein interactions. Nature 417:399–403
Wang S, Sim TB, Kim YS, Chang YT (2004) Tools for target identification and validation. Curr Opin Chem Biol 8:371–377
Weckwerth W (2003) Metabolomics in systems biology. Annu Rev Plant Biol 54:669–689
Wei GH, Liu DP, Liang CC (2004) Charting gene regulatory networks: strategies, challenges and perspectives. Biochem J 381:1–12
Wiechert W (2001) 13C metabolic flux analysis. Metab Eng 3:195–206
Winzeler EA, Shoemaker DD, Astromoff A, Liang H, Anderson K, Andre B, Bangham R, Benito R, Boeke JD, Bussey H et al. (1999) Functional characterization of the *S. cerevisiae* genome by gene deletion and parallel analysis. Science 285:901–906
Wodicka L, Dong H, Mittmann M, Ho MH, Lockhart DJ (1997) Genome-wide expression monitoring in *Saccharomyces cerevisiae*. Nat Biotechnol 15:1359–1367
Wohlschlegel JA, Yates JR (2003) Proteomics: where's Waldo in yeast? Nature 425:671–672
World Intellectual Property Organisation (2001a) WO Patent 0107567. Engineering of metabolic control. World Intellectual Property Organisation, Geneva
World Intellectual Property Organisation (2001b) WO Patent 0178652. Methods for drug discovery, disease treatment and diagnosis using metabolomics. World Intellectual Property Organisation, Geneva
Yao T (2002) Bioinformatics for the genomic sciences and towards systems biology. Japanese activities in the post-genome era. Prog Biophys Mol Biol 80:23–42
Yeang CH, Ideker T, Jaakkola T (2004) Physical network models. J Comput Biol 11:243–262
Yoon SH, Lee SY (2002) Comparison of transcript levels by DNA microarray and metabolic flux based on flux analysis for the production of poly-g-glutamic acid in recombinant *Escherichia coli*. Genome Informatics 13:587–588
Yoon SH, Han MJ, Lee SY, Jeong KJ, Yoo JS (2003) Combined transcriptome and proteome analysis of *Escherichia coli* during the high cell density culture. Biotechnol Bioeng 81:753–767

Zaragoza O, Lindley C, Gancedo JM (1999) Cyclic AMP can decrease expression of genes subject to catabolite repression in *Saccharomyces cerevisiae*. J Bacteriol 181:2640–2642
Zhu H, Snyder M (2003) Protein chip technology. Curr Opin Chem Biol 7:55–63
Zhu H, Bilgin M, Bangham R, Hall D, Casamayor A, Bertone P, Lan N, Jansen R, Bidlingmaier S, Houfek T et al. (2001) Global analysis of protein activities using proteome chips. Science 293:2101–2105
Zhu H, Bilgin M, Snyder M (2003) Proteomics. Annu Rev Biochem 72:783–812

2 Genome Evolution in Hemiascomycete Yeasts

L.J. Montcalm[1], K.H. Wolfe[1]

CONTENTS

[1] Department of Genetics, Smurfit Institute, University of Dublin, Trinity College, Dublin 2, Ireland

I. Introduction

The genome sequence of *Saccharomyces cerevisiae* was determined by an international consortium of more than 600 scientists (Goffeau et al. 1997). This landmark achievement took 7 years (1989–1996) and resulted in the first publication of the genome sequence of a eukaryote. Technology has advanced since then to the extent that, in 2003, the genomes of three other *Saccharomyces* species were sequenced in a few months and published together in a paper with only five authors (Kellis et al. 2003). This was quickly followed by the publications of genome sequences (either draft

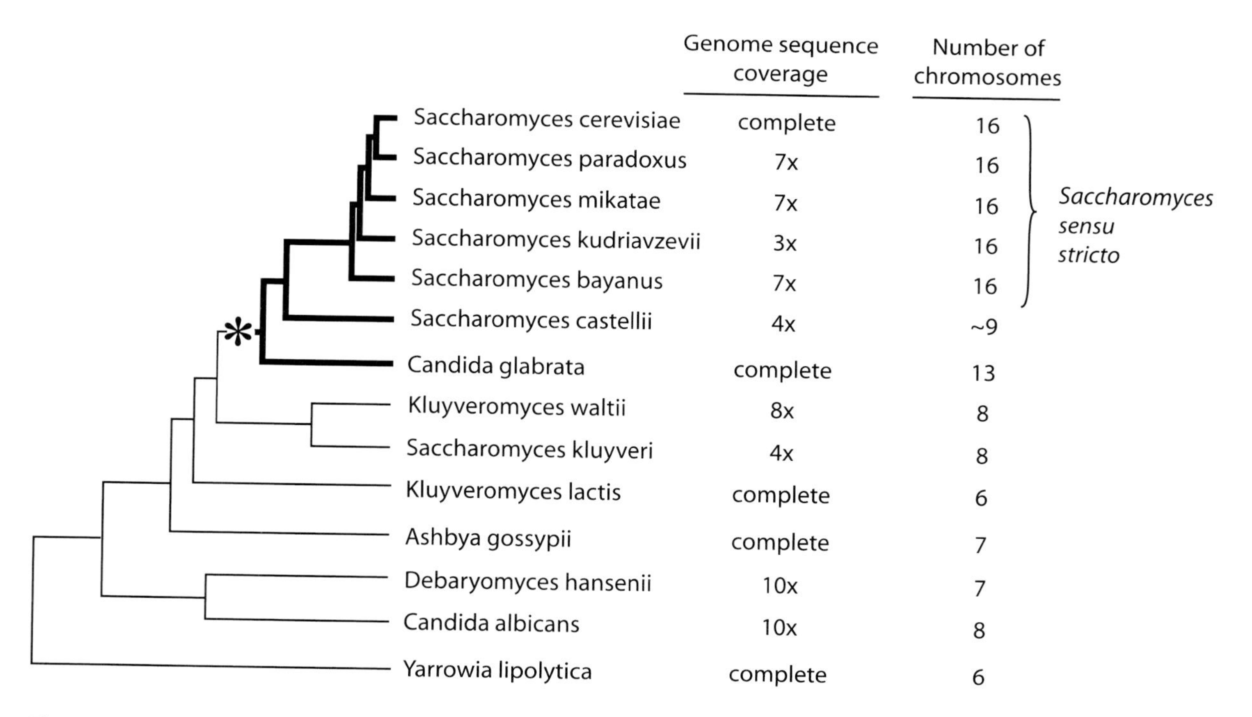

Fig. 2.1. Approximate phylogenetic relationships among the hemiascomycetes whose genomes have been sequenced. The *asterisk* indicates the likely position of the whole-genome duplication event (WGD), and the *thickened branches* show the post-WGD species

The Mycota XIII
Fungal Genomics
Alistair J.P. Brown (Ed.)

or complete) for several other yeasts, with the result that today genome data are available for 14 species in the class Hemiascomycetes (Fig. 2.1).

The phylogenetic relationships among the sequenced genomes are not yet fully established. The topology shown in Fig. 2.1 is based on the work of Kurtzman and Robnett (2003) who sequenced six genes from every species in the '*Saccharomyces* complex' of species, combined with data from Cai et al.'s (1996) 18S rRNA tree for the more deeply branching lineages, and Rokas et al.'s (2003) 106-gene phylogeny for the *Saccharomyces* sensu stricto species. In particular, it is not certain whether the three separate lineages leading to *Ashbya gossypii*, *Kluyveromyces lactis* and *K. waltii/S. kluyveri* as shown in Fig. 2.1 really are distinct lineages. The key internal branches have low bootstrap values in Kurtzman and Robnett's (2003) analysis, and in a recent analysis involving the same set of 106 genes there was a strongly supported clade comprising *S. kluyveri*, *A. gossypii* and *K. waltii*, with *K. lactis* outside (Hittinger et al. 2004). Comprehensive reconstruction of the phylogenetic relationships among these species, using the sequences of all the orthologous genes from all the sequenced genomes, remains to be done.

The lack of fossil records for hemiascomycetes makes it difficult to put an absolute timescale onto the phylogenetic tree in Fig. 2.1. Molecular clock estimates of the date for the divergence between *Candida albicans* and *S. cerevisiae* have ranged between 140 and 330 million years (Berbee and Taylor 1993; Pesole et al. 1995). It is perhaps more instructive to compare the average levels of protein sequence divergence between *S. cerevisiae* and other yeast species to the divergence between human and other animals (Dujon et al. 2004). By this yardstick, *C. glabrata* is slightly more divergent from *S. cerevisiae* than the pufferfish *Fugu rubripes* is from humans, and the most divergent species in Fig. 2.1, *Yarrowia lipolytica*, is about as divergent as the sea squirt (tunicate) *Ciona intestinalis*. Thus, the available genome data from the hemiascomycetes span a complete range of levels of sequence divergence. This will be further augmented in the years to come by the addition of more genome sequences from species closely related to some of the terminal taxa in Fig. 2.1 other than *S. cerevisiae*; for example, two species closely related to *C. albicans* (*C. dubliniensis* and *C. tropicalis*) are currently being sequenced, and so for any gene it will soon become possible to examine its evolution over short time periods in two independent groups of species: a *Saccharomyces* group and a *Candida* group.

II. Colinearity of Gene Order

One of the most striking features of yeast genome evolution is the degree to which the order of genes along chromosomes has been conserved between species. For the *Saccharomyces* sensu stricto species (Fig. 2.1), colinearity with *S. cerevisiae* approaches 100%. This conservation of gene order has proved very useful in allowing us to differentiate real genes from spurious ORFs, and in facilitating the detection of rapidly evolving loci (Brachat et al. 2003; Kellis et al. 2003; Zhang and Dietrich 2003).

Among the sensu stricto species, there are only a few places in the genome where colinearity is disrupted, and these disruptions fall into two types. First, there are a few differences in gene content between sensu stricto species, where a gene is present in one species but either absent or a pseudogene in another species (Bon et al. 2000; Fischer et al. 2001; Kellis et al. 2003). An example is the *XYZ3* locus beside the centromere of *S. cerevisiae* chromosome III, which is a pseudogene homologous to another gene (*DOM34*, a member of the pelota gene family) on chromosome XIV (Lalo et al. 1993). In the other sensu stricto species *S. paradoxus* and *S. bayanus*, the ortholog of *XYZ3* is an intact and apparently functional gene (Kellis et al. 2003; it is also a pseudogene in *S. mikatae*). Second, although all the sensu stricto species have 16 chromosomes, there have been several interchromosomal translocations during the evolution of these species (Fischer et al. 2000; Kellis et al. 2003), most of which are reciprocal translocations, where parts of two arms are reciprocally exchanged between two chromosomes. In addition, many intrachromosomal inversions have been mapped among the sensu stricto species. All the endpoints of inversions coincide with tRNA genes, and many of the reciprocal translocations have endpoints in Ty elements or other repeated sequences, suggesting that homologous recombination between these elements may have catalyzed the translocations.

As one moves out to more distantly related species, the colinearity of gene order begins to break down (Keogh et al. 1998; Llorente et al. 2000b; Wong et al. 2002). Between *S. cerevisiae* and *C. glabrata* (Dujon et al. 2004), only about 58% of genes have the same neighbors due to an increasing number of chromosomal rearrangements and loss of alternative copies of duplicated genes in the two species. Between *S. cerevisiae* and *A. gossypii*, there are about 500 breaks in synteny caused by interchromosomal rearrangements that have occurred since those species shared a common ancestor. Between *S. cerevisiae* and *C. albicans*, only about 9% of genes that are immediate neighbors in one species are also immediate neighbors in the other, and gene order is often seen to have become scrambled by the action of numerous inversions of segments of DNA containing a few genes (Seoighe et al. 2000).

Colinearity of gene order among species breaks down quite noticeably in the subtelomeric regions, which span the terminal 30–50 kilobase pairs of each chromosome (Dietrich et al. 2004; Dujon et al. 2004; Kellis et al. 2004). This is also apparent in comparisons among the *Saccharomyces* sensu stricto species (Kellis et al. 2003; Maciaszczyk et al. 2004), and in some cases even among different isolates of *S. cerevisiae* (Daran-Lapujade et al. 2003; Nomura and Takagi 2004). Subtelomeric regions are notable for the presence of some groups of co-functional genes – such as the *MAL* and *SUC* genes required for growth on particular carbon sources, or the *BIO* genes for synthesis of the enzyme cofactor biotin – and transcriptional regulation of these clusters may involve epigenetic mechanisms such as chromatin modification (Robyr et al. 2002; Halme et al. 2004). Subtelomeric regions are also notable for having high sequence similarity among different chromosomes, and for frequently containing damaged genes (i. e., recently formed pseudogenes; Harrison et al. 2002; Lafontaine et al. 2004). These parts of the genome may be 'evolution's cauldron': a sort of combined workshop and junkyard where dynamic DNA turnover rapidly creates new genes or groups of genes, epigenetic regulation is used as a way of testing these gene arrangements in a controlled manner, and ultimately natural selection either retains or discards them (Eichler 2001; Kent et al. 2003).

III. Genomic Evidence for Whole-Genome Duplication

Comparisons of gene order in the hemiascomycetes are complicated by the presence of a whole-genome duplication (WGD) that occurred during the evolution of some species in this group, including *S. cerevisiae* (Wolfe and Shields 1997). The most likely phylogenetic position of this event is shown by the asterisk in Fig. 2.1, and it divides the hemiascomycete species into two groups: 'pre-WGD' and 'post-WGD'. In this shorthand nomenclature, 'post-WGD' means a species that is a descendant of the WGD event (the species marked by thick branches in Fig. 2.1), and 'pre-WGD' means a species that separated from the *S. cerevisiae* lineage before the WGD event happened (thin branches in Fig. 2.1).

The WGD resulted in a transient doubling of the number of genes and chromosomes in an ancestral yeast, but this was followed quickly by the loss of many duplicated genes (Fig. 2.2). Today, the *S. cerevisiae* genome contains more than 500 pairs of duplicated genes (ohnologs) that were formed simultaneously by this event (Dietrich et al. 2004; Dujon et al. 2004; Kellis et al. 2004; Wolfe 2004). The functional consequences of the WGD in *S. cerevisiae* have been reviewed recently elsewhere (Kellis et al. 2004; Piskur and Langkjaer 2004; Wolfe 2004; Wong and Wolfe 2005) and will not be discussed here. Because there have been relatively few genomic rearrangements since the WGD, many of the ohnologs are still arranged in a pattern that betrays their origins: a series of genes on one chromosome has a series of homologs on another chromosome, generally with conserved order and orientation of the duplicated genes (Wolfe and Shields 1997). In between the duplicated genes, there are many singleton genes on each chromosome. These are loci that were originally duplicated as part of the WGD, but where one of the gene copies was subsequently lost. Most likely there was no evolutionary advantage to retaining both copies of the gene, and so there was no selection against a mutation that inactivated or deleted one of the copies. In *C. glabrata*, another post-WGD species, the number of genes that have been retained in duplicate is less than in *S. cerevisiae* (about 404 pairs), which means that when the *C. glabrata* genome is compared to itself, the duplicated blocks that can be detected are shorter and fewer than in *S. cerevisiae* (Dujon et al. 2004). The patterns of post-WGD gene losses that are seen in *S. cerevisiae*, *C. glabrata* and *S. castellii* are similar, which suggests that these three species are all descended from the same WGD event (D.R. Scannell, K.P. Byrne and K.H. Wolfe, unpublished data).

Comparing gene order between pre-WGD and post-WGD species reveals a 1:2 relationship (Keogh

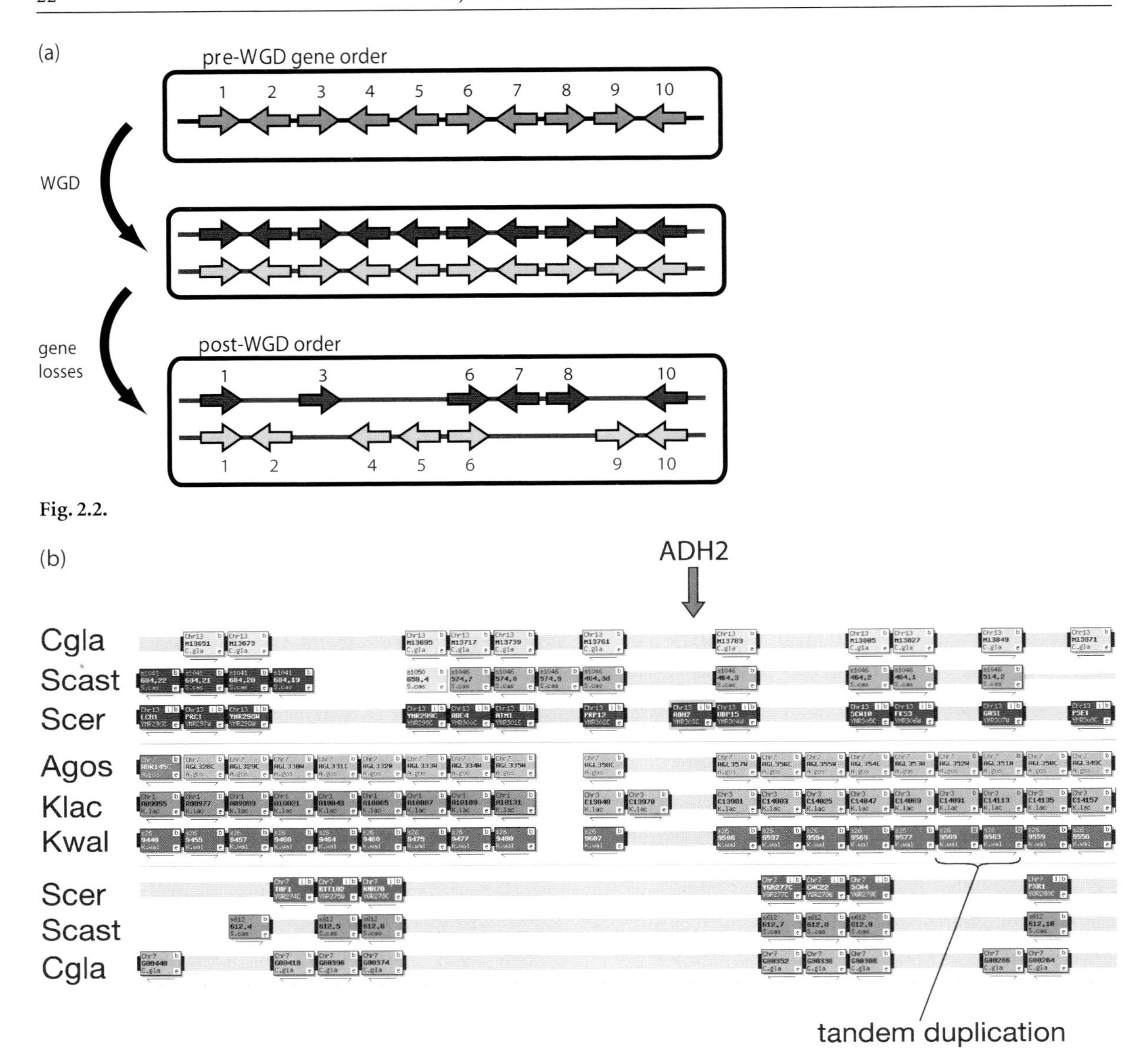

Fig. 2.2.

Fig. 2.3. (**a**) Illustration of our model of gene order evolution following whole-genome duplication (WGD). The *box at the top* shows a hypothetical region of chromosome containing ten genes numbered 1–10. After WGD, the whole region is briefly present in two copies. However, many genes subsequently return to single-copy state because there is no evolutionary advantage to maintaining both copies. In this example, only genes 1, 6 and 10 remain duplicated. However, the arrangement of these three homolog pairs in the post-WGD species (*bottom*) would be sufficient to allow the sister regions to be detected using that genome sequence alone. Also, the order of genes in sister regions in post-WGD species have well-defined relationships to the gene order that existed in the pre-WGD genome (*top*), which will also be similar to the gene order seen in any species that diverged from the WGD lineage before the WGD occurred (from Haber and Wolfe 2005, based on Keogh et al. 1998). (**b**) (next page) An example of gene order relationships between parts of two sister regions in the post-WGD species *S. cerevisiae* (*Scer*: chromosome XIII genes are darkly shaded and chromosome VII genes lightly shaded), *S. castellii* (*Scast*) and *C. glabrata* (*Cgla*), and the single homologous genomic region in the pre-WGD species *A. gossypii* (*Agos*), *K. lactis* (*Klac*) and *K. waltii* (*Kwal*). In this representation, each *rectangle* represents a gene, and homologs are arranged as *vertical columns. Arrows below the rectangles* show transcriptional orientation. *Shaded lines* connect genes that are adjacent but do not indicate the actual gene spacing on the chromosome. The *ADH2* gene, which is present only in *S. cerevisiae* (and other sensu stricto species not shown here), is highlighted, as is a tandem duplication shared by the pre-WGD but not the post-WGD species. This image is a screenshot from a Yeast Gene Order Browser (YGOB) currently under development in our laboratory (K.P. Byrne and K.H. Wolfe, unpublished data)

et al. 1998; Seoighe and Wolfe 1999). For any genomic region in a pre-WGD species such as *K. lactis*, there are two corresponding genomic regions in post-WGD species such as *S. cerevisiae* or *C. glabrata*, each of which contains a subset of the genes present in the region in the pre-WGD species (Fig. 2.2). The recent demonstrations that this 1:2 relationship between pre-WGD and post-WGD species extends across essentially the entire genomes, both of *S. cerevisiae* and of the pre-WGD species (Wong et al. 2002; Dietrich et al. 2004; Dujon et al. 2004; Kellis et al. 2004), provides direct proof that the WGD hypothesis is correct, despite the earlier controversy (Feldmann 2000; Llorente et al. 2000a, b; Souciet et al. 2000). A similar 1:2 gene order relationship is seen between the plants maize (which is an ancient polyploid) and rice (Lai et al. 2004). The only exceptions in yeasts to this 1:2 rule are in places where chromosomal rearrangements in one species make it difficult to track the synteny relationships (Fig. 2.2), and in the subtelomeric regions, where wholesale movement of genes appears to have taken place in every species. The WGD hypothesis has now met with widespread acceptance in the yeast community (Dujon et al. 2004; Piskur and Langkjaer 2004).

Two misunderstandings of the WGD hypothesis are often made and deserve some discussion. First, it is incorrect to think that the majority of duplicated genes in the *S. cerevisiae* genome were formed by the WGD event. In fact, only a few hundred pairs of duplicated genes in *S. cerevisiae* survive from the WGD (we originally identified 376 pairs in 1997, but by leveraging information from other genome sequences the number is now known to be about 551 pairs; Wolfe 2004; K.P. Byrne and K.H. Wolfe, unpublished data). Every gene in the genome was duplicated transiently, but at about 90% of locus pairs, one of the two copies of the gene was subsequently deleted again. Because the genome that underwent duplication was itself comprised of many multigene families, like any other eukaryotic genome, many *S. cerevisiae* genes have paralogs that were formed by duplications that happened independently of the WGD, but none of those duplications were multigene events involving large segments of chromosome. Second, there is nothing special about the 50% of the genome that we were able to arrange into 55 paired regions in our original analysis (Wolfe and Shields 1997). The hypothesis is that the whole genome underwent duplication, and so in principle every point in the genome should have a sister point somewhere else. The 55 paired regions that we originally reported are simply the parts of the genome that have retained the highest densities of duplicated genes, which made it possible to recognize the sisters.

IV. Pairs of Centromeres

Whole-genome duplication should have the effect of doubling the number of chromosomes, and chromosome number is likely to remain unchanged during the process of gene loss outlined in Fig. 2.2a. It has long been known from pulsed-field gel electrophoresis (PFGE) experiments that some yeasts, such as the *Saccharomyces* sensu stricto and *C. glabrata*, are characterized by having a large number of relatively small chromosomes as compared to other species (Sor and Fukuhara 1989; Vaughan-Martini et al. 1993), and it is now apparent that there is indeed an approximate doubling of the number of chromosomes in post-WGD species (Fig. 2.1). However, the arithmetic is not precise: the pre-WGD species have six, seven or eight chromosomes, and the post-WGD species have 16 (*Saccharomyces* sensu stricto), 13 (*C. glabrata*) or even nine (if a PFGE estimate for *S. castellii* is accurate; Petersen et al. 1999). How chromosome number can change during evolution is discussed below.

The number of chromosomes in a species is determined primarily by the number of centromeres. For the species whose genomes have been sequenced to completion or to high-coverage drafts, it is now possible to track the evolutionary history of each centromere (Fig. 2.4). Except in *S. cerevisiae* and *K. lactis* (Heus et al. 1993), the locations of centromeres have been predicted based on matches to the CDE I–CDE II–CDE III consensus in the genome sequence, rather than by laboratory experiments, but in each case there is only one match to the consensus per chromosome, and so the predictions are probably accurate.

Based on the identities of the protein-coding genes flanking the centromeres, there is a straightforward 2:1 relationship between the 16 centromeres of *S. cerevisiae* and the eight centromeres of *K. waltii* (Fig. 2.4; Kellis et al. 2004), which supports the hypothesis that the WGD involved an eight-chromosome ancestor turning into a 16-chromosome descendant (Keogh et al. 1998; Wong et al. 2002). *A. gossypii* has only seven centromeres, all of which are at positions that are also centromeres in *K. waltii* (Fig. 2.4).

If we assume that the ancestral number of chromosomes in the pre-WGD species was eight, then *A. gossypii* has lost the equivalent of the centromere of *K. waltii* chromosome 5 (*KwCEN5*). The genes flanking *KwCEN5* are found on two different chromosomes of *A. gossypii* and are not near centromeres. Only three of *K. lactis*'s six centromeres (*KlCEN2*, *KlCEN4* and *KlCEN5*) are in colinear relationships with *K. waltii* centromeres, both to the left and to the right of the centromere (Fig. 2.4). The other three *K. lactis* centromeres (*KlCEN1*, *KlCEN3* and *KlCEN6*) have synteny with a *K. waltii* centromere on one chromosome arm, but there is a rearrangement breakpoint immediately on the other side of the centromere. The most interesting aspect of these centromere comparisons among pre-WGD species is that there are two places in the *K. lactis* genome that show gene order conservation with the regions spanning centromeres in both *K. waltii* and *A. gossypii*, but which are not centromeres in *K. lactis* (marked by X symbols in Fig. 2.4). These 'non-centromeric' regions in *K. lactis* are the counterparts of *KwCEN7/AgCEN3* and *KwCEN2/AgCEN5*. They occur on *K. lactis* chromosomes 4 and 5 at sites hundreds of kilobases distant from the true centromeres of those chromosomes. If the ancestral number of chromosomes was eight and *K. lactis* underwent a reduction to six, then these are the two sites from which centromere function has been lost without further chromosomal rearrangement.

In the post-WGD species, at least 12 and probably all 13 of *C. glabrata*'s centromeres are directly orthologous to centromeres in *S. cerevisiae* (see legend to Fig. 2.4 for details of the one debatable case). The difference in chromosome number between *C. glabrata* and *S. cerevisiae* is again accounted for by the presence of 'non-centromeric' regions: three places in the *C. glabrata* genome that show conserved gene order with the regions flanking three centromeres in *S. cerevisiae* (*ScCEN10*, *ScCEN11* and *ScCEN14*), but which are not centromeres in *C. glabrata* (Fig. 2.4).

The above comparisons make it very probable that the ancestral number of chromosomes was eight, and the WGD doubled this to 16. The reduced numbers of chromosomes in *K. lactis*, *A. gossypii* and *C. glabrata* are probably due to chromosome fusions that resulted in the death of centromeres at the 'non-centromeric' sites described above (because chromosome fusions are only evolutionarily viable if one of the two centromeres becomes inactive, or if the chromosome breaks again). An ancestral number of eight seems more plausible than (say) seven, because under a model of a seven-chromosome ancestor undergoing WGD to produce a 14-chromosome descendant, there must subsequently have been a gain of one centromere in *K. waltii* and a gain of two centromeres in *S. cerevisiae* and, eerily, these centromeres must have been gained at equivalent positions on the three chromosomes. At the moment, there are no clear examples of centromeres being gained during hemiascomycete evolution other than by the WGD mechanism. In *C. albicans* there are eight chromosomes and their centromeres have been identified by biochemical methods (Sanyal et al. 2004), but the genes flanking these centromeres bear little or no relationship to the centromere groups identified in Fig. 2.4 (data not shown). This may just be a consequence of the larger evolutionary distance and more numerous rearrangements between *C. albicans* and the *S. cerevisiae* group of species (Seoighe et al. 2000), though the *C. albicans* gene order does seem to be even more rearranged than one might expect from its evolutionary distance alone.

V. A Model for Genome Duplication

Although it is clear that the genome became doubled in an ancestor of *S. cerevisiae* and the other post-WGD species, the exact details of what happened have not yet been worked out. It is convenient to refer to this event as a genome duplication, but more accurate to call it a polyploidization. Polyploidy can arise by either of two routes, depending on whether the two constituent genomes are identical (autopolyploidy) or different (allopolyploidy). It seems increasingly likely that the WGD event was an allopolyploidy, for several reasons. First, in an autopolyploid the two incoming sequences of every gene are identical, and so they cannot have diverged at all in function. This means that, in the immediate aftermath of an autopolyploidy event, the only selective force that can lead to the survival of both copies is selection for increased dosage of the locus. Although some loci, such as the cytosolic ribosomal protein genes of *S. cerevisiae*, probably have been retained due to selection for increased dosage, this explanation seems applicable only to a minority of the ohnologs that have been retained by *S. cerevisiae* (Wolfe 2004). Second, there are at least two documented instances of interspecies hybridizations that occurred during the

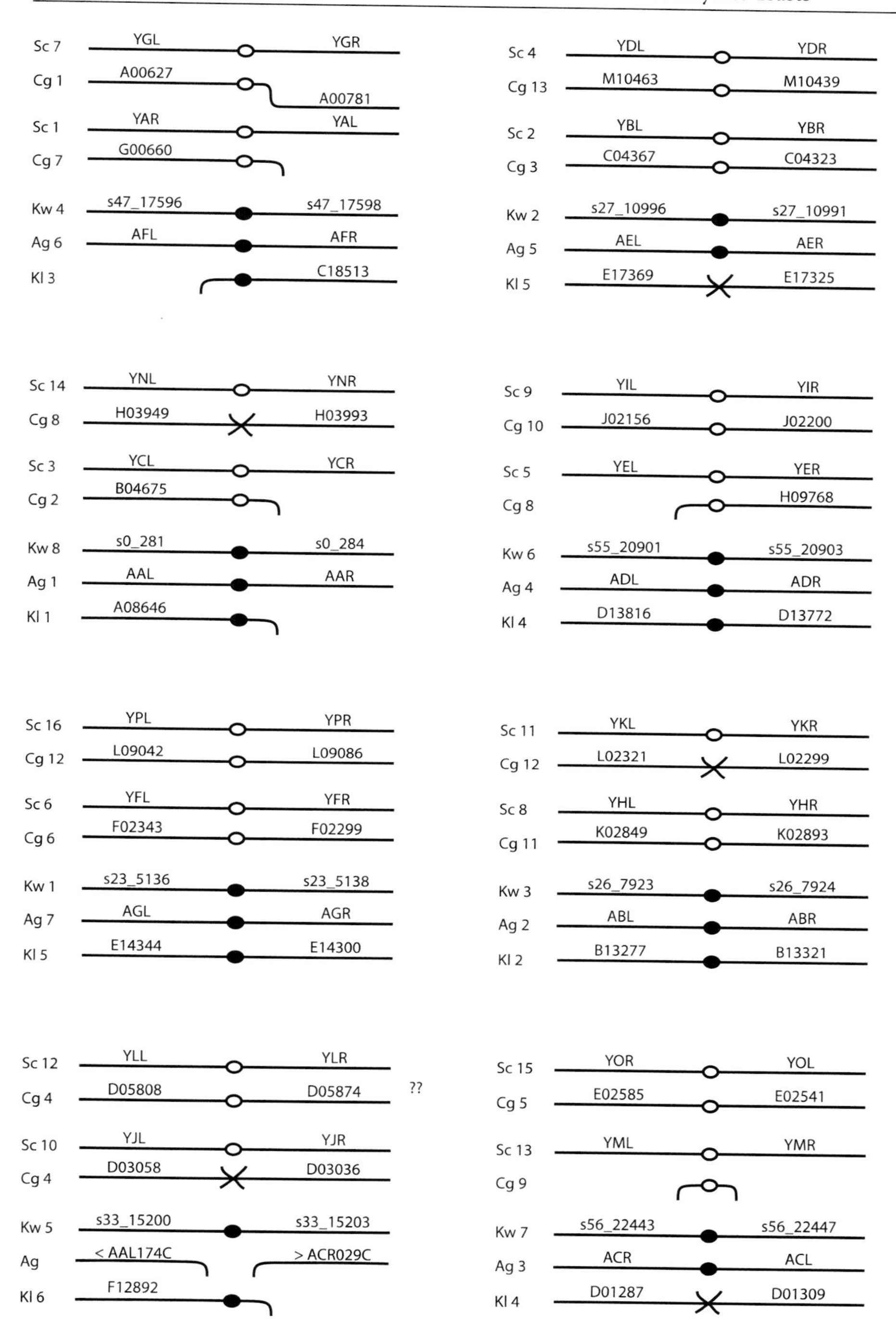

Fig. 2.4. Relationships among centromeric regions of *S. cerevisiae* (*Sc*), *C. glabrata* (*Cg*), *K. waltii* (*Kw*), *A. gossypii* (*Ag*) and *K. lactis* (*Kl*) chromosomes. Each of the eight panels shows a group of genomic regions that are related by virtue of their gene contents close to a centromere (or a similar non-centromeric region). *Closed circles* represent centromeres in pre-WGD species, and *open circles* represent centromeres of post-WGD species. *X symbols* indicate the absence of a centromere. Names indicate chromosome arms, or genes close to the centromere on each arm. *Hooked lines* indicate loss of relatedness. The assignment of the centromere of *C. glabrata* chromosome 9 to the group on the *bottom right* is less certain than for the other centromeres, and is based on the linkage of *CgCEN9* to the gene *CAGL0I08107g*, which is an ortholog of the genes *KLLA0D01243g* (*K. lactis*) and *s56_22439* (*K. waltii*) that are close to the corresponding regions in those species. Gene names in *K. lactis* and *C. glabrata* have been shortened by writing *A00627* instead of *CAGL0A00627g*, etc.

recent evolution of hemiascomycetes, which makes an older hybridization plausible, too. Hybridization between *S. cerevisiae*-like and *S. bayanus*-like ancestors formed the brewing yeast *S. pastorianus* (Hansen and Kielland-Brandt 1994, 2003), and hybridization between *Zygosaccharomyces rouxii* and a related species formed an allopolyploid that is used in the miso and soy sauce industries (Kinclova et al. 2001; J. Gordon and K.H. Wolfe, unpublished data). Third, the pattern of gene loss and retention in different post-WGD species is suggestive of allopolyploidy (D.R. Scannell, K.P. Byrne and K.H. Wolfe, unpublished data).

Here we propose a simple model for how an allopolyploidy could have resulted in the WGD (Fig. 2.5). Much of the genetics literature about polyploidy has been centered on studies of polyploid plants or animals (Coyne and Orr 2004), and yeasts differ somewhat from these organisms because they are able to grow mitotically as either haploids or diploids. In a yeast, allopolyploidy can occur if gametes from two different species fuse to form a hybrid zygote. Unless the two parental species are very closely related, the hybrid will be sterile because chromosomal rearrangements will prevent it undergoing meiosis (Delneri et al. 2003), and so it will grow only mitotically and may do so for many generations. One way for the hybrid to escape from this life of abstinence and regain the ability to reproduce sexually is if one allele at the *MAT* locus gets deleted spontaneously, which has the effect of turning a diploid into a haploid (Fig. 2.5). In other words, the *MAT* locus simply needs to be among the many duplicated loci that lost one copy in the aftermath of the WGD, during mitotic replication of the hybrid cell lineage. Once the *MAT* locus becomes single-copy, HO endonuclease will catalyze mating-type switching in mother cells. This will enable mating between mother and daughter cells, forming diploids with homogeneous genomes whose chromosome number will be the sum of the numbers of chromosomes in the parental species (Fig. 2.5).

It should be noted that the model sketched in Fig. 2.5 is also applicable to an autopolyploidization event, if a deletion of one *MAT* allele occurs in a normal diploid cell before it undergoes meiosis. However, for the reasons outlined above, we think that an allopolyploidization was more probable. It is also noteworthy that this model is plausible only for yeasts that have HO endonuclease, and that HO originated in yeasts only shortly before the WGD happened (Butler et al. 2004; see Sect. VII).

What is also surprising about the evolution of the yeast genome is that, despite the common occurrence of aneuploidy in industrial or laboratory strains (Hughes et al. 2000; Hansen and Kielland-Brandt 2003), and despite the frequency at which segmental duplications (single events that duplicate large regions of chromosome) can arise in laboratory experiments that select for increased copy number of a gene (Koszul et al. 2003), these types of event seem to have been quite rare during the natural evolution of yeasts. In the complete genomes of four *Saccharomyces* sensu stricto species, there is only one example of a segmental duplication other than in a subtelomeric region (Kellis et al. 2003).

VI. Pseudogenes, Relics and Null Alleles

It was originally thought that the *S. cerevisiae* genome contains very few pseudogenes, because the genome is compact and there is relatively little non-coding DNA. The *XYZ3* example mentioned above was one of the first descriptions of an *S. cerevisiae* pseudogene (Lalo et al. 1993), and there was so much doubt about its veracity that it was re-sequenced several times by several laboratories before it was finally accepted as real. More recently, careful analysis has revealed the existence of 120 pseudogenes in *S. cerevisiae* that are so degenerated that they have been called 'relics' (Lafontaine et al. 2004), of which about 40% are at internal (non-telomeric) sites in chromosomes. The severe truncation and extensive sequence divergence in the relics had the effect that, in many cases, they were found only through the painstaking use of thousands of dot-matrix plots to further investigate very short BLASTN hits between known genes and putatively relic-containing intergenic regions (Lafontaine et al. 2004).

Two other types of pseudogenes are notable. First, there are several examples in *S. cerevisiae* where a pair of homologous open reading frames (ORFs) exist, but one of these is much larger than the other. An early example of this is the *CDC4* gene on chromosome VI, which codes for a 779-residue protein and has a truncated homolog on chromosome V (*YER066W*, a 186-codon ORF; Yochem and Byers 1987). In cases such as this, it is often unclear whether the second locus is a pseudogene that contains a fortuitous ORF, or whether the shorter gene has a function. Second, there are several loci

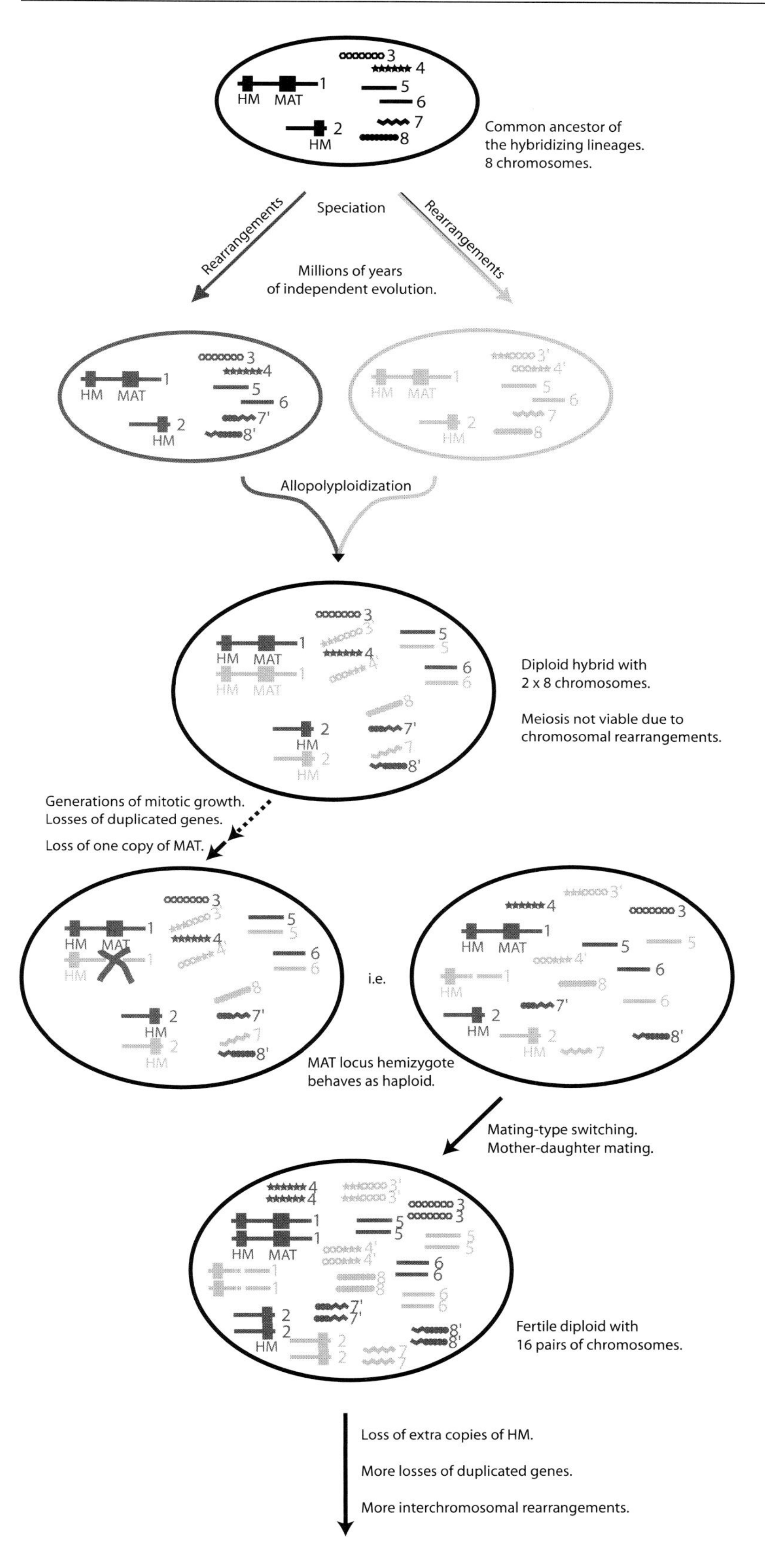

Fig. 2.5. A model for how an allopolyploidization could result in WGD, if one copy of the *MAT* locus becomes deleted after hybridization. *HM* refers to the *HML* and *HMR* cassettes of silent mating-type information, which are located on two different chromosomes in all species other than *S. cerevisiae* (Dujon et al. 2004)

where a gene in the reference *S. cerevisiae* genome sequence (from strain S288c) contains an internal stop codon, but the gene is known to be intact and functional in other laboratory strains of *S. cerevisiae* such as W303 or Σ1278b. Examples include *FLO8*, *CRS5*, *YOL153C* and *SDC25* (Damak et al. 1991; Liu et al. 1996; Dujon et al. 1997), and these are probably better described as null alleles in *S. cerevisiae* S288c, rather than pseudogenes. Some of these null alleles may have been selected for during the process of turning *S. cerevisiae* into a tractable laboratory organism (Hall and Linder 1993).

VII. Interspecies Differences in Gene Content

A. Genes Gained

One of the most striking differences among the sequenced hemiascomycete genomes is the extent to which tandemly duplicated genes are present in the genomes. Tandem repeats are quite rare in *S. cerevisiae* (<2% of genes), but in *Debaryomyces hansenii* about 10% of genes are part of a tandem array (Dujon et al. 2004). Species-specific tandem repeats are one of the most frequent causes of non-colinearity in genomic regions aligned between species (Fig. 2.2b). In *S. cerevisiae* one of the principal attributes of tandem arrays is that they can become further amplified under particular environmental conditions where there is strong selection for increased copy number of the gene – e.g., at the *CUP1* metallothionein locus (Fogel and Welch 1982) and the *PMR2* locus coding for a sodium ion pump (Wieland et al. 1995). Hence, it is possible that the tandemly repeated loci in other species are also involved in responses to environmental change.

Although there have been many expansions and contractions of gene families in different yeast species (Dujon et al. 2004), the great majority of genes in each hemiascomycete have orthologs in all the other species. One gene family expansion of great economic importance is the gene duplication that gave rise, specifically in the *Saccharomyces* sensu stricto lineage, to the *ADH2* gene by duplication of the *ADH1* alcohol dehydrogenase gene (Fig. 2.2b). *ADH2* codes for an enzyme that efficiently converts ethanol to acetaldehyde and is expressed only in aerobically growing cells (Wills 1976). This duplication is hypothesized to have enabled *S. cerevisiae* to adopt a lifestyle whereby it uses Adh1 to rapidly ferment carbon sources to make ethanol (a product that is toxic to many competing microorganisms), and then later uses Adh2 to respire the ethanol (Ashburner 1998; Benner et al. 2002). Without Adh2 it is unlikely that the sensu stricto yeasts would have evolved their vigorous fermentation phenotype.

There are a few interesting exceptions to the general rule of gene conservation across the hemiascomycetes – loci where either a clade of species has picked up a new gene by horizontal gene transfer, or where a species has lost a gene (or a set of genes) because they are no longer necessary due to a change in the lifestyle of the organism. Examples of these gene content differences are discussed below.

One clear case of horizontal gene transfer into a group of yeasts is *URA1*, the gene that in *S. cerevisiae* codes for dihydroorotate dehydrogenase (DHODase), the fourth step in the pathway of de novo pyrimidine biosynthesis (Gojkovic et al. 2004). There are two types of DHODase enzyme in eukaryotes, with very different sequences. The ancestral eukaryotic form of DHODase is mitochondrial and is found in plants, animals and most fungi. In contrast, the *URA1* DHODase present in *S. cerevisiae* is more closely related to DHODases of bacteria (specifically *Lactococcus lactis*) than to the ancestral eukaryotic enzyme, and is cytosolic in *S. cerevisiae*. Some yeasts, such as *S. kluyveri*, *K. lactis* and *K. waltii*, have kept genes for both types of DHODase (Dujon et al. 2004; Gojkovic et al. 2004; Kellis et al. 2004; Piskur and Langkjaer 2004). In *S. cerevisiae* and *S. castellii*, the ancestral gene has been lost and only the horizontally transferred bacterial *URA1* remains – an example of gene displacement. The driving force behind this shift may have been selective pressure on yeasts for increased ability to grow anaerobically (Gojkovic et al. 2004; Piskur and Langkjaer 2004). The mitochondrial DHODase uses a quinone as an electron acceptor. Hence, pyrimidine synthesis is coupled to mitochondrial respiration, whereas the cytosolic *URA1* enzyme instead uses fumarate as its electron acceptor, and fumarate is made via the glyoxylate cycle in the absence of respiration (Gojkovic et al. 2004). The adoption of *URA1* allows *S. cerevisiae* and *S. castellii* to make pyrimidines even in the absence of oxygen. Rather surprisingly, *C. glabrata* has done the opposite to *S. cerevisiae*: it lost the *URA1* ortholog and retained only the gene for the mitochondrial DHODase (*CAGL0M12881g*) – an example of failed horizontal gene transfer.

The gene for HO endonuclease, which catalyzes mating-type switching in *S. cerevisiae* by making a double-strand break at the *MAT* locus, is found only in species quite closely related to *S. cerevisiae* (it is present in all the post-WGD species and in one pre-WGD lineage represented by *Zygosaccharomyces rouxii*; Butler et al. 2004). The existence of HO has the consequence that natural isolates of *S. cerevisiae* grow as diploids, with only a very brief haploid phase in their life cycle. In contrast, most pre-WGD lineages such as *K. lactis* grow primarily as haploids; they switch mating type only rarely and they sporulate immediately after mating (Kurtzman and Fell 1998; Zonneveld and Steensma 2003). The evolutionary advantage conferred by HO is pseudohomothallism: a new, sexually reproducing population of *S. cerevisiae* can be established by just a single colonizing cell, whereas in a heterothallic species such a *K. lactis* it is necessary for two cells of opposite mating type to be present if the new population is to reproduce sexually. Thus, HO is likely to facilitate the dispersal of a species. Sexual reproduction is important for the long-term evolutionary success of a species because without recombination all chromosomes will accumulate some deleterious mutations by Muller's ratchet. Hence, mitotic (asexual) reproduction is not a stable long-term strategy (Berbee and Taylor 1993). Horizontal gene transfer is implicated in the origin of HO, because it is derived from an intein – a selfish genetic element found primarily in bacterial genomes – and because its closest homolog, the intein in the *S. cerevisiae* gene *VMA1*, has been horizontally transferred among yeast species (Gimble and Thorner 1992; Gimble 2000; Koufopanou et al. 2002; Okuda et al. 2003; see Haber and Wolfe 2005 for review).

A third possible candidate for horizontal gene transfer is *SIR1*, whose function is linked to that of HO (Fabre et al. 2005). The Sir1 protein's role is to mediate the recruitment of the silencer complex Sir2/Sir3/Sir4 to the silent *HM* cassettes of mating-type information. The Sir proteins alter the chromatin conformation of the *HM* cassettes so that they are both transcriptionally inactive and resistant to cleavage by HO endonuclease. Although Sir2, Sir3 and Sir4 have a broad phylogenetic distribution (Astrom and Rine 1998), Sir1 is restricted to almost the same set of species as those that have HO: it is present in the *Saccharomyces* sensu stricto species and *S. castellii*, but not in any of the studied pre-WGD species, except for *Z. rouxii* (Haber and Wolfe 2005). Every studied species has either both HO and *SIR1*, or neither, with the sole exception of *Candida glabrata*. Fabre et al. (2005) recently concluded that *HO* appeared in yeasts earlier than *SIR1*, but we disagree with their interpretation because (1) there is a *SIR1* homolog in *Z. rouxii* (which implies an older origin of the gene, and secondary loss in *C. glabrata*), and (2) the *K. lactis HO*-like sequence they describe does not seem to be a true ortholog of *HO* (it is an outlier to the *HO/VMA1* clade in phylogenetic trees, and instead clusters with the *S. castellii* intein-like gene Scas_542.2; data not shown).

B. Genes Lost

The converse of gene gain is gene loss. One of the benefits of complete genome sequencing is that it allows one to make definitive statements about what is absent from a genome, as well as what is present, and several examples have recently been found of complete biochemical pathways that have been lost in one or more hemiascomycete species.

S. kudriavzevii cannot grow on galactose as a sole carbon source, and analysis of its genome showed that seven *GAL* loci are all pseudogenes in this species (Hittinger et al. 2004). All the genes that are exclusively involved in galactose catabolism have become pseudogenes, but other genes that have additional roles in non-galactose aspects of metabolism (such as *GAL5*) are still intact. Loss of *GAL* activity in *S. kudriavzevii* is quite recent (<10 million years), because the *GAL* genes are all intact in other sensu stricto species. The *GAL* pathway has also been lost in *A. gossypii*, *K. waltii* and *C. glabrata*, by two or three additional loss events (Hittinger et al. 2004).

Similarly, the lactose assimilation pathway has been lost in many hemiascomycetes. In *K. lactis*, the ability to grow on lactose is conferred by *LAC4* (beta-galactosidase) and *LAC12* (lactose permease), two genes that are located beside each other in the genome (Chang and Dickson 1988; Poch et al. 1992). *LAC12* is a member of a large family of sugar permeases whose evolutionary history is difficult to trace. *LAC4* has no homolog in other hemiascomycete species except for *Debaryomyces hansenii*, but it also has homologs in several euascomycetes such as *Magnaporthe grisea* and *Neurospora crassa*. Given this phylogenetic distribution, it is likely that the *LAC4* gene has been lost from several hemiascomycete lineages, arguing against horizontal gene transfer from a bacterium into *K. lactis* as was suggested by Poch et al. (1992).

Other examples of pathway losses include the absence in *C. glabrata* of genes in the *DAL* (allantoin catabolism) pathway, and the *SNO* and *SNZ* genes involved in pyridoxine synthesis (Dujon et al. 2004). A three-gene pathway conferring the ability to grow on pyrimidines as a sole source of nitrogen is present in the pre-WGD species *S. kluyveri* (Gojkovic et al. 2000, 2001), *K. lactis* and *K. waltii*, but has been lost both in the post-WGD species and in *A. gossypii*. These examples serve to illustrate the principle that a gene will be maintained in an organism's genome only if there exists the natural selection to prevent its loss, and that a change in the ecology of the organism can result in a pathway becoming unnecessary. This seems to happen quite frequently in the case of genes involved in metabolic pathways that permit growth on particular substrates.

One gene loss whose significance is not yet fully understood is the loss of *MAT***a**2, a gene for an HMG-domain DNA-binding protein, from the *MAT* locus in post-WGD yeast species (Tsong et al. 2003; Butler et al. 2004). In *C. albicans*, *MAT***a**2 is a positive regulator of expression of **a**-specific genes in cells of mating type **a**. In *S. cerevisiae* no such activator is needed, because expression of **a**-specific genes occurs by default in the absence of the repressor molecules Mcm1-α2 (present in *MAT*α haploid cells) or **a**1-α2 (present in *MAT***a**/*MAT*α diploids) (Tsong et al. 2003; Scannell and Wolfe 2004). In *C. albicans* the α2 repressor has no effect on the expression of **a**-specific genes, unlike in *S. cerevisiae*. The *MAT***a**2 gene is present in all pre-WGD species (except *Pichia angusta*) but absent from all post-WGD species that have been studied (Butler et al. 2004).

Lastly, it is interesting that the gene for chitinase (*CTS1*) is absent from the *A. gossypii* genome (Dietrich et al. 2004). In mitotic growth of budding yeasts, chitinase degrades the chitin ring between mother and daughter, allowing the cells to separate. *A. gossypii* is unusual among hemiascomycetes in growing only filamentously – it forms multinucleate hyphae and has no yeast phase (Alberti-Segui et al. 2001). It is the only known hemiascomycete lacking a *CTS1* homolog.

VIII. Rapidly Evolving Loci

Most genes in hemiascomycete genomes have clear orthologs in other species that show both conservation of protein sequence and conservation of local gene order relationships (Fig. 2.2). In a few places in the genome, however, synteny information suggests that some genes in different species might be orthologs, even though there is little or no sequence similarity. These are likely to be loci with very rapid sequence evolution. An example is shown in Fig. 2.6: in *S. cerevisiae* a gene of unknown function, *YKL023W*, lies between *URA6* and *CDC16*. In every other hemiascomycete genome, there is also a putative gene in the interval between *URA6* and *CDC16*, and these genes are all transcribed toward *CDC16*. This suggests that the genes in the interval are orthologs, but they have no detectable sequence similarity to each other (in a BLASTP search, none of them hits any of the others with an *E*-value <1) and they range in size from 185 codons (*A. gossypii ACR169C*) to 408 codons (*K. lactis KLLA0E08008g*). *S. cerevisiae YKL023W* does have orthologs in the other *Saccharomyces* sensu stricto species, and it is among the fastest 5% of genes in terms of its rate of protein sequence evolution among the sensu stricto species (Kellis et al. 2003). This indicates that the genes in the *URA6–CDC16* interval in the other species are also very probably orthologs, but the combination of the gene's inherently fast rate of sequence change and the longer period of evolutionary time involved in comparisons outside the sensu stricto group has resulted in sequences that have diverged beyond the ability of BLAST to detect their relatedness. Other examples of rapidly evolving genes were identified by Kellis et al. (2003) in their comparison of sensu stricto genomes. One such locus, *YBR184W*, has a very high ratio of nonsynonymous to synonymous nucleotide substitutions (indicating low constraint on the amino acid sequence), and has no detectable homologs outside the sensu stricto species.

Gene pairs (ohnologs) that were formed by the WGD are also sometimes seen to be diverging very rapidly in *S. cerevisiae* or other post-WGD species (Kellis et al. 2004; Wolfe 2004). These pairs are strong candidates for having undergone divergence of function following the WGD, and several of them show asymmetrical evolution where one member of the pair has apparently retained an ancestral function while the faster-evolving copy has taken on a new function (Kellis et al. 2004).

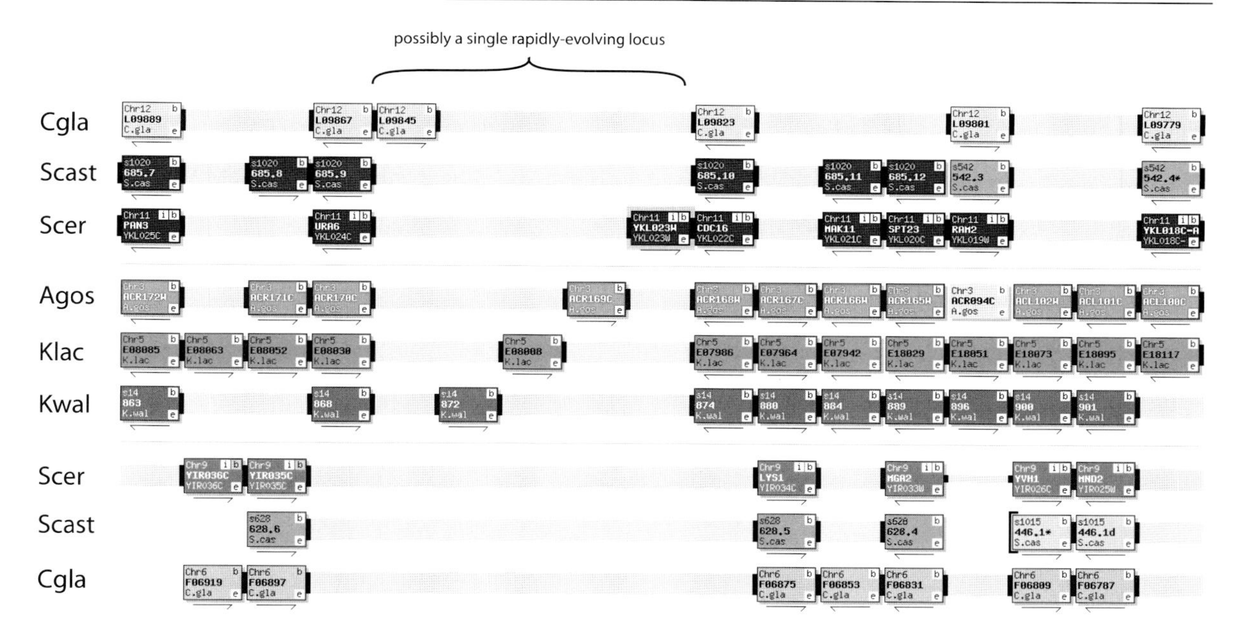

Fig. 2.6. Gene order relationships around the locus *YKL023W*. Although they are shown in separate columns because they do not have BLASTP hits to one another, the genes in the region marked by the *brace* may in fact be rapidly evolving orthologs: *C. glabrata* (*Cgla*), *S. castellii* (*Scast*), *S. cerevisiae* (*Scer*), *A. gossypii* (*Agos*), *K. lactis* (*Klac*) and *K. waltii* (*Kwal*). *S. castellii* also has a large ORF in the *URA6–CDC16* interval that is missing from this YGOB view (see Fig. 2.2b) because it was overlooked in the genome annotation (Cliften et al. 2003)

IX. Conclusions

S. cerevisiae is so entrenched as a model organism for the hemiascomycetes that it is worth pausing to consider whether, in the light of the new genomics data available, it is really a good representative of this group of yeasts. What the genomics data show is extensive conservation of gene content and extensive colinearity of gene order, and so from that standpoint *S. cerevisiae* is highly representative, at least of the post-WGD species. However, the genome data also show that there are many species-specific quirks. Overlaid onto the conserved core of genes that make up any hemiascomycete are species-specific attributes such as the presence of particular genes. In some cases – such as *LAC4* in *K. lactis*, or *ADH2* in *S. cerevisiae* – these attributes confer such a strong phenotype that they almost define the species, at least from a human perspective. In other cases – such as the loss of *CTS1* in *A. gossypii* – they may be correlated with a phenotypic change without being the root cause of it.

At the moment our knowledge of the genomes of most hemiascomycetes other than *S. cerevisiae* greatly outstrips our knowledge of their biology and natural ecosystems. Hence, it is likely that our appreciation of the genetic causes of the biological differences between these species is still in its infancy. In a sense, we are at the limits of the reductionist approach to biology: the cells of all hemiascomycetes work in much the same way, and we have already obtained a reasonable understanding of how they work through decades of research into the biology of *S. cerevisiae* and other model organisms. That is the beauty of the reductionist approach. What is perhaps most interesting, however, and what remains for the future, is what lies outside this reduced core – the rapidly evolving genes, the laterally transferred genes, the lost genes, and the expanded gene families – that make the proteome of each species unique and that enabled each species to evade extinction by carving out a foothold in the biosphere.

Acknowledgements. We thank Kevin Byrne, Devin Scannell, Jonathan Gordon and Simon Wong for discussion. Research in our laboratory is supported by Science Foundation Ireland. L.J. Montcalm was supported by the Plains of Abraham Trust.

References

Alberti-Segui C, Dietrich F, Altmann-Johl R, Hoepfner D, Philippsen P (2001) Cytoplasmic dynein is required to oppose the force that moves nuclei towards the hyphal tip in the filamentous ascomycete *Ashbya gossypii*. J Cell Sci 114:975–986

Ashburner M (1998) Speculations on the subject of alcohol dehydrogenase and its properties in *Drosophila* and other flies. Bioessays 20:949–954

Astrom SU, Rine J (1998) Theme and variation among silencing proteins in *Saccharomyces cerevisiae* and *Kluyveromyces lactis*. Genetics 148:1021–1029

Benner SA, Caraco MD, Thomson JM, Gaucher EA (2002) Planetary biology – paleontological, geological, and molecular histories of life. Science 296:864–868

Berbee ML, Taylor JW (1993) Ascomycete relationships: dating the origin of asexual lineages with 18S ribosomal RNA gene sequence data. In: Reynolds DR, Taylor JW (eds) The fungal holomorph: mitotic, meiotic and pleomorphic speciation in fungal systematics. CAB International, Wallingford, UK, pp 67–78

Bon E, Neuveglise C, Casaregola S, Artiguenave F, Wincker P, Aigle M, Durrens P (2000) Genomic exploration of the hemiascomycetous yeasts. 5. *Saccharomyces bayanus* var. *uvarum*. FEBS Lett 487:37–41

Brachat S, Dietrich FS, Voegeli S, Zhang Z, Stuart L, Lerch A, Gates K, Gaffney T, Philippsen P (2003) Reinvestigation of the *Saccharomyces cerevisiae* genome annotation by comparison to the genome of a related fungus: *Ashbya gossypii*. Genome Biol 4:R45

Butler G, Kenny C, Fagan A, Kurischko C, Gaillardin C, Wolfe KH (2004) Evolution of the *MAT* locus and its Ho endonuclease in yeast species. Proc Natl Acad Sci USA 101:1632–1637

Cai J, Roberts IN, Collins MD (1996) Phylogenetic relationships among members of the ascomycetous yeast genera *Brettanomyces*, *Debaryomyces*, *Dekkera*, and *Kluyveromyces* deduced by small-subunit rRNA gene sequences. Int J Syst Bacteriol 46:542–549

Chang YD, Dickson RC (1988) Primary structure of the lactose permease gene from the yeast *Kluyveromyces lactis*. Presence of an unusual transcript structure. J Biol Chem 263:16696–16703

Cliften P, Sudarsanam P, Desikan A, Fulton L, Fulton B, Majors J, Waterston R, Cohen BA, Johnston M (2003) Finding functional features in *Saccharomyces* genomes by phylogenetic footprinting. Science 301:71–76

Coyne JA, Orr HA (2004) Speciation. Sinauer, Sunderland, MA

Damak F, Boy-Marcotte E, Le-Roscouet D, Guilbaud R, Jacquet M (1991) *SDC25*, a *CDC25*-like gene which contains a RAS-activating domain and is a dispensable gene of *Saccharomyces cerevisiae*. Mol Cell Biol 11:202–212

Daran-Lapujade P, Daran JM, Kotter P, Petit T, Piper MD, Pronk JT (2003) Comparative genotyping of the *Saccharomyces cerevisiae* laboratory strains S288C and CEN.PK113-7D using oligonucleotide microarrays. FEMS Yeast Res 4:259–269

Delneri D, Colson I, Grammenoudi S, Roberts IN, Louis EJ, Oliver SG (2003) Engineering evolution to study speciation in yeasts. Nature 422:68–72

Dietrich FS, Voegeli S, Brachat S, Lerch A, Gates K, Steiner S, Mohr C, Pohlmann R, Luedi P, Choi S et al. (2004) The *Ashbya gossypii* genome as a tool for mapping the ancient *Saccharomyces cerevisiae* genome. Science 304:304–307

Dujon B, Albermann K, Aldea M, Alexandraki D, Ansorge W, Arino J, Benes V, Bohn C, Bolotin-Fukuhara M, Bordonné R et al. (1997) The nucleotide sequence of *Saccharomyces cerevisiae* chromosome XV. Nature suppl 387:98–102

Dujon B, Sherman D, Fischer G, Durrens P, Casaregola S, Lafontaine I, de Montigny J, Marck C, Neuvéglise C, Talla E et al. (2004) Genome evolution in yeasts. Nature 430:35–44

Eichler EE (2001) Recent duplication, domain accretion and the dynamic mutation of the human genome. Trends Genet 17:661–669

Fabre E, Muller H, Therizols P, Lafontaine I, Dujon B, Fairhead C (2005) Comparative genomics in hemiascomycete yeasts: evolution of sex, silencing and subtelomeres. Mol Biol Evol 22:856–873

Feldmann H (2000) Génolevures – a novel approach to 'evolutionary genomics'. FEBS Lett 487:1–2

Fischer G, James SA, Roberts IN, Oliver SG, Louis EJ (2000) Chromosomal evolution in *Saccharomyces*. Nature 405:451–454

Fischer G, Neuvéglise C, Durrens P, Gaillardin C, Dujon B (2001) Evolution of gene order in the genomes of two related yeast species. Genome Res 11:2009–2019

Fogel S, Welch JW (1982) Tandem gene amplification mediates copper resistance in yeast. Proc Natl Acad Sci USA 79:5342–5346

Gimble FS (2000) Invasion of a multitude of genetic niches by mobile endonuclease genes. FEMS Microbiol Lett 185:99–107

Gimble FS, Thorner J (1992) Homing of a DNA endonuclease gene by meiotic gene conversion in *Saccharomyces cerevisiae*. Nature 357:301–306

Goffeau A, Aert R, Agostini-Carbone ML, Ahmed A, Aigle M, Alberghina L, Allen E, Alt-Mörbe J, André B, Andrews S et al. (1997) The Yeast Genome Directory. Nature suppl 387:5–105

Gojkovic Z, Jahnke K, Schnackerz KD, Piskur J (2000) *PYD2* encodes 5,6-dihydropyrimidine amidohydrolase, which participates in a novel fungal catabolic pathway. J Mol Biol 295:1073–1087

Gojkovic Z, Sandrini MP, Piskur J (2001) Eukaryotic beta-alanine synthases are functionally related but have a high degree of structural diversity. Genetics 158:999–1011

Gojkovic Z, Knecht W, Zameitat E, Warneboldt J, Coutelis JB, Pynyaha Y, Neuveglise C, Moller K, Loffler M, Piskur J (2004) Horizontal gene transfer promoted evolution of the ability to propagate under anaerobic conditions in yeasts. Mol Genet Genomics 271:387–393

Haber JE, Wolfe KH (2005) Evolution and function of HO and VDE endonucleases in fungi. In: Belfort M, Derbyshire V, Stoddard B, Wood D (eds) Homing endonucleases and inteins. Springer, Berlin Heidelberg New York, Nucleic Acids and Molecular Biology vol 16, pp 161–175

Hall MN, Linder P (1993) The early days of yeast genetics. Cold Spring Harbor Laboratory Press, New York

Halme A, Bumgarner S, Styles C, Fink GR (2004) Genetic and epigenetic regulation of the *FLO* gene family generates cell-surface variation in yeast. Cell 116:405–415

Hansen J, Kielland-Brandt MC (1994) *Saccharomyces carlsbergensis* contains two functional *MET2* alleles similar to homologues from *S. cerevisiae* and *S. monacensis*. Gene 140:33–40

Hansen J, Kielland-Brandt MC (2003) Brewer's yeast: genetic structure and targets for improvement. Topics Curr Genet 2:143–170

Harrison P, Kumar A, Lan N, Echols N, Snyder M, Gerstein M (2002) A small reservoir of disabled ORFs in the yeast genome and its implications for the dynamics of proteome evolution. J Mol Biol 316:409–419

Heus JJ, Zonneveld BJ, Steensma HY, van den Berg JA (1993) The consensus sequence of *Kluyveromyces lactis* centromeres shows homology to functional centromeric DNA from *Saccharomyces cerevisiae*. Mol Gen Genet 236:355–362

Hittinger CT, Rokas A, Carroll SB (2004) Parallel inactivation of multiple *GAL* pathway genes and ecological diversification in yeasts. Proc Natl Acad Sci USA 101:14144–14149

Hughes TR, Roberts CJ, Dai H, Jones AR, Meyer MR, Slade D, Burchard J, Dow S, Ward TR, Kidd MJ et al. (2000) Widespread aneuploidy revealed by DNA microarray expression profiling. Nat Genet 25:333–337

Kellis M, Patterson N, Endrizzi M, Birren B, Lander ES (2003) Sequencing and comparison of yeast species to identify genes and regulatory elements. Nature 423:241–254

Kellis M, Birren BW, Lander ES (2004) Proof and evolutionary analysis of ancient genome duplication in the yeast *Saccharomyces cerevisiae*. Nature 428:617–624

Kent WJ, Baertsch R, Hinrichs A, Miller W, Haussler D (2003) Evolution's cauldron: duplication, deletion, and rearrangement in the mouse and human genomes. Proc Natl Acad Sci USA 100:11484–11489

Keogh RS, Seoighe C, Wolfe KH (1998) Evolution of gene order and chromosome number in *Saccharomyces*, *Kluyveromyces* and related fungi. Yeast 14:443–457

Kinclova O, Potier S, Sychrova H (2001) The *Zygosaccharomyces rouxii* strain CBS732 contains only one copy of the *HOG1* and the *SOD2* genes. J Biotechnol 88:151–158

Koszul R, Caburet S, Dujon B, Fischer G (2003) Eucaryotic genome evolution through the spontaneous duplication of large chromosomal segments. EMBO J 23:234–243

Koufopanou V, Goddard MR, Burt A (2002) Adaptation for horizontal transfer in a homing endonuclease. Mol Biol Evol 19:239–246

Kurtzman CP, Fell JW (1998) The yeasts, a taxonomic study. Elsevier, Amsterdam

Kurtzman CP, Robnett CJ (2003) Phylogenetic relationships among yeasts of the '*Saccharomyces* complex' determined from multigene sequence analyses. FEMS Yeast Res 3:417–432

Lafontaine I, Fischer G, Talla E, Dujon B (2004) Gene relics in the genome of the yeast *Saccharomyces cerevisiae*. Gene 335:1–17

Lai J, Ma J, Swigonova Z, Ramakrishna W, Linton E, Llaca V, Tanyolac B, Park Y-J, Jeong O-Y, Bennetzen JL, Messing J (2004) Gene loss and movement in the maize genome. Genome Res 14:1924–1931

Lalo D, Stettler S, Mariotte S, Slonimski PP, Thuriaux P (1993) Une duplication fossile entre les régions centromériques de deux chromosomes chez la levure. C R Acad Sci Paris 316:367–373

Liu H, Styles CA, Fink GR (1996) *Saccharomyces cerevisiae* S288C has a mutation in *FLO8*, a gene required for filamentous growth. Genetics 144:967–978

Llorente B, Durrens P, Malpertuy A, Aigle M, Artiguenave F, Blandin G, Bolotin-Fukuhara M, Bon E, Brottier P, Casaregola S et al. (2000a) Genomic exploration of the hemiascomycetous yeasts. 20. Evolution of gene redundancy compared to *Saccharomyces cerevisiae*. FEBS Lett 487:122–133

Llorente B, Malpertuy A, Neuveglise C, de Montigny J, Aigle M, Artiguenave F, Blandin G, Bolotin-Fukuhara M, Bon E, Brottier P et al. (2000b) Genomic exploration of the hemiascomycetous yeasts. 18. Comparative analysis of chromosome maps and synteny with *Saccharomyces cerevisiae*. FEBS Lett 487:101–112

Maciaszczyk E, Wysocki R, Golik P, Lazowska J, Ulaszewski S (2004) Arsenical resistance genes in *Saccharomyces douglasii* and other yeast species undergo rapid evolution involving genomic rearrangements and duplications. FEMS Yeast Res 4:821–832

Nomura M, Takagi H (2004) Role of the yeast acetyltransferase Mpr1 in oxidative stress: regulation of oxygen reactive species caused by a toxic proline catabolism intermediate. Proc Natl Acad Sci USA 101:12616–12621

Okuda Y, Sasaki D, Nogami S, Kaneko Y, Ohya Y, Anraku Y (2003) Occurrence, horizontal transfer and degeneration of VDE intein family in Saccharomycete yeasts. Yeast 20:563–573

Pesole G, Lotti M, Alberghina L, Saccone C (1995) Evolutionary origin of nonuniversal CUG(Ser) codon in some *Candida* species as inferred from a molecular phylogeny. Genetics 141:903–907

Petersen RF, Nilsson-Tillgren T, Piskur J (1999) Karyotypes of *Saccharomyces sensu lato* species. Int J Syst Bacteriol 49:1925–1931

Piskur J, Langkjaer RB (2004) Yeast genome sequencing: the power of comparative genomics. Mol Microbiol 53:381–389

Poch O, L'Hote H, Dallery V, Debeaux F, Fleer R, Sodoyer R (1992) Sequence of the *Kluyveromyces lactis* beta-galactosidase: comparison with prokaryotic enzymes and secondary structure analysis. Gene 118:55–63

Robyr D, Suka Y, Xenarios I, Kurdistani SK, Wang A, Suka N, Grunstein M (2002) Microarray deacetylation maps determine genome-wide functions for yeast histone deacetylases. Cell 109:437–446

Rokas A, Williams BL, King N, Carroll SB (2003) Genome-scale approaches to resolving incongruence in molecular phylogenies. Nature 425:798–804

Sanyal K, Baum M, Carbon J (2004) Centromeric DNA sequences in the pathogenic yeast *Candida albicans* are all different and unique. Proc Natl Acad Sci USA 101:11374–11379

Scannell DR, Wolfe K (2004) Rewiring the transcriptional regulatory circuits of cells. Genome Biol 5:206

Seoighe C, Wolfe KH (1999) Yeast genome evolution in the post-genome era. Curr Opin Microbiol 2:548–554

Seoighe C, Federspiel N, Jones T, Hansen N, Bivolarovic V, Surzycki R, Tamse R, Komp C, Huizar L, Davis RW et al.

(2000) Prevalence of small inversions in yeast gene order evolution. Proc Natl Acad Sci USA 97:14433–14437

Sor F, Fukuhara H (1989) Analysis of chromosomal DNA patterns of the genus *Kluyveromyces*. Yeast 5:1–10

Souciet J, Aigle M, Artiguenave F, Blandin G, Bolotin-Fukuhara M, Bon E, Brottier P, Casaregola S, de Montigny J, Dujon B et al. (2000) Genomic exploration of the hemiascomycetous yeasts. 1. A set of yeast species for molecular evolution studies. FEBS Lett 487:3–12

Tsong AE, Miller MG, Raisner RM, Johnson AD (2003) Evolution of a combinatorial transcriptional circuit: a case study in yeasts. Cell 115:389–399

Vaughan-Martini A, Martini A, Cardinali G (1993) Electrophoretic karyotyping as a taxonomic tool in the genus *Saccharomyces*. Antonie van Leeuwenhoek 63:145–156

Wieland J, Nitsche AM, Strayle J, Steiner H, Rudolph HK (1995) The *PMR2* gene cluster encodes functionally distinct isoforms of a putative Na+ pump in the yeast plasma membrane. EMBO J 14:3870–3882

Wills C (1976) Production of yeast alcohol dehydrogenase isoenzymes by selection. Nature 261:26–29

Wolfe K (2004) Evolutionary genomics: yeasts accelerate beyond BLAST. Curr Biol 14:R392–R394

Wolfe KH, Shields DC (1997) Molecular evidence for an ancient duplication of the entire yeast genome. Nature 387:708–713

Wong S, Wolfe KH (2005) Duplication of genes and genomes in yeasts. In: Sunnerhagen P, Piskur J (eds) Comparative genomics. Springer, Berlin Heidelberg New York, Topics in Current Genetics (in press) DOI 10.1007/b105770

Wong S, Butler G, Wolfe KH (2002) Gene order evolution and paleopolyploidy in hemiascomycete yeasts. Proc Natl Acad Sci USA 99:9272–9277

Yochem J, Byers B (1987) Structural comparison of the yeast cell division cycle gene *CDC4* and a related pseudogene. J Mol Biol 195:233–245

Zhang Z, Dietrich FS (2003) Verification of a new gene on *Saccharomyces cerevisiae* chromosome III. Yeast 20:731–738

Zonneveld BJM, Steensma HY (2003) Mating, sporulation and tetrad analysis in *Kluyveromyces lactis*. In: Wolf K, Breunig K, Barth G (eds) Non-conventional yeasts in genetics, biochemistry and biotechnology. Springer, Berlin Heidelberg New York, pp 151–154

3 Investigating the Evolution of Fungal Virulence by Functional Genomics

S. Ahmad[1], D.M. Soanes[1], M.C. Barooah[1], N.J. Talbot[1]

CONTENTS

Abbreviations: ATMT, *A. tumefaciens*-mediated transformation; COGs, clusters of orthologous groups; EST, expressed sequence tag; RAPD, random amplified polymorphic DNA; RFLP, restriction fragment linked polymorphism; REMI, restriction endonuclease-mediated DNA integration; TAG-KO, transposon-arrayed gene knockout

I. Introduction

Fungi are a morphologically diverse group of eukaryotes, most of which have a saprophytic life style and degrade dead organic matter for their source of nutrients. The fungal life style is exemplified by an ability to invade and colonise diverse substrates using cylindrical hyphal cells, and by the ability of these cells to secrete prodigious amounts of extracellular enzymes, with which complex polymeric substrates are degraded into simple sugars and amino acids for subsequent uptake by the growing fungus. A small number of fungal species, however, have acquired the ability to colonise living plant tissues and cause disease (Hammond-Kosack and Jones 1997).

Fungi infect a multitude of plant species, destroying harvests and causing immense economic damage to agriculture. Approximately 70% of major crop diseases are caused by fungi, resulting in crop losses which amount to 5%–10% of the total harvest of developed countries, and 40%–50% in developing countries each year. This is in spite of widespread modern farming practices, the use of fungicides, and extensive plant breeding programmes for resistance (Bowyer 1999). Although intensive efforts have been directed towards disease control, crop diseases still continue to threaten food security. Historically, serious deprivation has resulted directly from fungal and oomycete diseases. The Irish potato famine of 1845, for example, was caused by the oomycete pathogen *Phytophthora infestans*. More recently, a rice blast epidemic caused by the fungus *Magnaporthe grisea* resulted in the loss of 1090 tonne rice in Bhutan in 1995 (Thinlay et al. 2000).

The advent of genome sequencing efforts in filamentous fungi has provided a wealth of new information by which the characteristics of fungal species can be investigated and compared. At the time of writing in early 2005, there were 26 completed fungal genome sequences publicly available, ranging from unicellular yeasts, such as the model eukaryote *Saccharomyces cerevisiae* and many of its closest relatives, to morphologically complex, multicellular basidiomycete fungi, such as *Phanerochaete chrysosporium* and *Coprinus cinereus* (Table 3.1). This vast amount of information has allowed many new types of investigation into the evolution of fungal virulence to be carried out (Yoder and Turgeon 2001; Tunlid and Talbot 2002). In this review, we examine the use of the newly generated fungal genome sequences and expressed sequence tag collections of filamentous fungi as a resource for identifying conserved gene sequences, and as a means of learning more about the fascinating biology of this group of organisms, with a particular focus on the evolution of fungal pathogenicity.

[1] School of Biological and Chemical Sciences, University of Exeter, Washington Singer Laboratories, Perry Road, Exeter EX4 4QG, UK

The Mycota XIII
Fungal Genomics
Alistair J.P. Brown (Ed.)

Table. 3.1. Completed genome sequences of fungi which have been publicly released (early 2005)

Fungal species	Reference
Saccharomyces cerevisiae	Goffeau et al. (1996)
Saccharomyces bayanus	Kellis et al. (2003)
Saccharomyces kluyveri	http://www.genome.wustl.edu/projects/yeast/
Saccharomyces mikatae	Kellis et al. (2003)
Saccharomyces paradoxus	Kellis et al. (2003)
Saccharomyces kudriavzevii	http://www.genome.wustl.edu/projects/yeast/
Schizosaccharo myces pombe	Wood et al. (2002)
Kluyveromyces waltii	http://natchaug.labri.u-bordeaux.fr/Genolevures/
Kluyveromyces lactis	http://natchaug.labri.u-bordeaux.fr/Genolevures/
Candida albicans	http://www-sequence.stanford.edu/group/candida/
Candida glabrata	http://genopole.pasteur.fr/glabrata/
Yarrowia lipolytica	http://cbi.labri.fr/Genolevures/about.php
Neurospora crassa	Galagan et al. (2003), http://www.broad.mit.edu/annotation/fungi/neurospora/
Aspergillus nidulans	http://www.broad.mit.edu/annotation/fungi/aspergillus/
Fusarium graminearum	http://www.broad.mit.edu/annotation/fungi/fusarium/
Magnaporthe grisea	Dean et al. (2005), http://www.broad.mit.edu/annotation/fungi/magnaporthe/
Coccidioides immitis	http://www.broad.mit.edu/annotation/fungi/coccidioides_immitis/
Rhizopus oryzae	http://www.broad.mit.edu/annotation/fungi/fgi/
Debaromyces hansenii	http://natchaug.labri.u-bordeaux.fr/Genolevures/
Encephalitozoon cuniculi	http://www.genoscope.cns.fr/externe/English/Projets/Projet_AD/AD.html
Ashbya gossypii	Dietrich et al. (2004)
Pichia stipitis	http://www.jgi.doe.gov/sequencing/DOEmicrobes.html
Coprinus cinereus	http://www.broad.mit.edu/annotation/fungi/coprinus_cinereus/
Ustilago maydis	http://www.broad.mit.edu/annotation/fungi/ustilago_maydis/
Phanerochaete chrysosporium	Martinez et al. (2004)
Cryptococcus neoformans	http://www.broad.mit.edu/annotation/fungi/cryptococcus_neoformans

II. Identifying Virulence Factors in Plant Pathogenic Fungi

In order to cause plant disease, many species of filamentous fungi undergo a complex sequence of metabolic and developmental events which enable them to penetrate the host plant cuticle and invade living plant tissue (Tucker and Talbot 2001). These events include attachment of spores to the plant surface, germination of propagules on the plant, development of infection structures which mediate penetration of the host and, finally, colonisation of host tissue.

Recent advances in molecular genetic and genomic techniques have changed the way in which fungal biology is studied and has contributed greatly to our understanding of the strategies employed by pathogens to initiate plant disease. At the cellular level, improved microscopic techniques have helped to gain a better insight into the nature of fungal infection structures which mediate entry into plants (Howard 2001). Advances in immunological labelling procedures, for example, together with electron microscopy, have enabled visualisation of plant–fungal interactions with ever-greater resolution (Gold et al. 2001). The development of reporter gene technologies to enable visualisation of spatial patterns of gene expression during plant infection has also been an important advance. Green fluorescent protein from the jellyfish *Aequoria victoriae* is now widely used to study gene expression and protein localisation during fungal development (Lorang et al. 2001). Immunological procedures to target particular cellular targets have also helped to indicate the role of a given protein in pathogenicity. Antibodies have, for example, been used to inhibit cutinase, lipases and other cell wall-degrading enzymes in order to assess their roles in plant infection (Gold et al. 2001). In *Botrytis cinerea*, Commenil et al. (1998) have used anti-lipase antibodies to inhibit fungal colonisation. Immunolocalisation of pathogenicity factors is also commonly carried out (Tucker et al. 2004).

Arguably, the most powerful approaches to investigate fungal virulence, however, have been gene functional studies, involving the generation of targeted deletion mutants, or large-scale forward genetic screens. Insertional mutagenesis has been widely utilised to identify pathogenicity mutants (for reviews, see Gold et al. 2001; Talbot and Foster 2001; Tucker and Talbot 2001). Restriction enzyme-mediated DNA integration (REMI) is a process whereby DNA is introduced into a fungus in the presence of a restriction endonuclease which cleaves genomic DNA at numerous corresponding restriction sites in the genome, and also enhances the rate of DNA-mediated transformation. This procedure has been used extensively in the rice blast fungus *M. grisea*, and allowed characterisation of mutants which display various defects such as non-functional appressoria and non-invasive growth (Sweigard et al. 1998; Balhadère et al. 1999). The *PTH11* gene, which encodes a novel G-protein coupled receptor required for appressorium development (DeZwaan et al. 1999), and the *PDE1* gene which encodes a P-type ATPase involved in appressorium function (Balhadère and Talbot 2001) were both identified in REMI screens. The *Tox* locus, which is responsible for production of T-toxin and high virulence on T-cytoplasm maize of the corn pathogen *Cochliobolus heterostrophus*, was also tagged using REMI mutagenesis (Yang et al. 1996), which allowed characterisation of the polyketide synthase involved in T-toxin production (Yang et al. 1996). More recently, fungal transformation mediated through the plant pathogenic bacterium *Agrobacterium tumefaciens* has been used to transfer foreign genes into several fungal species and to carry out gene replacements (Bundock et al. 1995; de Groot et al. 1998). Problems associated with protoplast isolation in *Fusarium oxysporum* have been alleviated using *A. tumefaciens*-mediated transformation (ATMT; Mullins et al. 2001), and the high frequency of transformation achieved by ATMT has allowed rapid generation of large collections of insertional mutants in a variety of fungal pathogens, including most notably *M. grisea*. The fact that disrupted loci can be isolated readily by the inverse polymerase chain reaction (PCR) or 'step-down' PCR (Zhang and Gurr 2000) means that such libraries are an extremely valuable resource for pathogenicity gene discovery.

Another method for genome-wide mutagenesis studies in filamentous fungi, pioneered in the private sector, has been reported by Hamer et al. (2001). A transposon-arrayed gene knockout (TAG-KO) strategy has been developed to sequence fungal genomes in a rapid manner and to simultaneously create gene disruption cassettes for subsequent fungal transformation and mutant analysis. The process works by generation of sequencing templates using a bacterial transposon, such as Tn5, which also serve as gene disruption vectors for subsequent functional analysis of individual genes. The process has been used as a high-throughput means of identifying new targets for fungicide discovery in both *M. grisea* and the wheat blotch pathogen *Mycosphaerella graminicola* (Adachi et al. 2002).

Map-based cloning strategies have also been used to identify pathogenicity genes which were previously identified using chemical mutagenesis and classical genetics. This strategy uses RFLP and RAPD markers to identify mutant loci, and direct chromosome 'walks' using hybridisation from the neighbouring, linked DNA markers (Nitta et al. 1997). This approach was used, for example, to clone *AVR2-Pi-ta*. *PWL2*, and *AVR1-CO39* in *M. grisea* (Orbach et al. 2000). The difficulty of carrying out conventional mutagenesis studies and the inability to use complementation cloning in most fungi to identify genes, however, was a principal cause of the widespread use of reverse genetic approaches to identify pathogenicity genes.

Identifying genes in plant pathogenic fungi based on their expression pattern has often provided the first stage of such studies. Differential cDNA screening involves generating cDNA libraries, either derived from a particular growth stage of the pathogen which is observed during plant growth, or by exposing the organism to environmental conditions which favour expression of particular genes, such as starvation stress (Talbot et al. 1993). Differential cDNA screening in *M. grisea* led, for example, to the identification of *MPG1*, a hydrophobin-encoding gene found to be important in the development of spores and appressoria (Talbot et al. 1993, 1996). The *GAS1* and *GAS2* genes of *M. grisea*, which are also highly similar to genes expressed using plant infection by the obligate biotrophic pathogen *Blumeria graminis*, were also identified in this way (Xue et al. 2002). A purely deductive approach has also been used as a strategy for identifying virulence determinants in plant pathogenic fungi. Here, a key process is recognised, based on cell physiology or cell biology studies, which is predicted to influence fungal pathogenicity, and then systematically tested using targeted gene replacement. The prevalence of melanin in the appressorium cell wall of *M. grisea*, for instance, led to targeting of the dihydroxy-naphthalene (DHN) melanin biosynthetic pathway using standard mutagenesis. The application of this strategy led ultimately to the isolation of three genes, *ALB1*, *RSY1* and *BUF1*, which were found to be required for the melanisation of appressoria and essential for pathogenicity (Howard 1997). Similar genes were found to be essential for appressorium function in the anthracnose pathogen *Colletotrichum lagenarium* which also produces melanin-pigmented infection structures (Kubo et al. 1991, 1996). Genes involved in virulence-associated cell signalling have also predominantly been identified in pathogenic species based on homology to genes in *S. cerevisiae*, and then tested empirically to determine their role in pathogenesis. For example, the MAP kinase genes *UBC3* and *PMK1* from *Ustilago maydis* and *M. grisea* respectively were identified as functional homologues of the yeast *FUS3/KSS1* MAPK-encoding genes (Xu and Hamer 1996; Mayorga and Gold 1999).

In all of the reverse genetic strategies described above, it has been the advancement of DNA-mediated transformation techniques for phytopathogenic fungi which has permitted the testing of putative pathogenicity determinants (Mullins and Kang 2001). Transformation frequencies and, indeed, the rates of homologous recombination which lead to gene disruption vary greatly among species, but have so far been used to identify and validate approximately 70 virulence genes in a wide range of phytopathogenic species (Idnurm and Howlett 2001).

The availability of full genome sequences for a variety of phytopathogenic fungi, including *Magnaporthe grisea*, *Ustilago maydis* and *Fusarium graminearum* has, however, recently changed the research perspective from analysis of the functions of individual genes in fungal pathogenicity to investigation of the orchestrated action of large sets of genes during plant infection.

III. The Challenge of Comparative Genomics

The first years of the 21st century have seen the generation of a large number of genome sequences, including those of the major model eukaryotic organisms *Arabidopsis thaliana*, *Drosophila melanogaster*, *Mus musculus*, *Rattus norvegicus*, the mosquito *Anopheles gambiae* and the human genome (see Brown 2002). The vast amount of data generated by these projects has necessitated the development of new bioinformatic algorithms to investigate gene function and genome organisation. These include programmes to search for regions of similarity between two or more sequences (Altschul et al. 1990, 1997), to identify open reading frames (ORFs) and regulatory domains within genomic data, or functional motifs and domains in protein-coding sequences (see http://us.expasy.org/alinks.html). To be used effectively, genome sequence data need to be stored and presented in a form which can be readily accessed, queried and compared. The sequence of a single genome is, of course, informative in itself, but its value is greatly enhanced by comparison with other genomes. This can assist in the prediction and annotation of individual genes, and allow us to relate differences between species to their gene inventories. Such information is indicative of how evolutionary lineages have diverged from a common ancestor, and provide clues regarding how adaptation to a particular niche or situation has occurred in a given organism (Mushegian and Koonin 1996; Wei et al. 2002; Koonin 2003; Nobrega and Pennacchio 2004).

The first eukaryotic genome sequence, generated in 1996, was that of the ascomycete yeast *S. cerevisiae*, but it was not until 2002 that large-scale sequencing of fungal genomes was undertaken in earnest. As such, until that point, comparative genomics in filamentous fungi was restricted to the analysis of incomplete datasets, including most notably the use of collections of expressed sequence tags (ESTs), the uses of which are discussed in greater detail in Section VI.. The generation of DNA sequence information from *Neurospora crassa* (Galagan et al. 2003), *Phanerochaete chrysosporium* (Martinez et al. 2004) and *Schizosaccharomyces pombe* (Wood et al. 2002) and a number of other species (see Table 3.1) has for the first time allowed comparative genomic analysis to be carried out in diverse fungal species. Of particular importance in these studies has been the analysis of gene functions in fungi which are conserved in other species and yet are specific to the kingdom Fungi, or those which are conserved among ecologically related groups of fungal species such as pathogenic organisms. However, in order to understand the relationships between such genes, it is important first to be able to develop informatic resources to attempt to predict whether genes fulfil similar functions, or have diverged to fulfil distinct biological roles.

IV. Orthology and Paralogy

Homologous genes are described as those which are descended from a common ancestral gene. Homologues were further classified by Fitch (1970) into orthologous and paralogous genes. Orthologous genes are defined as those resulting from a speciation event, such that the gene history reflects the evolutionary history of the species in which it is found. Paralogous genes result from a gene duplication event, and so are homologous members of the same gene family which have descended side by side during an organism's history. These phrases are frequently used (Ouzounis et al. 2003), although not always with the same meaning. An important point of contention, for example, is whether the relationship between genes is indicative of their biological function. It is often assumed that orthologous genes carry out similar functions and that paralogues differentiate to perform different functions (Tatusov et al. 1996). However, this need not be true (Huynen et al. 1998; Gogarten and Olendzenski 1999) and cannot be assumed (Theiben 2002). In spite of these definitions, genes are often referred to as orthologues, meaning functionally equivalent genes, without any reference to speciation. Differentiating between orthologues and paralogues in genome sequences is valuable for identifying equivalent genes in different organisms, and also provides a means of investigating how the evolution of particular genes has occurred. This process is difficult, however, because orthologous genes exhibit a range of evolutionary rates, and therefore the degree of sequence similarity is not sufficient to distinguish orthologues from paralogues, because orthologous genes do not always show higher levels of sequence identity than do paralogues (Grishin et al. 2000; Hedges and Kumar 2003). For example, protein sequences of conserved proteins such as ubiquitins or histones in eukaryotes are typically 90%–98% identical, whereas the sequences of dihydroorotases (pyrimidine metabolic enzymes) are only typically 20%–30% identical, even when functionally equivalent. An example of the difficulty inherent in such studies is the report by Chen and Jeong (2000). A family of β-decarboxylating dehydrogenases was examined. Phylogenetic analysis indicated that genes encoding this family of enzymes diverged from a common ancestral gene and had since accumulated large numbers of sequence differences. Protein structural data and protein engineering, however, indicated that only a few amino acid substitutions were involved in substrate specificity determination and could therefore be considered reliable markers for determining orthology or paralogy. In the light of this information, the authors re-examined sequences within the main databases which were annotated as β-decarboxylating dehydrogenases, and reassigned a functional annotation for 26 proteins (Chen and Jeong 2000).

Notwithstanding these problems, to make comparisons between the gene inventories of different genomes it has become necessary to develop means of identifying, in particular, orthologous genes, based only on sequence data. Various methods are now used, most of which are based on searching for gene sequences which have a certain level of sequence similarity over a given proportion of their lengths. Tatusov et al. (1996) searched for orthologous gene pairs in the *Haemophilus influenzae* bacterial genome and 75% of the *Escherichia coli* genome. They defined an *E. coli* gene as an orthologue if it had a greater similarity, by at least several percentage points, to a *H. influenzae* gene than to any other *E. coli* gene. This analysis identified 1128 pairs of apparent orthologues which had identities varying in the range 18%–98%, with an average value of 59%. Approximately 50% of *E. coli* genes were identified as belonging to clusters of paralogues. Tatusov and colleagues (1997) recognised that divergence followed by duplication can lead to a tree of genes derived from the original ancestor; which they called clusters of orthologous groups (COGs). Subsequently, individual orthologues, or orthologous groups of paralogues, from three or more phylogenetic lineages were used to construct a COGs database of proteins from sequenced genomes, as a tool for functional annotation of novel or unknown proteins (Tatusov et al. 2000, 2001).

Phylogenomics is another method which uses orthology detection as a means of predicting conserved gene function (Eisen 1998). Homologues of a gene in question are identified and any regions of ambiguous homology are masked. Sequences are then aligned, and a phylogenetic tree constructed. The tree is then used to identify possible gene duplication events. In this way, Mushegian et al. (1998) compared orthologous genes of the human, *D. melanogaster*, *C. elegans* and *S. cere-*

visae genomes. They began by searching for homologous genes from each of the genomes. Their criteria for inferring orthology were that the sequences which showed the highest level of homology to one another (a 'reciprocal best hit' test) and showed similarity through the entire length of both genes were likely to be orthologous. This resulted in 42 sets of putative orthologous groups of genes. The sequences thus identified were aligned and trees constructed to detect phylogenetic relationships between the species. However, Xie and Ding (2000) re-examined these results and commented that using such a reciprocal best hit method can detect one-to-one orthologous relationships, but can miss one-to-many or many-to-many relationships where a gene in one species has more than one orthologue in another species. They also noted that where there are these more distant or complex relationships, orthologues might not retain the same function. To overcome this limitation, they used a more rigorous phylogenetic analysis, using a reconciled tree method to identify more complex gene relationships (Xie and Ding 2000).

A different approach was carried out by Bansal (1999) who used a bipartite graph matching and fuzzy-logic procedures for orthologue and paralogue differentiation. The process still used steps similar to those of the other methods, in that it involves homology searches to identify putative gene pairs, then uses the Smith-Waterman algorithm to determine the regions of homology between the genes. The bipartite graph matching was then used to group the gene pairs (Bansal 1999).

There are currently a number of shortcomings in all the available computational methods described above. Where the species in question are only distantly related, orthologous gene pairs may have diverged so much, for example, that it is not possible to accurately identify those relationships where the level of similarity may not be different to that of unrelated genes. Another reason why similarity searches may not reveal functional relationships is when a particular cellular function has been undertaken by the product of a non-orthologous gene. In a comparison of the *E. coli* and *H. influenzae* genomes, Koonin et al. (1996) identified 12 cases where essential cellular functions were encoded by non-orthologous genes. In addition, there need not be simple one-to-one relationships between orthologues (Xie and Ding 2000) – a gene may be orthologous to a number of other genes in another genome.

With these more complex and distant gene relationships, changes in gene function may occur, or putatively orthologous genes may be lost (Jensen 2001). New terms such as co-orthologue (Gates et al. 1999) and semi-orthologue (Sharman 1999) have been suggested to describe the relationships among duplicated gene and their single-copy orthologues or pro-orthologues. Using these computational methods, genome searches have revealed extensive gene conservation between phylogenetically distinct organisms. For example, 90% of rat genes have orthologues in both the mouse and human genomes (Rat Genome Sequencing Project Consortium 2004). However, systematic differences between whole sets of genes can also be apparent. Homologues of *S. cereviseae* genes were found to be on average almost 15% shorter in the microsporidian parasite *Encephalitozoon cuniculi*, which was proposed to reflect reduced protein–protein interactions in this organism (Katinka et al. 2001).

V. Comparative Genomic Analysis of Pathogens and Saprotrophs

The availability of full genome sequence information for a range of saprotrophic and pathogenic fungal species allows general questions about the evolution of fungal virulence to be asked for the first time. It is not yet known how genome structure and gene inventories have been modified in fungi during evolution of species to fulfil specific ecological niches. There are three mutually non-exclusive theories which have been proposed to explain the evolution of virulence in fungi (Tunlid and Talbot 2002).

- First, pathogenic fungi may have acquired sets of specific pathogenicity genes which are conserved among pathogens but are not present in closely related free-living organisms.
- Second, gene inventories in pathogens and non-pathogens may be fairly universal, but selection has occurred on regulatory genes such that pathogenicity is a consequence of particular spatio-temporal control of gene expression.
- Finally, pathogenicity may be the result of gene loss.

The study of bacterial pathogens has shown examples of each of these events in particular pathogens and it is clear, for example, that horizontal transfer of genes among pathogens has taken place (Wren 2000; Ochman et al. 2000). In eukaryotic organisms such as fungi, there is much less evidence of such events (Rosewich and Kistler 2000).

So far, there have been few systematic studies to address whether pathogenic fungi possess unique genes which are conserved only in pathogenic species. To carry out this type of analysis requires whole-genome homology searches, followed by a phylogenetic study to determine the evolutionary relationships between putatively conserved genes, in order to infer orthology, or at least functional relatedness. Once genes have been identified in this way, rigorous experimental analysis using targeted deletion in different pathogen species and cross-species complementation experiments would be required to confirm the presence of such conserved gene functions. The bioinformatic component of such a project is underway using the *M. grisea* genome and genome and EST information from 15 other pathogenic fungal species including *Mycosphaerella graminicola*, *Botrytis cinerea*, *F. graminearum*, *U. maydis* and the human pathogens *Candida albicans*, *Aspergillus fumigatus* and *Cryptococcus neoformans*. When compared to genome information from *S. cerevisiae*, *N. crassa*, *Schizosaccharomyces pombe* and *Aspergillus nidulans*, a set of 65 genes was found to be unique to the pathogenic species, with some individual genes conserved in up to five pathogenic species (S. Ahmad and N.J. Talbot, unpublished information). Some of these genes are now being functionally analysed.

Whole-genome comparisons between two taxonomically related fungal species such as *M. grisea* and *N. crassa* have also proved revealing in determining the characteristics of pathogenic fungi which may distinguish them from saprotrophic species (Dean et al. 2005). The rice blast fungus appears to posses approximately 10% more genes than does the saprotrophic *M. grisea*, including far more genes which are predicted to encode secreted proteins. In addition, there have been episodes of gene family expansion or maintenance in *M. grisea*. This analysis is constrained, however, by the influence which repeat-induced point mutation (RIP) has had on the genome of *N. crassa* (Galagan et al. 2003). RIP is a process by which duplicated genes are removed at the onset of meiosis, resulting in little opportunity for paralogous duplication and gene family expansion (Galagan et al. 2003; Borkovitch et al. 2004).

VI. Identifying Pathogen-Conserved Genes Using Incomplete Genome Data

In the absence of complete genomic sequence, single pass, partial sequencing of either 3′ or 5′ ends of complementary DNA (cDNA) clones to generate a set of expressed sequence tags (ESTs), offers a low-cost strategy to identify gene inventories. EST datasets are now available for a large number of phytopathogenic fungi, but generally in a flat file format with limited annotation. The largest collection of EST data from phytopathogenic fungi is housed at the consortium for Functional Genomics of Microbial Eukaryotes (COGEME) database, which contains 54,083 unique sequences from 18 species of fungi and oomycetes (Soanes et al. 2002). In the COGEME database, individual EST sequences were clustered to produce a set of consensus sequences (unisequences), to which putative functions were assigned based on similarity to sequences of known genes. The BlastX algorithm (Altschul et al. 1990) was used to query the NCBI non-redundant protein database, and the top five most similar sequences (with expectation values less than 1×10^{-5}) were retrieved for each unisequence. The cut-off value selected was more rigorous than the recommended value of 10^{-2} (Anderson and Brass 1998), below which matches are considered significant in 98% of cases. On the basis of these similarity scores, a putative product or function was assigned to each unisequence, but this remains highly speculative in some cases. Based on the assignments, the unisequences were classified by function according to a hierarchical scheme used by MIPS (Munich Information Centre for Protein Sequences; Mewes et al. 2004). This scheme groups gene products depending on the particular metabolic pathway or cellular process which they are predicted to be involved in. Approximately 44% of the unisequences in the COGEME database had no homology to sequences present in publicly available databases. Such genes are often termed orphans (ORFs of no known function) and are commonly found in the genomes of eukaryotic organisms. Orphan genes may represent genes whose phylogenetic distribution is restricted to certain evolutionary lineages or genes which rapidly diverge between closely related species (Siew and Fischer 2004). A comparison of ESTs from *N. crassa* with the *S. cerevisiae* genome indicated that there is a higher proportion of orphans in the *N. crassa* genome than in yeast (Braun et al. 2000). This may be due to the increased morphological complexity of the multicellular *N. crassa* when compared with *S. cerevisiae*, which is also far more intensively studied. However, it will be interesting in due course to investigate more closely related fungal species to determine whether the orphans in *N. crassa* include rapidly evolving genes.

Although the gene inventories contained in the COGEME EST database represent only a fraction of the transcriptome of each fungal species, careful analysis of the genes represented in these collections can give valuable information about metabolic pathways which may be present in different species of phytopathogenic fungi. A recent example of such a study examined amino acid biosynthesis genes within a group of fungi, including phytopathogenic species such as *M. grisea*, *My. graminicola*, *B. cinerea*, *F. graminearum* and the barley powdery mildew fungus *Blumeria graminis* (Giles et al. 2003). A relational database of information regarding amino acid biosynthetic pathways for each fungal species was generated and can be viewed at http://cogeme.ex.ac.uk/biosynthesis.html. The database was developed as a demonstration tool to show how pathway information can be gained from incomplete EST data. Amino acid biosynthesis pathways are reasonably well understood and documented (Braus 1991; Fritz 1997; Thomas and Surdin-Kerjan 1997). A relational database, with a web-interface offering analytic functions, was therefore developed to exploit the available fungal genomic data with a view to investigating the conservation of amino acid biosynthesis pathways of several pathogenic species. To determine whether an amino acid biosynthetic pathway was likely to be conserved within a particular fungal species based on the presence of a given number of EST sequences, Giles et al. (2003) utilised a Bayesian probability metric, which is given below.

$$\binom{n}{g} \frac{G!}{(G-g)!} \frac{(N-G)!}{(N-G-n+g)!} \frac{(N-n)!}{N!},$$

where N is the number of genes in the genome (assumed to be 10,000), G is the number of enzymes in the pathway, n is the number of genes in the sample dataset and g is the number of pathway enzymes in the sample dataset.

The main limitation in the analysis reported by Giles et al. (2003) results from the nature of the partial EST datasets stored in the COGEME database. These datasets are highly unlikely to meet the assumption required for validity of the probability calculations – that the data are drawn randomly from the entire genome. This is because the cDNA libraries from which the ESTs are derived are from different stages of fungal development. The unisequence sets created by clustering EST sequences do, however, have less sequence redundancy and may more closely meet the assumption the probability calculation is based on. Notwithstanding this limitation, the probability values yield information with the potential to provide early indications of the absence or presence of a particular amino acid pathway in an organism of interest. The data, for example, provided evidence for conservation of amino acid biosynthesis pathways in the Barley powdery mildew fungus *Blumeria graminis*. Since this fungus is an obligately pathogenic species which can grow only on its host plan (Kobayashi et al. 1991; Giese 1997), a requirement for amino acid biosynthesis might be predicted to be dispensable. The data presented by Giles et al. (2003) suggest that this is not the case and that the organism possesses amino acid biosynthetic genes in the same way as a free-living saprotroph or a facultative pathogen. The pathway prediction tool is adaptable and may of broader utility in determining the presence of other metabolic or signal transduction pathways in organisms for which only partial gene inventories are available.

VII. Exploring Gene Expression Patterns Using EST Datasets

When ESTs are generated randomly from non-normalised cDNA libraries, the frequency of a given EST is proportional to the abundance of the mRNA transcript corresponding to this gene, in the tissue from which the cDNA library was constructed (Audic and Claverie 1997). If ESTs are available which have been sequenced from a number of cDNA libraries, then consequently an expression profile can be constructed for each gene, showing EST frequency in libraries constructed from a variety of tissues, or experimental conditions. The sequencing of ESTs from a cDNA

A

Unigene ID: DNMag0391
Putative product / function: MPG1 class I hydrophobin

Transcript profile

Library	Mag02	Mag03	Mag04	Mag06	Mag07	Mag08	Mag10	Mag15
No. of ESTs[1]	35	0	92	18	27	8	9	402
% ESTs[2]	1.370936154	0	2.344546381	0.571247223	0.794818958	0.185787274	0.203573852	6.164698666

Details of each library (click on library name for further details, total number of ESTs sequenced from each library in brackets)
Mag02: 70-15 appressorium (2553).
Mag03: 70-15 mycelium grown in minimal medium (1466).
Mag04: CP987 mycelium grown in medium containing rice cell walls as sole carbon source (3924).
Mag06: Guy11 mycelium grown in complete medium (3193).
Mag07: Guy11 conidia (3405).
Mag08: Guy11 mycelium grown in nitrogen starvation medium (4310).
Mag10: *pmk1* germinated conidia (4421).
Mag15: Mixed mated culture (6521).

[1]Number of ESTs representing this unigene sequenced from each cDNA library.
[2]Number of ESTs representing this unigene shown as a % of the total number of ESTs sequenced from each library.

B

Pairwise comparison of transcript abundance between different libraries for this unigene

Library	Mag02	Mag03	Mag04	Mag06	Mag07	Mag08	Mag10	Mag15
Mag02		**0.999**	**0.994**	**0.998**	**0.96**	**0.999**	**0.999**	**0.999**
Mag03			**0.999**	**0.998**	**0.999**	0.8	0.8	**0.999**
Mag04				**0.999**	**0.999**	**0.999**	**0.999**	**0.999**
Mag06					0.7	**0.994**	**0.99**	**0.999**
Mag07						**0.999**	**0.999**	**0.999**
Mag08							0.1	**0.999**
Mag10								**0.999**

The values in this table are P values indicating the probability that differences in the number of EST transcripts, representing this unigene, sequenced from the two libraries are due to differential expression of the gene rather than chance sampling error. Significant differences in expression (P >= 0.95) are highlighted in bold (Audic and Claverie, 1997).

Fig. 3.1. Screen shots from the web-based front of the COGEME fungal EST database (http://cogeme.ex.ac.uk/transcript.html) showing a transcript profile for the gene which encodes the fungal hydrophobin *MPG1* (Talbot et al. 1993, 1996). **A** Output from the database shows the number of ESTs representing this unigene sequenced from each cDNA library, as well as the percentage of the total ESTs sequenced from each library which this represents. **B** This matrix shows *P*-values representing the probability that differences in EST abundance in pair-wise comparisons between each library are due to differential expression of the gene, rather than chance sampling error

A

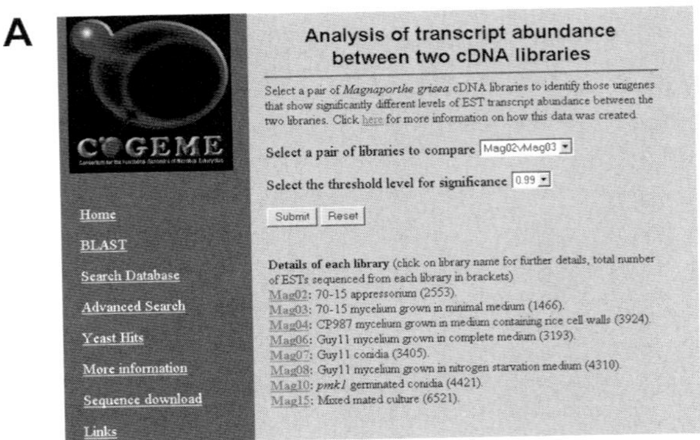

Fig. 3.2. Screen shots from the web-based front of the COGEME fungal EST database showing a comparison of EST transcript profiles between cDNA libraries which are derived from mRNA extracted from appressoria of a rice pathogenic strain of *Magnaporthe grisea*, 70-15 (Mag02), or hyphae cultured in standard minimal growth medium (Mag03). **A** The researcher can use this screen to specify which pair of libraries to compare and to set the minimum *P*-value for consideration of a difference in EST frequency as significant (0.99 or 0.95). **B** List of unigenes which are significantly more highly expressed in one library than in the other, split into those which are more highly expressed in Mag02 (*top*) and those which are more highly expressed in Mag03 (*bottom*)

library is, however, also a random event and therefore the frequency of sequencing a particular cDNA clone is subject to fluctuations due to chance sampling error. Consequently, the data produced in this way are inherently noisy, and more reliable results are produced only when large numbers of ESTs are sequenced from each library (Ewing et al. 1999). The presence of large EST collections for a variety of phytopathogenic fungi provides an opportunity to carry out this form of gene expression analysis, which is of particular value where resources such a microarrays are not yet available. An example of the strategy has been carried out with 958 unique genes (unisequences), which were well-represented by ESTs, from eight cDNA libraries of *M. grisea* derived from appressoria, conidia and mycelia grown either in complete medium, minimal medium, nitrogen starvation medium or with rice cell walls as the sole carbon source. This was done with wild-type strains, as well as with germinated conidia from a null mutant for the MAP kinase-encoding gene *PMK1* (Xu and Hamer 1996) and a mated culture library constructed from a mixture of asci, ascospores, perithecia and mycelium produced by crossing two different strains (D.M. Soanes and N.J. Talbot, unpublished data). A comparison of ESTs sequenced from two cDNA libraries was then carried out to identify which putative genes are differentially regulated between the two conditions or tissue types from which the libraries were constructed, as shown in Fig. 3.1. This information is available at http://cogeme.ex.ac.uk/transcript.html. A statistical analysis method, developed to identify whether the difference in EST sequence abundance between two libraries is significant (Audic and Claverie 1997), was used to test the robustness of the results scored. This test is based on the probability that the difference in frequency of ESTs between two cDNA libraries is due to random sampling, rather than to a difference in expression levels (Fig. 3.2). The theory links the threshold of selection of putatively regulated genes to the fraction of false positive clones one is willing to risk. This statistical method was used to produce a *P*-value for each pair of cDNA libraries for each unigene. The *P*-value indicates the probability that differences in the number of EST transcripts, representing each unigene, sequenced from the two libraries are due to differential expression of the gene, rather than chance sampling error (Audic and Claverie 1997).

VIII. Conclusions and Future Perspectives

The emergence of functional genomics has introduced a new era in the study of fungal pathogenicity (Yoder and Turgeon 2001). As illustrated above, genomic technologies allow us to acquire vast datasets which have obvious relevance to pathogenesis, but also provide challenges if they are to be analysed effectively. Molecular phylogeny has shown that pathogenic fungi are found in many taxonomic groups (including basidiomycetes and ascomycetes), which suggests that the pathogenic life style has evolved multiple times within the fungal kingdom. There are several experimental studies which have demonstrated that parasitic fungi have unique pathogenicity factors (as reviewed by Idnurm and Howlett 2001). There are fewer examples, however, of large-scale systematic screens for such genes from genome sequences, which provides a rich opportunity for future research. The alternative explanation, in which pathogenesis is due to specific differences in gene regulation, is exemplified by the presence

B

Comparison of transcript profiles of Mag02 and Mag03

Sequences significantly more highly expressed (P >= 0.99) in Mag02 than Mag03

Unigene ID	Putative product / function	Mag02[1]	Mag03[1]	P value[2]
DNMag0335	homologue of UVI-1 [Bipolaris oryzae], cell wall protein	30	0	0.999
DNMag0371	MAS3 protein	20	0	0.999
DNMag0391	MPG1 class I hydrophobin	35	0	0.999
DNMag0623	unknown	13	0	0.996
DNMag0815	unknown	11	0	0.991

Sequences significantly more highly expressed (P >= 0.99) in Mag03 than Mag02

Unigene ID	Putative product / function	Mag02[1]	Mag03[1]	P value[2]
DNMag0052	60S acidic ribosomal protein P2	0	7	0.999
DNMag0054	60S ribosomal protein L10	0	7	0.999
DNMag0166	calcineurin subunit B	0	9	0.999
DNMag0208	coproporphyrinogen oxidase, sixth step in heme biosynthetic pathway	0	7	0.999
DNMag0232	D-hydroxyacid dehydrogenase / D-lactate dehydrogenase	0	10	0.999
DNMag0293	glyceraldehyde 3-phosphate dehydrogenase	3	21	0.999
DNMag0438	pathogenesis related (SnodProt1)	0	8	0.999
DNMag0592	unknown	8	55	0.999
DNMag0612	unknown	2	55	0.999
DNMag0645	unknown	0	23	0.999
DNMag0650	unknown	0	18	0.999
DNMag0658	unknown	0	8	0.999
DNMag0673	unknown	0	7	0.999
DNMag0074	60S ribosomal protein L26	0	6	0.998
DNMag0260	extracellular chitinase	0	6	0.998
DNMag0560	translation elongation factor-1 gamma (EF-1-gamma)	0	6	0.998
DNMag0670	unknown	0	6	0.998
DNMag0079	60S ribosomal protein L30	1	7	0.996
DNMag0700	unknown	1	7	0.996
DNMag0023	40S ribosomal protein S12	0	5	0.995
DNMag0067	60S ribosomal protein L19	0	5	0.995
DNMag0070	60S ribosomal protein L22	0	5	0.995
DNMag0083	60S ribosomal protein L35	0	5	0.995
DNMag0107	adenosine kinase	0	5	0.995
DNMag0249	enolase (2-phosphoglycerate dehydratase)	0	5	0.995
DNMag0573	ubiquinol-cytochrome c reductase	0	5	0.995
DNMag0849	unknown	0	5	0.995
DNMag0850	unknown	0	5	0.995
DNMag0610	unknown	10	17	0.994
DNMag0036	40S ribosomal protein S25	1	6	0.99
DNMag0712	unknown	1	6	0.99

[1] Number of ESTs representing each unigene sequenced from each cDNA library.

[2] he P value indicates the probability that differences in the number of EST transcripts, representing each unigene, sequenced from the two libraries are due to differential expression of the gene rather than chance sampling error. (Audic and Claverie, 1997).

Fig. 3.2. (continued)

of conserved elements of signaling pathways for infection-related development in diverse phytopathogenic species. MAP kinase genes similar to the *PMK1*-encoded MAP kinase gene of *M. grisea* are present in a number of other plant pathogenic fungi, including *Botrytis cinerea* (Zheng et al. 2000), *Colletotrichum lagenarium* (Takano et al. 2000), *Cochliobolus heterostrophus* (Lev et al. 1999), *Fusarium oxysporum* (Di Pietro et al. 2001), *Pyrenophora teres* (Ruiz-Roldan et al. 2001) and *Ustilago maydis* (Mayorga and Gold 1999; Muller et al. 1999). In all cases, the corresponding MAPK genes have been disrupted and mutant strains found to be defective in pathogenicity. The extent to which conservation of the effector molecules which act downstream of these pathways are conserved in pathogen genomes is, however, not yet clear.

Widespread genome sequencing of filamentous fungi, including phytopathogenic species such as *M. grisea*, *U. maydis* and *F. graminearum*, is already providing key insights into the pathogenic life style, including the presence of a complex secreted proteome, and gene family expansions in genes associated with environmental perception and cell signalling (Dean et al. 2005). The contrast and comparison with genomes of closely related saprotrophic fungi such as *A. nidulans*, *N. crassa* and *C. cinereus* will prove very illuminating. One of the most significant informatics challenges will be in predicting whether conservation or divergence of particular biological functions has occurred between these fungal species. The availability of EST collections from a much larger group of phytopathogenic species is also important, not only in carrying out comparative genomics but also for examination of alternative splicing, gene expression analysis, genome annotation, and for determining whether microRNA-mediated gene regulation occurs in filamentous fungi as in other multicellular eukaryotes.

Finally, a large effort is still required to provide better mechanisms for gene functional analysis in filamentous fungi in order that the genes predicted to serve roles in pathogenesis can be readily tested for such a role, and the interplay between gene activities then investigated in detail. Drawing together these disparate, computational and empirically derived pieces of information into a cohesive and plausible model for the development of plant disease by fungi will be a key challenge for the next years.

Acknowledgements. Work on comparative genomics of fungi in our group is supported by the Biotechnology and Biological Sciences Research council (BBSRC) grant BBS/B/17174.

References

Adachi K, Nelson GH, Peoples KA, Frank SA, Montenegro-Chamorro MV, DeZwaan TM, Ramamurthy L, Shuster JR, Hamer L, Tanzer MM (2002) Efficient gene identification and targeted gene disruption in the wheat blotch fungus Mycosphaerella graminicola using TAGKO. Curr Genet 42:123–127

Altschul S, Gish W, Miller W, Myers EW, Lipman DJ (1990) Basic local alignment search tool. J Mol Biol 215:403–410

Altschul S, Madden TL, Schaffer AA, Zhang J, Zhang A, Miller W, Lipman DJ (1997) Gapped BLAST and PSI-BLAST: a new generation of protein database search programs. Nucleic Acids Res 25:3389–3402

Anderson I, Brass A (1998) Searching DNA databases for similarities to DNA sequences: when is a match significant? Bioinformatics 14:349–356

Audic S, Claverie JM (1997) The significance of digital gene expression profiles. Genome Res 7:986–995

Balhadère PV, Talbot NJ (2001) PDE1 encodes a P-type ATPase involved in appressorium-mediated plant infection by the rice blast fungus Magnaporthe grisea. Plant Cell 13:1987–2004

Balhadère PV, Foster AJ, Talbot NJ (1999) Identification of pathogenicity mutants of the rice blast fungus Magnaporthe grisea by insertional mutagenesis. Mol Plant-Microbe Interac 12:129–142

Bansal AK (1999) An automated comparative analysis of 17 complete microbial genomes. Bioinformatics 15:900–908

Borkovich KA, Alex LA, Yarden O, Freitag M, Turner GE, Read ND, Seiler S, Bell-Pedersen D, Paietta J, Plesofsky N et al. (2004) Lessons from the genome sequence of Neurospora crassa: tracing the path from genomic blueprint to multicellular organism. Microbiol Mol Biol Rev 68:1–108

Bowyer P (1999) Plant diseases caused by fungi: phytopathogenicity. In: Oliver RP, Schweizer M (eds) Molecular fungal biology. Cambridge University Press, pp 294–321

Braun EL, Halpern AL, Nelson MA, Natvig DO (2000) Large-scale comparison of fungal sequence information: mechanisms of innovation in Neurospora crassa and gene loss in Saccharomyces cerevisiae. Genome Res 10:416–430

Braus GH (1991) Aromatic amino acid biosynthesis in the yeast Saccharomyces cerevisiae: a model system for the regulation of a eukaryotic biosynthetic pathway. Microbiol Rev 55:349–370

Brown TA (2002) Genomes 2. Bios Scientific, Oxford Bundock P, Dendulkras A, Beijersbergen A, Hooykaas PJJ (1995) Trans-kingdom T-DNA transfer from Agrobacterium-tumefaciens to Saccharomyces cerevisiae. EMBO J 14:3206–3214

Chen R, Jeong SS (2000) Functional prediction: identification of protein orthologs and paralogs. Protein Sci 9:2344–2353

Commenil P, Belingheri L, Dehorter B (1998) Antilipase antibodies prevent infection of tomato leaves by Botrytis cinerea. Physiol Mol Plant Pathol 52:1–14
Dean RA, Talbot NJ, Ebbole DJ, Farman ML, Mitchell T, Orbach M, Thon M, Kulkarni R, Xu JR, Pan H et al. (2005) The genome sequence of the rice blast fungus Magnaporthe grisea. Nature 434:980–986
de Groot MJA, Bundock P, Hooykaas PJJ, Beijersbergen AGM (1998) *Agrobacterium tumefaciens*-mediated transformation of filamentous fungi. Nature Biotechnol 16:839–842
DeZwaan TM, Carroll AM, Valent B and Sweigard JA (1999) Magnaporthe grisea pth11p is a novel plasma membrane protein that mediates appressorium differentiation in response to inductive substrate cues. Plant Cell 11:2013–2030
Dietrich FS, Voegeli S, Brachat S, Lerch A, Gates K, Steiner S, Mohr C, Pohlmann R, Luedi P, Choi S et al. (2004) The Ashbya gossypii genome as a tool for mapping the ancient Saccharomyces cerevisiae genome. Science 304:304–307
Di Pietro A, Garcia-Maceira FI, Meglecz E, Roncero MIG (2001) A MAP kinase of the vascular wilt fungus Fusarium oxysporum is essential for root penetration and pathogenesis. Mol Microbiol 39:1140–1152
Eisen JA (1998) Phylogenomics: improving functional predictions for uncharacterized genes by evolutionary analysis. Genome Res 8:163–167
Ewing RM, Kahla AB, Poirot O, Lopez F, Audic S, Claverie JM (1999) Large-scale statistical analyses of rice ESTs reveal correlated patterns of gene expression. Genome Res 9:950–959
Fitch WM (1970) Distinguishing homologous from analogous proteins. Syst Zool 19:99–113
Fritz R, Lanen C, Colas V, Leroux P (1997) Inhibition of methionine biosynthesis in Botrytis cinerea by the anilinopyrimidine fungicide pyrimethanil. Pest Sci 49:40–46
Galagan JE, Calvo SE, Borkovich KA, Selker EU, Read ND, Jaffe D, Fitzhugh W, Mal J, Smirnov S, Purcell S et al. (2003) The genome sequence of the filamentous fungus Neurospora crassa. Nature 422:859–868
Gates MA, Kim L, Egan ES, Cardozo T, Sirotkin HI, Dougan ST, Lashkari D, Abagyan R, Schier AF, Talbot WS (1999) A genetic linkage map for zebrafish: comparative analysis and localization of genes and expressed sequences. Genome Res 9:334–347
Giese H, Hippe-Sanwald S, Somerville S, Weller J (1997) Erisyphe graminis. In: Carroll GC, Tudzynski P (eds) The Mycota, vol V, part B. Springer, Berlin Heidelberg New York, pp 55–78
Giles PM, Soanes DM, Talbot NJ (2003) A relational database for the discovery of genes encoding amino acid biosynthetic enzymes in pathogenic fungi. Comp Funct Genomics 4:4–15
Goffeau A, Barrell BG, Bussey H, Davis RW, Dujon B, Feldmann H, Galibert F, Hoheisel JD, Jacq C, Johnston M et al. (1996) Life with 6000 genes. Science 546:563–567
Gogarten JP, Olendzenski L (1999) Orthologs, paralogs and genome comparisons. Curr Opin Genet Dev 9:630–636
Gold SE, García-Pedrajas MD, Martínez-Espinoza AD (2001) New (and used) approaches to the study of fungal pathogenicity. Annu Rev Phytopathol 39:337–365
Grishin NV, Wolf YI, Koonin EV (2000) From complete genomes to measures of substitution rate variability within and between proteins. Genome Res 10:991–1000
Hamer L, Adachi K, Montenegro-Chamorro MV, Tanzer MM, Mahanty SK, Lo C, Tarpey RW, Skalchunes AR, Heiniger RW, Frank SA et al. (2001) Gene discovery and gene function assignment in filamentous fungi. Proc Natl Acad Sci USA 98:5110–5115
Hammond-Kosack KE, Jones JDG (1997) Plant disease resistance genes. Annu Rev Plant Physiol Plant Mol Biol 48:575–607
Hedges SB, Kumar S (2003) Genomic clocks and evolutionary timescales. Trends Genet 19:200–206
Howard RJ (1997) Breaching the outer barrier – cuticle and cell wall penetration. In: Carroll GC, Tudzynski P (eds) The Mycota, vol V. Plant relationships, part A. Springer, Berlin Heidelberg New York, pp 43–60
Howard RJ (2001) Cytology of fungal pathogens and plant-host interactions. Curr Opin Microbiol 4:365–373
Huynen M, Dandekar T, Bork P (1998) Differential genome analysis applied to the species-specific features of Helicobacter pylor. FEBS Lett 426:1–5
Idnurm A, Howlett BJ (2001) Pathogenicity genes of phytopathogenic fungi. Mol Plant Pathol 2:241–255
Jensen RA (2001) Orthologs and paralogs – we need to get it right. Genome Biol 1:21–23
Katinka MD, Duprat S, Cornillot E, Metenier G, Thomarat F, Prensier G, Barbe V, Peyretaillade E, Brottier P, Wincker P et al. (2001) Genome sequence and gene compaction of the eukaryote parasite Encephalitozoon cuniculi. Nature 414(6862):450–453
Kellis M, Patterson N, Endrizzi M, Birren B, Lander ES (2003) Sequencing and comparison of yeast species to identify genes and regulatory elements. Nature 423:241–254
Kobayashi I, Tanaka C, Yamaoka N, Kunoj H (1991) Morphogenesis of Erisyphe graminis conidia on artificial membranes. Trans Mycol Soc Jpn 187–198
Koonin EV (2003) Comparative genomics, minimal gene-sets and the last universal common ancestor. Nat Rev Microbiol 1:127–136
Koonin EV, Mushegian AR, Bork P (1996) Non-orthologous gene displacement. Trends Genet 12:334–336
Kubo Y, Nakamura H, Kobayashi K, Okuno T, Furasawa I (1991) Cloning of a melanin biosynthetic gene essential for appressorial penetration of Colletotrichum lagenarium. Mol Plant-Microbe Interact 4:440–445
Kubo Y, Takano Y, Endo N, Yasuda N, Tajima S, Furusawa I (1996) Cloning and structural analysis of the melanin biosynthesis gene SCD1 encoding scytalone dehydratase in Colletotrichum lagenarium. Appl Environ Microbiol 62:4340–4344
Lev S, Sharon A, Hadar R, Ma H, Horwitz BA (1999) A mitogen-activated protein kinase of the corn leaf pathogen Cochliobolus heterostrophus is involved in conidiation, appressorium formation, and pathogenicity: diverse roles for mitogen-activated protein kinase homologs in foliar pathogens. Proc Natl Acad Sci USA 96:13542–13547
Lorang JM, Tuori RP, Martinez JP, Sawyer TL, Redman RS, Rollins JA, Wolpert TJ, Johnson KB, Rodriguez RJ, Dickman MB et al. (2001) Green fluorescent protein is lighting up fungal biology. Appl Environ Microbiol 67:1987–1994

Martinez D, Larrondo LF, Putnam N, Gelpke MD, Huang K, Chapman J, Helfenbein KG, Ramaiya P, Detter JC, Larimer F et al. (2004) Genome sequence of the lignocellulose degrading fungus phanerochaete chrysosporium strain RP78. Nat Biotechnol 22:695–700
Mayorga ME, Gold SE (1999) A MAP kinase encoded by the ubc3 gene of Ustilago maydis is required for filamentous growth and full virulence. Mol Microbiol 34:485–497
Mewes HW, Amid C, Arnold R, Frishman D, Guldener U, Mannhaupt G, Munsterkotter M, Pagel P, Strack N, Stumpflen V et al. (2004) MIPS: analysis and annotation of proteins from whole genomes. Nucleic Acids Res 32:D41–44
Muller P, Aichinger C, Fedbrügge M, Kahmann R (1999) The MAP kinase Kpp2 regulates mating and pathogenic development in Ustilago maydis. Mol Microbiol 34:1007–1017
Mullins ED, Kang S (2001) Transformation: a tool for studying fungal pathogens of plants. Cell Mol Life Sci 58:2043–2052
Mullins ED, Chen X, Romaine P, Raina R, Geiser DM, Kang S (2001) Agrobacterium-mediated transformation of Fusarium oxysporum: an efficient tool for insertional mutagenesis and gene transfer. Phytopathology 91:173–180
Mushegian AR, Koonin EV (1996) A minimal gene set for cellular life derived by comparison of complete bacterial genomes. Proc Natl Acad Sci USA 93:10268–10273
Mushegian AR, Garey JR, Martin J, Liu LX (1998) Large-scale taxonomic profiling of eukaryotic model organisms: a comparison of orthologous proteins encoded by the human, fly, nematode, and yeast genomes. Genome Res 8:590–598
Nitta N, Farman ML, Leong SA (1997) Genome organization of Magnaporthe grisea: integration of genetic maps, clustering of transposable elements and identification of genome duplications and rearrangements. Theor Appl Genet 95:20–32
Nobrega MA, Pennacchio LA (2004) Comparative genomic analysis as a tool for biological discovery. J Physiol 554:31–39
Ochman H, Lawrence JG, Groisman EA (2000) Lateral gene transfer and the nature of bacterial innovation. Nature 405:299–304
Orbach MJ, Farrall L, Sweigard JA, Chumley FG, Valent B (2000) A telomeric avirulence gene determines efficacy for rice blast resistance gene Pi-ta. Plant Cell 12:2019–2032
Ouzounis CA, Coulson RMR, Enright AJ, Kunin V, Pereira-Leal JB (2003) Classification schemes for protein structure and function. Nat Rev Genet 4:508–519
Rat Genome Sequencing Project Consortium (2004) Genome sequencing of the Brown Norway rat yields insights into mammalian evolution. Nature 428(6982):475–476
Rosewich UL, Kistler HC (2000) Role of horizontal gene transfer in the evolution of fungi. Annu Rev Phytopathol 38:325–363
Ruiz-Roldan MC, Maier FJ, Schafer W (2001) PTK1, a mitogen-activated protein kinase gene is required for conidiation, appressorium formation, and pathogenicity of Pyenophora teres on barley. Mol Plant-Microbe Interact 14:116–125
Sharman AC (1999) Some new terms for duplicated genes. Semin Cell Dev Biol 5:561–563
Siew N, Fischer D (2004) Structural biology sheds light on the puzzle genomic ORFans. J Mol Biol 342:369–373
Soanes DM, Skinner W, Keon J, Hargreaves J, Talbot NJ (2002) Genomics of phytopathogenic fungi and the development of bioinformatic resources. Mol Plant-Microbe Interact 15:421–427
Sweigard JA, Carroll AM, Farrall L, Chumley FG, Valent B (1998) Magnaporthe grisea pathogenicity genes obtained through insertional mutagenesis. Mol Plant-Microbe Interact 11:404–412
Takano Y, Kikuchi T, Kubo Y, Hamer JE, Mise K, Furusawa I (2000) The Colletotrichum lagenarium MAP kinase gene CMK1 regulates diverse aspects of fungal pathogenesis. Mol Plant-Microbe Interact 13:374–383
Talbot NJ, Foster AJ (2001) Genetics and genomics of the rice blast fungus Magnaporthe grisea: developing an experimental model for understanding fungal diseases of cereals. Adv Bot Res 34:263–287
Talbot NJ, Ebbole DJ, Hamer JE (1993) Identification and characterisation of MPG1, a gene involved in pathogenicity from the rice blast fungus Magnaporthe grisea. Plant Cell 5:1575–1590
Talbot NJ, Kershaw MJ, Wakley GE, DeVries OMH, Wessels JGH, Hamer JE (1996) MPG1 encodes a fungal hydrophobin involved in surface interactions during infection-related development of Magnaporthe grise. Plant Cell 8:985–999
Tatusov RL, Mushegian AR, Bork P, Brown NP, Hayes WS, Borodovsky M, Rudd KE, Koonin EV (1996) Metabolism and evolution of Haemophilus influenzae deduced from a whole-genome comparison with Escherichia coli. Curr Biol 6:279–291
Tatusov RL, Koonin EV, Lipman DJ (1997) A genomic perspective on protein families. Science 278:631–637
Tatusov RL, Galperin MY, Natale DA, Koonin EV (2000) The COG database: a tool for genome-scale analysis of protein functions and evolution. Nucleic Acids Res 28:33–36
Tatusov RL, Natale DA, Garkavtsev IV, Tatusova TA, Shankavaram UT, Rao BS, Kiryutin B, Galperin MY, Fedorova ND, Koonin EV (2001) The COG database: new developments in phylogenetic classification of proteins from complete genomes. Nucleic Acids Res 29:22–28
Theiben G (2002) Secret life of genes. Nature 415:741
Thinlay X, Finckh MR, Bordeos AC, Zeigler RS (2000) Effects and possible causes of an unprecedented rice blast epidemic on the traditional farming system of Bhutan. Agric Ecosyst Environ 78:237–248
Thomas D, Surdin-Kerjan Y (1997) Metabolism of sulfur amino acids in Saccharomyces cerevisiae. Microbiol Mol Biol Rev 61:503–536
Tucker SL, Talbot NJ (2001) Surface attachment and pre-penetration stage development by plant pathogenic fungi. Annu Rev Phytopathol 39:385–417
Tucker SL, Thornton CR, Tasker K, Jacob C, Giles G, Egan M, Talbot NJ (2004) A fungal metallothionein is required for pathogenicity of Magnaporthe grisea. Plant Cell 16:1575–1588
Tunlid A, Talbot NJ (2002) Genomics of parasitic and symbiotic fungi. Curr Opin Microbiol 5:513–519

Wei L, Liu Y, Dubchak I, Shon J, Park J (2002) Comparative genomics approaches to study organism similarities and differences. J Biomed Inform 35:142–150

Wood V, Gwilliam R, Rajandream MA, Lyne M, Lyne R, Stewart A, Sgouros J, Peat N, Hayles J, Baker S (2002) The genome sequence of Schizosaccharomyces pombe. Nature 415:871–880

Wren BW (2000) Microbial genome analysis: Insights into virulence, host adaptation and evolution. Nat Rev Genet 1:30–39

Xie T, Ding DF (2000) Investigating 42 candidate orthologous protein groups by molecular evolutionary analysis on genome scale. Gene 261:305–310

Xu JR, Hamer JE (1996) MAP kinase and cAMP signalling regulate infection structure formation and pathogenic growth in the rice blast fungus Magnaporthe grisea. Genes Dev 10:2696–2706

Xue C, Park G, Choi W, Zheng L, Dean RA, Xu JR (2002) Two novel fungal virulence genes specifically expressed in appressoria of the rice blast fungus. Plant Cell 14:2107–2119

Yang G, Ross MS, Turgeon BG, Yoder OC (1996) A polyketide synthase is required for fungal virulence and production of polyketide T-toxin. Plant Cell 8:2139–2150

Yoder OC, Turgeon BG (2001) Fungal genomics and pathogenicity. Curr Opin Plant Biol 4:315–321

Zhang Z, Gurr SJ (2000) Walking in the unknown; a "step-down" PCR-based technique leading to direct sequence analysis of flanking genomic DNA. Gene 253:145–150

Zheng L, Campbell M, Murphy J, Lam S, Xu J-R (2000) The BMP1 gene is essential for pathogenicity in the gray mold fungus Botrytis cinerea. Mol Plant-Microbe Interact 13:724–732

Fungal Rythms and Responses

4 Circadian Rhythms, Photobiology and Functional Genomics in *Neurospora*

J.J. Loros[1], J.C. Dunlap[1]

CONTENTS

Abbreviations: EST, expressed sequence tag; MSUD, meiotic silencing of unpaired DNA; PKC, protein kinase C; PTGS, post-transcriptional gene silencing; RIP, repeat-induced point mutations; SNP, single nucleotide polymorphism

[1] Departments of Biochemistry and Genetics, Dartmouth Medical School, Hanover, NH 03755, USA

I. Introduction

Two widespread and tightly interconnected forms of cellular, tissue and organismal regulation are the ability to see environmental light in conjunction with the ability to tell the time of day. Chromophore-binding photoreceptive molecules facilitate the harvesting of photons, allowing an organism to respond to changes in light fluence. Biological rhythms provide organisms with the capacity to **anticipate environmental cycles** imposed by the Earth's rotation. The capability to gage time of day is called **circadian rhythmicity**, and the physiological and molecular basis of these rhythms has been an object of study for well over a century. Recent decades have seen the unraveling of this long mystery, and microbial systems have led the way in many regards. In organisms with clocks, which include most eukaryotes and the cyanobacteria, **most or all cells of the organism have their own molecular clock**. The cellular nature of circadian rhythmicity, now universally recognized as a characteristic of rhythms, was first described in microbes. Phase response curves to light played a central role in delineating the formal properties of biological oscillators, and were first determined in *Gonyaulax* (Hastings and Sweeney 1958), as was the concept of using phase response curves to define sensitive and insensitive phases to chemicals that facilitated the probing of the molecular nature of the oscillator (Hastings 1960). More recently, as studies utilized more genetic and molecular tools, the microbial system *Neurospora crassa* has pioneered the way, along with *Drosophila*, in finally describing **molecular feedback loops as the basis of rhythms** as we understand them today.

Work on *Neurospora* describing single gene rhythm mutants (Feldman and Waser 1971) occurred simultaneously with similar work in *Drosophila* (Konopka and Benzer 1971), as did

The Mycota XIII
Fungal Genomics
Alistair J.P. Brown (Ed.)

molecular cloning of the clock gene *frq* (Loros and Dunlap 2001), which provided the first case of rescue of a behavioral mutation by DNA-mediated transformation. Once such molecules were cloned and isolated, work in *Neurospora* was the first to move beyond a description of the system to the use of genetic engineering to manipulate the regulation of these genes (Aronson et al. 1994a). This led the way to establishing their roles in biological oscillators and proving that such feedback loops lay at the core of circadian rhythmicity. Molecular mechanisms for **light-induced phase shifting** and for **temperature entrainment** of circadian clocks were first determined in *Neurospora* (Crosthwaite et al. 1995; Liu et al. 1998), and many of these mechanisms show or have foreshadowed direct parallels in mammalian systems (Shigeyoshi et al. 1997). Paradigms used for understanding the rhythmic physiology of organisms have also been developed in *Neurospora*, including the view that **clock-regulated transcription** would provide a major means of clock output. This led to the first genome-wide screens for **clock-controlled genes** and the coining of the term "*ccg*" that has now become widely used to describe circadianly regulated genes and transcripts (Loros et al. 1989).

The ability of fungi to sense light using **photoreceptive molecules** is also common to most or all cells of the organism. The molecular basis of light sensing has also been the focus of intense study, now yielding to the combination of classic genetic and modern molecular, biochemical and genomic techniques. Indeed, the well-known **interplay between daily rhythms and light sensing** turned out to have profound interconnections at the molecular level. Many genes regulated by the clock are additionally regulated by light.

The **FREQUENCY (FRQ)** protein is the central negative element in the autoregulatory feedback at the core of the circadian clock in *Neurospora*. **White-collar-1 (WC-1)**, part of the **heterodimeric transcription factor** with **white-collar-2 (WC-2)** that drives *frq* expression and is thereby a central component of the clock feedback mechanism, is also the **photoreceptor for the circadian clock**, as well as all other **blue light-regulated genes** in *Neurospora*. This review will summarize much of what is presently known about the molecular bases of rhythms in *Neurospora*, with a focus on current, genomics-based approaches.

A. Fungi as Model Systems

Fungi, plants, and animals represent three main phylogenetic kingdoms within the eukaryotes. Fungi play huge roles, both positive and negative, in the economies of both industrialized nations and the world. Although fungi constitute the kingdom most closely related to Animalia (Sogin 1994; Simpson and Roger 2002), many are exceptionally tractable experimentally and are therefore universally used as model organisms for understanding all aspects of basic cellular regulation. These regulatory networks include cell cycle progression, gene expression, circadian timing, light sensing, recombination, secretion, and multicellular development. Additionally, mycorrhizal fungi, those that grow interdependently with the roots of plants, are crucial symbionts without which most trees and many grasses cannot live. Fungi also carry out most biomass turnover. Importantly, mycelial fungi rank with bacteria as the most serious of the human and plant pathogens. Because they are eukaryotes, treatment of opportunistic fungal infections in humans pose special risks, challenges and problems not posed by bacteria. The use of fungi in the manufacture of pharmaceutical products is a multibillion dollar per year industry. Penicillin and similar β-lactams, all produced by fungi, are the world's largest-selling antibiotics. It is estimated that 10%–35% of the world's food supply is lost each year due to fungal contamination, a loss of over US$ 200 billion per year. Within the United States alone, fungicide sales constitute more than a US$ 1 billion per year industry. The commercial production of chemicals by mycelial fungi constitutes an industry with a turnover of approximately US$ 35 billion per year. For example, import into the United States of some of these chemicals (such as citrate) costs in excess of US$ 1 billion annually. Industrial production of enzymes, largely by mycelial fungi, constitutes a US$ 1.5 billion per year industry. About 75% of the approximately 250,000 different species of fungi belong to the **Ascomycetes**, 90% being **mycelial fungi** and the remainder yeasts. The other 25% are Basidiomycetes, also mycelial, and having fruiting bodies known as "mushrooms". Much of the above economic impact is due to mycelial fungi yet, until recently, there has been amazingly little research funding and, therefore, effort devoted to their characterization.

B. *Neurospora* as a Research System

1. Overview

Of the mycelial fungi, *N. crassa* is neither a pathogen nor the most widely used industrial fungus. So, why study *Neurospora*? *Neurospora* has been studied for decades and is the best understood mycelial fungus. It is a saprophyte that displays both **asexual and sexual life cycles**, and it is easily maintained through both cycles in laboratory conditions. *Neurospora* exists vegetatively as an **incompletely septate syncytium**, growing equally well in simple liquid or on solid media of defined composition. It is nonpathogenic, and therefore easy and safe to work with, yet it is phylogenetically related to pathogens. Both asexual development and sexual differentiation are highly influenced by environmental factors including nutrient, light and temperature. *Neurospora* is typically haploid, undergoing only a transient diploid stage immediately prior to meiosis. Because of its interesting and diverse biology, simplicity of culture, facile genetics and rapid growth rate, *Neurospora* remains a widely used model that sustains a large and diverse research community.

The genetics of *Neurospora* is unparalleled within the mycelial fungi. It has the most identified genes, and the densest and most accurate genetic map. These attributes are a legacy of 70 years of effort that began concurrently with *Drosophila* genetics. Biochemical genetics as a field got its start from work in *Neurospora* (Beadle and Tatum 1945), and the ease of culture, rapid growth (mass doubling time of 140 min), and ease of harvest continue to support research. Sexual crosses are technically trivial and the **genetic generation time** (from progeny to progeny) is about 3 weeks. Sexual spores are stable for years at 4 °C, and asexual spores or mycelia can be stored for decades. Happily, molecular tools are advanced and being continually improved, as is typical in a vibrant research community. The first mycelial fungus to be transformed, shortly after yeast (Davis 2000), *Neurospora* transformation is routine at frequencies up to many thousands of transformants per microgram of DNA. A variety of selectable markers exist, and several regulatable promoters are regularly used to control expression of transgenes (e.g., Aronson et al. 1994a). As in animal cells, transformation is typically the result of ectopic insertion through non-homologous integration, although **targeted disruption by reciprocal homologous integration** is widely used to generate knockouts of known sequences via insertion of selectable markers (e.g., Aronson et al. 1994c) at frequencies of up to 90% of transformants (depending on the construct and the locus; 5%–10% is routine). A recent report demonstrates that homologous insertion can be achieved approaching 100% in Ku 70 and Ku 80 knockouts (Ninomiya et al. 2004). A number of available vectors are routinely used to target transformation to specific loci within the genome. Alternative methods for generating knockouts or knockdowns are also used with success. Most commonly used is **RIP (repeat-induced-point mutation**; Selker 1990), a rapid method whereby duplicated genes are detected in a parental strain during a sexual cross and mutated prior to meiosis. Knockdowns are also made through **quelling**, a **form of co-suppression** found in *Neurospora* (Cogoni and Macino 1997). These advances are described in greater detail below.

2. Biology of *Neurospora*

In *Neurospora*, as in many organisms, light has two primary roles. The first and most obvious one is to immediately and acutely regulate organismal responses such as pigment production. The second and more subtle response is to regulate the phase of the **biological clock**, the **internal circadian timer** that tells the organism what time of day it is and thereby modulates the organism's responses to many factors, including light. In most organisms, **light and the clock work** together to regulate many aspects of the life cycle, and *Neurospora* has proven to be an ideal model system for understanding the interplay between these two forms of regulation. Figure 4.1 provides an overview of the *Neurospora* life cycle and notes the places in the life cycle that are influenced by light and/or the clock.

Light signals influence many facets of both the sexual and asexual (vegetative) stages of life, but a good starting place is to consider the basic genetics and growth characteristics. In their natural habitat, sexual spores are activated by the heat from fires; *Neurospora* is classified as a Pyrenomycete. Cultures thus emerge after fires, and the organism spends most of its life growing vegetatively on the burned-over substrate. *Neurospora* grows vegetatively as a syncytium with incomplete cell walls separating cellular compartments, and comes in two mating types, **A** and **a**. Even genetically different strains of the same mating type can often fuse and intermingle their nuclei to form

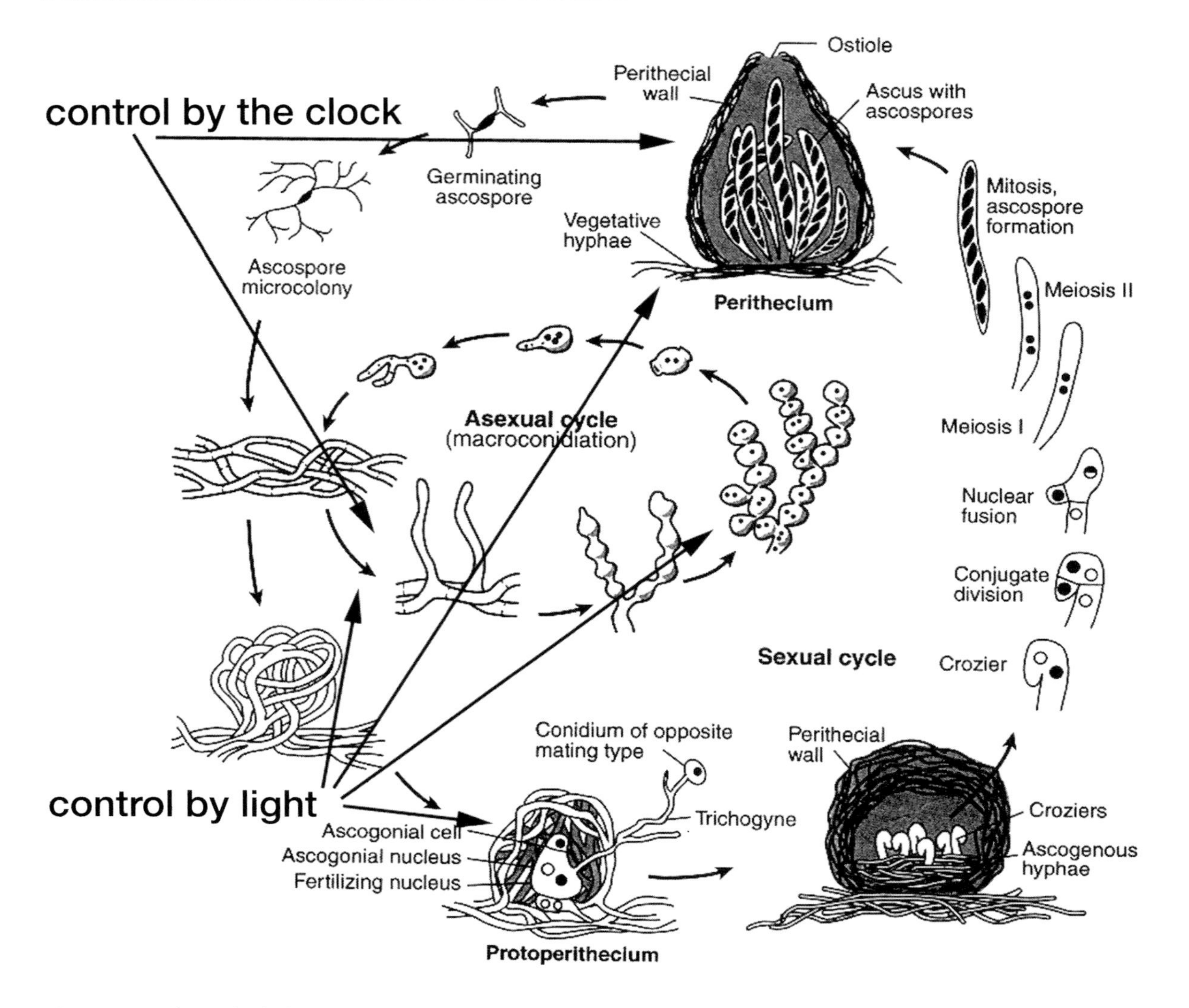

Fig. 4.1. Regulation by light and the clock in the life cycle of *Neurospora*. Many aspects of both the asexual vegetative life cycle and the sexual cycle are influenced by both the endogenous circadian clock and by ambient light. See text for detail. Adapted from Davis (2000) with permission

heterokaryons, but strains with different mating types never fuse. *Neurospora* can exist as a vegetative culture in two forms, as surface mycelia or it can elaborate aerial hyphae. The tips of aerial hyphae form morphologically distinct structures that act as asexual spores (called conidia) that are easily dispersed by wind. When nutrients become scarce, vegetative *Neurospora* of either mating type is able to induce sexuality by forming a **perithecium** (a **fruiting body**) that can be fertilized by a nucleus from a piece of mycelium or an asexual spore of the opposite mating type. Nuclei of each parent replicate in tandem and eventually fuse to make a transient diploid that immediately undergoes meiosis to produce an eight-spored ascus, each ascus containing the products of meiosis from a single diploid nucleus.

Acute effects of light during the asexual phase of the life cycle include the acute induction of conidiation, the developmental process leading to the production of the asexual spores. More conidia are produced and they are produced faster (Klemm and Ninneman 1978; Lauter 1996). Although pigmentation of the conidia is constitutive, as noted above in the historical descriptions of *Neurospora*, **carotenogenesis** in mycelia is light-induced (Harding and Shropshire 1980) and this response is quite rapid, being observable within the first 30 min after exposure to light. In a strain defective for light perception or transduction of the light signal, the production of white mycelia underlying yellow/orange conidia is seen; this screen proved to be of central importance in the genetic identification of the photoreceptors, as described below. Developmental responses such as **light-induced conidiation** are slower. Aerial hyphae are reported to display **phototropism** (e.g., Siegel et al. 1968), in that they preferentially form on the side of a dish near to light, but it is not clear to what extent this response is distinct from the overall light induction of conidiation. Also somewhat controversial are reports of changes in membrane conductivity (hyperpolarization and an increase in input resistance) in response to light (e.g., Potapova et al. 1984).

During the asexual cycle, shown in the center of the figure, the most global effect of light is to set the phase of the endogenous biological clock that acts to regulate a variety of aspects of the life cycle of the organism (Sargent and Briggs 1967; reviewed in Loros and Dunlap 2001). The clock controls the daily timing of a developmental switch that can initiate the morphological changes leading to conidiation. Light is used to set the phase of the clock, in that a light-to-dark transfer is interpreted as dusk and a dark-to-light transfer as dawn; the molecular basis of this response will become clear below in the chapter. Continued light also acts to suppress the expression of the clock. Conidiation (differentiation leading to asexual spore production) can be triggered by environmental signals including blue light, desiccation, and nutrient starvation, as well as by the endogenous circadian clock, in otherwise constant conditions. This asexual development involves a major morphological change that requires many novel gene products. Although the production of asexual spores is the best characterized light-phased circadian rhythm in *Neurospora*, other persisting rhythms at the physiological level have been described, which include the production of CO_2, lipid and diacylglycerol metabolism (e.g., Roeder et al. 1982; Lakin-Thomas and Brody 2000; Ramsdale and Lakin-Thomas 2000), a number of enzymatic activities (e.g., Hochberg and Sargent 1974; Martens and Sargent 1974), heat shock proteins (Rensing et al. 1987), and even growth rate (Sargent et al. 1966).

During the sexual phase of the life cycle, light again induces many and varied effects. The overall initiation of the sexual process is enhanced by light (Degli Innocenti and Russo 1983). This is particularly interesting since, once initiated, the process proceeds better in the dark. Not surprisingly, carotenogenesis of perithecial walls is light induced (Perkins 1988). Perhaps the most interesting photobiology in the sexual phase involves the behavior of mature perithecia once they are formed. The ejection of spores is induced by light, and the direction in which they are ejected is light regulated – that is, the tips ("beaks") of the perithecia display a distinct phototropism (Harding and Melles 1983). Finally, even in the dark, the number of spores shot from perithecia is regulated by the light-phased circadian clock.

The processes described above are all light regulated at face value, but *Neurospora* is also capable of a more subtle response to light, characteristic of further regulatory sophistication: *Neurospora* can respond to changes in the level of ambient light, a process known as **photoadaptation**. This response is manifested in two ways, both tied to the response to light at the molecular level. In the first, when the organism initially sees light, a response is triggered that peaks within 15–30 min, but this response generally decays within about 2 h. If the organism is exposed to light within this 2-h period, no additional response is seen. A related phenomenon is observed if the light remains on – the response decays but, interestingly, after the 2-h latency period the organism can respond again if the ambient level of light is increased.

3. *Neurospora* as a Genetic and Genomic System

The *Neurospora* genome is organized among seven chromosomes corresponding to the seven genetic linkage groups, although not as a one-to-one mapping since the chromosomes were defined cytologically, independently from the genetic accretion of the linkage groups. In any case, the entire genome comprises something approaching 2000 map units. General features of the genome are summarized in Table 4.1. There are about **11,000 genes**, a number that compares favorably with other genetic model systems – *Drosophila* (around 14,000 genes) and *Caenorhabditis elegans* (about 19,000 genes). Genes appear on average every 3.7 kb along the chromosome. The **average length of a gene is about 1.3 kb**, compared to about 1.1 kb for *Saccharomyces* and *Schizosaccharomyces*. **Introns are found in nearly all genes** and are evenly distributed, unlike *Saccharomyces* where most genes do not have introns and the few introns that exist tend to be clustered in the 5′ ends of genes. Due to whole-genome scanning mechanisms such as RIP that de-

Table. 4.1. General features of the *Neurospora crassa* genome (updated from Borkovich et al. 2004)

General	
Chromosomes	7
%G+C	50%
Protein-coding genes	10,620
Protein-coding genes>100 aa	9,200
Introns	17,118
tRNA genes	424
5S rRNA	74
Percent coding	44%
Average gene size	1673 bp (481 aa)
Median intron size	84 bp
Average intergenic distance	1953 bp
Identified by similarity to known sequences	1,336 (13%)
Conserved hypothetical proteins	4,606 (46%)
Predicted proteins (no similarity to known sequences)	4,140 (41%)
Current genome assembly (as of March 2005)	
Average contig length	273 kb
Total length of combined contigs	38,433,854 bp
Remaining sequence gaps	37

tect and eliminate duplications, there is a **dearth of repeats and gene families** relative to *Saccharomyces* (Galagan et al. 2003) and other genomes.

In general, *Neurospora* genes appear to be typical of those seen in "higher" eukaryotes. As noted above, most genes have several introns, and 5′ noncoding regions that may also contain introns precede coding regions. Promoters can be complex, in that multiple promoters can direct the expression of a single coding region, and they are regulated in the combinatorial manner typical of eukaryotes. A number of examples of **alternative splicing** are becoming known. There is a **distinct codon bias** in favor of C and against A in the third position. Additional details can be found in (Borkovich et al. 2004), in the paper describing the genomic sequence (Galagan et al. 2003), or on the web (http://www.genome.wi.mit.edu/annotation/fungi/neurospora/index.html).

Several independent estimates have suggested a surprising degree of genetic novelty in the *Neurospora* genome. Repeated estimates suggest that nearly **50 % of *Neurospora* genes have no homologs or orthologs** in GenBank (perhaps because of the 13 million sequences in GenBank, only about 270,000 come from ascomycetes). These are either novel genes with novel functions, or novel genes representing different ways of carrying out known functions; in either case, they will be of great interest. The 11,000 genes in the *Neurospora crassa* genome implies a complexity approaching that of *Drosophila* and *Caenorhabditis*, and about twice that found in yeasts, especially because of mechanisms in *Neurospora* that target and eliminate duplications (e.g., RIP). For this reason there are few gene families, and hence nearly all of the sequence complexity reflects actual diversity. *Neurospora* shares gene sequences with a variety of taxonomic groups, and so information from *Neurospora* will inform projects covering the breadth of biology, not just within the fungi. *Neurospora* is definitively not *Saccharomyces* with a larger genome. Instead, it is the gateway to an incredibly diverse group of organisms – the mycelial fungi – which are a cornerstone of our ecosystem and annually contribute over US$ 40 billion to the United States economy alone. Although both nonpathogenic and easy to manipulate, *Neurospora* is phylogenetically very closely allied and genetically syntenic, both with important animal and plant pathogens and with agriculturally and industrially important production strains (such as *Cochliobolus*, *Fusarium*, *Magnaporthe* and *Trichoderma*). Strong parallels have been noted in signaling pathways, photobiology, developmental regulation, and many aspects of metabolism including secondary metabolism, to name a few.

In a broader sense, there is great potential for information synergy, perhaps reflected in the fact that the *Neurospora* Genome site at the Whitehead Institute Center for Genome Research at MIT draws over 3500 hits per day. *Neurospora* is the best understood of the more than 250,000 species of mycelial fungi, and work in this system has historically paved the way to developments in other fungal species.

4. Genome Defense Mechanisms

As a first approximation, we normally expect the characteristics of an organism, or the characteristics of progeny arising from a cross between members of the same genus, to reflect the basic genetic composition of the organism. However, this is not always the case. One reason for this, as initially described by McClintock (1950), is that mobile genetic elements can gain access to genomes and then move about causing mutations both by their excision and their reinsertion into the genome. Perhaps because of its syncytial growth habit, fungi like *Neurospora* would be especially susceptible to such selfish DNA. As a result, three independent mechanisms have evolved to protect the genome and to prevent mobile genetic elements from passing through a cross into the genomes of progeny.

The first of these is RIP, the acronym for repeat-induced point mutations. Initially described by Selker and colleagues in 1987 (Selker et al. 1987), RIP describes a process in which, in the transient diploid cells that arise during a sexual cross, both genomes are scanned for duplicated sequences, i.e., for sequences present in more than one copy anywhere in the genome. If any are found, both copies are riddled with GC to AT transversions with the result that, typically, both copies are inactivated. There are regions of the genome that are protected from RIP, such as the ribosomal repeats in the nucleolus organizer region, but interestingly these cannot be targeted in transformation experiments.

A somewhat similar screening mechanism is known as **MSUD (meiotic silencing of unpaired DNA**; Shiu et al. 2001). Normally, of course, after karyogamy in a haploid, each gene is present in two copies that pair during the early stages of meiosis. MSUD is a process in *Neurospora*, akin to but not the same as transvection, wherein the presence of an unpaired gene results in the meiotic silencing of all copies of that gene in the genome. This is the basis of barrenness observed in crosses involving nonreciprocal translocation strains that would, if fruitful, generate segmental aneuploids. If the translocated region (present therefore in three copies) contains genes required for meiosis, then all three copies are silenced and the cross is aborted. Mutations in the *sad*-1 gene (Shiu et al. 2001) abrogate this phenomenon, suppressing a number of ascus-dominant mutations (*Round spore*, *Peak*, *Banana*) whose ascus phenotypes constitute the MSUD-mediated response to loss of the gene, and allowing diploid meiotic nuclei possessing a gene deletion in a meiosis-required gene (i.e., a single unpaired copy) to pass through crosses.

Whereas both RIP and MSUD are active only during the sexual cycle, *Neurospora* exhibits a third defense mechanism in vegetative cells. This is quelling, a **PTGS (post-transcriptional gene silencing)** that is in all ways akin to the double-stranded RNA-induced gene silencing seen in most eukaryotes and that has been studied in depth by Macino, Cogoni and colleagues (Cogoni and Macino 1994, 1997). As in most organisms, the process requires an RNA-dependent RNA polymerase, an Argonaute-like protein and a RecQ-like helicase (encoded by *qde*-1, *qde*-2, and *qde*-3), some of which were first identified as quelling defective mutants of *Neurospora*.

5. *Neurospora* Genome Projects

Currently, there are approximately 70 research laboratories in the US focusing on *Neurospora*, and an additional approximately 80 in the rest of the world, comprising a community of around 700 investigators (see http://www.fgsc.net/). In the US alone this represents more than 70 years of support and 500 graduate and postdoctoral training positions. Over the past 10 years there has been over US$ 60 million in funding to support *Neurospora* research. Readers may also appreciate that these numbers are likely to underestimate the actual effort, since they are based largely on self-reporting in surveys prepared for the NIH Non-Mammalian Models Workshop in 1999, and for the NSF grant that supported the genomic sequencing project. Compounding the inexactness is the fact that the *Neurospora* community is also growing rapidly, with at least five new laboratories starting within the past 3 years. The worldwide *Neurospora* community is tight-knit. Annual meetings tie the community together, and organism/system-specific interests are overseen by an elected *Neurospora* Policy Committee on which many of the investigators have served. In the early 1990s, interest in the genomics of *Neurospora* began to grow. In 1993 the Chair of the *Neurospora* Policy Committee formally initiated, in the interests of the community, a *Neurospora* genome focus of effort; with evolving leadership, this project is entering its ninth year. Efforts to physically map the genome (coordinated by the Committee to Order the *Neurospora* Genome) began in 1993. Independently, the first federally funded **EST (expressed sequence tag)** project began in 1995 at the University of New Mexico, and funds to physically map the genome were awarded to the University of Georgia in 1998. The initiative to raise funds to sequence the genome of *Neurospora* began in 1996 under the direction of the outgoing chair of the Policy Committee (http://gene.genetics.uga.edu/white_papers/ncrassa.html), as did a second, independently funded EST project (a collaboration between the University of Oklahoma and Dartmouth Medical School). Many members of the international community stepped forward to work toward this goal, the first success coming in 1998 with the award of DM 7.5 million in Germany for the complete sequencing of two chromosomes (http://mips.gsf.de/proj/neurospora/). In the US, a *Neurospora* Genomics Policy Committee (http://www.unm.edu/~ngp/WhitePaper.html) was organized. This led to an NSF grant for US$ 5.2 million that was awarded in 2000 for the completion of the archival, reference-quality (minimum tenfold coverage) **genomic sequence** that would serve as the **reference genome** for the mycelial fungi. These efforts, based at MIT (http://www.broad.mit.edu/annotation/fungi/neurospora/) , set the stage for further work.

The availability of whole genomic sequences has vastly accelerated the pace of research in eukaryotic model systems. However, to exploit this resource, research communities must (1) annotate the genome to extract the relevant information, (2) systematically disrupt the functions of the identified genes, (3) examine the regulation of the genes in different biological contexts, and finally (4) communicate this information to the scientific community at large, particularly to those studying similar problems in other systems. We may then integrate all these aspects of phenotype and regulation into a comprehensive portrait describing the biology of organisms. It is a tautology to state that the simplest organisms are the easiest to dissect and also reveal the least, and that the most complicated organisms, while the most information-rich, may be beyond the scope of current efforts. Yet, it is apparent from the extant genomic comparisons that conservation of important biological processes is the rule, and that simple models can inform more complex systems. The desirability of rich biology, coupled with the realistic need for approachable genetics recommended *Neurospora* for further analysis. Hence, a consortium of universities has come together in four projects to further work on the *Neurospora* genome. The institutions are Dartmouth Medical School, University of California at Riverside, University of California at Los Angeles, University of Missouri, MIT, the Oregon Health Sciences University, the University of California at Berkeley, and the University of New Mexico, with funding provided by the National Institutes of General Medical Sciences.

As further described on the Genome Project website (http://www.dartmouth.edu/%7Eneurosporagenome/), the first project is pursuing the systematic disruption of genes through targeted gene replacements, preliminary phenotyping of these strains, and their distribution to the scientific community at large. Project 1 will rely on bioinformatic support from Project 2. Through a primary focus on **annotation** and genomics, Project 2 is producing a platform for electronically capturing community feedback and data about the existing annotation, while building and maintaining a **database to capture and display information about phenotypes** that is relying on data from EST analyses (Project 4) to refine the gene structures. **Oligonucleotide-based microarrays** created in Project 3 are allowing **transcriptional profiling** of the nearly 11,000 distinguishable transcripts in *Neurospora*. This effort will provide a baseline analysis of gene expression under a variety of growth conditions, and later begin to analyze the global effects of loss of novel genes in strains created by Project 1. These data will be made available through the web via structures created in Project 2. Since alternative splicing, **alternative promoters**, and **long antisense transcripts** contribute widely to the overall complexity of expressed sequences in *Neurospora*, in Project 4, cDNA libraries are being generated from wild-type and related strains to document this complexity to aid in annotation in Project 2. Sequences from related strains have also driven assembly of an **SNP (single nucleotide polymorphisms)** map. Overall, this effort will help to anchor genomic exploration within the largely unexplored phylogenetic kingdom of the fungi.

These community-based efforts are providing a wonderful context for further work on circadian rhythms and **photobiology**. This remains an extremely active area of research; a search of [*Neurospora* AND circadian] yields over 100 publications since 2000, and genomics tools are well adapted for building on the general models that have derived from the past two decades of combined molecular and genetic analyses. As a backdrop for describing these ongoing studies, the present status of the *Neurospora* clock and light regulatory machineries will be described next.

II. Analysis of Circadian Rhythms in *Neurospora*

A. What is, and is not, a Circadian Rhythm

Circadian rhythms reflect the output of biological clocks; they can regulate cyclical biochemical, physiological, or behavioral functions and, in nature, have a period length of exactly 24 h. Under constant conditions in the laboratory, they run at their own inherent frequency, which is about one cycle per day – circadian. At 25 °C the period length of the *Neurospora* circadian clock is about 22 h. The phase of these rhythms is determined by the environmental light/dark cycle and temperature cycle, but they are not simply a reaction to the light/dark cycle. Rather, they represent the overt expression of an **endogenous and self-sustaining** timekeeping mechanism. The period of the cycle varies only a little, if at all, when the organism is examined on different media or at different temperatures (Dunlap et al. 2003), a characteristic known as **temperature, nutritional, or pH compensation**. The **ability to be reset by short light or temperature treatments** (called **entrainment**), the period length of about a day under constant conditions, and the compensation capacity are characteristics that distinguish circadian rhythms from cell cycle-regulated phenomena or from other ca. 24-h metabolic or developmental rhythms whose period lengths are often temperature or nutritionally dependent. These three characteristics of a true circadian rhythm define it as being circadian, and unite it with similar rhythms found ubiquitously in most eukaryotes and cyanobacteria.

In addition to circadian rhythms, organisms including *Neurospora* can express a variety of other rhythms of varying period lengths. These can often be observed as cycles in morphology (Feldman and Hoyle 1974) or in growth rate on exotic media (Lakin-Thomas 1996), but in some cases the rhythm presents at face value a bona fide circadian character (e.g., Loros 1984; Loros et al. 1986; Aronson 1994b). However, in all cases one or more of the canonical properties of circadian rhythmicity is lost (often temperature compensation), and so these rhythms are understood to have a different basis, not comparable to that known for circadian clocks.

B. Phylogenetic Conservation of Rhythms and Clock Components

All known eukaryotic clocks are based at least in part upon **transcriptional/translational feedback loops** that close within the cell (Dunlap 1999). These clocks all share similar components and organizational logic in their assembly and operation, although considerable diversity has evolved in the number and types of some of the components executing the necessary functions. A cartoon of the general scheme of the core feedback loops in the clock is seen in Fig. 4.2. In all eukaryotes, the **positive elements are heterodimers of two proteins that interact via PAS domains** to make a transcription factor. In *Neurospora*, the two are WC-1 and WC-2, in *Drosophila* CYC and CLK, and in mammals BMAL1 and CLOCK. WC-1 is a sequence and functional homolog of CLOCK. There is greater diversity amongst the **negative elements.** In *Neurospora*, the negative element is FRQ that acts as a dimer, in *Drosophila* it is PER and TIM as a complex or PER acting alone, and in mammals it is a complex of four

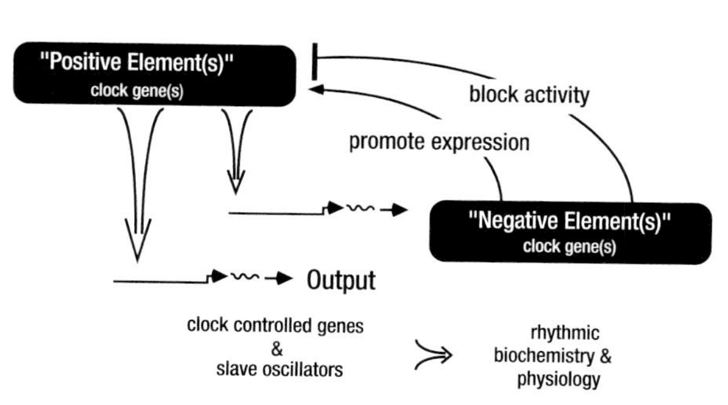

Fig. 4.2. A generalized view of the feedback loops associated with rhythmicity. Positive elements, heterodimers of PAS domain-containing proteins such as WC-1 and WC-2, act as transcription factors to drive expression of "negative elements" such as FRQ. Negative elements in turn feedback to attenuate the activity of their activators, thereby giving rise to a negative feedback loop that, with the proper delays, can oscillate. Negative elements also feed forward to promote the synthesis of their activators. One aspect of output occurs when positive elements activate expression of genes that do not participate in the clock loops. Adapted from Dunlap (1999)

proteins, two homologs of the *Drosophila* PER protein called PER1 and PER2, and two cryptochrome homologs, CRY1 and CRY2.

Outside of *Neurospora*, but within the fungi, relatively little is known about the molecular bases of rhythms. The *frq* gene has been identified in a number of relatives within the *Sordariaceae*, including several species of *Neurospora* and *Sordaria*. More broadly within the *Pyrenomycetes* and *Loculoascomyetes*, *frq* has been identified in diverse genera including organisms such as *Magnaporthe*, *Fusarium*, *Leptosphaeria*, and *Chromocrea* (Merrow and Dunlap 1994; Lewis and Feldman 1997; Lewis et al. 1997; J. Dunlap and J. Loros, unpublished data). Not surprisingly, the degree of sequence relatedness drops steeply with the degree of phylogenetic distance, but functional conservation of FRQ has been demonstrated by using the *Sordaria frq* gene to complement a loss-of-function mutation in *Neurospora* (Merrow and Dunlap 1994). All fungi examined to date have homologs to WC-1, the core element of the *Neurospora* clock showing the most sequence conservation with known clock components from mammals (Lee et al. 2000). A circadian rhythm has recently been reported in *Aspergillus nidulans* (Greene et al. 2003), an organism whose genome appears to have no close sequence homologs of FRQ.

C. Assays of Rhythmicity

To appreciate how genomics have been used to approach rhythms and photobiology in *Neurospora*, background is needed on how the clock is monitored. Several features have made *Neurospora* an attractive and successful model for the dissection of circadian timing. As noted above, in *Neurospora* the clock is controlling the potential to develop, rather than the developmental process itself. Once the switch is thrown, development can take either a long or a short time, hours to days, depending on the nutritional state and temperature of the culture. The cycle recurs once a day, such that a characteristic conidial banding pattern of developed or developing conidiophores can be readily followed in **hollow glass culture tubes called race tubes** (Ryan et al. 1943). Race tubes are glass tubes about a centimeter in diameter and 30–100 cm long that are bent up at a 45° angle at both ends to hold agar medium for growth (Fig. 4.3). As vegetative growth proceeds, CO_2 levels become elevated and can suppress conidiation, and therefore mask the rhythm. The CO_2 masking effect is alleviated by the *band* (*bd*) mutation (Sargent et al. 1966), and for this reason all laboratory stocks commonly used for circadian rhythm studies carry a mutation in the *bd* gene.

The *Neurospora* clock can also be assayed using **liquid cultures**, and this facilitates collection of materials for molecular study (Nakashima 1981; Perlman et al. 1981; Loros et al. 1989; Aronson et al. 1994a; Garceau et al. 1997). The clock runs normally in liquid cultures, and individual mycelial samples will retain their endogenous rhythmicity and phase when transferred from liquid to solid growth media. As an additional alternative, real-time analysis of rhythmic clock-controlled or core clock gene activity can be assayed in vivo using firefly luciferase as a reporter for gene expression (Mehra et al. 2002; Morgan et al. 2003).

D. Genetic Analysis of Rhythms

These assays have facilitated the isolation of strains in which the clock runs fast or slow, or not at all (such as *chr* and *frq*; Feldman 1967; Feldman and Hoyle 1973; Feldman et al. 1979; Gardner and Feldman 1980), or where both the period length and temperature compensation are affected. *prd*-1, *prd*-2, *prd*-3, *prd*-4, *cys*-9, *wc*-2 and other *frq* alleles affect both the period of the rhythm and temperature compensation (Feldman and Atkinson 1978; Feldman et al. 1979; Onai and Nakashima 1997; Collett et al. 2002). The *frequency* (*frq*) locus, which was identified multiple times (Feldman and Hoyle 1973; Gardner and Feldman 1980), can give rise to both **long** (24, 29 h) and **short** (16, 19 h) **period length mutants** as well as mutants lacking circadian rhythmicity (although they do retain non-circadian rhythms). Most genes associated with rhythms are identified through a single allele, and most have not yet been cloned. Hence their role in the clock, even to the extent of whether the effect is specific or pleiotropic, is not known. Occasionally, even for cloned genes, sequence data have not yielded much information as to the molecular mechanism of the clock. When *frq* was first cloned, it was evident that it did not have extended homology to any gene or protein in the GenBank database. It was only experimental manipulation of the gene that identified its central role in a feedback loop, as described below. On the other hand, there are a number of genes that have been identified in genetic screens that give rise to products of known biochemical function that can now be fitted into an internally consistent

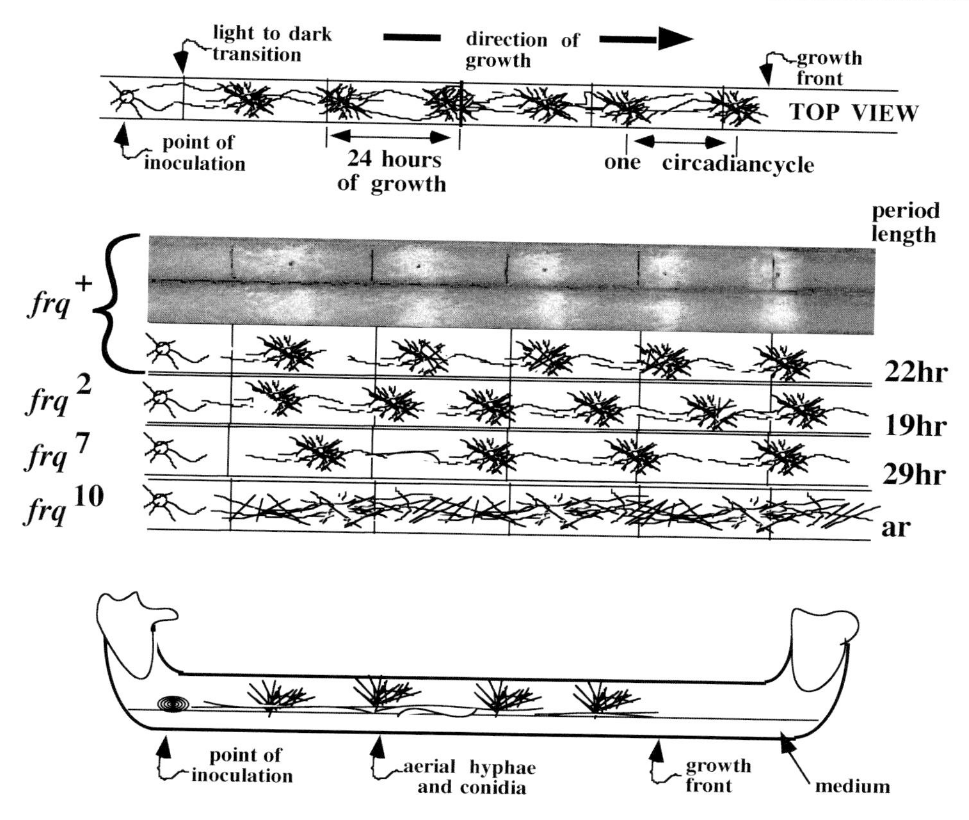

Fig. 4.3. The *Neurospora* conidiation rhythm can be viewed on race tubes. Cartoons of the race tube assay for monitoring the phenotypic expression of the *Neurospora* clock are shown at *top* and *bottom*. Conidia are inoculated at one end of a hollow glass tube about 40 cm long and 16 mm in diameter that is bent upward at both ends to accommodate an agar medium. Following growth for a day in constant light, the position of the growth front is marked and the culture is transferred to constant darkness. This light-to-dark transfer synchronizes the culture and sets the clock running from subjective dusk. The growth front can be marked every 24 h under red light. Positions of the easily visualized conidial bands (separated by undifferentiated surface mycelia), relative to the marked growth fronts, permit determination of both period length and phase of the rhythm. Photographs and cartoon depictions of race tubes representative frq^+ and *frq* mutant strains demonstrating genetic regulation of period length and overall rhythmicity are shown in the *middle*. A strain bearing the frq^2 allele shows a short period of approximately 19 h relative to the wild-type frq^+ period of 22 h. The frq^7 strain has a long period of about 29 h, and the frq^{10} gene replacement strain is arrhythmic

model for a feedback loop whose function is essential for normal operation of the *Neurospora* circadian clock under constant conditions. For example, phosphorylation has been established to play an important regulatory role in maintaining the kinetics of oscillations of the clock (Kloss et al. 1998; Price et al. 1998; Liu et al. 2000), and several kinases are found in Table 4.1. One of these, encoding casein kinase 2, is now known to be universally associated with all circadian clockworks, but was first identified by work in *Neurospora* (Yang et al. 2002). It is not surprising that several genes involved in mitochondrial function and energy metabolism have been identified as clock-affecting genes. What has emerged from the comparisons of clocks in other systems is a general pattern describing how gene products and genes act together in the circadian system.

E. The Mechanism of the Circadian Clock in *Neurospora*

1. The Molecular Clock Cycle

The white collar-1 (WC-1) and white collar-2 (WC-2) proteins, and both *frq* mRNA and FRQ are central components of the *Neurospora* circadian oscillator (Aronson et al. 1994a, b; Crosthwaite et al. 1997; Dunlap 1999; Collett et al. 2001; Fig. 4.4). WC-

1 and WC-2 heterodimerize to form a **white collar complex, WCC** (Ballario et al. 1996, 1998; Linden et al. 1997; Linden and Macino 1997; Talora et al. 1999; Denault et al. 2001; Cheng et al. 2002), and the WCC activates expression of the *frq* gene. Alternative splicing of *frq* mRNA (see below) can result in the production of **two distinct FRQ proteins** (Garceau et al. 1997) that feed back to block this activation. In the process of executing this feedback loop, both *frq* mRNA and FRQ protein are rhythmically expressed in a daily fashion, and FRQ protein acts to repress the abundance of its own transcript (Aronson et al. 1994a; Garceau et al. 1997; Merrow et al. 1997; Froehlich et al. 2003). Light and temperature, two of the most important environmental signals, reset the *Neurospora* clock by changing the levels of *frq* mRNA and FRQ protein (Crosthwaite et al. 1995; Liu et al. 1998), as further outlined below.

Figure 4.4 serves as a guide for following the progress of the *Neurospora* clock cycle through the day. By late night, FRQ protein has recently been degraded and *frq* RNA levels are low. WC-1 and WC-2 in the form of the WCC bind to the promoter of the *frq* gene at two sites (**LREs** or **light-responsive elements**) to drive the circadian rhythm in transcription of the *frq* gene (Froehlich et al. 2003). *frq* primary transcripts are spliced in a complex manner, and gradually by the early morning FRQ proteins appear (Garceau et al. 1997), dimerize (Cheng et al. 2001a), and soon enter the nucleus (Garceau et al. 1997; Luo et al. 1998) where they interact with the WCC (Cheng et al. 2001a; Denault et al. 2001; Merrow et al. 2001) so as to attenuate its activity (Froehlich et al. 2003). By midday, WCC activity reaches a nadir as FRQ levels rise (Lee et al. 2000), thereby turning down the expression of the *frq* gene. As FRQ appears, it is phosphorylated by casein kinase Ia, casein kinase II, and calcium/calmodulin-dependent kinase, actions that regulate the stability of FRQ (Liu et al. 2000; Gorl et al. 2001; Yang et al. 2001, 2002) as well as FRQ/WCC interactions (Yang et al. 2002). As a result of inhibited WCC, *frq* transcript levels begin to decline, although continued translation causes FRQ protein levels to continue to rise. Thus, *frq* mRNA levels peak in the midmorning (Aronson et al. 1994a; Crosthwaite et al. 1995), about 4–6 h before the peak of total FRQ in the afternoon (Garceau et al. 1997).

FRQ also exerts a second role in the cycle by promoting, through an unknown mechanism, the translation of WC-1 from existing *wc*-1 message (Lee et al. 2000; Merrow et al. 2001; Cheng et al. 2001b, 2002). The result is that WC-1 levels begin to rise even as **phosphorylation-promoted turnover of FRQ** begins. Thus, at close to the same time in a daily cycle, FRQ is promoting WC-1 synthesis to increase the level of the WCC while blocking activation of the *frq* promoter by the WCC. This creates a mass of WCC held inactive by FRQ (Lee et al. 2000; Denault et al. 2001; Froehlich et al. 2003). When the phosphorylation of FRQ finally triggers its turnover (Garceau et al. 1997; Liu et al. 2000), **FRQ-promoted synthesis of WC-1** is balanced by WC-1 degradation, and WC-1 levels peak in the night (Lee et al. 2000). The high WCC activity is released to initiate the next cycle and to maintain a robust amplitude in the feedback loop (Froehlich et al. 2003).

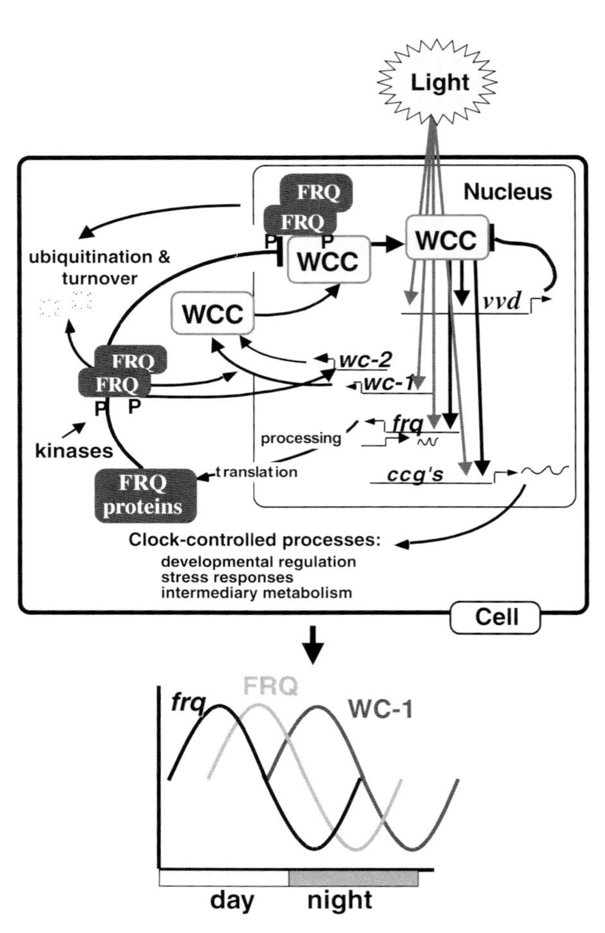

Fig. 4.4. Coupled feedback loops in the *Neurospora* circadian system. Known molecular components and their regulatory relationships in the *Neurospora* circadian clock are shown. *Lines ending in bars* connote negative regulation or repression. Arrows connote positive regulation. *P on FRQ* denotes phosphorylation. See text for details. Modified from Lee et al. (2000)

As described here, *frq* mRNA cycles with a peak during the **subjective day** (circadian daytime in constant conditions of darkness and temperature) and FRQ expression lags by about 4 h, peaking late during the day, near dusk. In constant conditions, therefore, subjective day is defined as the time when these components are at their highest levels, and night corresponds to the troughs in *frq* and FRQ. This information forms the basis of our understanding of how light adjusts the phase of the rhythm, a process called entrainment. Light acts through the WCC to rapidly induce *frq* expression (Fig. 4.4), resulting in a large and transient increase in *frq* and FRQ. This increase in *frq* is mediated by an *frq* antisense message. Depending on the time in the *frq*/FRQ cycle during which the light falls, it will either cause the *frq* peak to occur sooner or it will delay its decline, thereby advancing or delaying the clock.

2. The Principal Molecular Components of the Clock Cycle

a) *frq Transcripts and FRQ Proteins*

frq primary transcripts arise from multiple up-and downstream promoters (H. Colot, J. Loros and J. Dunlap, unpublished data), and are heavily spliced in a complex manner. Splicing determines whether long or short FRQ proteins are synthesized. A long antisense transcript arises from the region of *frq*. It is light induced, oscillates antiphase to sense *frq* transcripts, and appears to play a role in ensuring precise entrainment to light/dark cues (Kramer et al. 2003). In any case, two FRQ proteins are encoded by *frq*, a long form of 989 amino acids and a form lacking the first 100 amino acids (Garceau et al. 1997). Both forms can form **homo- and heterodimers** (Cheng et al. 2001a) and are needed for robust rhythmicity. Their expression pattern responds to ambient temperature. At low temperatures (<22 °C), the short form is preferentially required and less overall FRQ is needed, whereas at high temperatures (>26 °C) the large form is favored at higher overall levels (Liu et al. 1997).

b) *WC-1 and WC-2, Positive Elements in the Feedback Loop*

WC-1, a GATA-like transcription factor, is a 1167 amino acid protein regulated both transcriptionally and post-transcriptionally at the level of synthesis, and additionally through phosphorylation and protein–protein interactions (reviewed in Ballario and Macino 1997; Dunlap 2005; Dunlap and Loros 2005). WC-1 contains a polyglutamine activation tract, a single class 4 DNA-binding Zn-finger domain, two PAS domains that are important for protein–protein interactions and that are required for both light and clock functions (Lee et al. 2000), and also an additional, specialized subclass of PAS domains called **LOV** (for **light**, **oxygen**, and **voltage sensing**). WC-1 binds flavin adenine dinucleotide (FAD) as a chromophore. WC-2 is 530 aa in length and is related to WC-1 at the sequence level. It has a single PAS domain, as well as a single activation domain and Zn-finger domain. As noted above, WC-1 and WC-2 form a white collar complex (WCC) via their PAS domains, and bind as a complex to the light-responsive elements (LREs) in promoters of light-regulated genes such as *frq*.

Additionally, in *Neurospora* as in many circadian systems, the response of the system to light is regulated by the clock, a process referred to as **gating** and mediated in part by **vivid (VVD)**, **another photoreceptor** and member of the PAS protein superfamily (Heintzen et al. 2001; Schwerdtfeger and Linden 2003). Photoadaptation, a third light-modulating effect, is the temporary insensitivity of the organism to respond to a second light pulse or to an increase in light intensity. VVD also mediates this in part, perhaps by contributing in some way to the post-translational modification of WC-1 that leads to light-induced turnover of WC-1.

III. Photobiology in *Neurospora*

A. Overview

Light responses in *Neurospora* have been appreciated for well over 100 years and their analysis drove much of the early genetics on the organism: the **white collar alleles** were among the first to be generated in the organism. All known light responses in *Neurospora* are specific to **blue light**, and to date no responses to either red or far-red light have been documented, although many people have looked for them. Interestingly, the *Neurospora* genomic sequence has exposed a number of possible photoreceptors including several **bacteriophytochromes** and a **cryptochrome** (Borkovich et al. 2004; Dunlap and Loros 2004). When Macino and colleagues cloned the *white* *collar* genes (Linden et al. 1999), they speculated that these proteins might act together as the transcription factor that directly regulates light-inducible genes, with WC-1 being the actual photoreceptor. This foresight was verified when WC-1 was shown to be the photoreceptor for the circadian clock in *Neurospora* (Froehlich et al. 2002; He et al. 2002), and WC-1 and WC-2 were shown to interact in vivo at the *frequency* (*frq*) promoter (Froehlich et al. 2003). Independent of its role in the clock, the WCC is responsible for light regulation of the several percent of the genome that responds to light (Lewis et al. 2002).

B. Control of the WC-1 Photoreceptor

Neurospora contains many blue light-inducible, WCC-regulated genes, but not all are regulated in the same way. The initial burst of **light-induced gene expression** is transient whether light is delivered as a pulse or is continuous, and the speed and duration of the response is gene dependent. The response may peak at 15 min, as with the *frq* gene, or between 90 min and 2 h, as with the *eas* (*ccg*-2) gene. As noted above, WC-1 is **transiently hyperphosphorylated** in response to light, and this corresponds to the brief induction of gene expression in its targets as well as being correlated with turnover of the photoreceptor. WC-1 also induces its own expression in response to light (reviewed in Linden et al. 1999; Liu 2003).

FRQ promotes WC-1 expression, and so there is a rhythm in WC-1 protein levels that is not quite 180° out of phase with that of FRQ: WC-1 levels reach a peak in the night as FRQ is reaching its trough. WC-1 is also regulated via protein–protein interactions, and WC-1 and WC-2 have influence on each other's levels. Importantly, though, all these proteins are subject to phosphorylation events that appear to affect each protein's activities and interactions. Among the pertinent kinases is PKC (Arpaia et al. 1999), which forms a transient interaction with the Zn-finger domain of WC-1 that is lost immediately after light exposure, only to return 2 h later. Over-expression of a catalytically dead PKC increases WC-1 levels whereas constitutively active kinase results in decreased levels, consistent with a role of the kinase in determining both activity and stability (Franchi et al. 2005). Other data, however, suggest that PKC is not the only kinase involved.

IV. Temperature Effects on the Clock

The **ambient temperature** at which *Neurospora* is growing can influence its rhythmicity in three ways. (1) Steps up or down in temperature will reset the clock in a manner similar to light pulses. (2) If it is too hot or too cold, the clock will not run. (3) The period length of the clock is compensated so that it is roughly the same at any temperature within the physiological range of operation. This is the phenomenon known as "**temperature compensation**" mentioned above. Some temperature effects are mediated through the amount, and perhaps kind of FRQ protein made. As noted above, temperature influences *frq* splicing (H. Colot, J. Loros and J. Dunlap, unpublished data), and thereby both the total amount of FRQ and the ratio of the two FRQ forms. It was also noted above that the amount of FRQ in the nucleus determines time of day. However, because there is more overall FRQ at higher temperatures, and FRQ concentrations oscillate around higher levels at higher temperatures, the "time of day" associated with a given number of molecules of FRQ is different at different temperatures. At the peak at 25 °C, there are about 30 molecules of FRQ in the nucleus (Merrow et al. 1997), less than this number at 20 °C, and more at 30 °C. Thus, a shift in temperature corresponds literally to a step to a different time and a shift in the state of the clock, although initially no synthesis or turnover of components occur. After the step, relative levels of *frq* and FRQ are assessed in terms of the new temperature and they respond rapidly, completing the resetting (Liu et al. 1998).

V. Output from the Clock

A. Clock-Regulated Biology

By allowing anticipation to a changing environment, the functional usefulness of a clock is that it facilitates an organism's optimal performance according to the time of day. A major question in circadian biology, which work in *Neurospora* largely defined at the molecular level and to which it has contributed greatly, is how the clock delivers information to regulate downstream biological processes. Historically, the macroscopic asexual developmental process in *Neurospora*, macroconidiation, is the best understood circadian phenotype. Conidiation can be initiated by environmental signals including blue light, desiccation, and nutrient starvation as well as by endogenous signals from the circadian clock. It involves major morphological changes that require many novel gene products. Other physiological rhythms have been described including the production of CO_2, lipid and diacylglycerol metabolism (e.g., Roeder et al. 1982; Lakin-Thomas and Brody 2000; Ramsdale and Lakin-Thomas 2000), a number of enzymatic activities (e.g., Hochberg and Sargent 1974; Martens and Sargent 1974), heat shock proteins (Rensing et al. 1987), and even growth rate (Sargent et al. 1966).

A word of caution is appropriate here. Fungi express a variety of rhythms, many of which are not circadian according to the definition provided above. A major difficulty in the analysis of output pathways has been distinguishing circadian clock regulation from other metabolic or developmental regulations, many of which may oscillate. Because of the complex and coupled nature of cellular processes, assessing the roles of the circadian clock in daily physiological changes requires care and it is important to remember that not every rhythm is a circadian rhythm, nor is every rhythmic process controlling conidiation or other events necessarily a part of the circadian clock.

B. Clock-Controlled Genes – *ccgs*

A premise made in the *Neurospora* system (Loros et al. 1989) that has proven to be universally true in all circadian systems examined is that daily clock control of gene expression would be a major form of regulation controlling clock output. The first systematic screens for clock-regulated genes were performed in *Neurospora*, using **subtractive hybridization** to enrich for morning- and evening-specific mRNAs (Loros et al. 1989). The genes identified were called **clock-controlled genes** or **ccgs**, a term that is now used generically in the circadian literature. *Neurospora ccgs* have subsequently been identified using differential hybridization of time-specific libraries (Bell-Pedersen et al. 1996b), high-throughput cDNA sequencing (Zhu et al. 2001), and cDNA microarrays (Nowrousian et al. 2003). Most exhibit peak expression in the late night to morning and, additionally, many are regulated by light and other stimuli that initiate development.

Much can be learned about the specific roles of the clock in an organism's life through the isolation of rhythmic genes. Many processes in the biology of *Neurospora* are under clock regulation, including both sexual as well as asexual development, stress responses, and basic intermediary metabolism. The *eas* (*ccg*-2) gene encodes the *Neurospora* hydrophobin (Bell-Pedersen et al. 1992; Lauter et al. 1992), a class of small, highly expressed and exported hydrophobic proteins found forming the bundled rodlet layer on conidia and aerial hyphae. This hydrophobic coating allows fungal structures to emerge from a wet substrate into the air (Talbot 1999; Wessels 1999). The pheromone precursor gene *ccg*-4 (Bobrowicz et al. 2002) is clock-controlled, as are several other structural genes involved in development, such as the *ccg*-13, *con*-6 and *con*-10 genes (Lauter et al. 1992; Zhu et al. 2001). Trehalose synthase, encoded by the *ccg*-9 gene, catalyzes the synthesis of the disaccharide trehalose important in protecting cells from environmental stresses. Additionally, inactivation of *ccg*-9 results in altered conidiophore morphology and abolishes circadian control of conidiation, although the underlying clock continues to operate normally (Shinohara et al. 2002). The *ccg*-12 gene is allelic to the *Neurospora cmt* gene that encodes copper metallothionein (Bell-Pedersen et al. 1996b). Also, *ccg*-1 (*grg*-1), whose mRNA levels can reach 10% of the cell's total mRNA, is induced by heat shock and repressed by glucose (McNally and Free 1988; Lindgren 1994; Garceau 1996). This suggests clock control of a stress response. Induction of heat shock and general stress responses, including the induction of heat shock genes, have been associated with conidiation (Hafker et al. 1998), and possibly with resetting of the phase of the clock in *Neurospora* (Ruoff et al. 1999).

The first energy-harvesting enzyme in glycolysis, glyceraldehyde-3-phosphate dehydrogenase, is encoded by *ccg*-7. Its enzymatic activity had been shown to be rhythmic in developing cultures on solid medium (Hochberg and Sargent 1974) and is now known to cycle, along with transcript abundance, in a non-conidiating liquid culture system (Shinohara et al. 1998). Interestingly, GAPDH has subsequently been found to be clock regulated in other systems, including *Gonyaulax* (Fagan et al. 1999) and mammals (Iwasaki et al. 2004). **Circadian regulation of central metabolic functions** in the cell had not been previously documented, indicating a more global role in cell regulation than had been formerly appreciated.

C. Microarray Analysis of Light, Clock and Temperature Regulation

Genomic approaches involving microarray analysis have lent themselves well to examining the regulation of large subsets of a genome. Using these microarrays to understand clock and photobiological regulation has been a natural progression from past studies. *Neurospora* microarrays have been used to examine the overall extent of **circadian-regulated gene expression**, **light-regulated genes**, and **temperature-responsive genes** in relation to the clock and even for the identification of genes in a different species, an approach that should lend itself to the discovery of light-and clock-regulated genes in other fungal species.

In order to examine the extent of light-regulated gene expression under WC-1 control, microarrays containing 1764 clones were prepared (Lewis et al. 2002) from ESTs collected from three cDNA libraries representing the mycelial, asexual and sexual stages of the *Neurospora* life cycle (Nelson et al. 1997). With the release of the *Neurospora* genome sequence, the 1764 ESTs were found to represent 1343 unique genes. Target cDNAs made from dark-grown and light-exposed mycelia harvested between 30 min and 4 h after light exposure were hybridized to the probes. Approximately 3% of the genes, or 22 **probe spots**, showed a twofold or greater increase in expression with light exposure. The microarrays were then hybridized with **target cDNAs** generated from a *wc*-1 over-expressing strain containing the *wc*-1 gene under control of the inducible *qa*-2 promoter (Cheng et al. 2001b), to determine if increased levels of WC-1 was sufficient to induce light-inducible genes. In this case, about 7% or 65 unique genes displayed a twofold or greater increase in expression when *wc*-1 was highly expressed in the presence of inducer. Strikingly, there was a very low overlap (four out of 22) between the light-induced gene set and the genes induced by *wc*-1 over-expression. This suggested that most or all of the over-expressed set are direct or indirect targets of the WC-1 transcription factor that are independent of light, and that elevated levels of WC-1 are not sufficient for inducing light-responsive gene expression (Lewis et al. 2002). This is consistent with **WC-1's dark-specific role** in circadian clock gene regulation.

In other studies, the genes represented on the arrays were derived from a previously studied cDNA library (Bell-Pedersen et al. 1996b) whose inserts were extensively sequenced in collaboration with Bruce Roe at the University of Oklahoma (Zhu et al. 2001; http://www.genome.ou.edu/fungal.html). The libraries were constructed from non-developing mycelia grown in dark starvation conditions, and harvested at times equivalent to circadian morning and evening. In all, 13,000 cDNA clones were sequenced, yielding approximately 20,000 sequences that resolved to 1431 unique expressed sequence tags (ESTs). Roughly one-third of the ESTs were unique, the other two-thirds appearing in multiple clones. When compared to the National Center for Biotechnology Information database, nearly half (710 ESTs) showed significant matches to genes of either known or unknown function. Surprisingly, the other half of the ESTs did not display significant sequence similarity with any previously identified gene. Somewhat disappointingly, northern analysis of a randomly chosen subset of those ESTs that appeared in one but not both timed libraries found only a small fraction to be rhythmic (four out of 26 ESTs), suggesting that sequencing of 13,000 clones was not of a sufficient depth to uncover genes whose expression was limited to either the morning or the evening (Zhu et al. 2001). The information from this study yielded a UniGene set of about 1100 genes representing a little over 10% of the genome for array analysis.

Microarrays representing the 1100 genes were used to identify novel clock-controlled genes on a broad scale, in addition to examining the temperature control of genes. Within this set, about 2%–6% of genes on the arrays (depending on the **stringency of the data analysis**) were rhythmically expressed when the target RNA was harvested from cultures grown in constant dark and temperature (Nowrousian et al. 2003). These numbers are similar to that seen in other organisms; 6% in *Arabidopsis*, 1%–5% in *Drosophila*, and 2%–9% in mammals (Duffield 2003). In plants (Harmer et al. 2000), whole subsets of genes corresponding to metabolic pathways are rhythmically expressed. In *Neurospora* this was not the case, with rhythmic genes corresponding to isolated and, possibly, rate-limiting proteins such as GAPDH. Clustering of clock-regulated genes within the genome was also not found in this study or previously (Bell-Pedersen et al. 1996b; Nowrousian et al. 2003). Using the above study as a baseline, temperature shift-induced genes were examined in clock wild-type and in *frq*-null strains lacking a wild-type circadian clock. As shown with previous individual genes (Arpaia et al. 1993,

1995), light induction of gene expression does not require the clock, but unexpectedly and remarkably, all of the temperature-responsive genes identified did require a functional circadian clock for their temperature response (Nowrousian et al. 2003). Cultures grown under a 5 °C temperature cycle (12 h of 22° alternating with 12 h of 27°) revealed 14 genes as cyclically expressed. However, none of these were either cyclic or temperature inducible in the *frq*-null strain grown under the same conditions. Additionally, all 14 of these genes belonged to the clock-controlled genes identified in dark-grown cultures. This suggests an unanticipated role of the circadian circuitry as an **environmental temperature sensor**. There are important caveats to these conclusions, in that non-abundant transcripts are doubtlessly under-represented in EST-based libraries and more thorough analysis and larger gene sets should produce additional, as well as possibly a different proportion of, rhythmic genes. It is well known that on solid media, temperature-regulated conidial banding in *Neurospora* does not require *frq*, suggesting that there must be temperature-inducible or repressible genes yet to be identified or, a less likely option, that these developmental rhythms do not occur when mycelia are grown in liquid media.

An exciting and alternative explanation comes from microarray analyses by Correa et al. (2003) describing the identification of rhythmic genes from *frq*-null strains, whose phase is affected by loss of *frq* but not the ability to cycle. Using the arrays described previously (Lewis et al. 2002), clock-controlled genes were sought using target cDNA representing mRNAs collected from 22-h frq^{+} and 29-h frq^{7} strains. Each slide had two sets of the array and duplicate slides were hybridized, providing four replicate samples per gene to verify cycling, half with a 22-h periodicity and half with a 29-h periodicity. After filtering parameters were used to eliminate poorly expressed genes, 760 probes remained and three criteria for rhythmicity were applied: in both the wild-type and frq^{7} targeted arrays, (1) there must be at least one peak, as a night-peaking gene will do so only once during the 1.5-day harvest regime; (2) for those genes with two peaks, the period length must be between 18 and 30 h; and (3) the **peak-to-trough amplitude** must be a minimum of 1.5-fold (smaller than the generally twofold amplitude looked for by Nowrousian et al. 2003). This analysis yielded 145 *ccgs* out of the 760 expressing probes, suggesting that up to 20% of genes in rapidly growing or developing cultures might be rhythmic under these parameters. Previously identified *ccgs* 1, 2, 6, 7 and 12 were all among these 145. A subset of genes out of the 145 were subject to northern analysis and all shown to cycle, providing further validation, as had been done in previous studies. Examination of **time-of-day expression** showed peaks at all phases of the cycle. As had been found by previous studies, most peaked late night to early morning and, as expected, all the previously identified *ccgs* fell into these classes. The classification of genes from this and earlier *ccg* studies, according to known or predicted function based on the Whitehead sequencing project (Galagan et al. 2003; http://www.genome.wi.mit.edu/annotation/fungi/neurospora) is shown in Table 4.2. This revealed a wide variety of cellular processes to be under clock regulation. The largest classes included genes involved in metabolism, protein synthesis, cell signaling, and those of unknown function. BIOPROSPECTOR (http://www.bioprospector.stanford.edu) was used to look for possible **promoter elements**, thereby identifying an eight-nt element (TCTTGGCA) that appeared frequently in the late-night and early-day gene classes. This element matches the core sequence previously found to be necessary and sufficient for cycling of the morning-expressed *eas* (*ccg*-2) gene (Bell-Pedersen et al. 1996a).

Of the 145 genes, all but three displayed the appropriate period lengths of about 22 versus 29 h, depending on the strain from which the target was derived. The other three genes displayed the shorter period length in the microarrays, even when targeted with cDNA derived from the long period mutant. Surprisingly, when assayed via northern analysis, all three continued to cycle in

Table. 4.2. Functions of *Neurospora ccgs* (expanded from Correa et al. 2003 and Nowrousian et al. 2003)

Functional category	No. of *ccgs*
Cell division	1
Signaling/communication	16
Cell structure/cytoskeleton	8
Cell defense	4
Development	11
Gene regulation	5
Metabolism	42
Protein processing	10
Protein synthesis	33
Unclassified	50

a loss-of-function *frq* strain, suggesting control by an oscillator separate from the *frq*-based clock. Also intriguing, phasing differences in peak expression between the wild-type and *frq*^{7}-targeted microarrays were seen, indicating that the *frq*-based mechanism may additionally influence these genes. The idea that the *Neurospora* circadian system might involve multiple oscillators was not in itself novel (Loros 1984; Dunlap 2004), although molecular components or outputs of non-*frq* oscillators have not previously been found. More recently, the circadianly regulated *nit*-4 gene, encoding nitrate reductase, has been shown to participate in an autoregulatory feedback network and cycle in a non-temperature-compensated, and therefore non-circadian manner in the absence of *frq*. However, in a wild-type strain the *nit*-4 *ccg* is under clock control (Christensen et al. 2004).

D. Cross-Species Arrays

A successful first effort at cross-species microarray analysis has recently been attempted (Nowrousian et al. 2005). A UniGene set compiled from both EST sets described above was adopted for use in microarray analysis of gene expression for developmental mutants in *Sordaria macrospora*. Closely related to *Neurospora*, *Sordaria* is an excellent model for sexual development in the mycelial fungi, as it does not make asexual spores, thus allowing for clean interpretations of the function of genes identified as developmental mutations. These species share a high degree of sequence identity, with nucleic acid identity within coding regions averaging 89.5% (Nowrousian et al. 2004). Although under high-stringency hybridization conditions the overall signal intensity may be lower, the possibility of nonspecific hybridization also appeared to be lower, an important consideration for cDNA microarray hybridization analysis. In all, 2880 cDNA clones were spotted on the array, with 1420 giving hybridization signals significantly above background in five of six replicate hybridizations. Of these, 172 individual differentially regulated genes were found to be either up- or down-regulated in all three mutant strains used as targets, with more genes found to be regulated in a subset of the mutants, equaling about 12% of genes examined. These results point to the utility of **cross-species hybridizations** using *Neurospora* and other UniGene sets as they are developed for various fungal species. Importantly, cross-species microarray hybridizations could be useful for examining the photobiology and circadian rhythmicity of other fungal species.

VI. Conclusions

Steady advances in developing molecular technologies and characterizing the genome in the mycelial fungus *Neurospora crassa* have yielded a wealth of information about this important model research system. The sequence shows *Neurospora*, with approximately 11,000 genes, to be comparable in complexity to animal genomes such as those of the fruit fly and worm, with most genes having introns. *Neurospora*'s several genome scanning mechanisms has resulted in a compact genome with few repeated sequences or gene families, thereby underlining the diversity of the coding sequences. A remarkable 50% of the predicted genes show no significant homology with other sequences deposited in GenBank, although this may change soon with further sequencing of mycelial fungal genomes. A Program Project recently funded by the National Institutes of Health and awarded to a consortium of *Neurospora* research laboratories is currently undertaking the genome annotation, the global analysis of gene regulation via full-genome microarray chips, and the complete disruption of all coding regions of the *Neurospora* genome in order to further the aims of the scientific community at large.

Twenty years ago the circadian oscillatory system in *Neurospora* was imagined simply as three discrete components (input, oscillator, and output), and possibly to be based on a single feedback loop with outputs. Several genes in *Neurospora* were known to be light inducible through a possibly transcriptional mechanism. Although FRQ, WC-1 and WC-2, as well as their modifiers and regulators, are now known to be clock components, the observation that WC-1 also is the circadian blue-light photoreceptor for all genes has clearly united aspects of input and the clock. Another photoreceptor as well as output gene product, VVD, is known to act broadly to influence both the phase and abundance of many of the other clock-controlled genes, and additionally the *frq* gene. Non-circadian oscillators such as the nitrate reductase oscillator can proceed on their own or within the circadian system coordinated with the FRQ/WCC loop and perhaps other, as yet undescribed, feedback loop oscillators.

Most of the aspects of *Neurospora*'s molecular and cell biology closely resemble those of animal cells (Sogin 1994; Simpson and Roger 2002). Research in the past two decades on circadian rhythms has identified a number of parallels between clocks in animals models like *Drosophila* and mice, and in many cases has foreshadowed findings pertinent to understanding clocks in other systems. *Neurospora* has many assets as a model system for other mycelial fungi – its excellent genetics and molecular biology, and the ease with which it can be cultured and maintained. There is every reason to expect work on *Neurospora* to continue to inform research on fungi and animals in just the way a good model should.

Acknowledgements. We thank laboratory members for helpful suggestions on this manuscript. This work was supported by grants from the NIH MH44651, R37GM34985 and PO1 GM068087, NSF MCB-0084509 and the Norris Cotton Cancer Center.

References

Aronson B, Johnson K, Loros JJ, Dunlap JC (1994a) Negative feedback defining a circadian clock: autoregulation in the clock gene frequency. Science 263:1578–1584

Aronson BD, Johnson KA, Dunlap JC (1994b) The circadian clock locus frequency: a single ORF defines period length and temperature compensation. Proc Natl Acad Sci USA 91:7683–7687

Aronson BD, Lindgren KM, Dunlap JC, Loros JJ (1994c) An efficient method of gene disruption in *Neurospora* crassa and with potential for other filamentous fungi. Mol Gen Genet 242:490–494

Arpaia G, Loros JJ, Dunlap JC, Morelli G, Macino G (1993) The interplay of light and the circadian clock: independent dual regulation of clock-controlled gene ccg-2 (eas). Plant Physiol 102:1299–1305

Arpaia G, Loros JJ, Dunlap JC, Morelli G, Macino G (1995) The circadian clock-controlled gene ccg-1 is induced by light. Mol Gen Genet 247:157–163

Arpaia G, Cerri F, Baima S, Macino G (1999) Involvement of protein kinase C in the response of *Neurospora* crassa to blue light. Mol Genet Genomics 262:314–322

Ballario P, Macino G (1997) White collar proteins: PASsing the light signal in *Neurospora crassa*. Trends Microbiol 5:458–462

Ballario P, Vittorioso P, Magrelli A, Talora C, Cabibbo A, Macino G (1996) White collar-1, a central regulator of blue-light responses in *Neurospora crassa*, is a zinc-finger protein. EMBO J 15:1650–1657

Ballario P, Talora C, Galli D, Linden H, Macino G (1998) Roles in dimerization and blue light photoresponse of the PAS and LOV domains of *Neurospora* crassa WHITE COLLAR proteins. Mol Microbiol 29:719–729

Beadle GW, Tatum EL (1945) *Neurospora* II. Methods of producing and detecting mutations concerned with nutritional requirements. Am J Bot 32:678–686

Bell-Pedersen D, Dunlap JC, Loros JJ (1992) The *Neurospora* circadian clock-controlled gene, ccg-2, is allelic to eas and encodes a fungal hydrophobin required for formation of the conidial rodlet layer. Genes Dev 6:2382–2394

Bell-Pedersen D, Dunlap JC, Loros JJ (1996a) Distinct cis-acting elements mediate clock, light, and developmental regulation of the *Neurospora crassa* eas (ccg-2) gene. Mol Cell Biol 16:513–521

Bell-Pedersen D, Shinohara M, Loros J, Dunlap JC (1996b) Circadian clock-controlled genes isolated from *Neurospora crassa* are late night to early morning specific. Proc Natl Acad Sci USA 93:13096–13101

Bobrowicz P, Pawlak R, Correa A, Bell-Pedersen D, Ebbole DJ (2002) The *Neurospora crassa* pheromone precursor genes are regulated by the mating type locus and the circadian clock. Mol Microbiol 45:795–804

Borkovich K, Alex L, Yarden O, Freitag M, Turner G, Read N, Seiler S, Bell-Pedersen D, Paietta J, Plesofsky N et al. (2004) Lessons from the genome sequence of *Neurospora crassa*: tracing the path from genomic blueprint to multicellular organism. Mol Microbiol Rev 68:1–108

Cheng P, Yang Y, Heintzen C, Liu Y (2001a) Coiled coil mediated FRQ-FRQ interaction is essential for circadian clock function in *Neurospora*. EMBO J 20:101–108

Cheng P, Yang Y, Liu Y (2001b) Interlocked feedback loops contribute to the robustness of the *Neurospora* circadian clock. Proc Natl Acad Sci USA 98:7408–7413

Cheng P, Yang Y, Gardner KH, Liu Y (2002) PAS domain-mediated WC-1/WC-2 interaction is essential for maintaining the steady-state level of WC-1 and the function of both proteins in clock and light responses of *Neurospora*. Mol Cell Biol 22:517–524

Christensen M, Falkeid G, Hauge I, Loros JJ, Dunlap JC, Lillo C, Ruoff P (2004) A *frq*-independent nitrate reductase rhythm in *Neurospora crassa*. J Biol Rhythms 19:280–286

Cogoni C, Macino G (1994) Suppression of gene expression by homologous transgenes. Antonie Van Leeuwenhoek 65:205–209

Cogoni C, Macino G (1997) Isolation of quelling-defective (qde) mutants impaired in posttranscriptional transgene-induced gene silencing in *Neurospora crassa*. Proc Natl Acad Sci USA 94:10233–10238

Collett MA, Dunlap JC, Loros JJ (2001) Circadian clock-specific roles for the light response protein WHITE COLLAR-2. Mol Cell Biol 21:2619-2628

Collett MA, Garceau N, Dunlap JC, Loros JJ (2002) Light and clock expression of the *Neurospora* clock gene frequency is differentially driven by but dependent on WHITE COLLAR-2. Genetics 160:149–158

Correa A, Lewis ZA, Greene AV, March IJ, Gomer RH, Bell-Pedersen D (2003) Multiple oscillators regulate circadian gene expression in *Neurospora*. Proc Natl Acad Sci USA 100:13597–13602

Crosthwaite SC, Loros JJ, Dunlap JC (1995) Light-Induced resetting of a circadian clock is mediated by a rapid increase in frequency transcript. Cell 81:1003–1012

Crosthwaite SC, Dunlap JC, Loros JJ (1997) *Neurospora* wc-1 and wc-2: transcription, photoresponses, and the origins of circadian rhythmicity. Science 276:763–769

Davis RH (2000) *Neurospora*: contributions of a model organism. Oxford University Press, Oxford, UK

Degli Innocenti F, Russo VEA (1983) Photoinduction of perithecia in *Neurospora crassa* by blue light. Photochem Photobiol 37:49–51

Denault DL, Loros JJ, Dunlap JC (2001) WC-2 mediates WC-1-FRQ interaction within the PAS protein-linked circadian feedback loop of *Neurospora crassa*. EMBO J 20:109–117

Duffield GE (2003) DNA microarray analyses of circadian timing: the genomic basis of biological time. J Neuroendocrinol 15:991–1002

Dunlap JC (1999) Molecular bases for circadian clocks. Cell 96:271–290

Dunlap JC (2005) Blue light photoreceptors – beyond phytochromes and cryptochromes. In: Schaefer E (ed) Photomorphogenesis in plants and bacteria: function and signal transduction mechanisms. Kluwer, Dordrecht (in press)

Dunlap JC, Loros JJ (2004) The *Neurospora* circadian system. J Biol Rhythms 19:414–424

Dunlap JC, Loros JJ (2005) *Neurospora* photoreceptors. In: Briggs WR, Spudich J (eds) Handbook of photosensory receptors. Wiley, Berlin, pp 371–389

Dunlap JC, Loros JJ, Decoursey P (eds) (2003) Chronobiology: biological timekeeping. Sinauer, Sunderland, MA

Fagan T, Morse D, Hastings JW (1999) Circadian synthesis of a nuclear-encoded chloroplast glyceraldehyde-3-phosphate dehydrogenase in the dinoflagellate *Gonyaulax* polyedra is translationally controlled. Biochemistry 38:7689–7695

Feldman JF (1967) Lengthening the period of a biological clock in Euglena by cycloheximide, an inhibitor of protein synthesis. Proc Natl Acad Sci USA 57:1080–1087

Feldman JF, Atkinson CA (1978) Genetic and physiological characterization of a slow growing circadian clock mutant of *Neurospora crassa*. Genetics 88:255–265

Feldman JF, Hoyle M (1973) Isolation of circadian clock mutants of *Neurospora crassa*. Genetics 75:605–613

Feldman J, Hoyle MN (1974) A direct comparison between circadian and noncircadian rhythms in *Neurospora* crassa. Plant Physiol 53:928–930

Feldman JF, Waser N (1971) New mutations affecting circadian rhythmicity in *Neurospora*. In: Menaker M (ed) Biochronometry. National Academy of Sciences, Washington, DC, pp 652–656

Feldman JF, Gardner GF, Dennison RA (1979) Genetic analysis of the circadian clock of *Neurospora*. In: Suda M (ed) Biological rhythms and their central mechanism. Elsevier, Amsterdam, pp 57–66

Franchi L, Fulci V, Macino G (2005) Protein kinase C modulates light responses in *Neurospora* by regulating the blue light photoreceptor WC-1. Mol Microbiol 56(2):334–345

Froehlich AC, Loros JJ, Dunlap JC (2002) WHITE COLLAR-1, a circadian blue light photoreceptor, binding to the frequency promoter. Science 297:815–819

Froehlich AC, Loros JJ, Dunlap JC (2003) Rhythmic binding of a WHITE COLLAR containing complex to the frequency promoter is inhibited by FREQUENCY. Proc Natl Acad Sci USA 100:5914–5919

Galagan J, Calvo S, Borkovich K, Selker E, Read N, FitzHugh W, Ma L-J, Smirnov N, Purcell S, Rehman B et al. (2003) The genome sequence of the filamentous fungus *Neurospora crassa*. Nature 422:859–868

Garceau N (1996) Molecular and genetic studies on the *frq* and ccg-1 loci of *Neurospora*. PhD Thesis, Dartmouth Medical School, Hanover, NH

Garceau N, Liu Y, Loros JJ, Dunlap JC (1997) Alternative initiation of translation and time-specific phosphorylation yield multiple forms of the essential clock protein FREQUENCY. Cell 89:469–476

Gardner GF, Feldman JF (1980) The *frq* locus in *Neurospora crassa*: a key element in circadian clock organization. Genetics 96:877–886

Gorl M, Merrow M, Huttner B, Johnson J, Roenneberg T, Brunner M (2001) A PEST-like element in FREQUENCY determines the length of the circadian period in *Neurospora crassa*. EMBO J 20:7074–7084

Greene AV, Keller N, Haas H, Bell-Pedersen D (2003) A circadian oscillator in Aspergillus spp. regulates daily development and gene expression. Eukaryot Cell 2:231–237

Hafker T, Techel D, Steier G, Rensing L (1998) Differential expression of glucose-regulated (grp78) and heat-shock-inducible (hsp70) genes during asexual development of *Neurospora crassa*. Microbiology 144:37–43

Harding R, Melles S (1983) Genetic analysis of phototropism of *Neurospora crassa* perithecial beaks using white collar and albino mutants. Plant Physiol 72:745–749

Harding RW, Shropshire WJ (1980) Photocontrol of carotenoid biosynthesis. Annu Rev Plant Physiol 31:217–238

Harmer SL, Hogenesch JB, Straume M, Chang H-S, Han B, Zhu T, Wang X, Kreps JA, Kay SA (2000) Orchestrated transcription of key pathways in Arabidopsis by the circadian clock. Science 290:2110–2113

Hastings JW (1960) Biochemical aspects of rhythms: phase shifting by chemicals. Cold Spring Harbor Symp Quant Biol 25:131–143

Hastings JW, Sweeney BM (1958) A persistent diurnal rhythm of luminescence in *Gonyaulax* polyedra. Biol Bull 115:440–448

He Q, Cheng P, Yang Y, Wang L, Gardner K, Liu Y (2002) WHITE COLLAR-1, a DNA binding transcription factor and a light sensor. Science 297:840–842

Heintzen C, Loros JJ, Dunlap JC (2001) VIVID, gating and the circadian clock: the PAS protein VVD defines a feedback loop that represses light input pathways and regulates clock resetting. Cell 104:453–464

Hochberg ML, Sargent ML (1974) Rhythms of enzyme activity associated with circadian conidiation in *Neurospora crassa*. J Bacteriol 120:1164–1175

Iwasaki T, Nakahama K, Nagano M, Fujioka A, Ohyanagi H, Shigeyoshi Y (2004) A partial hepatectomy results in altered expression of clock-related and cyclic glyceraldehyde 3-phosphate dehydrogenase (GAPDH) genes. Life Sci 74:3093–3102

Klemm E, Ninneman H (1978) Correlation between absorbance changes and a physiological response induced by blue light in *Neurospora crassa*. Photochem Photobiol 28:227–230

Kloss B, Price JL, Saez L, Blau J, Rothenfluh A, Wesley CS, Young MW (1998) The *Drosophila* clock gene doubletime encodes a protein closely related to human casein kinase Ie. Cell 94:97–107

Konopka RJ, Benzer S (1971) Clock mutants of *Drosophila* melanogaster. Proc Natl Acad Sci USA 68:2112–2116

Kramer C, Loros JJ, Dunlap JC, Crosthwaite SK (2003) Role for antisense RNA in regulating circadian clock function in *Neurospora crassa*. Nature 421:948–952

Lakin-Thomas P (1996) Effects of choline depletion on the circadian rhythm in *Neurospora crassa*. Biol Rhythm Res 27:12–30

Lakin-Thomas PL, Brody S (2000) Circadian rhythms in *Neurospora crassa*. Proc Natl Acad Sci USA 97:256–261

Lauter F-R (1996) Molecular genetics of fungal photobiology. J Genet 75:375–386

Lauter F, Russo V, Yanofsky C (1992) Developmental and light regulation of eas, the structural gene for the rodlet protein of *Neurospora*. Genes Dev 6:2373–2381

Lee K, Loros JJ, Dunlap JC (2000) Interconnected feedback loops in the *Neurospora* circadian system. Science 289:107–110

Lewis M, Feldman JF (1997) Evolution of the frequency clock locus in ascomycete fungi. Mol Biol Evol 13:1233–1241

Lewis M, Morgan L, Feldman JF (1997) Cloning of (*frq*) clock gene homologs from the *Neurospora* sitophila and *Neurospora* tetrasperma, Chromocrea spinulosa and Leptosphaeria australiensis. Mol Gen Genet 253:401–414

Lewis ZA, Correa A, Schwerdtfeger C, Link KL, Xie X, Gomer RH, Thomas T, Ebbole DJ, Bell-Pedersen D (2002) Overexpression of White Collar-1 (WC-1) activates circadian clock-associated genes, but is not sufficient to induce most light-regulated gene expression in *Neurospora crassa*. Mol Microbiol 45:917–931

Linden H, Macino G (1997) White collar-2, a partner in blue-light signal transduction, controlling expression of light-regulated genes in *Neurospora crassa*. EMBO J 16:98–109

Linden H, Ballario P, Macino G (1997) Blue light regulation in *Neurospora crassa*. Fungal Genet Biol 22:141–150

Linden H, Ballario P, Arpaia G, Macino G (1999) Seeing the light: news in *Neurospora* blue light signal transduction. Adv Genet 41:35–54

Lindgren KM (1994) Characterization of ccg-1, a clock-controlled gene of *Neurospora crassa*. PhD Thesis, Dartmouth Medical School, Hanover, NH

Liu Y (2003) Molecular mechanisms of entrainment in the *Neurospora* circadian clock. J Biol Rhythms 18:195–205

Liu Y, Garceau N, Loros JJ, Dunlap JC (1997) Thermally regulated translational control mediates an aspect of temperature compensation in the *Neurospora* circadian clock. Cell 89:477–486

Liu Y, Merrow M, Loros JJ, Dunlap JC (1998) How temperature changes reset a circadian oscillator. Science 281:825–829

Liu Y, Loros J, Dunlap JC (2000) Phosphorylation of the *Neurospora* clock protein FREQUENCY determines its degradation rate and strongly influences the period length of the circadian clock. Proc Natl Acad Sci USA 97:234–239

Loros JJ (1984) Studies on *frq*-9, a recessive circadian clock mutant of *Neurospora crassa*. PhD Thesis, University of California Santa Cruz, Santa Cruz, CA

Loros JJ, Dunlap JC (2001) Genetic and molecular analysis of circadian rhythms in *Neurospora*. Annu Rev Physiol 63:757–794

Loros JJ, Richman A, Feldman JF (1986) A recessive circadian clock mutant at the *frq* locus in *Neurospora* crassa. Genetics 114:1095–1110

Loros JJ, Denome SA, Dunlap JC (1989) Molecular cloning of genes under the control of the circadian clock in *Neurospora*. Science 243:385–388

Luo C, Loros JJ, Dunlap JC (1998) Nuclear localization is required for function of the essential clock protein Frequency. EMBO J 17:1228–1235

Martens CL, Sargent ML (1974) Conidiation rhythms of nucleic acid metabolism in *Neurospora crassa*. J Bacteriol 117:1210–1215

McClintock B (1950) The origin and behavior of mutable loci in maize. Proc Natl Acad Sci USA 36(6):344–355

McNally M, Free S (1988) Isolation and characterization of a *Neurospora* glucose repressible gene. Curr Genet 14:545–551

Mehra A, Morgan L, Bell-Pedersen D, Loros J, Dunlap JC (2002) Watching the *Neurospora* clock tick. In: Abstr Vol Conf Society for Research on Biological Rhythms, 22–25 May 2002, Amelia Island, FL, pp 27

Merrow M, Dunlap JC (1994) Intergeneric complementation of a circadian rhythmicity defect: phylogenetic conservation of the 989 amino acid open reading frame in the clock gene frequency. EMBO J 13:2257–2266

Merrow M, Garceau N, Dunlap JC (1997) Dissection of a circadian oscillation into discrete domains. Proc Natl Acad Sci USA 94:3877–3882

Merrow M, Franchi L, Dragovic Z, Gorl M, Johnson J, Brunner M, Macino G, Roenneberg T (2001) Circadian regulation of the light input pathway in *Neurospora* crassa. EMBO J 20:307–315

Morgan LW, Greene AV, Bell-Pedersen D (2003) Circadian and light-induced expression of luciferase in *Neurospora crassa*. Fungal Genet Biol 38:327–332

Nakashima H (1981) A liquid culture system for the biochemical analysis of the circadian clock of *Neurospora*. Plant Cell Physiol 22:231–238

Nelson MA, Kang S, Braun E, Crawford M, Dolan P, Leonard P, Mitchell J, Armijo A, Bean L, Blueyes E et al. (1997) Expressed sequences form conidial,mycelial, and sexual stages of *Neurospora*. Fungal Genet Biol 21:348–363

Ninomiya Y, Suzuki K, Ishii C, Inoue H (2004) Highly efficient gene replacements in *Neurospora* strains deficient for nonhomologous end-joining. Proc Natl Acad Sci USA 101:12248–12253

Nowrousian M, Duffield GE, Loros JJ, Dunlap JC (2003) The frequency gene is required for temperature-dependent regulation of many clock-controlled genes in *Neurospora crassa*. Genetics 164:922–933

Nowrousian M, Wurtz C, Poggeler S, Kuck U (2004) Comparative sequence analysis of Sordaria macrospora and *Neurospora crassa* as a means to improve genome annotation. Fungal Genet Biol 41:285–292

Nowrousian M, Ringelberg C, Dunlap J, Loros J, Kück U (2005) Cross-species microarray hybridization to identify developmentally regulated genes in the filamentous fungus Sordaria macrospora. Mol Genet Genomics 273:137–149

Onai K, Nakashima H (1997) Mutation of the cys-9 gene, which encodes thioredoxin reductase, affects the circadian conidiation rhythm in *Neurospora crassa*. Genetics 146:101–110

Perkins DD (1988) Photoinduced carotenoid synthesis in perithecial wall tissue of *Neurospora crassa*. Fungal Genet Newslett 35:38–39

Perlman J, Nakashima H, Feldman J (1981) Assay and characteristics of circadian rhythmicity in liquid cultures of *Neurospora crassa*. Plant Physiol 67:404–407

Potapova T, Levina N, Belozerskaya T, Kritsky M, Chailakhian L (1984) Investigation of electrophysiological responses of *Neurospora crassa* to blue light. Arch Microbiol 137:262–265

Price JL, Blau J, Rothenfluh A, Adodeely M, Kloss B, Young MW (1998) double-time is a new *Drosophila* clock gene that regulates PERIOD protein accumulation. Cell 94:83–95

Ramsdale M, Lakin-Thomas PL (2000) sn-1,2-Diacylglycerol levels in the fungus *Neurospora crassa* display circadian rhythmicity. J Biol Chem 275:27541–27550

Rensing L, Bos A, Kroeger J, Cornelius G (1987) Possible link between circadian rhythm and heat shock response in *Neurospora crassa*. Chronobiol Int 4:543–549

Roeder PE, Sargent ML, Brody S (1982) Circadian rhythms in *Neurospora crassa*: oscillations in fatty acids. Biochemistry 21:4909–4916

Ruoff P, Vinsjevik M, Mohsenzadeh S, Rensing L (1999) The Goodwin Oscillator: on the importance of degradation reaction in the circadian clock. J Biol Rhythms 14:469–479

Ryan FJ, Beadle GW, Tatum EL (1943) The tube method for measuring the growth rate of *Neurospora*. Am J Bot 30:784–799

Sargent ML, Briggs WR (1967) The effect of light on a circadian rhythm of conidiation in *Neurospora*. Plant Physiol 42:1504–1510

Sargent ML, Briggs WR, Woodward DO (1966) The circadian nature of a rhythm expressed by an invertaseless strain of *Neurospora crassa*. Plant Physiol 41:1343–1349

Schwerdtfeger C, Linden H (2003) VIVID is a flavoprotein and serves as a fungal blue light photoreceptor for photoadaptation. EMBO J 22:4846–4855

Selker EU (1990) Premeiotic instability of repeated sequences in *Neurospora crassa*. Annu Rev Genet 24:579–613

Selker EU, Cambareri EB, Jensen BC, Haack KR (1987) Rearrangement of duplicated DNA in specialized cells of *Neurospora*. Cell 51:741–752

Shigeyoshi Y, Taguchi K, Yamamoto S, Takeida S, Yan L, Tei H, Moriya S, Shibata S, Loros JJ, Dunlap JC et al. (1997) Light-induced resetting of a mammalian circadian clock is associated with rapid induction of the mPer1 transcript. Cell 91:1043–1053

Shinohara M, Loros JJ, Dunlap JC (1998) Glyceraldehyde-3-phosphate dehydrogenase is regulated on a daily basis by the circadian clock. J Biol Chem 273:446–452

Shinohara ML, Correa A, Bell-Pedersen D, Loros JJ, Dunlap JC (2002) The *Neurospora crassa* clock-controlled gene-9 (ccg-9) encodes a novel form of trehalose synthase required for circadian-regulated conidiation. Eukaryot Cell 1:33–43

Shiu PKT, Raju N, Zickler D, Metzenberg R (2001) Meiotic silencing of unpaired DNA. Cell 107:905–916

Siegel RW, Matsuyama S, Urey J (1968) Induced macroconidia formation in *Neurospora crassa*. Experiencia 24:1179–1181

Simpson A, Roger AJ (2002) Eukaryotic evolution: getting to the root of the problem. Curr Biol 12:R691–R693

Sogin ML (1994) The origin of eukaryotes and evolution into major kingdoms. In: Bengston S (ed) Early life on earth. Nobel Symposium no 84. Columbia University Press, New York, pp 181–192

Talbot N (1999) Fungal biology. Coming up for air and sporulation. Nature 398:295–296

Talora C, Franchi L, Linden H, Ballario P, Macino G (1999) Role of a white collar-1-white collar-2 complex in blue-light signal transduction. EMBO J 18:4961–4968

Wessels JG (1999) Fungi in their own right. Fungal Genet Biol 27:134–145

Yang Y, Cheng P, Zhi G, Liu Y (2001) Identification of a calcium/calmodulin-dependent protein kinase that phosphorylates the *Neurospora* circadian clock protein Frequency. J Biol Chem 276:41064–41072

Yang Y, Cheng P, Liu Y (2002) Regulation of the *Neurospora* circadian clock by casein kinase II. Genes Dev 16:994–1006

Zhu H, Nowrousian M, Kupfer D, Colot H, Berrocal-Tito G, Lai H, Bell-Pedersen D, Roe B, Loros JJ, Dunlap JC (2001) Analysis of expressed sequence tags from two starvation, time-of-day-specific libraries of *Neurospora crassa* reveals novel clock-controlled genes. Genetics 157:1057–1065

5 Genomics of Protein Secretion and Hyphal Growth in *Aspergillus*

D.B. ARCHER[1], G. TURNER[2]

CONTENTS

Abbreviations: ER, endoplasmic reticulum; EST, expressed sequence tag; PPI, peptidylprolyl *cis-trans* isomerase; SRP, signal recognition particle; UPR, unfolded protein response

I. Introduction

Aspergillus is a genus which includes over 180 species of filamentous fungi (Pitt et al. 2000). However, this review will focus on those species for which there is relevant information on the secretion of proteins and hyphal growth, and especially where genomic approaches have been applied. Thus, the chapter will emphasise *A. nidulans*, *A. fumigatus*, *A. niger* and *A. oryzae* (and their close relatives). *A. nidulans* is often regarded as the model filamentous fungus for genetic studies and, indeed, it has proven to be an exceptionally useful and interesting fungus since the pioneering studies on *Aspergillus* genetics (Pontecorvo et al. 1953). Advances in *Aspergillus* genetics made with *A. nidulans* have been facilitated by it having a sexual cycle. *A. niger* and *A. oryzae* are asexual but are species which are used commercially for the production of enzymes, metabolites and fermented food products (Archer 2000). *A. fumigatus* is a common fungus in some environments (including compost heaps), but it can also be a serious allergen and human pathogen, particularly for immuno-compromised people. The genomes from all four of these *Aspergillus* species have recently been sequenced (Table 5.1), thus making formal genomics approaches to the study of protein secretion and hyphal growth more feasible. Selected data on the genomes of each *Aspergillus* species are included in Table 5.2. Each species has a G+C mole percentage of about 50%, and between 45 and 50% of the sequenced DNA is predicted to be coding sequence.

The *Aspergillus* genome sequences have been analysed using annotation software, and then annotated manually with input from scientists worldwide. The outputs from the sequencing and annotation efforts in *Aspergillus* follow the detailed analyses of the genome of *Neurospora crassa* (Galagan et al. 2003; Borkovich et al. 2004). There is now a growing resource of fungal genome sequence data which include yeasts and filamentous fungi. The genome sequence of *Phanerochaete chrysosporium*, a white-rot basidiomycete, has also recently been published (Martinez et al. 2004; Teeri 2004). That species has an extraordinary capacity for the depolymerisation and degradation of wood com-

[1] School of Biology, University of Nottingham, University Park, Nottingham NG7 2RD, UK
[2] Department of Molecular Biology and Biotechnology, University of Sheffield, Firth Court, Western bank, Sheffield S10 2TN, UK

The Mycota XIII
Fungal Genomics
Alistair J.P. Brown (Ed.)

Table. 5.1. Websites for *Aspergillus* genome information (modified after Archer and Dyer 2004)

Species	Website URL address
A. fumigatus	http://www.tigr.org/tdb/e2k1/afu1/
	http://www.sanger.ac.uk/Projects/A_fumigatus/
A. nidulans	http://www.broad.mit.edu/annotation/fungi/aspergillus/
A. niger	http://www.dsm.com/dfs/innovation/genomics/[a]
	http://www.integratedgenomics.com/products.html[a]
A. oryzae	Whole genome not yet publicly available[b]
	http://www.aist.go.jp/RIODB/ffdb/welcome.html
General *Aspergillus* sites	http://www.aspergillus.man.ac.uk/
	http://www.fgsc.net/aspergenome.htm
	http://www.aspergillus-genomics.org/
	http://www.genome.ou.edu/fungal.html
	http://www.cadre.man.ac.uk/

[a]These sites provide an entry point to the sequence data but access requires agreement being reached with the respective company

[b]Genome sequence data for *A. oryzae* are not yet available. A summary of the *A. oryzae* genome sequencing project and related topics has been published (Machida 2002)

Table. 5.2. Summary of information derived from the genome sequences of *Aspergillus species*[a]

	A. fumigatus[b]	*A. nidulans*[c]	*A. niger*[d]	*A. oryzae*[e]
Genome size (Mb)[f]	29	30	36	37
Predicted genes[f]	10,000	10,000	14,000	14,000
Genes with Pfam hits[f]	4,400	4,500	5,300	5,300

[a]This table was published in extended form elsewhere (Archer and Dyer 2004), and the data were obtained from the websites and contributions from William Nierman (TIGR, USA), James Galagan (Broad Institute, USA), Masa Machida (Tsukuba, Japan), and Gert Groot and Noel van Peij (DSM, The Netherlands)

[b]http://www.tigr.org/tdb/e2k1/afu1/

[c]http://www.broad.mit.edu/annotation/fungi/aspergillus/

[d]http://www.dsm.com/dfs/innovation/genomics/

[e]Not yet publicly available. Data from M. Machida

[f]Approximately. Current predicted numbers have been rounded

ponents, and provides a resource of secreted fungal enzymes which add to the range of activities available from the ascomycetous *Aspergillus* spp. which do not degrade wood.

The purpose of this chapter is to discuss the recent advances in protein secretion and hyphal development in *Aspergillus* spp. in the context of the newly available genome sequence data and their annotation. There have been many recent reviews published which have discussed, in particular, the secretion of proteins (and, especially, heterologous proteins) from *Aspergillus* spp. (Conesa et al. 2001; Punt et al. 2002; MacKenzie et al. 2004) but none of them had the advantage of comprehensive annotated genome sequences, even if some genome data were available. Our aims in this chapter are therefore to highlight areas where we have gained additional understanding from the genome sequences, and to indicate where we believe that post-genomic approaches can be used to extend our knowledge even further. It is not our attention to reiterate a review of protein secretion, and we refer to the cited reviews for those data. Other chapters in this volume address issues pertinent to this chapter, as does a recent summary of the move from genomics to post-genomics in *Aspergillus* (Archer and Dyer 2004).

II. Generic Genomic Methodologies

Putative homologues of many of the known genes of the secretory pathway have been found in *Aspergillus* species and in *N. crassa*, though functional analysis is still limited. Although many relevant techniques have been developed in the pre-genomics era, improvements will be needed to reflect the need to deal with the challenges of functional genomics.

Mutants defective in polar growth, septation, or with aberrant hyphal morphology have been selected from temperature-sensitive mutant screens in both *A. nidulans* and *N. crassa*, two of the major genetic models for filamentous fungi (Harris et al. 1994; Seiler and Plamann 2003). Some of the genes affected have been characterised, and they often turn out to be genes involved in secretion and polar growth in other species. In addition, genes are found for which no function has been previously described in any species (Gatherar et al. 2004; Pearson et al. 2004). This approach is complementary to the search for homologues of known secretion/polarity genes, and is essential if we are to understand how growth and secretion in filamentous fungi differs from the yeast paradigm.

We will discuss examples of how genes identified by both approaches may act in secretion and polar growth, with reference to Fig. 5.1. The Golgi apparatus of fungi is not as well characterised as that of mammalian cells, and is often referred to as the "Golgi-equivalent". However, we will trace the probable secretion pathway of fungi on the assumption of a similar system.

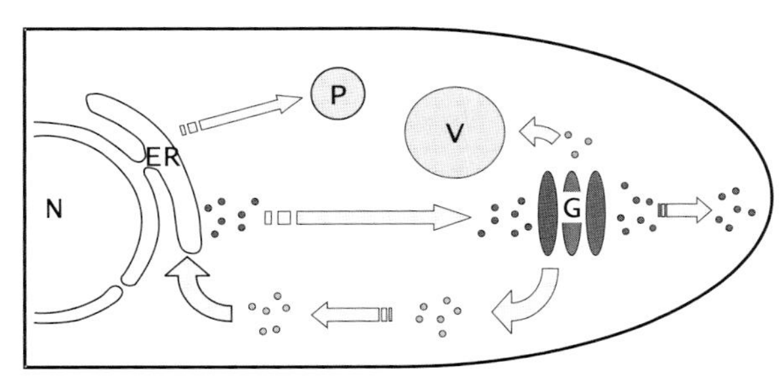

Fig. 5.1. A representation of the key sub-cellular compartments which are important in the secretion of proteins by a fungal hypha. *N* Nucleus, *ER* endoplasmic reticulum, *P* proteasome, *V* vacuole, *G* Golgi or equivalent. Polypeptides (co-translational) or proteins (post-translational) are directed to the ER by the N-terminal secretion signal. Protein folding, with the formation of disulphide bonds, is assisted in the lumen of the ER by chaperones and foldases before vectorial transport to the exterior of the cell. Post-translational modifications (e.g. glycosylation, protein processing) occur during this process. There are retrieval mechanisms for the recovery of ER-resident proteins, and proteins which do not satisfy the quality control checks for correct fold may be retro-translocated to the cytosol for degradation by the proteasome. The figure was produced by Adrian Watson, and is a modification of similar figures produced by several of our colleagues

Methods for transformation and gene manipulation of *Aspergillus* species have been fully reviewed elsewhere (Turner 1994; Riach and Kinghorn 1996; Brookman and Denning 2000), but the availability of several genome sequences presents new challenges for devising high-throughput approaches.

A. Targeted Gene Manipulation

Functional analysis of yeast genes is highly dependent on homologous recombination to target suitable vector constructs to genes of interest. This can be achieved easily in *Saccharomyces cerevisiae* using cassette vectors as templates for PCR, since the addition of short oligonucleotides (about 50 nt) at each end of the PCR product provides sufficient homology for gene targeting (reviewed in Wendland 2003). Homologous recombination in filamentous fungi varies with the species, but usually requires longer stretches of homologous sequence to achieve efficient targeting. A modified cassette template approach which uses recombinant PCR (fusion PCR; Krawchuck and Wahls 1999) can be used with *Aspergillus* species, and provides from 500 bp or more of flanking regions (Yang et al. 2004; Yu et al. 2004; Zarrin et al. 2005).

While this opens the way to more extensive genome analysis, it is not easy to detect the terminal phenotype of a deletion mutant if the gene is essential. Stable diploid strains can be constructed from all species of *Aspergillus* using the parasexual cycle but, unlike the case in *S. cerevisiae*, these are asexual diploids which cannot be haploidised by meiosis. This precludes the use of tetrad analysis of heterozygous mutants to observe the terminal phenotype. Though such diploids can be haploidised using drugs which cause mitotic non-disjunction, the recovery of a haploid segregant requires growth of a sector, and so failure to recover the mutant haploid is the only indication of lethality/severe growth impairment. Nevertheless, such approaches have been adopted to screen for essential genes in the opportunistic pathogen *A. fumigatus*, with a view to the identification of drug targets (Firon and d'Enfert 2002).

A similar approach for essential genes uses deletion in heterokaryons, where nuclei carrying both wild-type and deleted alleles can be detected by Southern blotting following deletion by transformation (Oakley et al. 1990). Failure to recover a pure, haploid deleted strain from the heterokaryon is taken as evidence for the gene being essential for viability. In some cases, it is possible to infer the terminal phenotype of the deletion mutant by plating out the mixture of uninucleate conidia which arise from the heterokaryotic mycelium.

Essential genes can be investigated in *A. nidulans* and *A. fumigatus* using the conditional promoter of the *alcA* (alcohol dehydrogenase) gene of *A. nidulans* (Waring et al. 1989). This gene is induced by ethanol, threonine and cyclopentanone, and repressed by sucrose and glucose (Felenbok et al. 2001). The *alcA* promoter can be used for rapid promoter exchange with any gene using a PCR-based approach (Fig. 5.2). For protein localisation, GFP and RFP have also been in use for some time (Fernandez-Abalos et al. 1998; Tavoularis et al. 2001; Su et al. 2004; Toews et al. 2004). Recently, the difficulty of visualising actin structures, important for cell biology studies, and which cannot be stained by phalloidin as in yeast, was overcome by using a GFP-tropomyosin fusion, the gene (*tpmA*) having been found by searching the genome sequence of *A. nidulans* (Pearson et al. 2004). Thus, the tools now exist for construction of a variety of PCR cassettes for rapid analysis of gene function and localisation. A valuable application of the PCR cassette approach in yeast has been the TAP-Tag system, used to investigate protein–protein interactions and to identify protein complexes on a genome-wide scale (Gavin et al. 2000), and this approach could now be extended to *Aspergillus* species.

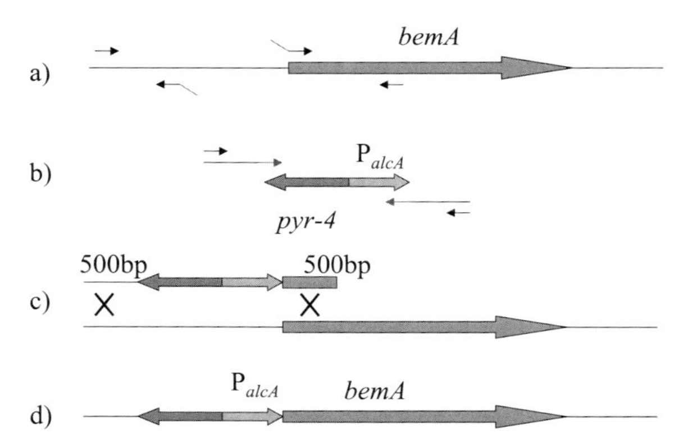

Fig. 5.2. Use of recombinant PCR for promoter replacement. **a** Sequences flanking the promoter region of the gene of interest are amplified. **b** A cassette carrying a selectable marker and conditional promoter is amplified using the fragments obtained in a, together with additional outer primers. **c** The linear fragment resulting from **b** is used to transform **a** Pyr^- strain of *A. nidulans* to Pyr^+. **d** The homologous integrant is identified by PCR and Southern analysis (Zarrin et al. 2005)

While multiple mutants can be constructed in *A. nidulans* using sexual crossing, this is more difficult in the asexual aspergilli. However, this has been partially overcome by using the BLASTER approach to gene disruption, where a two-way selectable marker such as the *pyr-4/pyrG* gene (orotidylate decarboxylase) is flanked by sequence repeats, and excision events can be selected on fluoro-orotic acid. This permits reuse of the same marker for further disruptions (d'Enfert 1996).

B. Other Approaches to Gene Disruption

Where disruption of genes is difficult using the direct transformation approach, an alternative, two-step method is to disrupt the gene of interest in a cosmid clone, and then take advantage of the large fungal DNA insert to provide very long stretches of homology to improve targeting during fungal transformation (Chaveroche et al. 2000; Langfelder et al. 2002). Homologous recombination with the clone insert can be achieved using an *E. coli* strain expressing phage λ Red functions. The *E. coli* strain carrying a cosmid with a known insert is transformed with a PCR product carrying selectable markers for *E. coli* (*zeo*–zeocin resistance) and the fungus (*pyrG*), and which has 50-bp flanking regions homologous to the target site within the cosmid insert. The recombinant cosmid is then used to transform the fungus.

Pseudo-random integration can be provided by transposon mutagenesis or random non-homologous integration (REMI). REMI has been attempted for gene function analysis but has its limitations, since it can lead to major rearrangements/deletions at the site of integration, and multiple site integration, complicating interpretation of the phenotype. Examples of the use of REMI include tagging of multidrug resistance genes in *A. nidulans* (de Souza et al. 2000), where only 36% of the transformants carried the integrating vector linked to the drug sensitivity phenotype. REMI approaches have also been used in *A. niger* and *A. oryzae* (Shuster and Connelley 1999; Yaver et al. 2000). It has also been used to validate signature-tagged mutagenesis in *A. fumigatus*, but the mutation frequency achievable by transformation was too low to be of practical use for the kind of large-scale screens carried out in prokaryotes (Brown et al. 2000).

Transposition within a bacterial host has been used by exploiting bacterial transposons for insertion into a clone, followed by identification of the clone with the insertion by PCR, and then transformation of the fungus (Jadoun et al. 2004). Fungal transposable elements are commonly found in the genome, but most are inactive. An active transposable element from *Fusarium oxysporum*, Fot1, has been used for insertional mutagenesis in *A. nidulans*. Most insertions did not occur in coding regions, limiting the use of the system (Nicosia et al. 2001), though a different transposable element from *F. oxysporum*, impala160, was applied more successfully to *A. fumigatus*, where essential genes were identified by insertional mutagenesis of a diploid strain (Firon et al. 2003).

Finally, *Agrobacterium* has been used successfully for plasmid transfer and genomic insertion of Ti-based vectors in a number of filamentous fungal species, including *A. awamori* and *A. niger*. Insertion can be by non-homologous or homologous integration (Groot et al. 1998; Gouka et al. 1999) but this approach has not yet been exploited as an efficient tagging method suitable for functional genomics.

C. Gene Identification by Complementation

Cloning genes by complementation has always been much more difficult in *A. nidulans* than in *S. cerevisiae*, mostly because of relatively low and variable transformation frequencies obtained with fungal protoplasts. The situation has improved recently with the generation of gene libraries made in AMA1-based replicating vectors, available from the Fungal Genetics Stock Centre (Osherov et al. 2002). As with *S. cerevisiae*, mutant screens and subsequent gene identification by complementation are still essential tools for investigating gene function, even in the post-genomic era. Recent examples from the genetic models *A. nidulans* (Gatherar et al. 2004; Pearson et al. 2004) and *N. crassa* (Seiler and Plamann 2003) show that we cannot expect to find all the components of polar growth and secretion required in filamentous fungi simply by comparative genomics.

D. Expression Cloning

If a cDNA library derived from a filamentous fungus is expressed in yeast, provided that a suitable enzyme assay is available, clones expressing novel enzymes can be rapidly identified. This approach was used to sample secreted enzymes of industrial interest from *Aspergillus aculeatus* (Dalboge 1997), and is applicable to species for which no genome sequence is available.

E. Transcriptomics

Transcriptome studies with *Aspergillus nidulans* are so far limited to glass slide arrays, with PCR products designed from partial EST libraries prior to completion of the genomic sequence. The slides have been validated by studying gene expression during a shift from ethanol to glucose (Sims et al. 2004a). Subsequently, further ESTs and some PCR products (using primers designed from the genome sequence) were added to the arrays. They were used in a comparative and functional genomics study to predict functions of *A. nidulans* genes and to enhance the automated annotation (Sims et al. 2004b). Those arrays have been used further in a transcriptomic study of recombinant protein secretion and the unfolded protein response (Sims et al. 2005). Although the arrays were predicted to represent up to ca. 30% of predicted ORFs, they proved useful in showing that the transcript levels of a wide range of genes were affected under secretion stress conditions. The value of expressing a heterologous protein, rather than chemicals, to induce stress was highlighted. No data have yet been reported for *A. fumigatus*, but glass slide arrays using 70mer oligonucleotides designed from all of the open reading frames (about 10,000) in the completed genome sequence have recently been constructed at TIGR (pfgrc.tigr.org).

The complete *A. niger* genome is represented on Affymetrix gene chips which are being used in collaborative studies with DSM (DSM Food Specialties, The Netherlands; Table 5.1). A generalised approach for studying the transcriptomics of the responses due to stress associated with protein secretion is presented in Fig. 5.3. *A. niger* ESTs have also been arrayed on slides (in separate studies by T. Goosen, TNO, Netherlands Organisation for Applied Research, Zeist, The Netherlands, and M. Gent, University of Manchester, UK, unpublished data). The arrays produced at TNO were hybridised with fluorescent dye-labelled target cDNA produced from RNA extracted from control and dithiothreitol-stressed cells of *A. niger*. Several up- and down-regulated transcripts were detected in response to dithiothreitol, and the results will be

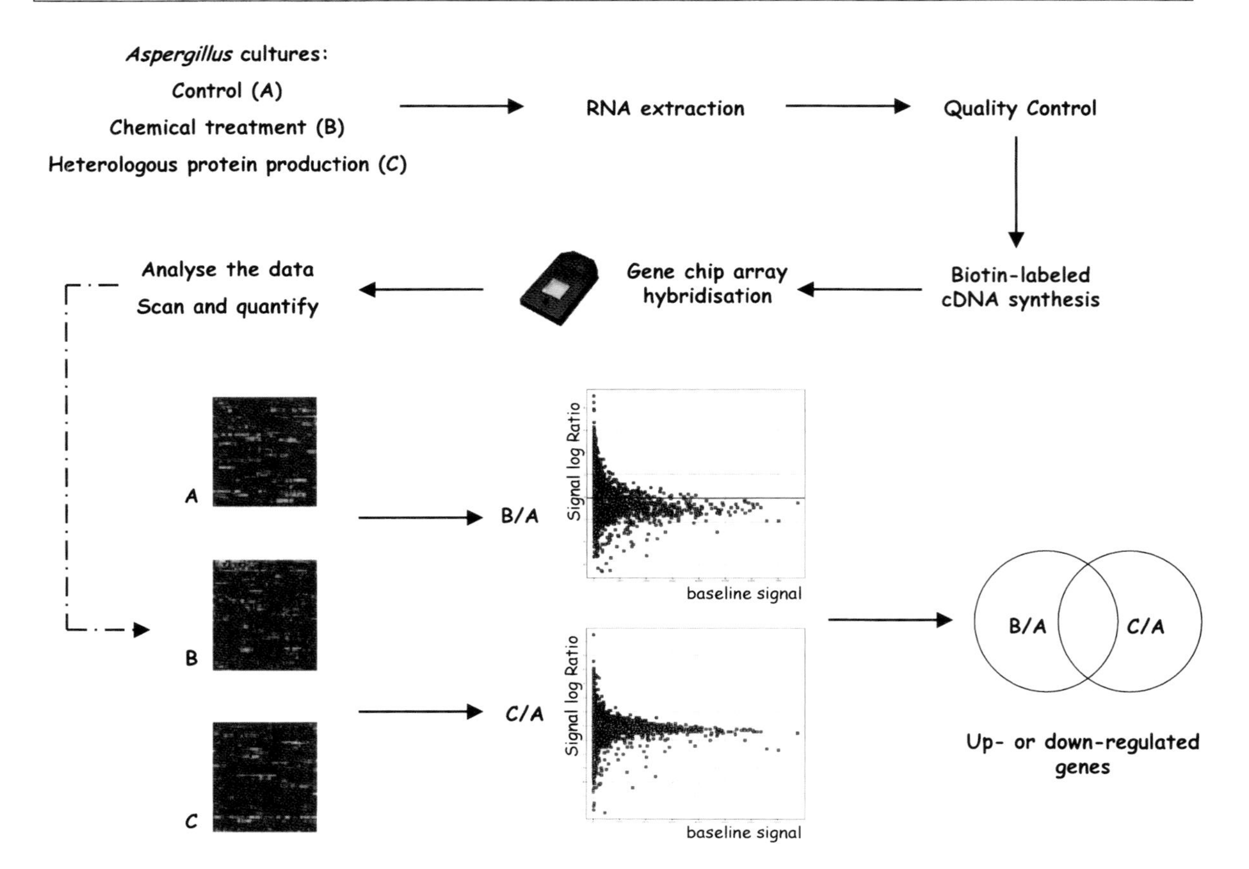

Fig. 5.3. An approach to the use of gene chips in transcriptomic studies of stress responses relevant to protein secretion by *Aspergillus*. The images were obtained using Affymetrix chips (from DSM, The Netherlands) of the *A. niger* genome, and we thank DSM for permission to reproduce them. We thank Thomas Guillemette for constructing the figure. The data from each oligonucleotide on each test chip are compared to that of the control. The compiled data are subsequently represented as a ratio of signal intensity, as well as a measure of intensity of the control signal for each oligonucleotide (baseline signal). Different stresses can then be compared to find overlapping responses. For example, the stress responses due to the reducing agent dithiothreitol include those which affect redox as well as those affecting protein secretion and the unfolded protein response. The ER stresses due either to strains secreting heterologous proteins or those exposed to other chemicals which affect the secretion of proteins (e.g. tunicamycin or brefeldin) may be separately analysed using gene chips and compared to those using other stress inducers. In addition, strains which are deleted for genes important in stress responses (e.g. the *hacA* gene of *Aspergillus*) may also be compared using this approach

published elsewhere. One interesting finding was that the transcriptional down-regulation of the *glaA* gene in response to dithiothreitol, reported elsewhere (Al-Sheikh et al. 2004), was confirmed.

F. Proteomics

Proteomic studies on *Aspergilli* and other filamenous fungi are at an early stage. The exoproteome of *A. flavus*, a species not yet fully sequenced, has been investigated by two-dimensional electrophoresis and MALDI-TOF mass spectrometry following induction of secreted proteins by the flavonoid rutin (Medina et al. 2004). Such an approach, combined with genome analysis, should prove useful for cataloguing the secreted protein repertoire of filamentous fungi, and monitoring changes resulting from mutation and growth conditions.

G. Metabolite Profiling

A novel approach to the identification of secondary metabolic pathway genes made use of transcriptional and metabolic profiling (Askenazi et al. 2003). Though a genome sequence was unavailable, a microarray of DNA fragments was constructed for transcriptome analysis. Employing a series of strains which had been previously engineered to give different yields of lovastatin, association analysis was used to determine gene expression patterns which correlated with lovastatin yield. This led to the identification of DNA fragments positively associated with lovastatin yield, which included the previously isolated lovastatin gene cluster. Additional genes associated with secondary metabolite production were also identified. This approach is especially useful where no complete genome sequence is available, which will remain the case for many of the estimated 100,000 or so known fungal species.

III. Secretion Genes in *Aspergillus* spp.

The proportion of predicted ORFs in the *Aspergillus* genomes which are directly or indirectly relevant to protein secretion and hyphal development is not accurately known. Certainly, annotations of the *Aspergillus* genome sequences indicate that there is a wide range of secreted proteins, and that the secretory system itself is complex and involves many proteins. The numbers of genes, predicted to encode proteins containing either signal peptides (cleaved secretion signal) or signal anchors (un-cleaved secretion signal, retaining the protein in a membrane) in the genomes of *A. fumigatus*, *A. nidulans* and *A. oryzae* is highly dependent on the confidence limit assigned to the prediction. Data for the predicted signal peptide/anchor proteins encoded in *A. fumigatus* are presented in Fig. 5.4. At the 95% confidence level of a correct assignment of a signal peptide, approximately 8% of the proteins encoded in the *A. fumigatus* genome are predicted to be secreted (4.5% for those proteins that have assigned Pfam numbers). This analysis indicates the significance of protein secretion to fungi and, in addition, that the secretion process also involves many proteins which have functions in maintaining and regulating the secretion process. The secretion of proteins is tightly integrated with other processes such as hyphal growth and branching (discussed in this chapter) and endocytosis (which, although not discussed here, is described in detail in *N. crassa* by Borkovich et al. 2004).

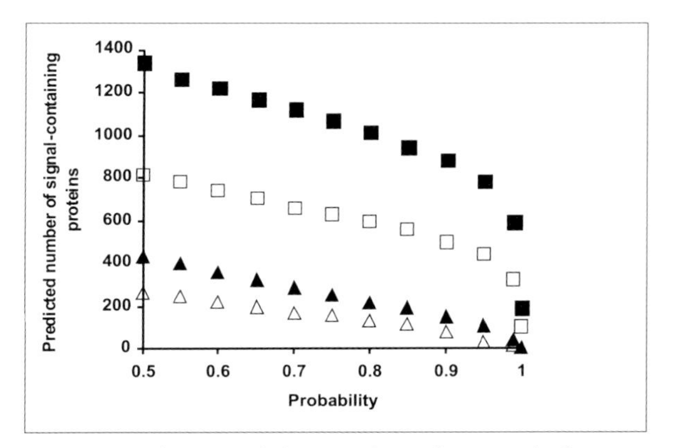

Fig. 5.4. Prediction of the number of N-terminal secretion signals and secretion anchors, using web-based programmes, in the genome of *A. fumigatus*. The analysis was done by J. Huang (TIGR, USA) and G. Robson (University of Manchester, UK). ■ Signal peptide, □ signal peptide+Pfam, ▲ signal anchor, △ signal anchor+Pfam (+Pfam indicates those proteins with an assigned Pfam description)

A. Entry of Proteins into the Secretory Pathway

In the yeast *S. cerevisiae*, proteins destined for entry into the secretory pathway can do so either co- or post-translationally whereas in mammals translocation appears to be, primarily at least, co-translational. The key difference between co- and post-translational translocation is that only co-translational translocation of proteins across the ER membrane requires the signal recognition particle (SRP) which recognises and binds to the signal (or leader) sequence at the N-terminus of proteins destined for entry into the secretory system. The post-translational translocation system is independent of the SRP and operates following the synthesis of proteins in the cytosol and release from the ribosome (Rapoport et al. 1999). The signal sequence binds to the so-called Sec complex which includes the protein-conducting channel (translocon) across the ER membrane for both the co- and post-translational systems. The composition and functions (Keenan et al. 2001), and the structure (Halic et al. 2004) of the eukaryotic ribonuclear-protein SRP, together with its role in co-translational translocation of proteins across the ER (Johnson and van Waes 1999), have been described.

Preliminary data suggested that both translocation systems operate in filamentous fungi (Conesa et al. 2001). The translocon is comprised in yeast of the trimeric Sec61p complex (Sec61p, homologue of mammalian Sec61α; Sss1p, homologue of mammalian Sec61γ; Sbh1p, homologue of mammalian Sec61β). The Sec complex of yeast includes the translocon and the tetrameric complex: Sec62p, Sec63p, Sec71p and Sec72p (Rapoport et al. 1999). Recent studies in yeast indicate that the signal peptide binds to both Sec61p and Sec62p in post-translational protein transport into the ER (Plath et al. 2004). In co-translational translocation, binding of the SRP arrests peptide elongation until the SRP docks with its receptor on the ER membrane, whereupon it is released (Keenan et al. 2001). For further detail of the SRP and its receptor, including the roles of GTPases, readers are referred to the cited reviews. For this chapter, it is sufficient to confirm that mining of the *Aspergillus* genome databases confirms the presence of homologues of genes necessary to support both the co- and post-translational translocation systems (D.B. Archer, unpublished data; A. Sims, personal communication; Robson et al. 2005), including signal peptidase for removal of the signal sequences. We do not know the functionality of the systems either with different secretory proteins or under different environmental conditions, and this will require detailed analysis.

B. Protein Folding in the ER

Proteins entering the ER lumen may be destined for export from the cell, targeted to other cellular destinations or retained within the lumen because of possession of a tetrameric C-terminal retention signal. Each protein undergoes assisted folding and modifications which may include structural isomerisation at prolines, the formation of disulphide bonds and glycosylation (initiated in the ER and continued in subsequent compartments of the secretory system). Assisted folding involves ER-resident chaperones whereas ER-resident foldases catalyse the isomerisation and formation of disulphide bonds. Glycosylation is catalysed by several different proteins, and subsequent passage of glycosylated proteins must pass scrutiny by a quality control system which employs calnexin, a chaperone bound to the ER membrane but active within the lumen of the ER. Quality control is a function of the ER which also operates for non-glycosylated proteins because exit from the ER and continued passage within the secretory system rely upon those proteins having attained a folded structure which is "stable" (discussed below) and does not expose hydrophobic regions. Those proteins which are judged by the quality control system to be irretrievably mis-folded (at least within kinetic parameters defined by the cell) are retro-translocated (dislocated) from the ER through the translocon and are ubiquitin-tagged for degradation by the proteasome. This is termed ER-associated degradation, ERAD. A closer inspection has revealed that the quality control system in yeast includes a series of checkpoints which determines the possible fates (e.g. proteasomal degradation or targeting to the vacuole) of mis-folded proteins (Vashist and Ng 2004). The core features of protein folding and modifications are found in filamentous fungi as well as yeasts and higher eukaryotes, albeit with minor differences, have been summarised previously in detail (Archer and Peberdy 1997; Conesa et al. 2001; MacKenzie et al. 2004) and will not be repeated here. The availability of the genome sequences for *Aspergillus* spp. has since permitted an in-depth comparison of the genes associated with each function.

Protein folding is assisted in the ER lumen by chaperones (e.g. BiP) and foldases (e.g. protein disulfide isomerase, PDI, and peptidyl-prolyl *cis-trans* isomerase, PPI). The PDI family (thioredoxin, PF00085) has been annotated for *A. fumigatus*, *A. nidulans* and *A. oryzae*, and each of these three species contains three genes encoding PDI family proteins characterised by thioredoxin domains (CGHC) and a C-terminal ER-retention signal, (K/H)DEL. Two of the PDI family proteins in each species have two thioredoxin domains, and one protein has one thioredoxin domain. These proteins have high similarity to proteins already described in *A. niger* (Jeenes et al. 1997; Ngiam et al. 1997, 2000; Wang and Ward 2000). The latter authors suggested that PrpA (one of the PDI family proteins) might be unique to filamentous fungi, and a very close homologue of the encoding gene is found in all of the *Aspergillus* sequences. BLAST analysis using the sequence without the thioredoxin domain showed significant alignment to other fungal genes (from *A. niger*, 1e-138, and *N. crassa*, 6e-91) but the next hits were to *Dictyostelium discoideum* (4e-26), *S. pombe* (2e-12) and *S. cerevisiae* (2e-11). Thus, PrpA and its homologues may be largely restricted to filamentous fungi. The *prpA* gene in *A. niger* is not essential for viability (Wang and Ward 2000). Transfer of electrons during oxidative protein folding (e.g. in the formation of disulfide bonds) involves Ero1p in *S. cerevisiae*. Homologues of this ER membrane protein are present in the *Aspergillus* genomes.

Most of the PPI family foldases predicted from the genome are cytosolic. The published sequence of *cypB* which encodes an ER-resident cyclophilin type of PPI in *A. niger* has been described by Derkx and Madrid (2001). The protein has an ER retention signal of HEEL, i.e. it diverges somewhat from the more usual (K/H)DEL signal, but it also has a predicted signal sequence strongly suggesting that the protein is ER-resident. A C-terminal HEEL motif is also found in the predicted CypB homologues in *A. fumigatus* and *A. oryzae*, but the *A. nidulans* homologue has HNEL. This protein also has a predicted signal sequence, so is likely to be ER-resident.

The *A. niger* calnexin gene sequence (Wang et al. 2003) has been described and homologues are found in the other *Aspergillus* genomes. No other calnexin homologues are present, confirming that there is no calreticulin homologue. Searches for the ER-resident Hsp70 protein (PF00012) identify BiP homologues in each genome as well as another ER-resident (judged by both the presence of a predicted signal sequence and a C-terminal ER retention motif) Hsp70 family member which has not been characterised to date in any filamentous fungus. These Hsp70 proteins may be homologues of the *S. cerevisiae* Lhs1p Hsp70 family member and may be involved in protection against oxygen deprivation as well as in assisting protein folding. The sequences of BiP homologues in the aspergilli are virtually identical to that described first in *A. niger* (van Gemeren et al. 1997). BiP is a key molecular chaperone resident in the ER but it also has important roles in the translocation of proteins into the ER (through its association with Sec63 in the Sec complex) and in regulating the unfolded protein response (UPR) through its association with the trans-membrane kinase IreA (see below).

The glycosylation of proteins will not be discussed here in detail. The enzymology of N-glycosylation has been discussed previously (Maras et al. 1999; Peberdy et al. 2001) and the analysis of the *N. crassa* genome has revealed many of the genes necessary (Borkovich et al. 2004). However, we do not know of a comprehensive mining of the *Aspergillus* genomes for genes predicted to be involved in the glycosylation of secretory proteins.

C. Post-ER Secretion

Secretory proteins are packaged in membrane-bound vesicles from the ER, to the Golgi (or functional equivalent in filamentous fungi), and thence to the plasma membrane (primarily at the hyphal tip). This vectorial transport of secretory proteins is common to all eukaryotes but is particularly interesting in the filamentous fungi because of the hyphal growth pattern. Secretion-related GTPases play a key role in this process, and they have been examined in *A. niger* before the genome sequences were available (Punt et al. 2001) and in *N. crassa* since its genome was sequenced (Borkovich et al. 2004). Five secretion-related GTPase-encoding genes were characterised in *A. niger* as members of the Rab/Ypt branch of the Ras superfamily of GTPases (Punt et al. 2001).

D. ER to Golgi

When ER-resident proteins are moved to the Golgi, their return to the ER (retrograde transport) seems to require COPI coated vesicles, and the *A. nidulans* COP-α homologue Sod^{VI}C is essential for the establishment and maintenance of polar growth (Whittaker et al. 1999).

Proteins folded correctly in the ER are recognised as appropriate cargo to be incorporated into COPII vesicles, which are moved to the cis-Golgi (entry point) on microtubules. Small GTPases of the RAB/Ypt/Sec4 family are involved in many of the fusion steps in vesicle targeting, including ER to Golgi, and between Golgi, vacuole, plasma membrane and endosome. Sar1 is involved in budding and docking COPII vesicles (ER to Golgi), and the *A. niger* homologue *SarA* is essential (Conesa et al. 2001). Docking of the vesicles to the cis-Golgi involves a transport protein particle (TRAPP I), which is a multisubunit complex. Binding of the vesicle to TRAPP I leads to activation of another small GTPase, Ypt1. Homologues of Ypt1 have been characterised from *A. niger* var *awamori* and *T. reesei*, and the latter was able to complement a yeast strain depleted for Ypt1 (Saloheimo et al. 2004). TRAPP II, which shares some proteins with TRAPP I, is involved in transport within the Golgi. Point mutations in a regulatory subunit of TRAPPII, Trs120 (HypA) result in wide, slow-growing hyphae with thick walls (Shi et al. 2004), indicating a polarity defect, and enzyme secretion is reduced. Deletion of *hypA* is not lethal but results in osmotically sensitive hyphae with an aberrant morphology.

Some proteins entering the Golgi are targeted to the vacuoles, and a large number of genes (VPS for vacuolar protein sorting) are required for this. Mutations in some of these genes also result in hyperbranching or other apparent polarity defects, giving abnormal hyphal morphology. These include *hbrA* of *A. nidulans* (a *VPS33* homologue and member of the Sec1 family; Memmott et al. 2001), and *digA* (*VPS18*; Geissenhoner et al. 2001).

E. Golgi to Plasma Membrane

For the final stages of secretion, vesicles exit the trans-Golgi and are transported to the plasma membrane, where they fuse to deliver their cargoes. Some of these secreted proteins are required for cell wall growth (budding in yeast, or hyphal elongation in filamentous fungi), and others are excreted into the surroundings, including the enormous range of hydrolytic enzymes typical of filamentous fungi (see below). It is not yet known how this differential targeting is achieved, but it presumably involves different types of vesicles. It is generally believed, with some evidence, that protein secretion occurs at hyphal tips (Archer and Perberdy 1997; Gordon et al. 2000).

Vesicle transport in yeast involves actin cables and myosin(s) which are in turn directed towards specific sites in the cell cortex (reviewed in Pruyne and Bretscher 2000a, b; Irazoqi and Lew 2004). This is achieved by cortical markers which define the point at which a bud will develop. The cortical markers result in localised assembly of protein complexes which nucleate the growth of actin cables, resulting in transport of vesicles to that point. Early cortical markers in yeast, which define the bud site, Bud3, Bud8 and Bud9, do not seem to have homologues in *Aspergillus* species, and alternative models have been proposed for how sites of polarised growth might be chosen in filamentous fungi (Harris and Momany 2004).

There is evidence from studies on mutants of *N. crassa* (Seiler et al. 1999; Riquelme et al. 2000) that in filamentous fungi, cytoplasmic microtubules, together with associated motor proteins, may play a greater role than in yeast in the long-range transport of vesicles to the tip region, where actin-myosin make the final delivery. Dynein mutations of *A. nidulans* (*nudA*), which abolish nuclear migration along the microtubules, do not prevent germ tube emergence, though hyphae are morphologically aberrant (Xiang et al. 1994).

The *myoA* gene of *A. nidulans* (a type I myosin) seems to be essential, and its down-regulation with the conditional *alcA* promoter led to severe polarity defects (McGoldrick et al. 1995). MyoA localises to hyphal tips, septa, and in cortical patches (Yamashita et al. 2000). As with hyphal tips, sites of septation also involve actin polarisation, and the *BNI1* homologue *sepA* was identified in *A. nidulans* via a mutant defective in septation (Harris et al. 1997). Interestingly, a GFP fusion with the secreted protein glucoamylase is also observed at both tips and septa in *A. niger* (Gordon et al. 2000). In yeast, the homologue of MyoA, Myo3, is associated with actin patches and endocytosis, and the exact role of different myosins in *A. nidulans* requires further study. A putative homologue of the type V myosin Myo2 is present in *A. nidulans*, but its function has not been tested. Distinguishing between actin cables and actin patches has not been possible until recently, since phalloidin does not stain *Aspergillus* actin, and the antibody staining does not permit differentiation. Recently, a GFP fusion was made with a tropomyosin homologue, and this actin-associated protein revealed some actin cables in the tip region (Pearson et al. 2004).

F. Polarity Establishment

Putative genes for much of the machinery of polarity establishment can be detected in the *Aspergillus* genomes, and some of these genes have been investigated for function, either in *Aspergillus* species or in related filamentous fungi. The GTPase Cdc42 plays a key role in regulating polarity establishment in all eukaryotes, together with its GEF Cdc24 and several GAPs. Cdc42 is recruited to the site of polar growth in yeast. Unlike *S. cerevisiae*, *Aspergillus* species also seem to possess a Rac homologue, a second Rho-type GTPase (Harris and Momany 2004). Though no functional analysis has been reported in *Aspergillus* species, a Rac homologue in *Penicillium marneffei* seems to contribute to polar growth (Boyce et al. 2003).

Other proteins which act with Cdc42 to establish polarity include Cla4, Ste20, Gic1, Gic2, and the scaffold proteins Bem1, Boi1 and Boi2. As in yeast, the first of two SH3 domains in the Bem1 homologue of *A. nidulans*, BemA, appears to be dispensable when expressed under the control of the *alcA* promoter, but down-regulation results in a severe polarity defect (Zarrin et al. 2005). These polarity establishment proteins are linked to the actin cables via the polarisome, a key component of which is the FH1/2 protein Bni1. The *A. nidulans* homologue of *BNI1*, *sepA*, was isolated and identified by complementation of a temperature-sensitive mutant (Harris et al. 1997). This mutant is defective in septum formation and shows a hyperbranching phenotype at the restrictive temperature, which may also be indicative of a polar growth defect.

G. The Spitzenkörper

A characteristic feature of filamentous fungi is a structure just behind the growing tip known as the Spitzenkörper (Reynaga-Pena et al. 1997), which can be readily observed by light microscopy, more easily in some fungi than in others, and which appears in electron micrographs to consist of a concentration of vesicles. It has been suggested that it acts as a vesicle supply centre, and this idea has been modelled mathematically. Sometimes, the body splits into two just before terminal branching is observed, and new Spitzenkörpers are observed to form at branch points. The molecular nature of the Spitzenkörper is not well defined, though it has been shown that a GFP–SepA (Bni1) fusion product localises not only to the membrane surface at the tip, but also to a region just behind the tip. Hence, this may be useful as a Spitzenkörper marker (Sharpless and Harris 2002). It further indicates that an element of the polarisome is associated with the Spitzenkörper, as well as with the point of growth. The availability of the genome sequence, and more rapid methods of GFP/RFP tagging should help to identify other components associated with this structure, and clarify its nature.

There will then remain many interesting questions about how the Spitzenkörper is assembled, and how it is positioned. Recent evidence is suggesting that cytoplasmic microtubules may be involved in the long-range movement of vesicles to the tip region and in positioning the Spitzenkörper, while short-range transport of vesicles from the Spitzenkörper region to the tip may require microfilaments (reviewed in Harris and Momany 2004). Immuno-staining of actin in *A. nidulans* shows its accumulation at the tip, and at newly forming septa. Recent use of a GFP–tropomyosin fusion, an actin-associated protein, has also revealed cable-like elements in the tip region (Pearson et al. 2004).

H. The Exocyst

Docking of secretory vesicles with the plasma membrane in yeast and mammalian cells requires a complex of eight subunits known as the exocyst (Fig. 5.5; reviewed in Lipschutz and Mostov 2002; Hsu et al. 2004), and putative homologues of these components can been found in the *Aspergillus* genome sequences (Table 5.3). Six of these proteins were discovered as products of the *SEC* genes of *S. cerevisiae*, and mutations in these genes led to defects in secretion. While this indicates conserved functions in the secretory pathway, functional analysis will be necessary to reveal the location of this structure in filamentous fungi. It might be expected that this structure would be located on the plasma membrane at growing tips. Docking of vesicles involves interaction with the exocyst component Sec15, and a vesicle-associated GTPase Sec4 (Guo et al. 1999). Sec4 is a member of the Rab family of GTPases, involved in regulation of vesicle traffic. While a gene encoding a putative Sec4 homologue, SrgA, has been deleted in *A. niger* (Punt et al. 2001), this was not lethal, as is the case in *S. cerevisiae*. Deletion resulted in

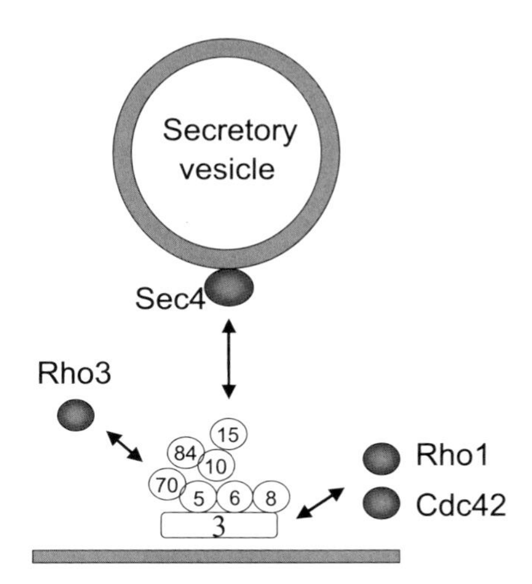

Fig. 5.5. Vesicle docking and the exocyst. *S. cerevisiae sec* mutants are deficient in secretion. Sec3, 5, 6, 8, 10, 15, and Exo70 and Exo84 are proteins of the exocyst, a 19.5S complex associated with the plasma membrane at sites of secretion and cell growth. Rho3, Rho1, Cdc42 and Sec4 are GTPases which interact with components of the exocyst complex. Docking of a secretory vesicle involves an interaction between Sec4 and Sec15. The complex also interacts with microtubules and actin filaments. Secretory vesicles are delivered via actin cytoskeleton (modified from Guo et al. 1999; Lipschutz and Mostov 2002)

Table. 5.3. Percent amino acid identities between exocyst components of *A. nidulans* and of selected other fungi (courtesy of Steve Baldwin)

Exocyst subunit	*Aspergillus nidulans*	*Aspergillus fumigatus*	*Magnaporthe grisea*	*Gibberella zeae*	*Ustilago maydis*	*Neurospora crassa*	*Saccharomyces cerevisiae*
Sec3	100	69	39	39	24	39	19
Sec5	100	70	49	50	25	49	21
Sec6	100	80	51	50	22	50	19
Sec8	100	72	42	42	19	42	20
Sec10	100	74	51	50	31	53	25
Sec15	100	86	61	62	34	63	20
Exo70	100	78	44	45	22	45	23
Exo84	100	76	47	46	24	45	19

aberrant hyphal morphology, indicating some role in hyphal growth, but this observation also emphasises that we cannot assume that what is true for yeast is true for *Aspergillus*. Also, it shows the need for further investigation of this structure and its relationship to hyphal growth and secretion.

Studies in yeast have also shown interactions of exocyst components with other polarity-associated genes, including Rho1 and Cdc42 with Sec3. RhoA, the *A. nidulans* homologue of Rho1, involved in cell wall integrity control by the MAP kinase pathway, also appears to be involved in polarity, branching and cell wall deposition (Guest et al. 2004). While comparative genomics is leading to the rapid identification of components of the secretion pathway in filamentous fungi, it is clear that additional components are present where there is no obvious homologue in *S. cerevisiae*, and a combination of mutant screens and investigation of protein–protein interactions will be required to go beyond the yeast paradigm.

IV. Secreted Proteins from *Aspergillus*

A. Native Proteins

The secretion of proteins is a key feature of the lifestyle of *Aspergillus* spp. and many other filamentous fungi. Enzymes are secreted from hyphal tips for the mobilization of carbon and energy sources from polymeric organic materials, and this capacity underpins the highly successful ecological niche of many filamentous fungi as primary degraders of waste organic matter. This capacity is exploited commercially for the production of enzymes with applications in food, detergents and in many other industrial sectors (Archer 2000). As already mentioned, a substantial proportion of all proteins encoded by *Aspergillus* spp. are predicted to have either secretion signals or secretion anchors. It is not our intention to list all proteins likely to be secreted from *Aspergillus*, but the availability of the genome sequences now provides an unprecedented opportunity for data mining and to gain sequences of particular enzymes or classes of enzymes. *Aspergillus* spp. are noted and exploited commercially for their capacity to secrete (gluco)amylases, proteases and pectinases, but there is a far wider range of enzymes secreted.

A. fumigatus is not a commercial organism, but it retains an efficient system for the secretion of enzymes which are key to its ecology, its growth, its allergenicity and probably to its pathogenicity. *A. fumigatus* conidia are widely dispersed in air and their environmental sources are not fully catalogued. It is likely that *A. fumigatus* is commonly distributed in nature and its capacity for growth on a wide variety of substrates is predicted from the annotated genome sequence. *A. fumigatus* is known to be a common component of composts and is considered to be the major compost-associated biohazard (Beffa et al. 1998). Temperature gradients are a feature of composts and the thermotolerance of *A. fumigatus*, relative to many other fungi, together with the degradative capacity due to versatility in the range of enzymes secreted by *A. fumigatus*, probably explains its success in composts.

Each *Aspergillus* sp. contains genes encoding a wide range of different proteinases, including the fungal-specific class of aspartyl-endopeptidases (PF01828; A. Sims, personal communication) which are thermostable, active at low pH and pepstatin-insensitive. Each *Aspergillus* sp. has genes encoding a wide range of glycosyl hydrolases, (phospho)lipases and pectinases which ensure that those fungi are capable of degrading the organic polymers available. Many secreted enzymes have been described from *Aspergillus* spp., either biochemically or in relation to the encoding genes and their regulation. From detailed analysis of the *Aspergillus* genomes, *A. fumigatus* contains genes encoding more cellulose-binding domains/modules (fungal type) than do *A. oryzae* and *A. nidulans* combined (Robson et al. 2005). Of the 14 cellulose-binding domains assigned as PF00734 in *A. fumigatus*, only one gene appears to encode a cellulose-binding domain attached to a protein without clear annotation while the remainder of the cellulose-binding domains are attached to glycosyl hydrolases (families 5, 6, 7, 45, 61) as well as one feruloyl esterase and a cutinase. Not all of the predicted cellulose-binding domain-containing proteins also contain clear signal peptides or signal anchors (G. Robson, personal communication) but that may reflect errors in automated annotations. There are several other glycosyl hydrolases without obvious cellulose-binding domains as well as three cutinases lacking a cellulose-binding domain in *A. fumigatus*.

The degradation of pectins is particularly complex due to the heterogeneous nature of the polymer. For the complete mineralisation of pectin, a wide range of enzymes is required to ensure that both the backbone and side chains are degraded. Pectinolytic enzymes have been well described from *A. niger* (de Vries and Visser 2001) and, more recently, some expression profiling of 26 pectinolytic genes has been reported when *A. niger* was grown on pectin and pectin-derived carbon sources (de Vries et al. 2002). Hybridised dot-blots were used in this study because, at that time, genome-wide arrays were not available. The approach did lead to the distinction of subsets of pectinolytic genes in relation to their regulation. The available genomes, allied to gene arrays/chips, will extend the ability to identify the complete set of pectinolytic genes and will facilitate investigations of their regulation in the wider context of co-regulated genes. Similar analyses will now be possible with genes important in the degradation of cellulose and starch by *Aspergillus* spp.

B. Heterologous Enzymes

A. niger and *A. oryzae* (and their close relatives *A. awamori* and *A. sojae* respectively) have been developed as cell factories for the secreted production of heterologous proteins. Host-vector systems have been described and many different heterologous proteins expressed, with yields ranging from just detectable to commercial levels (MacKenzie et al. 2004). The natural hosts from which the heterologous genes were taken have been from many sources but, in the main, the secreted yields have been better when a fungal gene is expressed in another fungal host. Yield and authenticity of product are the key factors in using fungi as cell factories, and the strategies developed to optimise protein production have been summarised several times (Conesa et al. 2001; Punt et al. 2002; MacKenzie et al. 2004) and will not be recapitulated here.

A summary of the strategies employed most successfully in the secreted production of heterologous proteins is:

- use a strong, native promoter
- optimise (may involve maximizing, but not necessarily) the gene copy number
- integrate genes at a transcriptionally active locus
- use a fungal host of low proteolytic activity, or minimize the induction of proteases by appropriate cultural technology
- express a gene fusion whereby the target protein is fused downstream of a carrier protein (e.g. glucoamylase). Cleavage of the two components of the fusion protein may be effected in vivo by engineering a dibasic amino acid endoproteolytic cleavage site at the fusion junction.

There are several pertinent areas which remain under- investigated. For example, there is known to be an impact of glycosylation on secreted yields of chymosin (Ward 1989), but this effect has not been taken very far in other systems. There is a limited range of promoters in current use but the genome sequences should provide other options quite readily, even if they will require testing. Promoters have not, in the main, been modified for optimised activity by, for example, improving core promoter activity (Minetoki et al. 1998; Liu et al. 2003). Promoters regulated by carbon catabolite repression are in common use and strains defective in carbon catabolite repression can be useful, although specific mutation of responsive sites in the promoters is not yet widely adopted. Knowledge of proteases from genomic information may prove to be very useful in strain construction, although it may be more efficient to examine the regulatory systems so that families of protease genes can be made ineffective. Where increasing gene copy number beyond a critical point becomes ineffective, increased expression of necessary transcription factors may overcome a limitation imposed by the titration of the native levels of these factors.

The carrier protein system is highly effective in improving secreted yields of heterologous proteins. The strategy depends on the efficient and accurate cleavage of the target protein from the carrier by the furin-type of endoprotease resident in the Golgi (Jalving et al. 2000; Punt et al. 2003). We know from a growing body of results that the proteolytic processing is not always faithful or efficient (Spencer et al. 1998; MacKenzie et al. 1998; Punt et al. 2003; Clemente et al. 2004) and that it probably depends on the context in which the dibasic amino acid pair is presented to membrane-bound protease. To improve the efficiency of cleavage, it is useful to design the context appropriately (Spencer et al. 1998) and this approach has been effective in the secretion of humanized antibodies from *A. niger* (Ward et al. 2004).

C. The Stress of Protein Secretion

The process of protein secretion is largely conserved amongst all eukaryotic cells. A key element of the secretion process is achieving the correct fold of proteins in the lumen of the endoplasmic reticulum (ER), so that the cell's natural quality control systems are satisfied and then, the folded proteins can continue the secretion process. The natural quality control systems within the secretory pathway of eukaryotic cells ensure that only properly folded protein is secreted (Ellgaard and Helenius 2003). How cells discriminate between properly folded versus mis-folded protein, and how they selectively retain and target the latter for degradation are challenging questions which remain to be answered, although more information is now becoming available. Improperly folded proteins (i.e. those which have not attained their native fold) induce the unfolded protein response (UPR) and other ER stress responses in yeasts and filamentous fungi (Travers et al. 2000; Ngiam et al. 2000; Patil and Walter 2001; Al-Sheikh et al. 2004; Sims et al. 2005). The unfolded protein response mediates the transcriptional up-regulation of genes encoding ER-resident chaperones and foldases, and up-regulates the ER-associated degradation (ERAD) following retro-translocation of proteins from the ER to the cytoplasm, where those proteins are tagged by ubiquitin and targeted for degradation by the proteasome. The UPR is mediated by the bZIP transcription factor HacA in *Aspergillus* (Hac1p in *S. cerevisiae*). However, some ER stress responses appear to be independent of Hac1p and the ER membrane protein Ire1p which senses unfolded proteins and initiates the signal from the ER (Schröder et al. 2003). Hac1p is the translation product from a spliced HAC1 mRNA in yeast (reviewed in Patil and Walter 2001), and analogous sensing by Ire1p and mRNA processing occurs in *Aspergillus*, albeit with differences in the detail (Saloheimo et al. 2003). Sensing of, and the responses to, ER stress in eukaryotes has been summarised recently by Zhang and Kaufman (2004), Rutkowski and Kaufman (2004), and Schröder and Kaufman (2005). There are alternative possible fates for degradation of mis-folded proteins, depending upon which checkpoints within the secretion pathway have been passed. On that basis, properly folded proteins are those which have passed scrutiny by a series of checkpoints (Vashist and Ng 2004).

Secretion of at least some heterologous proteins induces the UPR in *Aspergillus* (e.g. Wiebe et al. 2001; Mulder et al. 2004) and, as with studies in yeast, chemicals such as dithiothreitol and tunicamycin also induce the UPR in *Aspergillus*, with the dithiothreitol response being very strong. The genome sequences, together with gene chips/arrays, provide the opportunity to assess the global transcriptional responses to ER stress in *Aspergillus*. This has been achieved already with *S. cerevisiae* where there is a platform for comparative studies with *Aspergillus*. In *S. cerevisiae*, nearly 400 genes were transcriptionally up-regulated due directly or indirectly to the UPR (Travers et al. 2000), indicating that the effect is broad and has an impact on many aspects of the cellular metabolism. Although the UPR leads to a selective up-regulation of transcription of genes encoding proteins which combat stress due to mis-folded proteins, it may also be beneficial to homeostasis if genes which encode secretory proteins are selectively transcriptionally down-regulated, thus reducing the protein traffic in the secretory pathway. It has recently been shown that ER stress may lead to the transcriptional down-regulation of genes encoding at least some secreted proteins, but not non-secreted proteins in filamentous fungi (Pakula et al. 2003; Al-Sheikh et al. 2004) and probably in plants (Martinez and Chrispeels 2003). This response may be independent of the UPR, but that has to be verified. The response may compensate for the apparent lack of translational attenuation initiated by ER stress. In mammalian cells, translational attenuation occurs and is mediated through the ER trans-membrane protein PERK (reviewed recently by Kaufman 2004).

The Hac1p transcription factor of yeast (HacA in *A. niger*) is essential for the UPR. Hac1p binds to the UPR element in sensitive promoters, and leads to the UPR-mediated transcriptional up-regulation of genes encoding chaperones and foldases. Rüegsegger et al. (2001) reported translational regulation of *HAC1* mRNA upon induction of the UPR in *S. cerevisiae*. They showed that the un-spliced *HAC1* mRNA is associated with polyribosomes even though it does not lead to the synthesis of full-length protein. The expression of the un-spliced *HAC1* mRNA was already known to be post-transcriptionally regulated by an intron (and this block was removed upon excision of the intron by a non-conventional splicing event) but the mechanism was not fully appreciated. Rüegsegger et al. (2001) showed that the intron base-paired with a region of the 5′ UTR to cause stalling. Hence, despite being associated with polyribosomes, no Hac1p protein was produced. Saloheimo et al. (2003) have confirmed the existence of some RNA secondary structures in the *Hac* mRNAs from *A. nidulans* and *T. reesei* which may regulate *Hac* mRNA translation in these fungi. It has also been confirmed that HacA mediates the UPR in *A. niger* and that it up-regulates its own transcription (Mulder et al. 2004). It is possible to delete the *hacA* gene from *A. niger* (H. Mulder, personal communication) to facilitate studies on HacA dependency (as was used, for example, in *S. cerevisiae* by Travers et al. 2000). The constitutive up-regulation of *hacA* enhances the secreted yield of at least two heterologous proteins by *A. niger* (Valkonen et al. 2003). Other studies indicate that there may be an optimal level of individual foldases and chaperones for maximized secretion of any given heterologous protein (Moralejo et al. 2001; Lombrana et al. 2004) by *Aspergillus* spp. Genomic-based technologies will provide the means to establish the difference in responses which occur with the secretion of different heterologous proteins, enabling some predictive rules to be established. One factor determining the likelihood of efficient secretion of a protein is its thermodynamic stability (Kowalski et al. 1998a, b; Song et al. 2002; Kjeldsen and Pettersson 2003). Hence, gene chips can be used to investigate whether there are changes at the level of ER stress with proteins of different stabilities.

A link between nitrogen availability and the UPR has been reported in yeast by Schröder et al. (2000). Their study revealed that nitrogen levels, which themselves regulate the rate of overall translation, also lead to effects on the UPR. The splicing of *HAC1* mRNA occurs only in nitrogen-rich conditions and, therefore, UPR was suggested to play a role in nitrogen sensing. This study also showed that the range of events emanating from activation of the UPR was broader than previously appreciated. Indeed, the active form of Hac1p, which mediates the UPR, is a negative regulator of differentiation responses to nitrogen starvation, meiosis and pseudohyphal growth (Schröder et al. 2000, 2004; Kaufman et al. 2002). Repression of the early meiotic genes was shown to involve an association of Hac1p with the histone deacetylase complex in *S. cerevisiae* (Schröder et al. 2004), thus extending our knowledge of the roles and mechanisms of action of the UPR. However, we do not yet know how far analogies between yeast and *Aspergillus* can be taken. Global genomic approaches in *S. cerevisiae* have also examined other responses, e.g. to carbon sources (Kuhn et al. 2001) and other nutritional and environmental factors (Gasch et al. 2000), and can now be explored in *Aspergillus*. The genome sequences provide a basis not only for global studies using gene chips/arrays but also for proteomic studies, including protein–protein interactions in *Aspergillus*, analogous to studies in other systems which include *S. cerevisiae* (Uetz et al. 2000; Gavin et al. 2000; Han et al. 2004). Most studies with *Aspergillus* protein secretion and ER stress responses have involved a limited number of gene targets and these have been extensively reviewed (Punt et al. 2001; MacKenzie et al. 2004). A recent suggestion that the entire secretory pathway of *T. reesei* might be transcriptionally induced by ER stress (Saloheimo et al. 2004) begs investigation in *Aspergillus* which will be achievable with gene chips/arrays.

D. Polar Growth and Secretion

Aspergillus species grow in a highly polarised manner, typical of all filamentous fungi (Trinci et al. 1994). The uninucleate, haploid, asexual spore or conidium germinates to give a growth tube following a short period of isotropic growth. The hypha then enters a duplication cycle of elongation by tip growth, mitoses, septation, and lateral branch formation. Though many components and features of the mechanism used for budding in yeasts are conserved in *Aspergillus* and other filamentous fungi, hyphal compartments are multinucleate, septum formation is not followed by cell separation, and hyphal elongation does not cease during mitosis (Riquelme et al. 2003). The challenge is to understand how yeast homologues, together with gene products specific to filamentous fungi, regulate this peculiar form of growth.

A number of genes involved in polar growth and secretion have been identified prior to the completion of the genome sequences using mutant screens. In addition, some genes have been found using degenerate PCR-based approaches to screen for putative homologues of yeast polarity and secretion genes, and in some cases these have been tested for function. The completion of the genome sequences and subsequent annotation of open reading frames greatly expands the opportunities for identification of putative homologues of known secretion/polar growth genes. However, the application of the gene targeting approaches outlined above is essential to confirm function, and such work has not yet been carried out systematically on a genome-wide scale. At this time, it is possible to point out some of these putative homologues. While yeast is used as the paradigm for polar growth and secretion, and homologues of many of the relevant yeast genes can be found in filamentous fungi, this is not always the case. Some yeast polarity genes have no obvious fungal homologues, and some genes required for polar growth in fungi have no obvious yeast homologues. This emphasises the differences as well as the similarities in polar growth and secretion mechanisms.

E. Hyphal Branching and Polarity

Since filamentous fungi, unlike *S. cerevisiae*, are able to maintain more than one axis of polar growth, many questions remain about how this is achieved. While polar growth is maintained at an established tip, polarity has to be established in germinating conidia, and during branch formation from a sub-apical compartment. This suggests distinct mechanisms of polarity establishment and maintenance.

During germination, asexual spores (conidia) first grow isotropically, and then establish a point of growth for germ tube emergence. Later, a second germ tube emerges from the conidium (Harris 1999). Within growing hyphae, septation results in sub-apical compartments which each contain several nuclei, probably arrested in G_1 (Fiddy and Trinci 1976). A point of growth for a lateral branch is then established, and a new hypha emerges, resembling events during conidial germination.

One question is how these growth points in conidia and sub-apical compartments are established. *S. cerevisiae* defines the point of growth with cortical markers, and some of the BUD gene products required for this give yeast its characteristic patterns of lateral and apical budding. No clear homologues of *BUD3*, *BUD8* and *BUD9* were found in the *A. nidulans* genome, and Harris and Momany (2004) have proposed three models by which polarity might be established, including a stochastic model. In this model, establishment of a random site for polar growth would lead to suppression of other potential sites. Detailed observations on the site of lateral branch emergence (Fiddy and Trinci 1976) might favour this model, at least in *A. nidulans*. Emergence of a lateral branch results in the nuclei in the compartment re-entering the mitotic cycle. Thus, establishing the signalling networks which link polar growth with the mitotic cycle, already well studied in *S. cerevisiae*, should also be a fruitful area for investigation aided by the genome data.

F. Branching Frequency

Mutations leading to an increased branching frequency are relatively common, and result in more compact colonies on solid media. These mutants can arise during commercial fungal fermentations, and have consequences for culture viscosity, oxygen limitation, and product formation (Wiebe et al. 1996; Bocking et al. 1999). Since secretion of enzymes is believed to occur at the tips, the question arises as to whether a higher density of tips could lead to improved secretion of a biotechnological product. However, if hyperbranching results from defects in polar growth and secretion, then the opposite might be true. Some attempts have been made to answer this question, with contradictory results (Lee et al. 1998; Bocking et al. 1999).

In a limited number of cases, the genetic basis of hyperbranching mutations has been determined, indicating that they result from mutations in genes associated with polar growth of vesicle traffic. A series of temperature-sensitive hyperbranching mutants (Hbr) were isolated in *A. nidulans*, and two genes, *hbrA* and *hbrB*, isolated by complementation cloning. The *hbrA* gene encodes a homologue of the vacuolar sorting protein VPS33/SLP1 (Memmott et al. 2001), and *hbrB* encodes a protein of unknown function whose sequence is conserved only in filamentous fungi (Gatherar et al. 2004). The *cot*-1 mutant of *Neurospora crassa* results in a hyperbranching phenotype, and the gene responsible encodes a serine-threonine protein kinase (Yarden et al. 1992). Homologues of this gene have been found in many eukaryotes, including *CBK1* in *S. cerevisiae*. Full down-regulation of *hbrB* (Gatherar et al. 2004) and the *cot*-1 homologue *cotA* (Turner et al. 2004; Zarrin et al. 2005), using the conditional *alcA* promoter, leads to swollen condia (which cannot easily germinate) as well as swollen hyphae. Moderate down-regulation leads to hyperbranching. This might indicate that impairment of polar growth at hyphal tips leads indirectly to more frequent lateral branching. Many of these mutants appear to have swollen hyphae, which may result in a critical compartment volume being attained with shorter, broader compartments. This in turn would lead to an increased tip/length ratio, without changing the number of branches per compartment.

It would be interesting to identify mutations which lead to increases in branches per compartment, since this might help distinguish between the stochastic model and the cortical marker model.

V. Conclusion

The current availability of genome sequences for three *Aspergillus* spp. (*A. nidulans*, *A. fumigatus* and *A. oryzae*) has undoubtedly opened up new and quicker approaches to explore the molecular basis of protein secretion and hyphal growth. Genomes of other *Aspergilli* have also either been sequenced or this is underway (e.g. *A. niger*, *A. flavus*, *A. terreus*; see *Aspergillus* Genome Resources at http://www.ncbi.nlm.nih.gov/projects/genome/guide/aspergillus/) and, together with genome data from other fungal species, there is a major resource for post-genomic studies with fungi. Hyphal development and the secretion of proteins are important facets of the lifestyles of saprophytic and pathogenic (*A. fumigatus*) and commercially important (*A. oryzae*, *A. niger*) *Aspergillus* spp. but the molecular tools for exploiting the genome sequences are best developed in the sexual 'model' species, *A. nidulans*. We therefore anticipate that there will continue to be a role for *A. nidulans* as a model organism when investigating the biology of the less amenable *Aspergilli*.

Sequencing the genomes of several fungal species has shown that major, costly projects can be successfully completed in this manner. The annotation of *N. crassa* and *Aspergillus* genomes involved a large part of the fairly small international fungal research community working together. That is a strength which should be fostered to make the most rapid progress in the post-genomic era in order to advance the development of tools (e.g. improved vectors and methodologies for tagging, gene regulation, knockouts, better genome-wide arrays) which can then underpin specific projects in individual laboratories. A NIGMS-sponsored programme project for functional analysis of the *N. crassa* genome has begun recently (http://www.dartmouth.edu/~neurosporagenome/).

Most proteins are secreted from fungi at the hyphal tips. This form of polar development and protein secretion is unique, and occurs in filamentous fungi which have the other unique property of being comprised of multinucleate cells. The fungi are ubiquitous in most environments, cause disease and are exploited commercially. Fungi therefore provide a system of immense attraction for scientific exploration, and the genome sequences provide an unprecedented opportunity for these studies.

Acknowledgements. Genomic data for *A. fumigatus* was provided by The Institute for Genomic Research (www.tigr.org/tdb/e2k1/afu1) and The Wellcome Trust Sanger Institute (www.sanger.ac.uk/Projects/A_fumigatus); genomic data for *A. nidulans* was provided by The Broad Institute (http://www.broad.mit.edu/annotation/fungi/aspergillus/); genomic data for *A. oryzae* was provided by The National Institute of Advanced Industrial Science and Technology (http://oryzae.cbrc.jp/ and www.bio.nite.go.jp/dogan/Top). Coordination of analyses of these data was enabled by an international collaboration involving more than 50 institutions from 10 countries and coordinated from Manchester, UK (www.cadre.man.ac.uk and www.aspergillus.man.ac.uk). The authors thank the Biotechnology and Biological Sciences Research Council for supporting their research.

References

Al-Sheikh HM, Watson AJ, Lacey GA, Punt P, MacKenzie DA, Jeenes DJ, Pakula T, Penttilä M, Alcocer MJC, Archer DB (2004) Endoplasmic reticulum stress leads to the selective transcriptional downregulation of the glucoamylase gene in Aspergillus niger. Mol Microbiol 53:1731–1742

Archer DB (2000) Filamentous fungi as microbial cell factories for food use. Curr Opin Biotechnol 11:478–483
Archer DB, Dyer PS (2004) From genomics to post-genomics in Aspergillus niger. Curr Opin Microbiol 7:499–504
Archer DB, Peberdy JF (1997) The molecular biology of secreted enzyme production by fungi. CRC Crit Rev Biotechnol 17:273–306
Askenazi M, Driggers EM, Holtzman DA, Norman TC, Iverson S, Zimmer DP, Boers ME, Blomquist PR, Martinez EJ, Monreal AW et al. (2003) Integrating transcriptional and metabolite profiles to direct the engineering of lovastatin-producing fungal strains. Nat Biotechnol 21:150–156
Beffa T, Staib F, Fischer JL, Lyon PF, Gumowski P, Marfenina OE, Dunoyer-Geindre S, Georgen F, Roch-Susuki R, Gallaz L et al. (1998) Mycological control and surveillance of biological waste and compost. Med Mycol 36 suppl I:137–145
Bocking SP, Wiebe MG, Robson GD, Hansen K, Christiansen LH, Trinci AP (1999) Effect of branch frequency in Aspergillus oryzae on protein secretion and culture viscosity. Biotechnol Bioeng 65:638–648
Borkovich KA, Alex LA, Yarden O, Freitag M, Turner GE, Read ND, Seiler S, Bell-Pedersen D, Paietta J, Plesofsky N et al. (2004) Lessons from the genome sequence of Neurospora crassa: tracing the path from genomic blueprint to multicellular organism. Microbiol Mol Biol Rev 68:1–108
Boyce KJ, Hynes, MJ, Andrianopolous A (2003) Control of morphogenesis and actin localization by the Penicillium marneffei RAC homolog. J Cell Sci 116:1249–1260
Brookman JL, Denning DW (2000) Molecular genetics in Aspergillus fumigatus. Curr Opin Microbiol 3:468–474
Brown JS, Aufauvre-Brown A, Brown J, Jennings JM, Arst H Jr, Holden DW (2000) Signature-tagged and directed mutagenesis identify PABA synthetase as essential for Aspergillus fumigatus pathogenicity. Mol Microbiol 36:1371–1380
Chaveroche MK, Ghigo JM, d'Enfert C (2000) A rapid method for efficient gene replacement in the filamentous fungus Aspergillus nidulans. Nucleic Acids Res 28:E97
Clemente A, MacKenzie DA, Jeenes DJ, Domoney C (2004) The effect of variation within inhibitory domains on the activity of pea protease inhibitors from the Bowman-Birk class. Prot Exp Purif 36:106–114
Conesa A, Punt PJ, van Luijk N, van den Hondel CAMJJ (2001) The secretion pathway in filamentous fungi: a biotechnological view. Fungal Genet Biol 33:155–171
Dalboge H (1997) Expression cloning of fungal enzyme genes; a novel approach for efficient isolation of enzyme genes of industrial relevance. FEMS Microbiol Rev 21:29–42
d'Enfert C (1996) Selection of multiple disruption events in Aspergillus fumigatus using the orotidine-5′-decarboxylase gene, pyrG, as a unique transformation marker. Curr Genet 30:76–82
Derkx PMF, Madrid SM (2001) The foldase CYPB is a component of the secretory pathway of Aspergillus niger and contains the endoplasmic reticulum retention signal HEEL. Mol Genet Genomics 266:537–545
de Souza CC, Goldman MH, Goldman GH (2000) Tagging of genes involved in multidrug resistance in Aspergillus nidulans. Mol Gen Genet 263:702–711
de Vries RP, Visser J (2001) Aspergillus enzymes involved in degradation of plant cell wall polysaccharides. Microbiol Mol Biol Rev 65:497–522
de Vries RP, Jansen J, Aguilar G, Parenicova L, Joosten V, Wulfert F, Benen JAE, Visser J (2002) Expression profiling of pectinolytic genes from Aspergillus niger. FEBS Lett 530:41–47
Ellgaard L, Helenius A (2003) Quality control in the endoplasmic reticulum. Nat Rev Mol Cell Biol 4:181–191
Felenbok B, Flipphi M, Nikolaev I (2001) Ethanol catabolism in Aspergillus nidulans: a model system for studying gene regulation. Prog Nucleic Acid Res Mol Biol 69:149–204
Fernandez-Abalos JM, Fox H, Pitt C, Wells B, Doonan JH (1998) Plant-adapted green fluorescent protein is a versatile vital reporter for gene expression, protein localization and mitosis in the filamentous fungus, Aspergillus nidulans. Mol Microbiol 27:121–130
Fiddy C, Trinci APJ (1976) Mitosis, septation, branching and the duplication cycle in Aspergillus nidulans. J Gen Microbiol 97:169–184
Firon A, d'Enfert C (2002) Identifying essential genes in fungal pathogens of humans. Trends Microbiol 10:456–462
Firon A, Villalba F, Beffa R, d'Enfert C (2003) Identification of essential genes in the human fungal pathogen Aspergillus fumigatus by transposon mutagenesis. Eukaryot Cell 2:247–255
Galagan JE, Calvo SE, Borkovich KA, Selker EU, Read ND, Jaffe D, FitzHugh W, Ma LJ, Smirnov S, Purcell S et al. (2003) The genome sequence of the filamentous fungus Neurospora crassa. Nature 422:859–868
Gasch AP, Spellman PT, Kao CM, Carmel-Harel O, Eisen MB, Storz G, Botstein D, Brown PO (2000) Genomic expression programs in the response of yeast cells to environmental changes. Mol Biol Cell 11:4241–4257
Gatherar IM, Pollerman S, Dunn-Colemann N, Turner G (2004) Identification of a novel gene hbrB required for polarized growth in Aspergillus nidulans. Fungal Genet Biol 41:463–471
Gavin A-C, Bösche M, Krause R, Grandi P, Marzioch M, Bauer A, Schultz J, Rick JM, Michon A-M, Cruciat C-M et al. (2000) Functional organization of the yeast proteome by systematic analysis of protein complexes. Nature 415:141–147
Geissenhoner A, Sievers N, Brock M, Fischer R (2001) Aspergillus nidulans DigA, a potential homolog of Saccharomyces cerevisiae Pep3 (Vps18), is required for nuclear migration, mitochondrial morphology and polarized growth. Mol Genet Genomics 266:672–685
Gordon CL, Khalaj V, Ram AF, Archer DB, Brookman JL, Trinci AP, Jeenes DJ, Doonan JH, Wells B, Punt PJ et al. (2000) Glucoamylase: green fluorescent protein fusions to monitor protein secretion in Aspergillus niger. Microbiology 146:415–426
Gouka RJ, Gerk C, Hooykaas PJ, Bundock P, Musters W, Verrips CT, Groot MJ de (1999) Transformation of Aspergillus awamori by Agrobacterium tumefaciens-mediated homologous recombination. Nat Biotechnol 17:598–601

Groot MJ de, Bundock P, Hooykaas PJ, Beijersbergen AG (1998) Agrobacterium tumefaciens-mediated transformation of filamentous fungi. Nat Biotechnol 16:839–842
Guest GM, Lin X, Momany M (2004) Aspergillus nidulans RhoA is involved in polar growth, branching, and cell wall synthesis. Fungal Genet Biol 41:13–22
Guo W, Roth D, Walch-Solimena C, Novick P (1999) The exocyst is an effector for Sec4p, targeting secretory vesicles to sites of exocytosis. EMBO J 18:1071–1080
Halic M, Becker T, Pool MR, Spahn CMT, Grassucci RA, Frank J, Beckmann R (2004) Structure of the signal recognition particle interacting with the elongation-arrested ribosome. Nature 427:808–814
Han J-DJ, Bertin N, Hao T, Goldberg DS, Berriz GF, Zhang LV, Dupuy D, Walhout AJM, Cusick ME, Roth FP et al. (2004) Evidence for dynamically organized modularity in the yeast protein–protein interaction network. Nature 430:88–93
Harris SD (1999) Morphogenesis is coordinated with nuclear division in germinating Aspergillus nidulans conidiospores. Microbiology 145:2747–2756
Harris SD, Momany M (2004) Polarity in filamentous fungi: moving beyond the yeast paradigm. Fungal Genet Biol 41:391–400
Harris SD, Morrell JL, Hamer JE (1994) Identification and characterization of Aspergillus nidulans mutants defective in cytokinesis. Genetics 136:517–532
Harris SD, Hamer L, Sharpless KE, Hamer JE (1997) The Aspergillus nidulans sepA gene encodes an FH1/2 protein involved in cytokinesis and the maintenance of cellular polarity. EMBO J 16:3474–3483
Hsu SC, TerBush D, Abraham M, Guo W (2004) The exocyst complex in polarized exocytosis. Int Rev Cytol 233:243–265
Irazoqui JE, Lew DJ (2004) Polarity establishment in yeast. J Cell Sci 117:2169–2171
Jadoun J, Shadkchan Y, Osherov N (2004) Disruption of the Aspergillus fumigatus argB gene using a novel in vitro transposon-based mutagenesis approach. Curr Genet 45:235–241
Jalving R, van de Vondervoort PJ, Visser J, Schaap PJ (2000) Characterization of the kexin-like maturase of Aspergillus niger. Appl Environ Microbiol 66:363–368
Jeenes DJ, Pfaller R, Archer DB (1997) Isolation and characterisation of a novel stress-inducible PDI-family gene from Aspergillus niger. Gene 193:151–156
Johnson AE, van Waes MA (1999) The translocon: a dynamic gateway at the ER membrane. Annu Rev Cell Dev Biol 15:799–842
Kaufman RJ (2004) Regulation of mRNA translation by protein folding in the endoplasmic reticulum. Trends Biochem Sci 29:152–158
Kaufman RJ, Scheuner D, Schröder M, Shen X, Lee K, Liu CY, Arnold SM (2002) The unfolded protein response in nutrient sensing and differentiation. Nat Rev Mol Cell Biol 3:411–421
Keenan RJ, Freymann DM, Stroud RM, Walter P (2001) The signal recognition particle. Annu Rev Biochem 70:755–775
Kjeldsen T, Pettersson AF (2003) Relationship between self-association of insulin and its secretion efficiency in yeast. Prot Exp Purif 27:331–337
Kowalski JM, Parekh RN, Wittrup KD (1998a) Secretion efficiency in Saccharomyces cerevisiae of bovine pancreatic trypsin inhibitor mutants lacking disulfide bonds is correlated with thermodynamic stability. Biochemistry 37:1264–1273
Kowalski JM, Parekh RN, Mao J, Wittrup KD (1998b) Protein folding stability can determine the efficiency of escape from endoplasmic reticulum quality control. J Biol Chem 273:19453–19458
Krawchuk MD, Wahls WP (1999) High efficiency gene targeting in Schizosaccharomyces pombe using a modular, PCR-based approach with long tracts of flanking homology. Yeast 15:1419–1427
Kuhn KM, DeRisi JL, Brown PO, Sarnow P (2001) Global and specific translational regulation in the genomic response of Saccharomyces cerevisiae to a rapid transfer from a fermentable to a nonfermentable carbon source. Mol Cell Biol 21:916–927
Langfelder K, Gattung S, Brakhage AA (2002) A novel method used to delete a new Aspergillus fumigatus ABC transporter-encoding gene. Curr Genet 41:268–274
Lee IH, Walline RG, Plamann M (1998) Apolar growth of Neurospora crassa leads to increased secretion of extracellular proteins. Mol Microbiol 29:209–218
Lipschutz JH, Mostov KE (2002) Exocytosis: the many masters of the exocyst. Curr Biol 12:R212–R214
Liu L, Liu J, Qiu RX, Zhu XG, Dong ZY, Tang GM (2003) Improving heterologous gene expression in Aspergillus niger by introducing copies of protein-binding sequence containing CCAAT to the promoter. Lett Appl Microbiol 36:358–361
Lombrana M, Moralejo FJ, Pinto R, Martin JF (2004) Modulation of thaumatin secretion by Aspergillus awamori by modification of bipA gene expression. Appl Environ Microbiol 70:5145–5152
Machida M (2002) Progress of Aspergillus oryzae genomics. Adv Appl Microbiol 51:81–106
MacKenzie DA, Kraunsoe JAE, Chesshyre JA, Lowe G, Komiyama T, Fuller RS, Archer DB (1998) Aberrant processing of wild-type and mutant bovine pancreatic trypsin inhibitor secreted by Aspergillus niger. J Biotechnol 63:137–146
MacKenzie DA, Jeenes DJ, Archer DB (2004) Filamentous fungi as expression systems for heterologous proteins. In: Kuck U (ed) The Mycota II. Genetics and biotechnology, chap 15. Springer, Berlin Heidelberg New York, pp 289–315
Maras M, van Die I, Contreras R, van den Hondel CA (1999) Filamentous fungi as production organisms for glycoproteins of biomedical interest. Glycoconj J 16:99–107
Martinez IM, Chrispeels MJ (2003) Genomic analysis of the unfolded protein response in Arabidopsis shows its connection to important cellular processes. Plant Cell 15:561–576
Martinez D, Larrondo LF, Putnam N, Gelpke MDS, Huang K, Chapman J, Helfenbein KG, Ramaiya P, Detter JC, Larimer F et al. (2004) Genome sequence of the lignocellulose degrading fungus Phanerochaete chrysosporium strain RP78. Nat Biotechnol 22:695–700
McGoldrick CA, Gruver C, May GS (1995) myoA of Aspergillus nidulans encodes an essential myosin I re-

quired for secretion and polarized growth. J Cell Biol 128:577–587
Medina ML, Kiernan UA, Francisco WA (2004) Proteomic analysis of rutin-induced secreted proteins from Aspergillus flavus. Fungal Genet Biol 41:327–335
Memmott SD, Pollerman S, Burrow S, Dunn-Colemann N, Turner G (2001) A vacuolar protein sorting gene affects hyphal branching and vacuole biogenesis in Aspergillius nidulans. Genbank Accession AY064215.1 (http://www.ncbi.nih.gov/)
Minetoki T, Kumagai C, Gomi K, Kitamoto K, Takahashi K (1998) Improvement of promoter activity by the introduction of multiple copies of the conserved region III sequence, involved in the efficient expression of Aspergillus oryzae amylase-encoding genes. Appl Microbiol Biotechnol 50:459–467
Moralejo FJ, Watson AJ, Jeenes DJ, Archer DB, Martín JF (2001) A defined level of protein disulfide isomerase expression is required for optimal secretion of thaumatin by Aspergillus awamori. Mol Genet Genomics 266:246–253
Mulder HJ, Saloheimo M, Penttilä M, Madrid S (2004) The transcription factor HACA mediates the unfolded protein response in Aspergillus niger, and up-regulates its own transcription. Mol Genet Genomics 271:130–140
Ngiam C, Jeenes DJ, Archer DB (1997) Isolation and characterisation of a gene encoding protein disulfide isomerase, pdiA, from Aspergillus niger. Curr Genet 31:133–138
Ngiam C, Jeenes DJ, Punt PJ, van den Hondel CA, Archer DB (2000) Characterization of a foldase, protein disulfide isomerase A, in the protein secretory pathway of Aspergillus niger. Appl Environ Microbiol 66:775–782
Nicosia MGL, Brocard-Masson C, Demais S, Van AH, Daboussi MJ, Scazzocchio C (2001) Heterologous transposition in Aspergillus nidulans. Mol Microbiol 39:1330–1344
Oakley BR, Oakley CE, Yoon Y, Jung MK (1990) Gamma-tubulin is a component of the spindle pole body that is essential for microtubule function in Aspergillus nidulans. Cell 61:1289–1301
Osherov M, Mathew NJ, May GS (2002) Polarity-defective mutants of Aspergillus nidulans. Fungal Genet Biol 31:181–188
Pakula TM, Laxell M, Huuskonen A, Uusitalo J, Saloheimo M, Penttilä M (2003) The effects of drugs inhibiting protein secretion in the filamentous fungus Trichoderma reesei: evidence for down-regulation of genes that encode secreted proteins in the stressed cells. J Biol Chem 278:45011–45020
Patil C, Walter P (2001) Intracellular signaling from the endoplasmic reticulum to the nucleus: the unfolded protein response in yeast and mammals. Curr Opin Cell Biol 13:349–356
Pearson CL, Xu K, Sharpless KE, Harris SD (2004) MesA, a novel fungal protein required for the stabilization of polarity axes in Aspergillus nidulans. Mol Biol Cell 15:3658–3672
Peberdy JF, Wallis GLF, Archer DB (2001) Protein secretion by fungi. Appl Mycol Biotechnol 1:73–114
Pitt JI, Samson RA, Frisvad JC (2000) List of accepted species and their synonyms in the family Trichocomaceae. In: Samson RA, Pitt JI (eds) Integration of modern taxonomic methods for Penicillium encoding protein and Aspergillus classification. Harwood, The Netherlands, pp 9–10
Plath K, Wilkinson BM, Stirling CJ, Rapoport TA (2004) Interactions between Sec complex and prepro-alpha-factor during posttranslational protein transport into the endoplasmic reticulum. Mol Biol Cell 15:1–10
Pontecorvo G, Roper JA, Hemmons LM, MacDonald KD, Bufton AWJ (1953) The genetics of Aspergillus nidulans. Adv Genet 5:141–238
Pruyne D, Bretscher A (2000a) Polarization of cell growth in yeast. J Cell Sci 113:365–375
Pruyne D, Bretscher A (2000b) Polarization of cell growth in yeast. J Cell Sci 113:571–585
Punt PJ, Seiboth B, Weenink XO, van Zeijl C, Lenders M, Konetschny C, Ram AFJ, Montijn R, Kubicek CP, van den Hondel CAMJJ (2001) Identification and characterization of a family of secretion-related small GTPase-encoding genes from the filamentous fungus Aspergillus niger: a putative SEC4 homologue is not essential for growth. Mol Microbiol 41:513–525
Punt PJ, van Biezen N, Conesa A, Albers A, Mangnus J, van den Hondel CAMJJ (2002) Filamentous fungi as cell factories for heterologous protein production. Trends Biotechnol 20:200–206
Punt PJ, Drint-Kuijvenhoven A, Lokman BC, Spencer JA, van der Kamp EMC, Jeenes D, Archer DB, van den Hondel CAMJJ (2003) The role of the Aspergillus niger furin-type protease gene pclA in processing of fungal proproteins and fusion proteins. Evidence for alternative processing of recombinant (fusion-) proteins. J Biotechnol 106:23–32
Rapoport TA, Matlack KES, Plath K, Misselwitz B, Staeck O (1999) Posttranslational protein translocation across the membrane of the endoplasmic reticulum. Biol Chem 380:1143–1150
Reynaga-Pena CG, Gierz G, Bartnicki-Garcia S (1997) Analysis of the role of the Spitzenkorper in fungal morphogenesis by computer simulation of apical branching in Aspergillus niger. Proc Natl Acad Sci USA 94:9096–9101
Riach MBR, Kinghorn JR (1996) Genetic transformation and vector developments in filamentous fungi. In: Bos CJ (ed) Fungal genetics: principles and practice. Marcel Dekker, New York, pp 209–233
Riquelme M, Gierz G, Bartnicki-Garcia S (2000) Dynein and dynactin deficiencies affect the formation and function of the Spitzenkorper and distort hyphal morphogenesis of Neurospora crassa. Microbiology 146:1743–1752
Riquelme M, Fischer R, Bartnicki-Garcia S (2003) Apical growth and mitosis are independent processes in Aspergillus nidulans. Protoplasma 222:211–215
Robson GD, Huang J, Wortmann J, Archer DB (2005) A preliminary analysis of the process of protein secretion, and the diversity of putative secreted hydrolases encoded in Aspergillus fumigatus: insights from the genome. Med Mycol Supp 1 43:S41–S47
Rüegsegger U, Leber JH, Walter P (2001) Block of HAC1 mRNA translation by long-range base pairing is released by cytoplasmic splicing upon induction of the unfolded protein response. Cell 107:103–114
Rutkowski DT, Kaufman RJ (2004) A trip to the ER: coping with stress. Trends Cell Biol 14:20–28
Saloheimo M, Valkonen M, Penttilä M (2003) Activation mechanisms of the HACI-mediated unfolded protein

response in filamentous fungi. Mol Microbiol 47:1149–1161

Saloheimo M, Wang H, Valkonen M, Vasara T, Huuskonen A, Riikonen M, Pakula T, Ward M, Penttila M (2004) Characterization of secretory genes ypt1/yptA and nsf1/nsfA from two filamentous fungi: induction of secretory pathway genes of Trichoderma reesei under secretion stress. Appl Environ Microbiol 70:459–467

Schröder M, Kaufman RJ (2005) ER stress and the unfolded protein response. Mutation Res 569:29–63

Schröder M, Chang JS, Kaufman RJ (2000) The unfolded protein response represses nitrogen-starvation induced developmental differentiation in yeast. Genes Dev 14:2962–2975

Schröder M, Clark R, Kaufman RJ (2003) IRE1- and HAC1-independent transcriptional regulation in the unfolded protein response of yeast. Mol Microbiol 49:591–606

Schröder M, Clark R, Liu CY, Kaufman RJ (2004) The unfolded protein response represses differentiation through the RPD3-SIN3 histone deacetylase. EMBO J 23:2281–2292

Seiler S, Plamann M (2003) The genetic basis of cellular morphogenesis in the filamentous fungus Neurospora crassa. Mol Biol Cell 14:4352–4364

Seiler S, Plamann M, Schliwa M (1999) Kinesin and dynein mutants provide novel insights into the roles of vesicle traffic during cell morphogenesis in Neurospora. Curr Biol 9:779–785

Sharpless KE, Harris SD (2002) Functional characterization and localization of the Aspergillus nidulans formin SEPA. Mol Biol Cell 13:469–479

Shi X, Sha Y, Kaminskyj S (2004) Aspergillus nidulans hypA regulates morphogenesis through the secretion pathway. Fungal Genet Biol 41:75–88

Shuster JR, Connelley MB (1999) Promoter-tagged restriction enzyme-mediated insertion (PT-REMI) mutagenesis in Aspergillus niger. Mol Gen Genet 262:27–34

Sims AH, Robson GD, Hoyle DC, Oliver SG, Turner G, Prade RA, Russell HH, Dunn-Coleman NS, Gent ME (2004a) Use of expressed sequence tag analysis and cDNA microarrays of the filamentous fungus Aspergillus nidulans. Fungal Genet Biol 41:199–212

Sims AH, Gent ME, Robson GD, Dunn-Coleman NS, Oliver SG (2004b) Combining transcriptome data with genomic and cDNA sequence alignments to make confident functional assignments for Aspergillus nidulans genes. Mycol Res 108:853–857

Sims AH, Gent ME, Lanthaler K, Dunn-Coleman NS, Oliver SG, Robson GD (2005) Transcriptome analysis of recombinant protein secretion by Aspergillus nidulans and the unfolded-protein response in vivo. Appl Environ Microbiol 71:2737–2747

Song Y, Sakai J, Saito A, Usui M, Azakami H, Kato A (2002) Relationship between the stability of lysozymes mutated at the inside hydrophobic core and secretion in Saccharomyces cerevisiae. Nahrung 46:209–213

Spencer JA, Jeenes DJ, MacKenzie DA, Haynie DT, Archer DB (1998) Determinants of the fidelity of processing glucoamylase-lysozyme fusions in Aspergillus niger. Eur J Biochem 258:107–112

Su W, Li S, Oakley BR, Xiang X (2004) Dual-color imaging of nuclear division and mitotic spindle elongation in live cells of Aspergillus nidulans. Eukaryot Cell 3:553–556

Tavoularis S, Scazzocchio C, Sophianopoulou V (2001) Functional expression and cellular localization of a green fluorescent protein-tagged proline transporter in Aspergillus nidulans. Fungal Genet Biol 33:115–125

Teeri TT (2004) Genome sequence of an omnipotent fungus. Nat Biotechnol 22:679–680

Toews MW, Warmbold J, Konzack S, Rischitor P, Veith D, Vienken K, Vinuesa C, Wei H, Fischer R (2004) Establishment of mRFP1 as a fluorescent marker in Aspergillus nidulans and construction of expression vectors for high-throughput protein tagging using recombination in vitro (GATEWAY). Curr Genet 45:383–389

Travers KJ, Patil CK, Wodicka L, Lockhart DJ, Weissman JS, Walter P (2000) Functional and genomic analyses reveal an essential coordination between the unfolded protein response and ER-associated degradation. Cell 101:249–258

Trinci APJ, Wiebe MG, Robson GD (1994) The mycelium as an integrated entity. In: Esser K, Lemke PA (eds) The Mycota I. Growth, differentiation and sexuality. Springer, Berlin Heidelberg New York, pp 175–193

Turner G (1994) Vectors for genetic manipulation. In: Martinelli SD, Kinghorn JR (eds) Aspergillus: 50 years on, chap 24. Elsevier, Amsterdam, pp 641–665

Turner G, Safaie M, John S, Dunn-Coleman N (2004) Control of polar growth in Aspergillus nidulans by the cotA kinase. GenBank Accession AY620243.1

Uetz P, Giot L, Cagney G, Mansfield TA, Judson RS, Knight JR, Lockshon D, Narayan V, Srinivasan M, Pochart P et al. (2000) A comprehensive analysis of protein–protein interactions in Saccharomyces cerevisiae. Nature 403:623–627

Valkonen M, Ward M, Wang H, Penttilä M, Saloheimo M (2003) Improvement of foreign protein production in Aspergillus niger var. awamori by constitutive induction of the unfolded protein response. Appl Environ Microbiol 69:6979–6986

van Gemeren IA, Punt PJ, Drint-Kuyvenhoven A, Broekhuijsen MP, van Hood A, Beijersbergen A, Verrips CT, van den Hondel CAMJJ (1997) The ER chaperone-encoding bipA gene of black Aspergilli is induced by heat shock and unfolded proteins. Gene 198:43–52

Vashist S, Ng DTW (2004) Misfolded proteins are sorted by a sequential checkpoint mechanism of ER quality control. J Cell Biol 165:41–52

Wang H, Ward M (2000) Molecular characterisation of a PDI-related gene prpA in Aspergillus niger var awamori. Curr Genet 37:57–64

Wang H, Lambert J, Morlon E, Archer DB, Peberdy JF, Ward M, Jeenes DJ (2003) Isolation and characterization of a calnexin homologue, clxA, from Aspergillus niger. Mol Genet Genomics 268:684–691

Ward M (1989) Production of calf chymosin by Aspergillus awamori. In: Hershberger CL, Queener SW, Hegeman G (eds) Genetics and molecular biology of industrial microorganisms. American Society for Microbiology, Washington, DC, pp 288–294

Ward M, Lin C, Victoria DC, Fox BP, Fox JA, Wong DL, Meerman HJ, Pucci JP, Fong RB, Heng MH et al. (2004) Characterization of humanized antibodies secreted by Aspergillus niger. Appl Environ Microbiol 70:2567–2576

Waring RB, May GS, Morris NR (1989) Characterization of an inducible expression system in Aspergillus nidulans using alcA and tubulin-coding genes. Gene 79:119–130

Wendland J (2003) PCR-based methods facilitate targeted gene manipulations and cloning procedures. Curr Genet 44:115–123

Whittaker SL, Lunness P, Milward KJ, Doonan JH, Assinder SJ (1999) sodVIC is an alpha-COP-related gene which is essential for establishing and maintaining polarized growth in Aspergillus nidulans. Fungal Genet Biol 26:236–252

Wiebe MG, Blakebrough ML, Craig SH, Robson GD, Trinci AP (1996) How do highly branched (colonial) mutants of Fusarium graminearum A3/5 arise during Quorn myco-protein fermentations? Microbiology 142:525–532

Wiebe MG, Karandikar A, Robson GD, Trinci APJ, Candia JLF, Trappe S, Wallis G, Rinas U, Derkx PMF, Madrid S et al. (2001) Production of tissue plasminogen activator (t-PA) in Aspergillus niger. Biotechnol Bioeng 76:164–174

Xiang X, Beckwith SM, Morris NR (1994) Cytoplasmic dynein is involved in nuclear migration in Aspergillus nidulans. Proc Natl Acad Sci USA 91:2100–2104

Yamashita RA, Osherov N, May GS (2000) Localization of wild type and mutant class I myosin proteins in Aspergillus nidulans using GFP-fusion proteins. Cell Motility Cytoskeleton 45:163–172

Yang L, Ukil L, Osmani A, Nahm F, Davies J, De Souza CP, Dou X, Perez-Balaguer A, Osmani SA (2004) Rapid production of gene replacement constructs and generation of a green fluorescent protein-tagged centromeric marker in Aspergillus nidulans. Eukaryot Cell 3:1359–1362

Yarden O, Plamann M, Ebbole DJ, Yanofsky C (1992) cot-1, a gene required for hyphal elongation in Neurospora crassa, encodes a protein kinase. EMBO J 11:2159–2166

Yaver DS, Lamsa M, Munds R, Brown SH, Otani S, Franssen L, Johnstone JA, Brody H (2000) Using DNA-tagged mutagenesis to improve heterologous protein production in Aspergillus oryzae. Fungal Genet Biol 29:28–37

Yu JH, Hamari Z, Han KH, Seo JA, Reyes-Dominguez Y, Scazzocchio C (2004) Double-joint PCR: a PCR-based molecular tool for gene manipulations in filamentous fungi. Fungal Genet Biol 41:973–981

Zarrin M, Leeder A, Turner G (2005) A rapid method for promoter exchange in Aspergillus nidulans using recombinant PCR. Fungal Genet Biol 42:1–8

Zhang K, Kaufman RJ (2004) Signaling the unfolded protein response from the endoplasmic reticulum. J Biol Chem 279:25935–25938

6 The Genomics of Stress Response in Fission Yeast

B.T. Wilhelm[1], J. Bähler[1]

CONTENTS

Abbreviations: CESR, core environmental stress response; MAPK, MAP kinase; MAPKK, MAP kinase kinase; MMS, methyl methanesulfonate; SESR, specific environmental stress response

I. Introduction

The study of the molecular mechanisms behind stress responses in model organisms has been a rich field of research that has defined crucial aspects of cellular behaviour. Most traditional studies in this field have relied on a combination of biochemical and genetic techniques to highlight the role of a particular gene or pathway. While such approaches have the advantage of providing a well-defined set of behaviours for a particular protein, the results are often not directly comparable with other experiments performed under different conditions. The ability therefore to combine all available published data into a coherent and conclusive, overall cellular response is often difficult, if not impossible. New experimental approaches developed within the last decade have already revolutionized the way in which molecular biology is performed.

Most notably, the advent of microarrays has allowed a shift in experimental paradigms to a complementary and powerful approach for studying not only stress response, but virtually any cellular behaviour from expression differences to transcription factor binding. The ability to globally and simultaneously survey the expression level of an entire genome not only facilitates the study of transcriptional regulation but also, through the use of time-course studies, enables a regulatory network to be established. Because of their small genomes (relative to other eukaryotes) and well-defined cellular behaviours, fungi remain one of the most useful model systems for studying the transcriptional behaviour at the level of the whole genome. The other obvious advantage of surveying the entire genome using this approach is that, in addition to displaying the responses that would be predicted, whole-genome studies also highlight those changes which are not expected. Indeed, the ability to discover novel effects without a priori knowledge of the system is a key strength of microarray studies.

The use of microarrays and other genome-wide technologies for experimentation, often generically described as genomics (or functional genomics) studies, inherently generates a large amount of data. Typically, the most challenging part in these experiments is to extract biologically meaningful information from the huge mass of data. In any case, such data represent a valuable and reusable resource when new discoveries are made, and the existing datasets can be reanalysed to create connections between novel and established pathways and behaviours. However, while the accumulation of data represents a benefit, significant difficulties

[1] Wellcome Trust Sanger Institute, Hinxton, Cambridge CB10 1SA, UK

The Mycota XIII
Fungal Genomics
Alistair J.P. Brown (Ed.)

still exist in making these data available to the wider scientific community. The current situation is similar in many respects to the early days of genome sequencing before databases and common storage formats for DNA sequences were firmly established, and also before there was a requirement to submit such experimental data to databases as a condition of scientific publication. Although microarray data are clearly more complex than sequence data and standardisation remains a significant challenge, efforts to support a new MIAME reporting standard for microarray experiments, and the establishment of microarray databases, should prove highly beneficial for the field (Brazma et al. 2001, 2003; Spellman et al. 2002).

The focus of this chapter will be the genomics of stress response in the fission yeast *Schizosaccharomyces pombe*, highlighting some of the published work in this field as well as discussing the similarities and differences with the other major experimental free-living yeast, *Saccharomyces cerevisiae*. Fission yeast is a single-celled fungus that belongs to the phylum *Ascomycota* and grows as an elongating, rod-shaped cell. Prior to its introduction as a laboratory model organism, it was used in the production of millet beer in eastern Africa. The complete genome of *S. pombe* has been sequenced and published (Wood et al. 2002); it is comprised of three large chromosomes totalling approximately 13.8 Mb and a small (20 kb) mitochondrial genome. There are somewhat fewer than 5000 genes (either predicted or experimentally confirmed) in the genome although, like the genes of most other organisms whose genomes have been sequenced, a majority of these do not yet have defined cellular functions (Wood et al. 2002; Bähler and Wood 2003).

Although often compared to *S. cerevisiae*, more than 1 billion years of evolutionary divergence separates the two yeasts, according to recent estimates (Heckman et al. 2001). Fission yeast has similar experimental advantages, including a low-complexity genome, as budding yeast, which has been widely used to pioneer functional genomic approaches (Delneri et al. 2001; Kumar and Snyder 2001). Although it is impossible to declare either yeast to be the "best" experimental system, the case for studying *S. pombe* as a complementary model for the biology of multicellular eukaryotes is strong. Numerous aspects of not only the structure of the genome (large chromosomes, genes with introns, large centromeres) but also the functioning of the genome (large-number homologs for conserved cellular functions, etc.) are more similar between *S. pombe* and higher eukaryotes than is the case for *S. cerevisiae* (Forsburg 1999; Sunnerhagen 2002). As such, and although much cellular behaviour has first been studied in *S. cerevisiae*, there is still significant value in performing similar experiments in *S. pombe*. Evidence supporting this argument can be seen in previous genomic studies performed on *S. pombe* which clearly demonstrate that, despite superficial similarities in some characteristics, the two yeasts have undergone significant independent evolution in the billion years since their last common ancestor (Mata et al. 2002; Chen et al. 2003; Mata and Bähler 2003; Rustici et al. 2004).

II. Stress Response Pathways in Fission Yeast

Before discussing the role that genomics approaches are beginning to have on the study and understanding of stress response in fission yeast, it is useful to provide a brief overview of what was known before the application of newer techniques to the investigation. The study of stress response in fission yeast has benefited from earlier related research in other organisms. The discovery of two related stress-activated protein kinases (SAPKs) in budding yeast (Brewster et al. 1993) and mammals (Han et al. 1994) suggested that this pathway represented an evolutionarily conserved way of signalling stress. The subsequent discovery of a functional homolog of the Hog1/p38 protein, Sty1/Spc1, in *S. pombe* confirmed that a similar pathway exists in fission yeast (Shiozaki and Russell 1995; Degols et al. 1996). However, the activity of Hog1 functions specifically during osmotic stress, whereas Sty1 functions more broadly in response to many stresses (Toone and Jones 1998). Similarly, in mammalian cells the SAPK pathway is involved in controlling a variety of processes (e.g. immune response, development, apoptosis, cell proliferation, and DNA repair; Verheij et al. 1996; Wilkinson et al. 1996; Molnar et al. 1997; Takenaka et al. 1998; Nebreda and Porras 2000). Moreover, in both mammalian cells and fission yeast, but not in budding yeast, the SAPK pathway activates AP-1-like transcription factors (Toone and Jones 1999). Fission yeast therefore provides a valuable model to study this central regulatory pathway. Figure 6.1 shows a summary of the stress response pathway based on published work in the

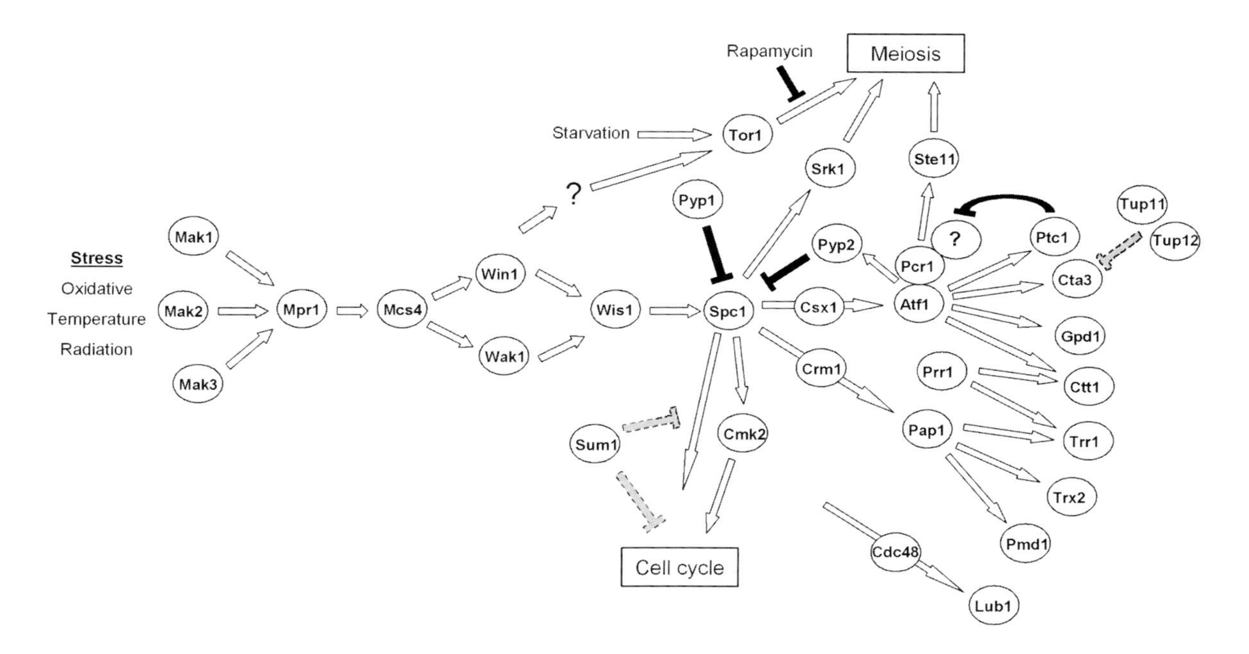

Fig. 6.1. An overview of the stress signalling pathway in fission yeast. The stress signalling cascade in *S. pombe* is shown with the various proteins being represented by named ovals. Open arrows indicate activating interactions, closed lines indicate a repressive or inhibitory interaction, and shaded dotted lines indicate interactions which are unclear. Question marks represent either unknown proteins in a pathway or the possible existence of such a protein

field. It is important to note that in some cases, the connections shown have not yet been clearly defined and therefore represent targets of further investigation, as outlined in Sect. IV below.

The initial proteins in the cascade involved in the sensing of extracellular changes are poorly defined, but the Mak proteins (also known as Phk proteins) have been shown to be initial targets for activation during oxidative, temperature and radiation stress (Nguyen et al. 2000; Aoyama et al. 2001; Buck et al. 2001). A "phosphorelay system" has been defined involving the Mak proteins, Mpr1 and Mcs4, where the sensor kinase exhibits a histidine-kinase activity and the response regulator contains a phospho-accepting aspartate in its receiver domain. The phosphorylation signal is therefore passed from sensor to acceptor and then goes on to activate either Win 1 or Wak1, depending on the level and type of stress, which results in the phosphorylated activation of Wis1, a MAPKK. The AP-1-like transcription factor Pap1 can also directly sense oxidative stress leading to nuclear accumulation (Toone et al. 1998; Castillo et al. 2003; Vivancos et al. 2004).

As illustrated in Fig. 6.1, the MAPK Sty1 is a central branch point for the stress signalling cascade in fission yeast. Despite the role Sty1 plays in stress response, it was originally cloned as a suppressor of lethality caused by the loss of protein phosphatase 2C activity (Millar et al. 1995). Subsequent research confirmed a connection for this protein in both pathways. Sty1 is inhibited by the Pyp1/2 proteins (Millar et al. 1995) where Pyp2 is transcribed as a result of Sty1 activity via Atf1 (Wilkinson et al. 1996), thus forming a negative feedback loop.

Many of the proteins involved in the stress response pathway, either as regulators such as Sty1 or as up-regulated targets such as Srk1 and Ste11, have been shown to be involved in normal cell growth processes and checkpoints (Shiozaki and Russell 1995, 1997; Gachet et al. 2001). The connection between stress response and effects on cell growth would appear fairly intuitive; cells presumably will have evolved in such a way as to block processes such as mitosis or meiosis if the metabolic environment is unsuitable. However, the evolution of secondary roles for these proteins under non-stressed growth is less clear. One simple explanation for the existence of proteins with multiple functions is that it is a means of allowing the organism to "do more with less", whereby each dual role could have evolved in either direction (i.e. assume a new role in the stress response for a protein normally involved in mitotic growth, or the reverse). Regardless of the origins of this complex behaviour, the net effect is that, because the stress response involves proteins with multiple roles under different conditions, it is not at all straightforward to unambiguously dissect the role of the proteins in the stress pathway.

III. Applications of Genomics Techniques to Stress Response

As discussed above, while the field of stress response in *S. pombe* is quite well established, the application of genomic approaches in this regard is still just beginning. Initial genomics studies have been published in the past 2–3 years and, although the amount of data currently available is not yet broadly comparable to that of other organisms, it does nevertheless provide some useful insight into the general effects of stress. The use of microarray-based studies of stress response could be classified into one of three partially overlapping categories – broadly descriptive experiments, genome-wide screening experiments, or hypothesis-driven experiments – examples of each of these types of investigations are discussed below. Although the assignment of an experiment to one of these classes is rather arbitrary, it is probably nonetheless fair to say that work in the field is now beginning to move beyond the most straightforward types of investigations (broadly descriptive) to those which are, arguably, more interesting (hypothesis driven). That being said, much "obvious" work still needs to be done, as this will provide an invaluable framework of reference for all other experiments.

A. Identifying Targets of Signalling Pathways

One of the most straightforward applications of microarray technology is to conduct a screen for expression phenotypes under certain conditions. This approach was used by Smith et al. (2002) in their investigation of the stress-induced signalling cascade in fission yeast. The Sty1/Spc1 MAPK had already been identified as a key component of the *S. pombe* stress response. As discussed above, this protein is activated by Wis1 and in turn activates at least two transcription factors, Atf1 and Pap1. Analysis of the phenotypes of *atf1* and *pap1* deletion mutants suggested that, although these factors are major effectors of the stress signal, there were effects which were specific to the *sty1* deletion mutant, suggesting that there were other downstream targets of Sty1. A microarray screen of putative regulatory genes expressed after various stresses identified the *srk1* (Sty1-regulated kinase 1) gene as being up-regulated under several stress conditions, and these results were confirmed with conventional northern blot analysis (Smith et al. 2002). Based on its similarity to the Rck2 kinase in budding yeast, which has been shown to be involved in regulating stress response through interaction with the Hog1 protein (Bilsland-Marchesan et al. 2000), the function of Srk1 was further investigated. Srk1 is in the same complex as Sty1 and, upon stress induction, the protein translocates to the nucleus (Smith et al. 2002).

This study demonstrates the basic utility of microarray experimentation in identifying hundreds of potential targets for further investigation. While this approach is often used, the study by Chen et al. (2003) discussed below represents a more comprehensive attempt to characterize the global transcriptional response to various stresses in fission yeast.

B. Identification of Core and Specific Stress Response Genes

In one of the most comprehensive genomics studies to date on the *S. pombe* stress response, the transcriptional responses to five major stresses were examined (heat shock, oxidative stress, osmotic stress, exposure to heavy metals, and DNA damage; Chen et al. 2003). This study not only analysed the commonalities between genes that were activated in different stresses, but also looked at the dynamics of the response in order to characterize the differences between transient versus sustained stress behaviour.

By using five different types of stress, Chen and co-workers were able to identify genes whose expression is altered either in specific stress responses or in all stress types. Genes whose expression was up- or down-regulated in all or most of the five conditions tested were called core environmental stress response (CESR) genes, while those that were specific to a single stress were called specific environmental stress response (SESR) genes. The study of these two groups of genes has provided useful information regarding the "stress pathway", which nicely complements what is already known regarding stress response in fission yeast.

Using conservative criteria to determine CESR genes, we generated a list of 140 genes which were up-regulated (the entire list can be found at http://www.sanger.ac.uk/PostGenomics/S_pombe; see also Fig. 6.2) and 106 genes which were down-regulated. As mentioned below, it is worth noting that a much higher number of genes were consistently up-regulated across all stress conditions tested, but failed to exceed the two-fold cut-off used in the study. The size of the subset of CESR genes should therefore be seen as a fairly conservative estimate. Of the remaining induced genes, surprisingly few belonged to classes of genes traditionally associated with stress such as heat shock proteins and proteins involved in DNA repair. Although antioxidant genes were expressed, many others were involved in carbohydrate metabolism, signalling, or transcriptional regulation. The significance of the up-regulation of these genes is not clear. Furthermore, the fact that some pseudogenes were also induced raises three possibilities: (1) that there are binding sites for stress-induced transcription factors present in the promoters of these genes, which in some cases transcribe genes which are evolutionary remnants; (2) that normal transcriptional regulation or maintenance of transcriptional repression (particularly of transposable elements and pseudogenes) are lost during the stress response; or (3) that the induction of pseudogenes has some regulatory function, as is emerging in some cases (e.g. Hirotsune et al. 2003). It is important to note that the functions of 80% of the genes in the CESR group are uncharacterized and, thus, deciding whether they directly play a role in stress response is not possible without further experiments.

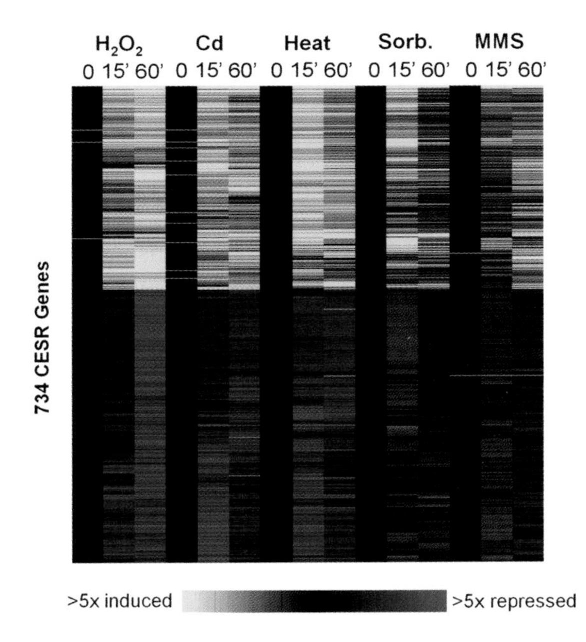

Fig. 6.2. CESR gene induction and repression during various stresses. Data from expression changes during five different stress conditions measured at 0, 15 and 60 min after stress (Chen et al. 2003). The five stresses tested include oxidative damage with hydrogen peroxide (H_2O_2), heavy metal exposure with cadmium (Cd), heat, osmotic stress with sorbitol (*Sorb.*) and DNA damage through an alkylating agent (*MMS*). The 314 induced CESR genes are depicted in *yellow* and the 420 repressed CESR genes in *blue*

The set of genes that did not exhibit universal expression changes across all stress conditions tested formed a continuum of partially overlapping and stress-specific groups. The largest observed overlap between SESR genes was in the heat and oxidative stress conditions where over 200 non-CESR genes were shared. At the other end of the spectrum, the number of genes specifically induced by only a single stress condition was generally low, with a high of 56 for oxidative stress and a low of two for DNA alkylation damage caused by MMS. Interestingly, significant overlaps exist between induced non-CESR genes in the case of MMS treatment and oxidative damage, suggesting they share a common regulator, as has been proposed by work in *S. cerevisiae* (Gasch et al. 2001). Additionally, it is not surprising that the magnitude of the stress response would depend in some cases on the strength of the initial stress, as discussed below, leading to variation in the size of gene sets induced. This dataset provides a framework for comparing the transcriptional responses in *S. pombe* cells to other external factors; for example, cisplatin induces a gene expression program that partially overlaps with the responses to both cadmium and MMS (Gatti et al. 2004). Recent microarray papers also revealed several genetic factors which lead to a stress response: significant overlaps were reported between CESR genes and (1) genes derepressed in *tup11* mutants (Fagerström-Billai and Wright 2005), (2) genes derepressed in silencing mutants (Hansen et al. 2005), and (3) genes induced in cells with shortening telomeres or circular chromosomes (Mandell et al. 2005).

Of the CESR genes that were found to be down-regulated, there was no striking pattern in terms of the classification of these genes, although enrichment for transcripts coding for proteins involved in protein synthesis, transcription and cellular signalling was noted. Indeed, only few genes (11) were found when using the same requirements for up-regulated genes being present in four of the five stresses. The requirement had to be lowered to three of five conditions to obtain a set similar in size to that of the up-regulated genes. This suggests that genes actively repressed during stress may be more specifically tailored to the individual stress. This notion is supported by the observation that the degree of transcriptional repression, but not activation, was lower in two of the stress conditions (MMS and sorbitol), indicating that strong transcriptional repression is not a mandatory feature of all stress responses. Alternatively, the strength of the stress in two cases may have been strong enough to induce a positive transcriptional response but too weak to interrupt normal cell proliferation and growth. This influence of response thresholds on transcriptional profiles has already been noted in some signalling pathways, and is also supported by own recent unpublished work.

To investigate this aspect of the stress response, a follow-up study of this work is currently underway (Chen et al., unpublished data). This new study has focussed on oxidative stress, which has been implicated in the etiology of several human diseases including cancer (Halliwell et al. 1992; Halliwell and Gutteridge 1999) and is also a major contributor to aging (Finkel and Holbrook 2000). Reactive oxygen species have deleterious effects when present in excess, but they are also critical for cell proliferation by modulating the activity of various signalling pathways (Torres 2003). The regulation of gene expression in response to various oxidants (hydrogen peroxide, menadione and t-butylhydroperoxide) and to different doses of oxidants has been carefully dissected. As suggested from previous low-scale analyses (Quinn et al. 2002), strong differences in the responses to low and high doses of oxidative stress are evident (adaptive vs. acute response). At least two signalling pathways are used to regulate these distinct responses, with some cross-talk occurring between pathways. A small core set of genes induced independently of oxidant and dose have also been identified, and these genes are all dependent on Pap1 expression. These data will help to understand how cells orchestrate and fine-tune their transcriptional programs in response to various degrees of oxidative stress.

One final aspect examined in the main global study by Chen et al. (2003) was the role of temporal aspects for gene expression in fission yeast. By examining global expression profiles at 15 and 60 min after treatment, it was possible to identify differences in transient and sustained changes in expression induced by different stresses. While sorbitol and MMS treatments induced rapid but transient stress responses, other stresses induced sustained stress responses lasting more than 60 min (Fig. 6.2). The origin of these differences is not clear, although several possible explanations exist. The ability of the cells to sense or respond to the different stresses may vary, as both sorbitol and MMS have a relatively weak effect on gene expression relative to the other stresses. Alternatively, the variation might reflect differences in how the cell deals with the stress or the time required for a stress to become effective. In such a scenario, transcription factors activated by DNA damage stress may remain active until all the DNA damage has been repaired, even though the initial stress has been removed.

As mentioned above, this study remains the most comprehensive global analysis of stress response in *S. pombe* to date. Despite this fact, the large amount of data generated and the limited space in typical journal articles, combined with a lack of available data regarding the function of unknown proteins, mean that there is still a wealth of information about stress response to be uncovered from these experiments. This prospect is further discussed at the end of this chapter in Sect. IV.

A second global study of the transcriptional stress response of fission yeast to ionising radiation provides additional insight into both general and specific stress responses (Watson et al. 2004). In these experiments, yeast cultures were exposed to 500 Gy of gamma radiation and mRNA samples were taken for microarray analysis. Approximately 200 genes were identified whose expression changed more than two fold. Of these, almost 70% were part of the fission yeast CESR genes identified in the earlier study by Chen et al. (2003). Of those genes which were not part of the CESR group, only a minority had any defined role in stress whereas the rest were mostly of metabolic or unknown function. While this may be unexpected (although similar findings have been made in *S. cerevisiae*; see references cited in Watson et al. 2004), it also highlights the power of using a microarray-based approach, as some of the transcripts coding for proteins of unknown function represent conserved proteins which may play a vital role, and which may have been overlooked in a "targeted" investigation.

Interestingly, the number of CESR genes induced by the stress regulator Sty1 during ionising radiation treatment was much lower (39%) than that in the other stresses tested (70%–90%). In addition, analysis of the dynamics of stress-induced gene expression through grouping of genes into either a fast or slow response group showed that a majority of the fast CESR induced genes are dependent on Sty1, Rad3 or both, while Chk1 was responsible for regulating the expression of more than double the number of slow CESR genes that were dependent on Sty1 or Rad3. The notion that Rad3 and Sty1 are the only stress regulators is also not supported by the finding that, in both *rad3* and *sty1* deletion mutants, *cta1* is still strongly up-regulated.

The microarray study also is consistent with the fact that Rad3 and the budding yeast protein Mec1 are structurally related to the mammalian ATM gene. The model that oxidative damage underlies some aspects of the phenotype of humans lacking ATM (resulting in the disease Ataxlia Telangiectasia) could be supported if ATM, through its similarity to the yeast stress regulators, can also regulate mammalian CESR genes (Watson et al. 2004). Such an intriguing possibility not only highlights the conservation of regulatory functions in stress responses across an enormous evolutionary timescale, but also underscores the value in using "simple" model organisms to gain insight into more complex eukaryotic organisms.

C. Identification of Mediating Factors of Global Response

In addition to providing direct evidence for the effects of stress on the fission yeast transcriptome, genomics approaches for studying stress response have also been able to help characterize events which mediate the response and, in some cases, tailor the cellular response to the stress. Previous reports had demonstrated that the stress-induced proteins Atf1 and Pap1 are responsible for a wide variety of stress responses and for low level H_2O_2/ROS damage respectively. While both of these proteins are also known to be regulated by Sty1, the mechanism by which a distinction is made between activation of either of these proteins was unknown. In a screen of genes involved in oxidative but not osmotic stress, a gene was identified that encodes the RNA-binding protein Csx1 which was essential for the survival of *S. pombe* under oxidative stress (Rodriguez-Gabriel et al. 2003). This work further demonstrated that Csx1 was capable of binding to *atf1* mRNA during oxidative stress, and that this binding strongly increased the stability of *atf1* mRNA. The stabilizing effect of Csx1 enables the Atf1-dependent expression of Pyp2, which in turn dephosphorylates Sty1 (see Fig. 6.3). This interesting demonstration of post-transcriptional regulation was not a general effect, as the transcription factors Pap1 and Prr1, which are also induced by oxidative stress, were unaffected by Csx1 levels. The application of microar-

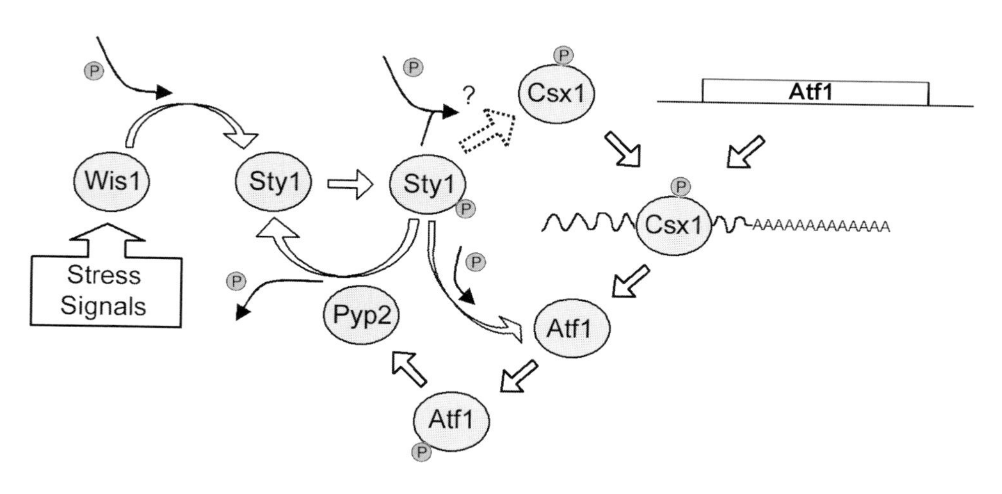

Fig. 6.3. Model of negative feedback regulation of Sty1. The proteins involved in the regulation circuit of Sty1 signalling in stress response are represented by *named ovals*. *Curved open arrows* in conjunction with *closed arrows* and *shaded circles* represent either phosphorylation or dephosphorylation events whereas the other arrows denote either binding or other predicted interactions. The *atf1* gene and mRNA are represented by a *rectangle* and *wavy line* respectively

rays to answer the question of the extent of Csx1 regulation was enlightening. Through comparisons of mRNA expression levels in various deletion strains, it was shown that virtually all genes whose expression was regulated by Atf1 are also dependent on Csx1. In addition, almost 40 genes were noted whose expression is dependent on Csx1, but not on Atf1 or its upstream regulator, Sty1.

Two other examples where microarray information led to insight in specific stress regulation were recently published. The transcription factor Zip1, which belongs to the same family as Atf1 and Pap1, was shown to be essential upon cadmium exposure and to mediate the global response to cadmium including growth arrest; in normally growing cells, this stress response is suppressed by constitutive degradation of the Zip1 protein (Harrison et al. 2005). In a second recent paper, microarray data revealed an unexpected connection between the sterol biosynthetic pathway and the fission yeast response to hypoxia: the cells monitor oxygen-dependent sterol synthesis as an indirect measure of oxygen supply to launch a hypoxic response when appropriate (Hughes et al. 2005).

As demonstrated by these studies, microarrays represent a powerful tool not only for investigating gene regulation in a direct sense, but also for rapidly putting new discoveries such as indirect regulation through mRNA stability into the context of known cellular behaviours, and for the unbiased discovery of unexpected connections.

IV. Correlations Between Budding and Fission Yeast Stress Responses

The large number of global studies of stress response as well as other cellular behaviours in budding yeast has allowed a comparison of the transcriptional programs of these similar yet divergent yeasts. As with fission yeast, there is a large set of genes (about 900) whose expression is stereotypically up-regulated (300 genes) or down-regulated (600 genes) in response to a broad range of cellular stresses (Gasch et al. 2000; Causton et al. 2001). Among the budding yeast core stress genes that contain fission yeast orthologs, many are also stereotypically induced or repressed during stress in fission yeast (Table 6.1; Chen et al. 2003). In contrast to the single signalling pathway used by fission yeast, *S. cerevisiae* responds to diverse stresses through a number of different receptor/signalling pathways to activate a similar set of genes as does *S. pombe* (Rep et al. 1999; Gasch et al. 2000; Alexandre et al. 2001; Causton et al. 2001). An additional common feature, the fact that the strength of the initial stress is proportional to the magnitude of the observed expression changes, indicates that the ability to fine-tune the stress response is either a feature which was present in the last common ancestor of the two yeasts or that it is simply under strong positive selective pressure. Given the differences in the regulation of the stress response in the two organisms, the latter seems a more likely possibility. For genes whose expression is influenced only in specific stress conditions, the use of a variety of stress-specific signalling pathways in budding yeast provides an easy explanation for stress-specific responses. In fission yeast, there is no clear mechanism yet known for the activation of such SESR genes, although such a process should exist if the available experimental data are to be satisfactorily explained.

In fission yeast, the global study of various stresses discussed in Sect. II highlighted the fact that the induction of the CESR genes is regulated through the SAPK pathway. This pathway is conserved in both budding yeast as well as in mammals (Fig. 6.4), indicating it probably had an ancient origin. In addition, the idea that the SAPK pathway is

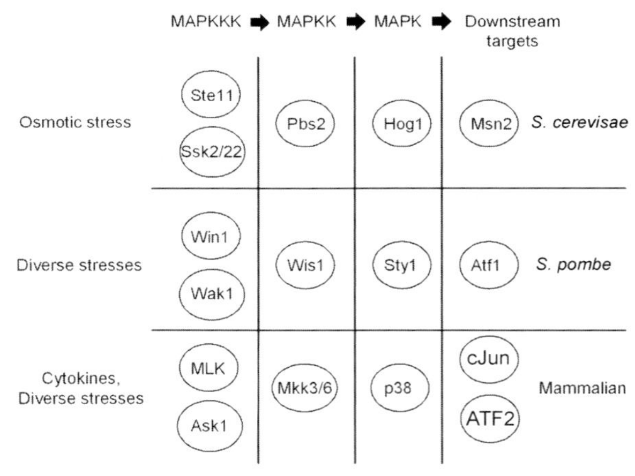

Fig. 6.4. Conservation of the SAPK signalling pathway amongst species. The components of the SAPK signalling pathways of two yeasts and mammals are shown. The *far left-hand column* indicates the initial signal which triggers the cascade while some of the downstream targets are shown in the *far right-hand column*. For the sake of clarity, only one simplified form of the signalling pathway for mammals is shown

Table. 6.1. Core set of stress-induced genes in budding and fission yeasts[a]

S. cerevisiae ortholog	*S. pombe* ortholog	Annotation
HSP12	*hsp9*	Heat shock protein, chaperone activity
HSP104	SPBC16D10.08c	Heat shock protein, chaperone activity
HSP26	*hsp16*	Heat shock protein
HSP78	SPBC4F6.17c	AAA family ATPase, mitochondrial chaperone
GLO1	*glo1*	Glyoxalase I, glutathione metabolism
GLO2	SPCC13B11.03c	Hydroxyacylglutathione hydrolase, glyoxalase II family
RIB5	*rib5*	Riboflavin synthase
YKL151C	SPCC61.03	Carbohydrate kinase family
MRF1	*mrf1*	Zinc-binding dehydrogenase
YJR096W	SPAC2F3.05c	Aldo/keto reductase, oxidoreductase
BSD2	SPAC328.07c	Metal homeostasis protein, putative membrane transporter
PNC1	SPBC365.20c	Pyrazinamidase/nicotinamidase
ATG3	SPBC3B9.06c	Involved in autophagy
TPS2	SPAC3G6.09c	Glycosyl transferase family, trehalose-phosphate synthase
SYM1	SPAC4G9.14	Member of Mpv17 and PMP22 family, peroxisomal protein
SRA1	*cgs1*	cAMP-dependent protein kinase (regulatory subunit)
YNL274C	SPACUNK4.10	Putative hydroxyacid dehydrogenase
YBR056W	*exg3*	Glucan 1,3-beta-glucosidase
GPD1	*gpd2*	Glycerol-3-phosphate dehydrogenase
GRX1	*grx1*	Glutaredoxin
GTT1	*gst3*	Glutathione transferase
TPS1	*tps1*	Trehalose-phosphate synthase
UBC8	*ubc8*	Ubiquitin conjugating enzyme
UGA2	SPAC1002.12c	Succinate-semialdehyde dehydrogenase
UBP15	*ubp22*	Protease involved in deubiquitination
YDL124W	SPAC19G12.09	Aldo-keto reductase family
YAK1	SPAC1E11.03	Protein kinase, negative regulator of cell growth
AYR1	SPAC23D3.11	1-acyl dihydroxyacetone phosphate reductase
FOX2	SPAC4H3.08	Involved in peroxisomal fatty acid beta-oxidation
PRB1	*prb1*	Vacuolar protease
MAG1	*mag1*	DNA-3-methyladenine glycosylase, involved in DNA repair
MOH1	SPAP691.02	Putative zinc-binding protein
SOD1	*sod1*	Cu, Zn-superoxide dismutase
YNL200C	SPAC15A10.05c	Conserved protein, unknown function
OXR1	SPAC8C9.16c	Conserved protein, unknown function
YJR008W	SPAC4H3.04c	Conserved protein, unknown function
YMR090W	SPBC216.03	Unknown function
YHR087W	SPBC21C3.19	Unknown function
YOL032W	SPBC21H7.06c	Unknown function
MSC1	*ish1*	Unknown function
YNL115C	SPAC23C11.06c	Unknown function
YLL023C	SPBC1539.04	Unknown function
YLR149C	SPCC4G3.03	Unknown function
YNL305C	SPCC576.04	Unknown function, similarity to apoptotic inhibitory molecule

[a]Among the genes which were stereotypically induced under stress conditions in either of two budding yeast studies (Gasch et al. 2000; Causton et al. 2001), we identified the *S. pombe* orthologs which were also induced as CESR genes in fission yeast (Chen et al. 2003)

derived from an ancestral one is also supported by findings that the overlap between genes shared in stress response is much higher than for other cellular processes such as the cell cycle and meiosis (Mata et al. 2002; Rustici et al. 2004). Interestingly, the functional role of this pathway appears to have somewhat diverged in the three species. Whereas this pathway now primarily regulates only osmotic stress response in *S. cerevisiae* via Hog1 (Cano and Mahadevan 1995), in mammals the pathway is part of a much more complex signalling network controlling normal development and function in addition to stress (reviewed in Roux and Blenis 2004). The adaptation of the SAPK signalling pathway in different species therefore represents an interesting example of evolutionary experimentation.

V. Unanswered Questions and Future Directions

Although the use of microarrays for global studies on stress response in yeast has extended our knowledge in this field, it has not yet been able to provide a framework for other aspects of yeast biology which overlap with, and definitely influence, the stress response. These areas of convergence therefore represent potential further avenues of investigation that would not only add knowledge to how yeast deal with stress, but also allow for a more integrated view of the organism which is the holy grail of systems biologists.

A. Correlations Between Transcription and Translation

Despite their powerful potential for investigation, microarrays have often been faulted as measuring only an intermediary in cell biology, namely mRNA, rather than surveying the net effect (i.e. the translated protein). While this criticism is factually valid, the impact of the implied error of this approach – that measured mRNA levels do not generally correlate with protein levels – has not been conclusively ascertained. In any case, it will ultimately be necessary to integrate data from different levels of regulation to gain a comprehensive picture of gene expression control. The study of post-transcriptional and translational regulation is an active and emerging field and, despite the technical difficulties related to this research, several groups have used similar techniques to assess the role of translational regulation (Pradet-Balade et al. 2001).

Arava and co-workers, looking at budding yeast, studied the number of ribosomes associated with various transcripts through velocity sedimentation on a sucrose gradient (Arava et al. 2003). Although the experiments were not performed in *S. pombe*, this analysis has nevertheless validated an experimental procedure for global investigations of this phenomenon. The results indicated that virtually all (>99%) mRNAs examined had a majority of their transcripts associated with one or more ribosomes and, of the small group for which this was not true, neither unusual sequence motifs nor overrepresentation by functional class were found. Despite this fact, several transcripts that are known to be regulated at the level of translation were found to be associated with only a single ribosome, supporting the use of this methodology to identify transcript targets that are translationally regulated. In addition, the study revealed a surprising finding that ribosome density decreases with increasing ORF length, although the significance of these data is not clear.

A more recent study involved a similar fractionation approach but also used details of the mRNA distribution between polysome fractions along with isotope-coded affinity tag (ICAT) to perform quantitative proteomics (MacKay et al. 2004). As in the previous study (Arava et al. 2003), most transcripts (about 80%) are associated with polysomes while those which were not, including *S. cerevisiae* genes such as *GCN4* and *HAC1*, are known to be translationally regulated. When growing cells were exposed to mating pheromone and compared to untreated cells, 816 genes of 3874 assayed were observed to change at least two fold in their protein levels. Of these, 617 (about 75%) were solely the result of changes in transcript levels, while the remaining 199 showed alterations in their translational efficiencies. Although the conclusions from this study seem to indicate that only a small number of genes showed regulation at the level of translational regulation in response to mating pheromone, it is important to note the limitations of this approach. As the authors mention, small ORFs which show enrichment in monosome fractions may not be translationally regulated, as their distribution would suggest, but may simply be limited in their polysome profile due to their short transcript length.

Along similar lines, a third paper dealing with *S. pombe* but bypassing this issue of measuring mRNA intermediates has recently been published, looking at the fission yeast response to cadmium exposure through amino acid coded mass tagging (AACT), liquid chromatography and tandem mass spectrometry (LCMS/MS; Bae and Chen 2004). Using this approach based on protein measurements, it was demonstrated that 161 proteins were up- or down-regulated (with threshold values of 1.3- and 0.78-fold respectively) in response to cadmium exposure. As approximately 400 genes were found to be induced (>two fold) under cadmium stress in earlier studies (Chen et al. 2003), this would suggest that the majority of alterations in gene transcription do not strongly influence protein concentration. These studies, however, are not directly comparable, as the experimental stress conditions used were not identical and additionally, the mRNA studies indicated that Cd^{2+} response is transient and decreases before the 1-h time point but the proteomic study did not begin observations until 1 h. It is therefore possible that faster changes in protein concentrations, due to rapid turnover of proteins, might have been overlooked. In addition, only 1133 proteins (or ~25% of all annotated genes) were actually identified in the translational study, and thus a more comprehensive study would be required for a conclusive comparison.

These findings, which generally show a limited effect of post-transcriptional regulation of mRNA on protein concentration, should not be taken to indicate that translational regulation does not represent an integral facet of overall expression regulation, along with mRNA stability and transcriptional regulation, and protein turnover. Indeed, as so little work has been done to study the global effects of this phenomenon, and even less related to stress response, it is quite possible that its importance continues to be significantly underrated. There are therefore many potential genomics-based experiments that could be performed, using approaches similar to those employed by Arava et al. (2003) and MacKay et al. (2004) where RNA in polysome preparations is hybridised on a microarray, to assess the extent of post-transcriptional regulation under different stress conditions.

B. Chromatin Remodelling in Response to Stress

There has been an enormous amount of literature published in the last decade that has highlighted the role of chromatin structure as it relates to gene expression across a wide range of eukaryotes, including fission yeast. Some of the first research related to the acetylation of histone tails created a conceptual link between changes in chromatin structure which could be correlated with transcriptional activation (Garcia-Ramirez et al. 1995). Since that time, a huge variety of histone modifications have been identified in addition to acetylation, including phosphorylation (Nowak and Corces 2000), methylation (Strahl et al. 1999), ubiquitination (Sun and Allis 2002) and sumoylation (Shiio and Eisenman 2003), and functional roles defined for a number of them (reviewed in Jenuwein and Allis 2001; Morales et al. 2001; Li 2002; Vermaak et al. 2003). This expanding body of research has resulted in the theory of a "histone code" which regulates gene expression epigenetically in all eukaryotic genomes. In fission yeast in particular, histone modifications have been shown to be involved in connecting seemingly unrelated pathways, including stress response.

The silent mating-type locus in fission yeast contains three genes, *mat1P/M*, *mat2P* and *mat3M*, where the latter two genes are transcriptionally silenced and serve as recombination material when mating-type switching occurs (Beach and Klar 1984). An investigation of this locus revealed that the transcriptionally silenced region contained histones enriched in histone H3 with methylated lysine residue 9, whereas outside this region the histones were enriched in H3 with methylated lysine residue 4 (Noma et al. 2001). Furthermore, this study demonstrated that the deletion of the inverted repeats found in the region allowed the histone modification pattern to spread along the chromosome, thereby interfering with the expression of neighbouring genes. Most intriguingly, a more recent investigation has shown that these histone modifications are actually targeted to the locus by the RNAi machinery as well as by the direct binding of the two well-characterized, stress-induced transcription factors Atf1 and Pcr1 (Jia et al. 2004).

Two recent publications also provide support for the idea that regulation of gene expression at the level of histone modifications may be a universal feature of many metabolic pathways. The first study relates to the Hog1 pathway in budding yeast. As discussed above, Hog1 is selectively involved in response to osmotic stress. In studies to characterize the connection between the transcriptional responses from MAPK signalling and histone modifications, De Nadal et al. (2004) demonstrated that Hog1 directly interacts with Rpd3, a histone deacetylase that acts mainly on histone H3 and H4 tails. The deletion of this histone deacetylase (HDAc) resulted in the reduction in expression of 80%–90% of the 222 genes previously upregulated in response to osmotic stress. The interaction of Hog1 and Rpd3 was shown to be specific for osmoresponsive promoters, and required for recruitment of the Rpd3–Sin3 deacetylase complex to these promoters.

The second study involves the well-known fission yeast stress proteins Atf1 and Pcr1. Through analysis of the chromatin structure surrounding the area of a CRE-like sequence in *ade26-M26* that is bound by the two factors, it was demonstrated that osmotic stress, but not other stresses, induces an increase in the number of micrococcal nuclease sites in the region (Hirota et al. 2004; Yamada et al. 2004). Through additional deletion experiments, evidence was presented for the involvement of Tup11 and Tup12 in regulating chromatin structure in order to tailor transcriptional responses to specific stresses (Greenall et al. 2002). Recent microarray experiments revealed that Tup11 and Tup12 have both overlapping and specialized functions in stress response regulation (Fagerström-Billai and Wright 2005). Another recent microarray study points to a potential link between stress response and chromatin remodelling: many of the genes which were derepressed in silencing mutants were activated during environmental stresses (Hansen et al. 2005), although further work is required to test whether the absence of silencing proteins directly causes this correlation or whether it reflects a secondary effect. These experiments, as well as others not discussed here, suggest that there is an important connection between stress signalling and alterations in chromatin structure leading to transcriptional reprogramming.

While these studies highlight a clear role for chromatin modification in stress response in both budding and fission yeast, there are still a vast number of histone modifications whose role is unknown. Given the wide variety of roles already defined for known histone modifications, it seems extremely likely that others could be found to be related to stress response.

C. Mapping Genomic Binding Sites of Stress Proteins

The recent emergence of a technique that combines chromatin immunoprecipitation and microarray analysis, referred to as ChIP-on-chip, has already found many uses (Ren et al. 2000; Iyer et al. 2001; Sun et al. 2003). The process involves treating cells with a cross-linking agent (usually formaldehyde) and then immunoprecipitating this DNA, after shearing, with a specific antibody against a DNA interacting protein. The DNA which is bound to the protein by the cross-linking can be recovered and used for microarray analysis. This provides a practical way to globally map the DNA binding sites for any protein that either has a suitable antibody available or has been epitope-tagged.

Many of the traditional studies of molecular pathways, such as intracellular signalling involved in stress response, often rely on indirect information. For instance, a deletion mutant may indicate that a regulatory factor lies upstream of another known transcription factor, but not whether it is immediately upstream. With the development of ChIP, it has been possible to detect the presence of a specific protein in a specific genomic location (such as the promoter of a target gene). This approach is useful but still has the severe limitations of requiring a priori knowledge of a candidate protein to immunoprecipitate, and being able to practically survey only select regions of the genome. With the powerful combination of ChIP and microarrays, it is now possible, in a series of experiments, to map the binding sites of all stress-related transcription factors across the genome. Such a collection of information, along with information already published, would make it possible to piece together a virtually complete transcriptional network of stress response. Studies of the sort in budding yeast have already been applied to define a closed loop of transcription factors that serially regulate the cell cycle (Simon et al. 2001; Lee et al. 2002). A similar approach in fission yeast under stress would likely provide great insight into the framework which previous research has established.

D. Examining Subtle Changes Associated with Stress

Lastly, all of the studies published to date have, for practical but arbitrary reasons, used a cut-off of two-fold change in expression as the level defined as unregulated. The use of such a threshold is related to the "noise" inherent in microarray experiments. It is clear, however, that many of the more subtle changes in expression that are observed will probably have an effect in concert with those more dramatic changes. These unanalysed changes also represent a potential wealth of knowledge regarding stress response. However, the effort necessary to define with certainty the changes in expression increases significantly as one looks at progressively smaller changes.

In addition to the actual experimental conditions being examined, gene expression patterns are also modulated in response to more subtle physiological variations caused by differences in growth media, temperature, and cell density. Notably, the set of previously identified CESR genes (Chen et al. 2003) show adjustments to growth conditions. For example, CESR-induced genes are expressed at a slightly higher level in minimal medium compared to full medium, and at 25 °C compared to 30 °C, and they show increasing expression once cells reach densities higher than about 10^7 cells/ml (when they are still exponentially growing). Thus, cells adapt and fine-tune their gene expression programs to adjust to even slight differences in the environment. These data indicate that a transcriptional response to, for example, environmental stress response is not an "all-or-nothing" effect but can be continuously adjusted to different needs. The sensitivity of microarray analysis can therefore be used to reveal such subtle cellular responses and complex phenotypes which are not apparent with other methods (Burns et al., unpublished data).

VI. Conclusions

The advantages provided by the use of genomics approaches to the study of fission yeast biology and, as discussed in this chapter, of stress response are numerous. It is also quite clear that while such global approaches are useful as a sort of screening tool by themselves, their maximum value can be seen when genomics studies are coupled with a well-designed, hypothesis-driven experiment, based on knowledge from traditional experimental approaches. The difference between these two approaches is equivalent to the difference between having a map that shows cities and one that shows both the cities and the roads connecting them. While it might be possible to identify interesting-looking cities without any other knowledge, knowing the roads which connect them all together provides a much deeper understanding of the underlying "biology". It is clear that genomics studies in *S. pombe* and other organisms are only at their beginning, and global approaches will reach their full potential in the coming years.

Acknowledgements. B. Wilhelm is supported by a Sanger post-doctoral fellowship, and core funding for J. Bähler is provided by Cancer Research UK. We wish to express our gratitude to Dr. J. Mata and Dr. J.-R. Landry for critical reading of this manuscript.

References

Alexandre H, Ansanay-Galeote V, Dequin S, Blondin B (2001) Global gene expression during short-term ethanol stress in *Saccharomyces cerevisiae*. FEBS Lett 498:98–103

Aoyama K, Aiba H, Mizuno T (2001) Genetic analysis of the His-to-Asp phosphorelay implicated in mitotic cell cycle control: involvement of histidine-kinase genes of *Schizosaccharomyces pombe*. Biosci Biotechnol Biochem 65:2347–2352

Arava Y, Wang Y, Storey JD, Liu CL, Brown PO, Herschlag D (2003) Genome-wide analysis of mRNA translation profiles in *Saccharomyces cerevisiae*. Proc Natl Acad Sci USA 100:3889–3894

Bae W, Chen X (2004) Proteomic study for the cellular responses to Cd^{2+} in *Schizosaccharomyces pombe* through amino acid-coded mass tagging and liquid chromatography tandem mass spectrometry. Mol Cell Proteomics 3:596–607

Bähler J, Wood V (2003) The genome and beyond. In: Egel R (ed) The molecular biology of Schizosaccharomyces pombe. Springer, Berlin Heidelberg New York, pp 13–23

Beach DH, Klar AJ (1984) Rearrangements of the transposable mating-type cassettes of fission yeast. EMBO J 3:603–610

Bilsland-Marchesan E, Arino J, Saito H, Sunnerhagen P, Posas F (2000) Rck2 kinase is a substrate for the osmotic stress-activated mitogen-activated protein kinase Hog1. Mol Cell Biol 20:3887–3895

Brazma A, Hingamp P, Quackenbush J, Sherlock G, Spellman P, Stoeckert C, Aach J, Ansorge W, Ball CA, Causton HC et al. (2001) Minimum information about a microarray experiment (MIAME) - toward standards for microarray data. Nat Genet 29:365–371

Brazma A, Parkinson H, Sarkans U, Shojatalab M, Vilo J, Abeygunawardena N, Holloway E, Kapushesky M, Kemmeren P, Lara GG et al. (2003) ArrayExpress - a public repository for microarray gene expression data at the EBI. Nucleic Acids Res 31:68–71

Brewster JL, de Valoir T, Dwyer ND, Winter E, Gustin MC (1993) An osmosensing signal transduction pathway in yeast. Science 259:1760–1763

Buck V, Quinn J, Soto Pino T, Martin H, Saldanha J, Makino K, Morgan BA, Millar JB (2001) Peroxide sensors for the fission yeast stress-activated mitogen-activated protein kinase pathway. Mol Biol Cell 12:407–419

Cano E, Mahadevan LC (1995) Parallel signal processing among mammalian MAPKs. Trends Biochem Sci 20:117–122

Castillo EA, Vivancos AP, Jones N, Ayte J, Hidalgo E (2003) *Schizosaccharomyces pombe* cells lacking the Ran-binding protein Hba1 show a multidrug resistance phenotype due to constitutive nuclear accumulation of Pap1. J Biol Chem 278:40565–40572

Causton HC, Ren B, Koh SS, Harbison CT, Kanin E, Jennings EG, Lee TI, True HL, Lander ES, Young RA (2001) Remodeling of yeast genome expression in response to environmental changes. Mol Biol Cell 12:323–337

Chen D, Toone WM, Mata J, Lyne R, Burns G, Kivinen K, Brazma A, Jones N, Bähler J (2003) Global transcriptional responses of fission yeast to environmental stress. Mol Biol Cell 14:214–229

Degols G, Shiozaki K, Russell P (1996) Activation and regulation of the Spc1 stress-activated protein kinase in *Schizosaccharomyces pombe*. Mol Cell Biol 16:2870–2877

Delneri D, Brancia FL, Oliver SG (2001) Towards a truly integrative biology through the functional genomics of yeast. Curr Opin Biotechnol 12:87–91
De Nadal E, Zapater M, Alepuz PM, Sumoy L, Mas G, Posas F (2004) The MAPK Hog1 recruits Rpd3 histone deacetylase to activate osmoresponsive genes. Nature 427:370–374
Fagerström-Billai F, Wright APH (2005) Functional comparison of the Tup11 and Tup12 transcriptional corepressors in fission yeast. Mol Cell Biol 25:716–727
Finkel T, Holbrook NJ (2000) Oxidants, oxidative stress and the biology of ageing. Nature 408:239–247
Forsburg SL (1999) The best yeast? Trends Genet 15:340–344
Gachet Y, Tournier S, Millar JB, Hyams JS (2001) A MAP kinase-dependent actin checkpoint ensures proper spindle orientation in fission yeast. Nature 412:352–355
Garcia-Ramirez M, Rocchini C, Ausio J (1995) Modulation of chromatin folding by histone acetylation. J Biol Chem 270:17923–17928
Gasch AP, Spellman PT, Kao CM, Carmel-Harel O, Eisen MB, Storz G, Botstein D, Brown PO (2000) Genomic expression programs in the response of yeast cells to environmental changes. Mol Biol Cell 11:4241–4257
Gasch AP, Huang M, Metzner S, Botstein D, Elledge SJ, Brown PO (2001) Genomic expression responses to DNA-damaging agents and the regulatory role of the yeast ATR homolog Mec1p. Mol Biol Cell 12:2987–3003
Gatti L, Chen D, Beretta GL, Rustici G, Carenini N, Corna E, Colangelo D, Zunino F, Bähler J, Perego P (2004) Global gene expression of fission yeast in response to cisplatin. Cell Mol Life Sci 61:2253–2263
Greenall A, Hadcroft AP, Malakasi P, Jones N, Morgan BA, Hoffman CS, Whitehall SK (2002) Role of fission yeast Tup1-like repressors and Prr1 transcription factor in response to salt stress. Mol Biol Cell 13:2977–2989
Halliwell B, Gutteridge JMC (1999) Free radicals in biology and medicine. Oxford University Press, Oxford Halliwell B, Gutteridge JM, Cross CE (1992) Free radicals, antioxidants, and human disease: where are we now? J Lab Clin Med 119:598–620
Han J, Lee JD, Bibbs L, Ulevitch RJ (1994) A MAP kinase targeted by endotoxin and hyperosmolarity in mammalian cells. Science 265:808–811
Hansen KR, Burns G, Mata J, Volpe TA, Martienssen RA, Bähler J, Thon G (2005) Global effects on gene expression in fission yeast by silencingand RNA interference machineries. Mol Cell Biol 25:590–601
Harrison C, Katayama S, Dhut S, Chen D, Jones N, Bähler J, Toda T (2005) SCFPof1-ubiquitin and its target Zip1 transcription factor mediate cadmium response in fission yeast. EMBO J 24:599–610
Heckman DS, Geiser DM, Eidell BR, Stauffer RL, Kardos NL, Hedges SB (2001) Molecular evidence for the early colonization of land by fungi and plants. Science 293:1129–1133
Hirota K, Hasemi T, Yamada T, Mizuno KI, Hoffman CS, Shibata T, Ohta K (2004) Fission yeast global repressors regulate the specificity of chromatin alteration in response to distinct environmental stresses. Nucleic Acids Res 32:855–862
Hirotsune S, Yoshida N, Chen A, Garrett L, Sugiyama F, Takahashi S, Yagami K, Wynshaw-Boris A, Yoshiki A (2003) An expressed pseudogene regulates the messenger-RNA stability of its homologous coding gene. Nature 423:91–96
Hughes AL, Todd BL, Espenshade PJ (2005) SREBP pathway responds to sterols and functions as an oxygen sensor in fission yeast. Cell 120:831–842
Iyer VR, Horak CE, Scafe CS, Botstein D, Snyder M, Brown PO (2001) Genomic binding sites of the yeast cell-cycle transcription factors SBF and MBF. Nature 409:533–538
Jenuwein T, Allis CD (2001) Translating the histone code. Science 293:1074–1080
Jia S, Noma K, Grewal SI (2004) RNAi-independent heterochromatin nucleation by the stress-activated ATF/CREB family proteins. Science 304:1971–1976
Kumar A, Snyder M (2001) Emerging technologies in yeast genomics. Nat Rev Genet 2:302–312
Lee TI, Rinaldi NJ, Robert F, Odom DT, Bar-Joseph Z, Gerber GK, Hannett NM, Harbison CT, Thompson CM, Simon I et al. (2002) Transcriptional regulatory networks in *Saccharomyces cerevisiae*. Science 298:799–804
Li E (2002) Chromatin modification and epigenetic reprogramming in mammalian development. Nat Rev Genet 3:662–673
MacKay VL, Li X, Flory MR, Turcott E, Law GL, Serikawa KA, Xu XL, Lee H, Goodlett DR, Aebersold R et al. (2004) Gene expression analyzed by high-resolution state array analysis and quantitative proteomics: response of yeast to mating pheromone. Mol Cell Proteomics 3:478–489
Mandell JG, Goodrich KJ, Bähler J, Cech TR (2005) Expression of a recQ helicase homolog affects progression through crisis in fission yeast lacking telomerase. J Biol Chem 280:5249–5257
Mata J, Bähler J (2003) Correlations between gene expression and gene conservation in fission yeast. Genome Res 13:2686–2690
Mata J, Lyne R, Burns G, Bähler J (2002) The transcriptional program of meiosis and sporulation in fission yeast. Nat Genet 32:143–147
Millar JB, Buck V, Wilkinson MG (1995) Pyp1 and Pyp2 PTPases dephosphorylate an osmosensing MAP kinase controlling cell size at division in fission yeast. Genes Dev 9:2117–2130
Molnar A, Theodoras AM, Zon LI, Kyriakis JM (1997) Cdc42Hs, but not Rac1, inhibits serum-stimulated cell cycle progression at G1/S through a mechanism requiring p38/RK. J Biol Chem 272:13229–13235
Morales V, Giamarchi C, Chailleux C, Moro F, Marsaud V, Le Ricousse S, Richard-Foy H (2001) Chromatin structure and dynamics: functional implications. Biochimie 83:1029–1039
Nebreda AR, Porras A (2000) p38 MAP kinases: beyond the stress response. Trends Biochem Sci 25:257–260
Nguyen AN, Lee A, Place W, Shiozaki K (2000) Multistep phosphorelay proteins transmit oxidative stress signals to the fission yeast stress-activated protein kinase. Mol Biol Cell 11:1169–1181
Noma K, Allis CD, Grewal SI (2001) Transitions in distinct histone H3 methylation patterns at the heterochromatin domain boundaries. Science 293:1150–1155
Nowak SJ, Corces VG (2000) Phosphorylation of histone H3 correlates with transcriptionally active loci. Genes Dev 14:3003–3013

Pradet-Balade B, Boulme F, Beug H, Mullner EW, Garcia-Sanz JA (2001) Translation control: bridging the gap between genomics and proteomics? Trends Biochem Sci 26:225–229

Quinn J, Findlay VJ, Dawson K, Millar JB, Jones N, Morgan BA, Toone WM (2002) Distinct regulatory proteins control the graded transcriptional response to increasing H(2)O(2) levels in fission yeast *Schizosaccharomyces pombe*. Mol Biol Cell 13:805–816

Ren B, Robert F, Wyrick JJ, Aparicio O, Jennings EG, Simon I, Zeitlinger J, Schreiber J, Hannett N, Kanin E et al. (2000) Genome-wide location and function of DNA binding proteins. Science 290:2306–2309

Rep M, Reiser V, Gartner U, Thevelein JM, Hohmann S, Ammerer G, Ruis H (1999) Osmotic stress-induced gene expression in *Saccharomyces cerevisiae* requires Msn1p and the novel nuclear factor Hot1p. Mol Cell Biol 19:5474–5485

Rodriguez-Gabriel MA, Burns G, McDonald WH, Martin V, Yates JR III, Bähler J, Russell P (2003) RNA-binding protein Csx1 mediates global control of gene expression in response to oxidative stress. EMBO J 22:6256–6266

Roux PP, Blenis J (2004) ERK and p38 MAPK-activated protein kinases: a family of protein kinases with diverse biological functions. Microbiol Mol Biol Rev 68:320–344

Rustici G, Mata J, Kivinen K, Lio P, Penkett CJ, Burns G, Hayles J, Brazma A, Nurse P, Bähler J (2004) Periodic gene expression program of the fission yeast cell cycle. Nat Genet 36:809–817

Shiio Y, Eisenman RN (2003) Histone sumoylation is associated with transcriptional repression. Proc Natl Acad Sci USA 100:13225–13230

Shiozaki K, Russell P (1995) Cell-cycle control linked to extracellular environment by MAP kinase pathway in fission yeast. Nature 378:739–743

Shiozaki K, Russell P (1997) Stress-activated protein kinase pathway in cell cycle control of fission yeast. Methods Enzymol 283:506–520

Simon I, Barnett J, Hannett N, Harbison CT, Rinaldi NJ, Volkert TL, Wyrick JJ, Zeitlinger J, Gifford DK, Jaakkola TS et al. (2001) Serial regulation of transcriptional regulators in the yeast cell cycle. Cell 106:697–708

Smith DA, Toone WM, Chen D, Bähler J, Jones N, Morgan BA, Quinn J (2002) The Srk1 protein kinase is a target for the Sty1 stress-activated MAPK in fission yeast. J Biol Chem 277:33411–33421

Spellman PT, Miller M, Stewart J, Troup C, Sarkans U, Chervitz S, Bernhart D, Sherlock G, Ball C, Lepage M et al. (2002) Design and implementation of microarray gene expression markup language (MAGE-ML). Genome Biol 3:research0046

Strahl BD, Ohba R, Cook RG, Allis CD (1999) Methylation of histone H3 at lysine 4 is highly conserved and correlates with transcriptionally active nuclei in Tetrahymena. Proc Natl Acad Sci USA 96:14967–14972

Sun ZW, Allis CD (2002) Ubiquitination of histone H2B regulates H3 methylation and gene silencing in yeast. Nature 418:104–108

Sun LV, Chen L, Greil F, Negre N, Li TR, Cavalli G, Zhao H, Van Steensel B, White KP (2003) Protein-DNA interaction mapping using genomic tiling path microarrays in Drosophila. Proc Natl Acad Sci USA 100:9428–9433

Sunnerhagen P (2002) Prospects for functional genomics in *Schizosaccharomyces pombe*. Curr Genet 42:73–84

Takenaka K, Moriguchi T, Nishida E (1998) Activation of the protein kinase p38 in the spindle assembly checkpoint and mitotic arrest. Science 280:599–602

Toone WM, Jones N (1998) Stress-activated signalling pathways in yeast. Genes Cells 3:485–498

Toone WM, Jones N (1999) AP-1 transcription factors in yeast. Curr Opin Genet Dev 9:55–61

Toone WM, Kuge S, Samuels M, Morgan BA, Toda T, Jones N (1998) Regulation of the fission yeast transcription factor Pap1 by oxidative stress: requirement for the nuclear export factor Crm1 (Exportin) and the stress-activated MAP kinase Sty1/Spc1. Genes Dev 12:1453–1463

Torres M (2003) Mitogen-activated protein kinase pathways in redox signaling. Front Biosci 8:d369–391

Verheij M, Bose R, Lin XH, Yao B, Jarvis WD, Grant S, Birrer MJ, Szabo E, Zon LI, Kyriakis JM et al. (1996) Requirement for ceramide-initiated SAPK/JNK signalling in stress-induced apoptosis. Nature 380:75–79

Vermaak D, Ahmad K, Henikoff S (2003) Maintenance of chromatin states: an open-and-shut case. Curr Opin Cell Biol 15:266–274

Vivancos AP, Castillo EA, Jones N, Ayte J, Hidalgo E (2004) Activation of the redox sensor Pap1 by hydrogen peroxide requires modulation of the intracellular oxidant concentration. Mol Microbiol 52:1427–1435

Watson A, Mata J, Bähler J, Carr A, Humphrey T (2004) Global gene expression responses of fission yeast to ionizing radiation. Mol Biol Cell 15:851–860

Wilkinson MG, Samuels M, Takeda T, Toone WM, Shieh JC, Toda T, Millar JB, Jones N (1996) The Atf1 transcription factor is a target for the Sty1 stress-activated MAP kinase pathway in fission yeast. Genes Dev 10:2289–2301

Wood V, Gwilliam R, Rajandream MA, Lyne M, Lyne R, Stewart A, Sgouros J, Peat N, Hayles J, Baker S et al. (2002) The genome sequence of Schizosaccharomyces pombe. Nature 415:871–880

Yamada T, Mizuno KI, Hirota K, Kon N, Wahls WP, Hartsuiker E, Murofushi H, Shibata T, Ohta K (2004) Roles of histone acetylation and chromatin remodeling factor in a meiotic recombination hotspot. EMBO J 23:1792–1803

7 Programmed Cell Death and Apoptosis in Fungi

M. Ramsdale[1]

CONTENTS

Abbreviations: cAMP, adenosine 3′,5′ cyclic monophophate; CDRE, calcineurin-dependent response element; Cyt *c*, cytochrome *c*; dsDNA, double-stranded DNA; ER, endoplasmic reticulum; EM, electron microscopy; FACS, fluorescence activated cell sorter; PCD, programmed cell death; PI, propidium iodide; PMN, piece-meal microautophagy of the nucleus; ROS, reactive oxygen species; STRE, general stress response element; TCA, tricarboxylic acid; TUNEL, terminal deoxynucleotidyl transferase-mediated dUTP nick-end labelling; VDAC, voltage-dependent anion channel; YRE, Yap1 response element

I. Introduction

Programmed cell death (PCD) has been described in nearly all of the major life forms including eubacteria (Yarmolinsky 1995; Hochman 1997; Engelberg-Kulka and Glaser 1999; Matsuyama et al. 1999; Lewis 2000), plants (Lam 2004), protists (Ameisen 1996) and animals (Golstein et al. 2003). To date, PCD has not been described in the archaebacteria, but this may reflect a failure to examine this issue, rather than its absence. The ubiquitous nature of PCD hints either at its ancient

[1] Aberdeen Fungal Group, School of Medical Sciences, Institute of Medical Sciences, University of Aberdeen, Foresterhill, Aberdeen AB25 2ZD, UK

The Mycota XIII
Fungal Genomics
Alistair J.P. Brown (Ed.)

origin, or that it is an aspect of the life of both unicellular and multicellular organisms which is so important that it has evolved many times over. In either case, it is apparent that it plays an important part in the life histories of most organisms.

Differences in the nature of death responses are often reflected in basic differences in the cell biology of the organisms under consideration. Whilst many have questioned that PCD as a notion has any relevance to the biology of unicellular organisms (Fraser and James 1998), it is now becoming apparent that at a deep-rooted level, related molecules are taking part in the cell death decisions of organisms as diverse as bacteria, yeast, plants, worms, flies and man. The study of these molecules and the contributions they make to cell death processes in fungi is the key focus of this review.

To date, only a small number of studies have utilized genomic technologies such as transcript profiling in assessments of fungal cell death (e.g. Allen et al. 2003; Orlandi et al. 2004). Furthermore, no studies have yet been published which use proteomic technologies to investigate fungal PCD. Consequently, this review largely focuses on what we do know about the biology and genetics of fungal programmed cell death, in the hope of stimulating interest in this field amongst the wider mycological community. Both genomics and transcript profiling have been used successfully in the dissection of the molecular events which accompany programmed cell death responses in animals, and these technologies could be very usefully employed to reveal the molecular basis of such events in fungi. Identifying the endogenous molecular switches which trigger and execute PCD responses in animals has been of enormous interest to the wider medical research community. Knowledge of the similarities and differences between animal and fungal PCD might be conveniently exploited in the search for novel antifungal agents or targets. In order to appreciate what is known about the fungal death response, it is meaningful to start with an overview of cell death in animals.

A. Programmed Cell Death in Model Organisms

In animals, various forms of PCD are used to eliminate cells during the course of development, to eradicate cells which have undergone cancerous transformations, or which are damaged by microbial infection, mutagens, oxidants and heat (Metzstein et al. 1998). Type I mammalian PCD responses (including apoptosis) are defined by a set of morphological and biochemical hallmarks which include a rapid loss of mitochondrial membrane potential, externalisation of the phospholipid phosphatidylserine on the plasma membrane, DNA fragmentation and cell shrinkage. Later stages of apoptosis include the blebbing of the plasma membrane and the enclosure of organelles in membrane-bound sacs. In numerous situations, cell suicide involves type II, or autophagic, cell death which is characterized by the targeting of organelles (golgi, polyribosomes and ER) to the lysosome prior to the onset of nuclear pyknosis.

At a molecular level, these events are mediated by either internal (intrinsic) or external (extrinsic) cell suicide programs. In the intrinsic death pathway, death signals induce the release of mitochondrial proteins such as cytochrome *c* (Cyt *c*; Liu et al. 1996) and Smac/Diablo (Du et al. 2000; Verhagen et al. 2000), through a poorly understood mechanism which may involve the formation of pores by pro-apoptotic BCL-2-family proteins and/or the opening of the permeability transition pore. Once in the cytosol, Smac/Diablo releases inhibitor-of-apoptosis-protein (IAP)-mediated inhibition of procaspases, and Cyt *c* forms a complex with Apaf-1 and procaspase-9. Activation of procaspases-9 results in an amplification of the caspase cascade, cleavage of a range of intracellular substrates and ultimately cell death (Liu et al. 1996; Li et al. 1997; Zou et al. 1997). In the extrinsic pathway, signals are mediated by death receptors of the TNF receptor superfamily which can activate the caspase cascade more directly. Caspases are regarded as the key to apoptotic death responses in mammals because their activation represents a point of no return.

Early bioinformatic screens of fungal genomes revealed very few pro- or anti-apoptotic sequences which might be expected to participate in a fungal PCD response (Fraser and James 1998). Whilst this was initially disappointing, it does increase the feasibility of developing fungus-specific drug therapies. As a result of more complex bioinformatic screens and the functional analysis of many genes in fungi, it appears that some overlap may exist. Indeed, caspase-related, BCL-2-related and AIF-like molecules have recently been identified (Madeo et al. 2002a). Furthermore, it is now apparent that suicide pathways exist in animals which are caspase-independent and involve the translocation of AIF from the mitochondrion to the nucleus, particularly in response to reactive oxygen species (ROS)-induced stress (Lockshin and Zakeri 2004). Evidently, there is more than one way to die.

II. Fungal Cell Death Responses

Death events in fungi can be instigated by harsh physical and chemical insults, or can be developmentally regulated. The organised disintegration of yeast cells and hyphae is linked with phenomena such as autophagy and autolysis, aging/senescence (both chronological and replicative), somatic and sexual incompatibility, hyphal interference, spore discharge mechanisms and the sculpturing of complex tissues during fruit-body formation.

A. Autophagy

Autophagy has been described in numerous physiological forms of cell death, including during insect development (e.g. insect metamorphosis), mammalian embryogenesis and human neurodegenerative disease (Cataldo et al. 1995; Anglade et al. 1997; Migheli et al. 1997). Starvation-induced sorocarp formation in *Dictyostelium discoideum* has also been linked to autophagic cell death (Cornillon et al. 1994). Autophagy has also been demonstrated in filamentous fungi, both during nutrient starvation (Dementhon et al. 2003) and also as part of the heterokaryon incompatibility response (Pinan-Lucarre et al. 2003). Autophagy is distinct from apoptosis but does appear to be required for its efficient implementation (Klionsky and Emr 2000; Uchiyama 2001).

Since autophagy appears to be highly conserved in animals, plants and fungi (Klionsky and Emr 2000), studies of autophagy in yeast have been important in elucidating the molecular processes involved with the onset and completion of autophagic processes. Autophagy involves the sequestration of organelles in a double membrane-enclosed vesicle termed the autophagosome (Klionsky and Ohsumi 1999). The autophagosomes, under the control of the cytoskeleton, fuse to the vacuolar membrane and then discharge their contents into the vacuolar lumen. Several mutants defective in autophagy have been identified.

Signalling of autophagic events requires the TOR2 and Ras/protein kinase A/cAMP pathways as well as Apg1 (a serine/threonine protein kinase) in complex with Apg13, Apg17, Vac8 and Cvt9 (Kamada et al. 2000). Protein conjugation requires Apg5, Apg7, Apg10 and Apg12, which are involved with ubiquitin-like interactions maintained in a complex with Apg16 (a coiled-coil protein which binds to Apg5). Autophagosome size is determined by Aut2 and Aut7, whilst docking of the autophagosomes requires Vam3, Vam7, Vps11, Vps16, Vps18, Vps33, Vps39 and Vps41, along with a functional Rab GTPase encoded by YPT7. Degradation of autophagosome contents is mediated by a lipase (Cvt18), the protease Prb1 and requires a number of vacuolar ATPases (*VMA*-encoded genes; see Klionsky and Emr 2000).

Piecemeal microautophagy of the nucleus (PMN) has recently been described in *Saccharomyces cerevisiae* by Roberts et al. (2003). Whilst the nucleus has not previously been considered a target for autophagy because it is an essential organelle, it appears that nuclear blebs are pinched from the nucleus into the vacuolar lumen where they can be degraded. PMN permits selective degradation of nonessential nuclear components whilst preserving chromatin, which is selectively excluded from the blebs. PMN occurs at nucleus–vacuole (NV) junctions, which are formed through physical interactions between Vac8 in the vacuole membrane and Nvj1 in the nuclear envelope. The link between PMN and death responses needs further exploration, but it could be important as a model for autophagy and apoptotic blebbing in higher eukaryotes.

Both Ras and Tor signals may be involved with the regulation of fungal autophagic death responses. Budovskaya et al. (2004) showed that the Ras/protein kinase A signalling pathway plays an important role in regulating the onset of autophagy in *Saccharomyces cerevisiae*. Increased signalling through Ras leads to a block in autophagy, and also inhibits the cytoplasm to vacuole targeting (CVT) pathway which is responsible for the delivery of a specific subset of vacuolar proteins in growing cells. These findings contrast with the situation in mammals where overexpression of Ras results in caspase-independent autophagic cell death (Chi et al. 1999), emphasizing the differences between mammals and fungi. TOR signalling is required for normal autophagy and stationary-phase entry (Cutler et al. 2001), and is also linked to the senescence phenotypes of *Podospora anserina* (Dementhon et al. 2003).

B. Autolysis

In *Penicillium chrysogenum*, carbon-limited cultures undergo autolysis which is associated with the production of ROS (Sámi et al. 2001). Autolysis is also a striking feature of foraging mycelia, which reallocate resources from one part of the mycelium so that they can invest it in the growth of another part (Cooke and Rayner 1984). Autolysis is clearly linked to nutritional deprivation in fungi, and may represent one extreme of the autophagic death response described above.

Mousavi and Robson (2003) demonstrated that autolytic cultures of *Aspergillus fumigatus* displayed several characteristic features of apoptotic cells, including annexin-V binding to the outer plasma membrane and TUNEL staining of nuclear material. Moreover, they were able to show that autolysis was an active process which could be inhibited by the addition of cycloheximide to cultures, and that the apoptotic phenotype itself could also be abolished by the addition of the pan-caspase inhibitor Z-FAD-fmk. Caspase-1- and -8-like activities (but not caspase-3-like activity) could be detected in autolysing cultures using *p*-nitroaniline-tagged peptide substrates, emphasizing the link to animal apoptosis.

C. Aging

The biology and genetics of aging/senescence in filamentous fungi have recently been reviewed in this series and elsewhere (Osiewacz 2004; Osiewacz and Hamann 2006); it is therefore not necessary to duplicate these efforts here. A number of studies on yeasts have, however, highlighted the close link between programmed cell death responses and aging and are therefore briefly covered in this review.

Two types of aging are recognized in yeasts – replicative aging and chronological aging. The death of aged cells is considered 'useful' because it limits the spread of deleterious mutations and other 'acquired' characteristics such as oxidized proteins which preferentially segregate in the mother cells (Lai et al. 2002).

Replicative aging is measured as the number of cell divisions which can be undertaken by a mother cell before it ceases to divide. Typically, any one mother cell can divide about 25–35 times before the onset of senescence (Jazwinski 1993). Genetic alterations and environmental changes which increase levels of ROS lead to a shortening of the lifespan of mother cells. Deletion of superoxide dismutases or catalases (Barker et al. 1999; Wawryn et al. 1999), along with exposure to oxygen at elevated or reduced partial pressures or the addition of glutathione, all have the 'expected' effects on lifespan (Nestelbacher et al. 2000), consistent with the oxygen theory of aging (Harman 1962).

Laun et al. (2001) found that aged mother cells of *S. cerevisiae* contained ROS of mitochondrial origin, and that older cells displayed markers of apoptosis including positive TUNEL labeling, annexin-V binding and chromatin fragmentation. The accumulation of mini-rDNA circles which has been described in aged yeasts (Sinclair and Guarente 1997) could also be attributed to ROS-induced alterations in the activity of DNA damage repair genes – a hypothesis which should be readily testable with the large bank of DNA damage repair mutants available in yeast.

Chronological aging sets a limit to the time a culture of cells can remain viable after the onset of the stationary phase of growth. Terminal cells in aged yeast populations exhibit phenotypic characteristics reminiscent of those seen during mammalian apoptosis, and chronological longevity of yeast cultures is temporarily extended by deleting the yeast metacaspase YCA1 (Laun et al. 2001; Herker et al. 2004).

D. Somatic Incompatibility

Many fungi undergo a vegetative or heterokaryon incompatibility response (Leslie and Zeller 1996; Glass et al. 2000) which has been likened to apoptosis in animals (Aylmore and Todd 1986; Ainsworth et al. 1990; Jacobson et al. 1998; Ramsdale 1998). PCD in this situation is linked to the recognition of non-self (indicated by allelic differences at so-called vegetative or heterokaryon incompatibility loci), which may operate to prevent non-self takeover reactions (Rayner 1991) or to prevent the spread of harmful genetic elements such as viruses (Hartl et al 1975; Anagnostakis and Day 1979; Debets et al 1994; van Diepeningen et al. 1997). Although plasmids, mitochondria and viruses can be spread between incompatible mycelia, this usually occurs at a much reduced frequency when compared to compatible reactions (Cortesi et al. 2001; Biella et al. 2002).

During incompatibility responses, hyphae progress through a series of stages marked by cytoplasmic condensation and granulation, vacuolization, and shrinkage of the plasma membrane away from the hyphal wall. Somatic incompatibility is often associated with the generation of ROS (possibly through the action of phenoloxidase enzyme systems such as laccase) and an increase in proteolytic activity (Esser and Blaich 1994). EM studies of rejection responses reveal nuclear degradation (which is often restricted to one partner in an interaction; Aylmore and Todd 1986) and the accumulation of vesicular bodies. Marek et al. (2003) examined the cytology of the heterokaryon incompatibility responses and senescence in *Neurospora crassa* and were able to demonstrate TUNEL-positive nuclei in transformants containing incompatible *het-c* alleles.

A significant number of heterokaryon incompatibility loci have now been cloned and identified (see Glass et al. 2000 for a review). Data about the molecular events associated with somatic incompatibility have been reviewed by Esser (2006).

E. Meiotic Death, Spore Discharge and Spore Germination

Meiotic apoptosis in fungi was first noted in *spo11* mutants of *Coprinus cinereus* (Celerin et al. 2000) and later studied in more detail by Lu et al. (2003). White-cap mutants of *Coprinus cinereus* display three classes of meiotic defect which can be distinguished cytologically. In all three groups, nuclei undergo normal meiotic metaphase I, but later arrest with condensed chromatin and other apoptotic phenotypes. In some mutants, plasmolysis and cytoplasmic shrinkage can also be observed in basidia, along with condensed chromatin, dispersed fragmented nuclei and TUNEL-positive intracellular bodies.

The triggering of the death response appAmut Bmutears to be linked to cell cycle controls. Examination of *Amut Bmut* strains kept in continuous light revealed that death was triggered downstream of diplotene-metaphase I, and that the checkpoint sensor was only activated once the meiotic nuclei had entered the metaphase I (pseudo-G2/M) checkpoint. In sporulation mutants, death was triggered at the tetrad stage, probably at the G1/S phase transition. This situation contrasts strikingly with that in animals, where apoptosis arising from defects in meiosis usually occurs at the time of the defect, rather than at a particular checkpoint later on in the cell cycle (Lu et al. 2003).

In *Coprinus cinereus*, meiotic arrest could also be detected in wild-type dikaryons, showing that it may have adaptive value in the elimination of arrested basidia, thereby promoting the more efficient utilization of resources (Money 2003). Whilst meiotic apoptosis has been observed in *spo11* mutants of mice (Baudat et al. 2000), within fungi it seems only to play a role in the basidiomycetes, since homologous sporulation mutants in *N. crassa*, *S. cerevisiae* and *Schizosaccharomyces pombe* arrest but do not appear to die (Lu et al. 2003).

Raju and Perkins (2000) described PCD as a part of the normal development of the ascospores of the homothallic fungus *Coniochaete tetraspora*. In this fungus, immature asci contain eight uninucleate ascospores. Typically, however, two ascospore pairs survive and two pairs then degenerate. Since the determinant of ascospore death segregates 1:1 in a Mendelian fashion, every ascus must originate from a diploid nucleus which is heterozygous for the factor. For this to occur in self-fertile homokaryons, one of the two nuclei entering a zygotic partnership must become epigenetically marked at each generation to create a new 'allele', or a programmed genomic rearrangement must occur prior to the onset of each meiotic event.

PCD has not been overtly linked with spore discharge, but it is evident that in some situations the death of support cells is essential for the dispersal of others (Moore 2004). In the rusts, terminal aeciospores are separated by suspensor cells which sacrifice their own future reproductive potential, a phenomenon which is potentially regulated by a PCD mechanism. Autolysis of fruit bodies of 'ink-caps' is another clear example (Umar and van Griensven 1997).

PCD may also be an integral part of the mechanics of spore germination and differentiation. Development of turgor pressure in appressoria of *Magnaporthe grisea* is reliant upon the death of the germinating spore. Genomic disparity between nuclei in spores can also lead to post-germination mortality in several basidiomycetes, e.g. *Heterobasidion annosum* and *Stereum hirsutum* (Ramsdale and Rayner 1996; Ramsdale 1998), which has the potential to weed out poor genetic combinations in natural populations.

F. Development and Tissue Formation

Higher fungi differentiate a number of tissue types during somatic and sexual development. PCD has been implicated in stipe cavitation and gill formation in a number of mushrooms (Lu 1991; Umar and van Giensven 1997, 1998). Examination of developing gills in primordia of *Coprinus cinereus* reveals extensive cellular degeneration within the gill cavities, but not within the gill domains (Lu 1991), which is associated with the activity of numerous hydrolytic enzymes. Degenerating hyphae contain multivesicular bodies and residual organelles within the vacuoles. In the final stages of degeneration, cellular debris becomes scattered within the intercellular spaces as a result of the lysis of the dead cells. In *Agaricus bisporus*, morphogenetic cell death is required for the development of gill splits in developing basidiocarps and mycelial cords. Morphogenetic cell death is associated with a loss of nuclei, cytoplasmolysis, mitochondrial swelling, rupture of the cell membrane and dissolution of the hyphal walls. Severely damaged organelles become enclosed by mucilaginous material in the inter-hyphal space, along with clusters of ribosomes, polygonal crystals, and glycogen rosettes (Umar and van Griensven 1997). Cell death often contributes to the formation of cribiforme cavities and gelatinous matrices which, in some species such as the jelly fungi, constitute the most significant part of the fruiting structure (Moore 1965). Studies performed on a variety of fungi (*Stropharia rugoso-annulata*, *Coprinus domesticus*, *Psathyrella candolleana*, *Tremella mesenterica*, *Otidea onotica* and *Peziza ostracoderma*) reveal changes which support the view that PCD plays a key role in different stages of fungal fruit-body development (Umar and van Griensven 1997).

III. Stimulating Fungal Apoptosis

The cytological changes associated with mammalian type I PCD, or apoptosis, have now been observed in many fungi following a variety of lethal stimuli. In *S. cerevisae*, apoptosis can be induced by treatment with harsh chemical agents or physical insults. A number of physiological stimuli such as mating pheromone and starvation can also induce apoptosis-like changes. Expression of heterologous pro-apoptotic regulators from mammals (Bax, Bak) or nematodes (Ced4) can also bring about apoptosis-like cell death in a number of yeasts. An increasing number of mutations are being identified which result in apoptotic cell death in yeasts (see sections below). Indeed, it is hard to find any lethal mutations which do not result in the expression of some or all markers of apoptotic cells, suggesting that apoptotic cell death may be a feature common to all 'slow-killing' responses in yeast.

Apoptotic markers which have been observed in fungi include externalisation of phosphatidylserine, enhanced dsDNA accumulation, nuclear condensation and fragmentation, cell shrinkage, membrane blebbing, mitochondrial permeability transition pore opening, release of Cyt *c*, cleavage of caspase substrates, and ROS accumulation (Madeo et al. 2002a).

A. Physical and Environmental Insults

1. UV Irradiation

Del Carratorre et al. (2002) examined the response of *S. cerevisiae* after exposure to UV light at doses of 90–120 J m^{-2}. Typically, 40%–70% of cells were TUNEL-positive. Exposure to higher doses killed cells, but apoptotic phenotypes were not detected in these cases. Apoptotic features observed included nuclear condensation and an accumulation of cells with a sub-G1 DNA content (as has been reported in mammalian apoptosis; Telford et al. 1992).

2. Sugar

In the absence of nutrients to support growth, sugars, including glucose at 2%, can kill yeast cells (Granot et al. 2003). Sugar-induced cell death is accompanied by nuclear fragmentation, DNA and RNA degradation, and cytoplasmic shrinkage. Most cells also stain with PI, showing a necrotic component to the death response after 24-h incubation. Cell death is associated with the production of ROS and significantly, ROS scavengers such as ascorbic acid can prevent the killing. Sugar-induced cell death commonly gives rise to petite mutants which presumably have avoided the apoptotic response by virtue of their inability to produce ROS.

Soloca et al. (2003) have described killing of yeasts by sugars at very high (60%) concentrations which can be accompanied by changes characteristic of apoptosis – including the activation of a caspase-like activity. The relationship of this response to that at much lower doses is not clear but it may be osmotic, rather than metabolic in origin.

3. Weak Acids

Weak acids such as acetic, sorbic, pentanoic and butyric acid have been shown to kill yeasts, including *Zygosaccharomyces bailii* and *S. cerevisiae* (Fernandes et al. 1997; Ludovico et al. 2001, 2003). Exposure of *S. cerevisiae* at pH 3.0 to acetic acid doses of 20–80 mM produced cells which stained positive by TUNEL, were able to bind annexin and exclude propidium iodide, and showed nuclear condensation and margination. Cells treated at 120–200 mM displayed hallmark features of necrosis including PI staining and a complete loss of internal organisation, as revealed by transmission electron microscopy. Similar observations were made in *Z. baillii*, albeit at higher doses of acetic acid (consistent with its reputation as an acid-tolerant species). In *Z. bailii*, an extensive reorganisation of mitochondria was observed which had not been seen in *S. cerevisiae*. FACs analysis, using low levels of rhodamine 123, revealed a reduction in $\Delta\Psi m$ (membrane depolarization), and EM studies indicated alterations in the number and structure of mitochondrial cristae, the presence of myelinic bodies, and some mitochondrial swelling (but not lysis). Although a temporal analysis was not presented, the proportions of cells displaying apoptotic attributes at different doses suggest that chromatin condensation precedes phosphatidylserine exposure which precedes or is coincident with dsDNA breakage.

In *S. cerevisiae*, death at low (pro-apoptotic) doses, but not at high (pro-necrotic) doses, of acetic acid can be inhibited by cycloheximide, indicating an active aspect to the death response. Furthermore, the failure of cycloheximide to prevent necrosis in *S. cerevisiae* indicates that cells can directly enter a necrotic state. This contrasts with the situation in *Candida albicans* where necrotic cells treated with acetic acid are typically derived secondarily from prior apoptotic events (Phillips and Ramsdale, unpublished data).

4. Salt

Salt treatment (1.5 M NaCl) causes a time-dependent reduction in viability which is accompanied by hallmarks of an apoptotic PCD response (Huh et al. 2002). Changes are apparent within 40 min of the onset of the treatment, and include nuclear fragmentation, TUNEL staining, vacuolization and eventually cell lysis.

Salt sensitivity was found to be dependent upon the P-type ATPase (ENA1). A mutant of calcineurin (a PP2B phosphatase) which regulates Na^+ homeostasis and mediates salt tolerance (Mendoza et al. 1994) was found to be hypersensitive to NaCl. Calcineurin affects Na^+ homeostasis via the transcriptional activator *CRZ1* (which binds to CDREs), and activates a number of genes including *ENA1*. The mammalian anti-apoptotic protein Bcl-2 was able to abrogate the increased salt sensitivity of a *cnb1* mutant, but had no effect on the NaCl sensitivity of a *hog1* mutant. This suggests that the primary apoptotic effect is ionic, rather than osmotic.

Mutants with defects in components of the permeability transition pore complex (*vdac1*, *vdac2*, *vdac3* and *aac1*, *aac2* or *aac3*) and electron transport chain (*atp4*) did not affect salt sensitivity. It might therefore be concluded that Cyt *c* release is not important for the initiation of salt-induced yeast apoptosis, since this requires an intact permeability transition pore. Alternatively, such results may indicate that salt triggers apoptosis through both Cyt *c*-dependent and Cyt *c*-independent pathways. Without further data, it will not be possible to distinguish between these possibilities.

Yeast cells with deletions in the *SRO7/SOP1*-encoded homolog of a *Drosophila* tumour suppressor gene (*l(2)gl*) also show enhanced sensitivity to salt stress (Wadskog et al. 2004). Mutant *sro7* cells exposed to NaCl exhibit a number of markers of apoptosis including ROS accumulation, DNA breakage, and nuclear fragmentation. In yeast, *SRO7* is a multicopy suppressor of *rho3* and is cold-sensitive; when combined with a mutation in *sni2*, it has an exocytic defect, accumulating post-Golgi vesicles. Other salt-sensitive mutants including *gpd1* and *ckb1* were not associated with apoptotic changes, showing that the effects are specific to the *sro7*-sensitive response. According to Wadskog (2003, cited in Wadskog et al. 2004), the salt sensitivity of *sro7* is due to the defective targeting of the *ENA1* ATPase to the plasma membrane.

Deletion of *YCA1* in *sro7* and wild-type cells prevented the apoptotic changes and protected the cells against the primary rapid loss of viability, although the cells were still salt-sensitive. By contrast, *sro77* null strains exhibited salt-induced death which was independent of *YCA1* but which was still associated with the development of an apoptotic phenotype. This supports the idea of at least two independent apoptosis pathways in yeast. YCA1 activity increased in the presence of salt in wild-type cells, and was increased further in an *sro7* backgound. A direct interaction between *sro77* and the yeast metacaspase can be inferred from a number of two-hybrid interaction studies (Uetz et al. 2000; Drees et al. 2001). However, the finding that the caspase activity levels were not enhanced in the *sro77* strain does not necessarily support the authors' conclusion that *sro7* and *sro77* exert antagonistic effects on caspase activation.

5. Hydrogen Peroxide

Hydrogen peroxide is a constant threat to microorgansims, which have developed a number of antioxidant resistance mechanisms to counteract its effects. Despite this, exposure of *S. cerevisiae*, *C. albicans* and *A. fumigatus* to low doses of hydrogen peroxide induces killing with all of the classical attributes of mammalian apoptosis (Madeo et al. 1999; Phillips et al. 2003; Mousavi and Robson 2004). At higher doses, cells die with features more reminiscent of necrosis. Annexin and TUNEL-positive cells are observed, along with the margination of chromatin at the nuclear envelope and/or the fragmentation of nuclei. In *S. cerevisiae*, nuclear condensation is visible after only 30-min treatment (Madeo et al. 1999). Some cells show plasma-membrane vesicles which are similar to the 'blebs' linked with apoptosis in mammalian cells. A strain lacking *GSH1* (therefore, depleted in glutathione) exhibits a similar death response, which can be exacerbated by the addition of low doses of hydrogen peroxide or abrogated by the exogenous application of glutathione.

Co-incubation of cells with cycloheximide almost eliminates the TUNEL straining of cells treated with 3 mM H_2O_2 and largely prevents cell death, as revealed by a clonogenic assay, indicating that apoptotic death requires the active participation of the cell machinery.

6. Mating Pheromone

In yeasts, cells of α-mating type produce alpha-factor which triggers **a**-type cells to mate. It has been known for some time that prolonged exposure to high doses of mating pheromone is toxic (Kurjan 1993). Severin and Hyman (2002) set out to assess if the process could be genetically programmed. The addition of alpha-factor pheromone to **a**-type cells induces the production of ROS in about 30% of cells after 1.5-h exposure. However, no ROS are produced when alpha-factor is added to α-cells. TUNEL analysis revealed dsDNA damage in pheromone-treated cells, and FACS analysis the accumulation of cells with a sub-G1 DNA content; phloxin B staining confirmed that the cells were dying.

Deletion of *STE20*, the mating factor receptor, abolished pheromone killing of **a**-type cells and the accumulation of ROS (see section below on signalling). Generation of ρ^- cells using ethidium bromide abolished pheromone killing, too, indicating a requirement for mitochondrial DNA. Ethidium bromide treatment did not, however, prevent schmoo formation, showing that although the cells could still respond to pheromone, they were specifically blocked in the PCD response.

Chloroquine treatment of mixed **a**- and α-type cells (which prevented cell fusion) induced high levels of killing which could be eliminated by the addition of cyclosporin A. Thus, mating-induced cell agglutination, in the absence of cell fusion, may lead to PCD under normal conditions. Overall, as Severin and Hyman proposed, the altruistic suicide of yeast cells unable to mate after a prolonged period of contact might be beneficial for the community as a whole.

Yeasts lacking Cyt *c* were still able to produce ROS in response to pheromone, but were unable to complete their PCD response. Confusingly, inhibition of permeability transition pore formation with cyclosporin A was able to prevent both ROS generation and cell death. Since cyclosporin A also inhibits the calcineurin/calmodulin signalling pathway, *CMD1* mutants were also tested. In a *cmd1-6* mutant, ROS production and PCD was accelerated. Since yeast mutants defective in calmodulin/calcineurin are hypersensitive to pheromone (Cyert and Thorner 1992; Moser et al. 1996), it seems likely that this system normally inhibits PCD.

B. Antifungal Agents and Cytotoxic Drugs

1. Amphotericin B

The primary mode of action of the antifungal drug amphotericin B is well established. Amphotericin B binds to sterols, such as ergosterol, disrupting the osmotic integrity of the fungal membrane and resulting in the leakage of intracellular ions and metabolites. Liao et al. (1999) examined physiological changes in *C. albicans* cells treated with amphotericin B, and described three patterns of death. Death was always accompanied by a drop in ATP level but could be subdivided on the basis of plasma-membrane integrity and mitochondrial membrane potential. Later, Phillips et al. (2003) found that nonviable cells which were able to exclude propidium iodide and produce ROS after amphotericin B treatment corresponded to an apoptotic subpopulation of cells. Normal treatment of systemic candidiasis may therefore reduce infection loads by initiating apoptosis of this pathogen.

Protoplasts of *A. fumigatus* treated with 0.25 $\mu g\,ml^{-1}$ amphotericin B result in about 15% cells staining positive for annexin V, rising to 50% when treated with 0.5–1 $\mu g\,ml^{-1}$ (Mousavi and Robson 2004). TUNEL labelling also increased to 65% at 0.5 $\mu g\,ml^{-1}$ amphotericin B, but was lower (at around 20%) when treatment was 1 $\mu g\,ml^{-1}$. PI staining (indicative of necrosis) was less than 20% at 0.25 and 0.5 $\mu g\,ml^{-1}$ amphotericin B, but increased to 85% at 1 $\mu g\,ml^{-1}$. Pre-incubation of cells with cycloheximide prevented the appearance of the apoptotic markers, but was not able to prevent the formation of PI-positive staining at 1 $\mu g\,ml^{-1}$. Taken together, this suggests that low fungicidal doses of amphotericin B are pro-apoptotic and higher doses are pro-necrotic.

2. Pradimicin A

Pradimicin A, a mannose-binding antifungal antibiotic, induced apoptosis-like cell death in *S. cerevisiae* (Hiramoto et al. 2003). Nuclear fragmentation and DNA damage can be observed in yeast cells by DAPI and TUNEL staining. Accumulation of reactive oxygen species (ROS) was also detected in dying cells. Moreover, pradimicin-induced cell death and the accumulation of ROS could be prevented with a free-radical scavenger, suggesting some dependency between the two.

3. Osmotin

The pathogenesis-related protein osmotin induces apoptosis-like cell death in *S. cerevisiae* (Narasimhan et al. 2001). FACS analysis of DNA content revealed a high level of cells with a sub-G1 DNA content in osmotin-treated cells (surprisingly, some sub-G1 contents were also seen in untreated cells, albeit at a considerably lower level). A number of abnormal mitochondrial morphologies were associated with osmotin-induced death, but these appear to be consequences, rather than causes of PCD, since oligomycin treatment, ATP4 deletion or Bcl-2 expression have no effect on killing. A temporal analysis of the death response showed that ROS production preceded dsDNA breaks which, in turn, preceded nuclear fragmentation.

4. Sphingosines and Ceramides

Phytosphingosine and dihydrospingosine treatment of *Aspergillus nidulans* is followed by the production of ROS, which is in turn followed by nuclear condensation and apoptosis (Cheng et al. 2003). Pulse field gel electrophoresis also revealed extensive DNA degradation and TUNEL-positive cells. Exposure to phytosphingosine or dihydrosphingosine for only a few minutes was sufficient in this respect, showing that cells were committed very early on to undergo a PCD response.

Whilst phytosphingosine-induced nuclear condensation was not inhibited by ROS scavengers and was also independent of the fungal metacaspase *casA*, overexpression of *casA* on its own was sufficient to induce apoptosis. The nuclear condensation component of the apoptotic response was independent of both condensin (as revealed by phytosphingosine-induced apoptosis in a $smcB^{cut14}$ mutant) and of mitosis per se (since it still occurred in temperature-sensitive *nimA* mutants). Nuclear fragmentation, but not condensation, must require protein synthesis, given that it can be inhibited with cycloheximide. Oligomycin, which inhibits mitochondrial function, abolished ROS production and also abolished nuclear condensation, suggesting that mitochondria play an essential role in phytosphingosine-mediated apoptosis.

5. Translation Inhibitors

Phenanthroline and other metal 'phen' complexes kill both mammalian and *C. albicans* cells (Coyle et al. 2004). Mammalian cells show hallmarks of apoptosis, whilst *C. albicans* cells accumulate ROS and show elevated oxygen consumption rates. Nuclear disruption (enlargement and crescent-shaped bodies) was observed after treatment of yeast cells with $[Ag_2(phen)_3(mal)].\ 2H_2O$, $[Mn_2(phen)_3(mal)].\ 2H_2O$ and $[Cu_2(phen)_3(mal)].\ 2H_2O$, and cytoplasmic shrinkage after $[Cu_2(phen)_3(mal)].\ 2H_2O$ treatment. $[Ag_2(phen)_3(mal)].2H_2O$, $[Mn_2(phen)_3(mal)].\ 2H_2O$, but not $[Cu_2(phen)_3(mal)].\ 2H_2O$, resulted in a non-specific form of DNA damage. Whilst it may be that such differences

Table. 7.1. Genes and proteins with potential link to endogenous apoptotic cell death responses in fungi

Gene or protein	Organism[a]	Description	Death response	References
AIF1/YNR074c	Sc	Putative apoptosis inducing factor (Apaf)	Mitochondrial protein with nuclease activity which translocates to nucleus during apoptosis	Wissing et al. (2004)
ASF1/CIA1	Sc	Nucleosome assembly factor, histone chaperone	Defect induces apoptosis	Yamaki et al. (2001)
ATP10	Sc	Mitochondrial inner mebrane protein required for F_1F_0 ATP synthetase	Enhanced expression during apoptosis, and required for apoptotic death	Yamaki et al. (2001), Ludovico et al. (2002)
BCY1	Sc	Regulatory subunit of PKA	Deletion enhances osmotin-induced apoptosis	Narasimhan et al. (2001)
casA	An	*Aspergillus nidulans* metacaspase	Overexpression accelerates apoptosis; deletion inhibits apoptosis	Cheng et al. (2003)
CDC13	Sc	Single stranded DNA-binding protein required for telomere replication	Inactivation induces caspase activity	Qi et al. (2003)
CDC48	Sc	ER-located ATPase required for retrotranslocation of ubiquitinated proteins	Defect induces apoptosis	Madeo et al. (1997)
CDC6	Sc	Origin licensing protein	Degraded during apoptosis	Blanchard et al. (2002)
CDR1/2	Ca	Multidrug resistance transporter; in-to-out floppase	Involved with translocation of phosphatidylserine	Smrti et al. (2002)
CMD1	Sc	Master regulator of calcium responses; calmodulin	Defect accelerates pheromone-induced apoptosis	Severin and Hyman (2002)
CNB1	Sc	Regulatory subunit B of calcineurin	Defect accelerates salt-induced apoptosis	Huh et al. (2002)
CRZ1	Sc	Stress-linked transcription factor regulated by calcineurin	Defect enhances sensitivity to antifungals and may be involved with salt sensitivity	Edlind et al. (2002), Huh et al. (2002)
CYC3	Sc	Cytochrome *c* lyase attaches heme to apo-Cyt *c*	Required for acetic acid-induced apoptosis	Ludovico et al. (2002)
CYC8	Sc	Transcriptional co-repressor with Tup1	Multicopy suppressor of *cdc13*-induced apoptosis	Qi et al. (2003)
CYT1/Cyt c	Sc	Cytochrome c1, mitochondrial or cytoplasmic	Mitochondrion to cytoplasm translocation during apoptosis, lack of Cyt *c* blocks pheromone-induced apoptosis	Yamaki et al. (2001), Severin and Hyman (2002), Ludovico et al. (2002)
DCP1	Sc	mRNA processing, decapping	Defect induces apoptosis	Mazzoni et al. (2003)
DCP2	Sc	mRNA processing, decapping	Defect induces apoptosis	Mazzoni et al. (2003)
DGA1	Sp	Acyl-CoA:diacylglycerol acyltransferase	Mutant accumulates DAG which stimulates apoptosis	Zhang et al. (2003)
ENA1	Sc	P-type ATPase required for salt tolerance	Defect enhances salt-induced apoptosis	Huh et al. (2002)
GCS1	Ca	γ-glutamylcysteine synthetase	Defect promotes apoptosis	Baek et al. (2004)
LSM1	Sc	mRNA processing, decapping and decay	Defect induces apoptosis	Mazzoni et al. (2003)
LSM6	Sc	mRNA processing, splicing, decapping and decay	Defect induces apoptosis	Mazzoni et al. (2003)

[a] An, *Aspergillus nidulans*; Ca, *Candida albicans*; Cc, *Coprinus cinereus*; Mr, *Mucor racemosus*; Sc, *Saccharomyces cerevisiae*; Sp, *Schizosaccharomyces pombe*

Table. 7.1. (continued)

Gene or protein	Organism[a]	Description	Death response	References
MEC1	Sc	Protein kinase required for DNA damage checkpoints	Caspase activation is dependent on Mec1	Qi et al. (2003)
MRas1	Mr	Ras-protein; GTP-binding protein with homology to mammalian Ras-oncoproteins	Required for lovastatin-induced apoptosis	Roze and Linz (1998)
MRE11	Sc	Component of MRX exonuclease required for double-strand DNA repair	Antagonizes Mec1-dependent apoptosis	Qi et al. (2003)
NMA111 (YNL123W)	Sc	Omi/Htr2A-like serine protease	Defect induces apoptosis	Fahrenkrog et al. (2004)
ORC2	Sp	Origin licensing protein	Defect induces apoptosis	Burhans et al. (2003)
PARP	An	Poly-ADP-ribose polymerase	Cleaved during PCD	Thrane et al. (2004)
PCA1	Sp	*S. pombe* metacaspase	Not required for DAG-mediated apoptosis	Zhang et al. (2003)
PKB/PI3K	Sc	Protein kinase B signalling	Inhibitors of PKB/PI3K block apoptosis	Jeon et al (2002)
PLH1	Sp	Phospholipid diacylglycerol acyltransferase	Mutant accumulates DAG which stimulates apoptosis	Zhang et al. (2003)
RAD9	Sp	DNA damage checkpoint protein	Partially suppresses apoptosis induced by defect in *orc2-1*, similarity to mammalian Bcl-2-family proteins	Burhans et al. (2003)
RAS1	Ca	Ras-protein; GTP-binding protein with homology to mammalian Ras-oncoproteins	Required for acetic acid-induced apoptosis	Ramsdale (unpublished data)
RAS2	Sc	Ras-protein; GTP-binding protein with homology to mammalian Ras-oncoproteins	Required for osmotin-induced apoptosis	Narasimhan et al. (2001)
Rho^0/ρ^0	Sc	Strain lacking mitochondrial DNA	Suppresses acetic acid- and *cdc13*-mediated apoptosis	Ludovico et al. (2002)
RSM23	Sc	Mitochondrial ribosomal protein of small subunit, DAP-3 homolog	Null mutation blocks Yca1-induced apoptosis	Madeo et al. (2002b)
SCH9	Sc	Serine/thronine kinase	Defect blocks aging-dependent apoptosis	Fabrizio et al. (2004)
Spo11	Cc	Whitecap mutant	Defect produces apoptotic basidiospores	Celerin et al. (2000)
SRO7/SOP1	Sc	Yeast homolog of *Drosophila* tumour suppressor; Suppressor of *rho3* defect	Interacts with Yca1; defect enhances salt-induced apoptosis in a Yca1-**dependent** fashion	Wadskog et al. (2004)
SRO77	Sc	Yeast homolog of *Drosophila* tumour suppressor; Suppressor of *rho3* defect	Interacts with Yca1; defect enhances salt-induced apoptosis in a Yca1-**independent** fashion	Wadskog et al. (2004)
STM1	Sc	DNA repair (quadruplex nucleic acid-binding) protein, acts with *cdc13*	Accumulation promotes apoptosis, deletion reduces H_2O_2-induced apoptosis	Ligr et al. (2001)
UBP10	Sc	Ubiquitin-specific protease	Defect induces apoptosis	Orlandi et al. (2004)
WWM1	Sc	WW domain containing protein involved with desiccation response	Growth defect of mutant suppressed by Yca1; physical interaction with Yca1	Szallies et al. (2002)
YCA1/MCA1	Sc	Putative cysteine protease – yeast metacaspase	Deletion inhibits apoptosis; overexpression enhances H_2O_2-induced apoptosis	Madeo et al. (2002b), Wadskog et al. (2004)

[a] An, *Aspergillus nidulans*; Ca, *Candida albicans*; Cc, *Coprinus cinereus*; Mr, *Mucor racemosus*; Sc, *Saccharomyces cerevisiae*; Sp, *Schizosaccharomyces pombe*

Table. 7.1. (continued)

Gene or protein	Organism[a]	Description	Death response	References
YDR333c	Sc	Unknown function	Multicopy suppressor of *cdc13*-induced apoptosis	Qi et al. (2003)
YME1	Sc	AAA-protease located in mitochondrial inner membrane required for degradation of misfolded proteins	Activated in response to drop in Cyt *c* oxidase	Manon et al. (1997)

[a] An, *Aspergillus nidulans*; Ca, *Candida albicans*; Cc, *Coprinus cinereus*; Mr, *Mucor racemosus*; Sc, *Saccharomyces cerevisiae*; Sp, *Schizosaccharomyces pombe*

manifest themselves as a result of differences in the timing of these events, the production of ROS in the absence of DNA damage (seen after $[Cu_2(phen)_3(mal)].2H_2O$ treatment) could imply that ROS play a primary role as signals of apoptosis but do not act directly as DNA damaging agents.

Killing by these translational inhibitors may involve a reduction in the levels of cytochrome *b* and *c* and the associated uncoupling of respiratory function. Such uncoupling may then contribute to the formation of pro-apoptotic ROS. McCann et al. (2000) found that phenanthrolines lower reduced:oxidized glutathione ratios, consistent with a pro-oxidant role for the effects of these drugs.

6. Histatins

Histatins are short peptides found in the oral cavity which have been shown to display both fungistatic and fungicidal activity against a range of *Candida* species. Several workers have independently suggested that histatins might induce PCD (Helmerhorst et al. 2001; Phillips et al. 2003), not least because of the delay after treatment starts before killing is observed. However, Wunder et al. (2004) concluded that histatins did not induce apoptosis in *C. albicans*, since they could not find any evidence of DNA laddering or the release of cytochrome *c* when isolated mitochondria were treated with histatin 5. Although elevated levels of ROS were detected following the exposure of cells to histatin 5 or the intracellular expression of an Hst 5 construct, the reduction in viability of *SOD1/2* mutants of *S. cerevisiae* or SOD1 mutants of *C. albicans* was the same as that of wild-type strains. It was therefore concluded that ROS production merely accompanies the death response, but does not contribute significantly to its fungicidal activity.

Whilst the results of Wunder et al. (2004) can be taken to suggest that the primary mode of action of histatins is not the direct release of cytochrome *c* from mitochondria, their failure to detect PCD at all might be simply due to the fact that they did not look for classical markers of apoptosis in intact cells. Isolated mitochondria might not be expected to respond to apoptotic stimuli in a natural manner when removed from the cellular environment.

Several other antifungal drugs also appear to operate independently of apoptotic mechanisms. Treatment of protoplasts of *A. fumigatus* with itraconazole or of *A. nidulans* with aureobasidin A, amphotericin B and itraconazole does not appear to induce apoptosis (Cheng et al. 2003; Mousavi and Robson 2004). Furthermore, fluconazole toxicity has been shown to be independent of known apoptotic mechanisms in yeast, though this was tested only in the context of a failure of Bcl-2 to inhibit killing, and the paper addressed only growth inhibition and did not look at death characteristics per se (Kontoyiannis and Murray 2003).

C. Lethal Mutations

In the sections which follow, an overview is given of the current state of our knowledge of the molecular events underlying programmed cell death responses in fungi. A summary of the significant genes which have been implicated in fungal programmed cell death responses is provided in Table 7.1.

1. *CDC48*

CDC48 encodes a protein of the AAA family which has been implicated in homotypic membrane fusion and in a complex with Npl4 and Ufd1. It participates in the retro-translocation of ubiquitinated proteins from the ER into the cytosol for targeted degradation by the proteasome (Elkabetz et al. 2004). Expression of a temperature-sensitive mutant *cdc48*^{S565G} at the non-permissive temperature leads to phosphatidylserine exposure, margination of chromatin, DNA fragmentation, and the formation of cell fragments (Madeo et al. 1997).

The mammalian orthologue of *CDC48*, p97/VCP (Shirogane et al. 1999), and the *C. elegans* homologue, *Mac-1* (Wu et al. 1999), have now also been implicated in apoptosis, making *CDC48* the first apoptotic regulator identified in yeast (Frohlich and Madeo 2001).

2. Cell Division Cycle and DNA Repair Mutants

Cdc13 is a multifunctional protein involved with telomere replication and protection and which, when defective, activates the DNA damage checkpoint. Inactivation of *CDC13* has also been reported to induce metacaspase activity, ROS production and phosphatidylserine exposure in yeast (Qi et al. 2003).

Multicopy suppressors of these responses were found when a temperature-sensitive *cdc13-1* mutant strain was transformed with a yeast genomic library and a mammalian HeLa cDNA library. Two clones were identified in the yeast screen – a fragment of *YDR333C* and an antisense clone of *CYC8*. Screening of the HeLa cell cDNA library revealed a third clone, *MTCO3*. All three of these genes have a role to play in mitochondrial functioning, implicating the mitochondrion in the death response – a conclusion further supported by the discovery that *cdc13-1*-driven changes are suppressed in a ρ^0 strain. *CDC13*-initiated apoptotic signals were also found to be dependent upon the ATM homolog *MEC1* but not *TEL1*, and could be enhanced in an *mre11* null strain. MRE11 is part of the RMX complex which exhibits exonuclease activity. The enhanced death of *mre11* might therefore be due to a failure to activate the appropriate DNA damage repair function of the complex.

Induction of apoptosis-like cell death in the yeasts *S. cerevisiae* and *S. pombe* also requires other functional cell cycle checkpoint pathways (Burhans et al. 2003). In a screen of ROS production by checkpoint mutants, the highest levels were detected amongst alleles affecting proteins required for the initiation of replication (e.g. *orc2*, *orp2 orp5*, *cdc18* and *dfp1*). In general, mutants defective in dsDNA break repair such as *pku70*, *lig4*, *rad32* and *rad50* or damage bypass and replication fork stability *srs2*, *rqh1* showed very little ROS production.

Apoptotic phenotypes can be detected in mutant yeast strains bearing conditional defects in the DNA replication initiation complex. When an *orc2-1* strain is shifted to a non-permissive temperature, cells produce condensed chromatin and fragmented nuclei, and accumulate ROS (Watanabe et al. 2002; Burhans et al. 2003). Significantly, the defect can be partially suppressed by deletion of *RAD9* (a yeast homologue of BCL-2-family proteins) or by mutation of *MEC1*, which indicates that the killing is dependent on functional DNA damage checkpoints.

Burhans et al. (2003) proposed that the *orc2-1* defect induces cell death when the number of replication forks drops below a threshold required to activate cell cycle checkpoints and/or rescue stalled forks. Loss of functional CDC6 can result in an unforeseen disassembly of pre-replication complexes, and has been implicated in both yeast and mammalian apoptosis (Weinberger et al. 1999; Blanchard et al. 2002).

Elimination of both Chk1 and Cds1 in *S. pombe* was found to be required to generate ROS in the presence of hydroxyurea (HU). Both kinases activate pathways which converge upon Cdc2, a cyclin-dependent kinase. A mutant version of Cdc2 (*cdc2Y15F*) which is constitutively active displays extensive ROS production (even in the absence of HU), suggesting that the deregulation of cdc2 and associated premature entry into M phase is the major source of ROS. In mammals, deregulation of cdk1 (the homologue of *S. pombe* Cdc2) is also a strong apoptotic signal (Castedo et al. 2002).

The relationship between apoptosis and DNA replication is clearly complex. It is apparent that some apoptosis pathways require checkpoint proteins, others are inhibited by them, and yet others are independent. The genetic tractability of yeast will help in the elucidation of these pathways and their mechanisms.

3. *ASF1/CIA1*

Disruption of the histone chaperone *ASF1/CIA1* in *S. cerevisiae* produces cells with a slow-growth phenotype which is associated with G2/M-phase cell cycle arrest, and an accumulation of enlarged dead cells which contain fragmented nuclei with condensed chromatin (Yamaki et al. 2001). Using lucifer yellow and quinacrine mustard as markers, the acidity of the vacuoles was found to decrease whilst that of the cytoplasm increased. Membrane potentials of both vacuoles and mitochondria were found to be lower in the mutant than in the wild-type. Studies on spheroplasts and isolated mitochondria revealed that death was linked to the release of Cyt *c*.

Microarray analysis revealed that several genes encoding components of the vacuolar proton pump (*VMA7*, *VMA8*, *VMA10*, *VMA21* and *VMA22*) were down-regulated in comparison to controls, and that expression of the mitochondrial proton pump genes *ATP1* and *ATP5* was enhanced. The human homolog of *ASF1*/*CIA1* has been shown to interact with hCCG1, a key component of TFIID which is involved in the regulation of mammalian apoptosis (Sekiguchi et al. 1995). Holstege et al. (1998) reported that disruption of yeast *CCG1* influences *VMA* gene expression in a similar manner to that seen in *ASF1* nulls, and so the link between them certainly deserves further attention.

4. mRNA Processing Mutants

Saccharomyces cerevisiae cells expressing a truncated Lsm4 protein from *Kluyveromyces lactis* and strains with deletions of *LSM1*, *LSM6*, *DCP1* or *DCP2* display phenotypic markers of apoptosis including chromatin condensation, DNA fragmentation, and ROS accumulation (Mazzoni et al. 2003). All of these strains have defects in aspects of pre-mRNA splicing, mRNA capping and RNA degradation. Whilst only a slight induction of apoptosis was observed for *DHH1* and *PAT1*, the overall behaviour of these mutants is indicative of a role for normal mRNA decapping in the prevention of apoptosis. It was proposed therefore that stabilization of mRNAs could lead to the abnormal accumulation of pro-apoptotic proteins which, in turn, trigger the apoptotic response.

IV. Death-Related Signalling Pathways

Understanding of fungal PCD responses will require the elucidation of the signal pathways which transmit information about the status of the cell and the environment to the suicide machinery. Both pharmacological approaches and functional genetic analyses have revealed some of the components of the death-related signal transduction pathways, although there is still a great deal of research to be done.

A. Ras and cAMP

Both pro- and anti-apoptotic functions for Ras have been identified in fungi, mirroring its complexity in the determination of cell proliferation, differentiation and apoptosis in animal models.

Lovastatin, an indirect inhibitor of Ras prenylation, can induce an apoptosis-like cell death response in hyphae but not yeasts of *Mucor racemosus* (Roze and Linz 1998). Lovastatin treatment alters the processing of MRas1, blocks the accumulation of MRas3 protein, and causes the loss of an MRas1/p20 protein complex. Exogenous application of dibutyryl cAMP (which can initiate morphogenesis from hyphal to yeast-like growth) prevents the lovastatin-induced cell death response, further implicating the Ras pathway. Some cross-talk between the Ras and phosphoinositide 3-OH kinase signalling pathways can be suspected, however, because the addition of wortmannin in combination with dibutyryl cAMP produces a synergistic response, completely blocking cell death (even though wortmannin alone has no effect).

In *S. cerevisiae*, ROS production, the expression of antioxidant proteins and the apoptotic response which follows treatment with osmotin, is partially dependent upon an induced suppression of the *RAS2*/cAMP pathway. *RAS2*^{G19V}, a dominant active allele of *RAS2*, increased sensitivity to osmotin whereas a null mutant decreased sensitivity. The response was linked specifically to the Ras-protein kinase A, rather than to the Ras-MAP kinase pathway, because the effects of the dominant active allele were also seen in a *ste20* background. Consistent with osmotin-induced Ras-protein kinase A signalling, a *bcy1* null, with a constitutively active protein kinase A activity, significantly increased sensitivity to osmotin.

De-repression of STRE-dependent transcriptional responses in *ras2* mutants (Stanhill et al. 1999) could account for the elevated resistance of *RAS2* nulls to stress treatments, including osmotin. During osmotin-induced PCD, both STRE-element and YRE-reporter constructs are repressed (Narasimhan et al. 2001), and so it might be argued that the balance between stress and apoptotic signals determines cell fate. Although osmotin-induced PCD in wild-type cells is associated with a suppression of STRE signalling, this situation cannot be generalized to other PCD responses because, at similar levels of apoptosis-like cell death, hydrogen peroxide still stimulates STRE-reporter expression (Narasimhan et al. 2001). It may be concluded, however, that osmotin stimulates pro-apoptotic ROS production via the activation of the *RAS2*/cAMP pathway which, in turn, inhibits YRE (Yap1-dependent) and STRE-mediated antioxidant stress responses.

The Ras pathway has been strongly linked to morphogenetic signals in a number of fungi (D'Souza and Heitman 2001). One common theme is the finding of an interrelationship between cell death and fungal morphogenesis/differentiation. As highlighted by Narasimhan et al. (2001), overexpression of plant defence molecules induces hyper-branching or formation of spiral hyphae (Epple et al. 1997), and growth inhibition per se can also be linked to hyphal branching (Garcia-Olmedo et al. 1998; Ali and Reddy 2000). A study of the link between death and morphogenesis in *C. albicans* showed that it could not be attributed to any of the known signalling pathways (*EFG1*, *RIM101*, *TEC1*, *CPH1*) which contribute to morphogenesis (Phillips et al. 2003). *RAS1* deletion in *C. albicans* is able to decelerate, or even prevent PCD induced by acetic acid (Phillips et al. 2003; Phillips and Ramsdale, unpublished data), just as it is in an *S. cerevisiae ras2* strain. By contrast, *ste12* knockouts in *S. cerevisiae* are defective in their apoptotic response, but disruption of the homologous downstream effectors of the Ras pathway in *C. albicans* (*CPH1* or *EFG1*) has no effect.

B. Phosphoinositide 3-OH Kinase and Protein Kinase B Signalling

The phosphoinositide 3-OH kinase and protein kinase B signalling pathways have been shown to protect a variety of cell types against apoptosis (Mathieu et al. 2001; von Gise et al. 2001). Jeon et al. (2002) observed that ROS-induced apoptosis in yeast could be stimulated by the addition of wortmannin or LY294002, or inhibited by the addition of PIP_2. Since protein kinase B activity declines during ROS-induced cell death, and protein kinase B is activated by phosphorylation, it may be that phosphoinositide 3-OH kinase itself effects the primary response. In mammals, phosphoinositide 3-OH kinase inhibits caspase-3 (Tessier et al. 2001), delaying the onset of p53-mediated apoptosis. Further work needs to be undertaken to examine whether such an interaction is responsible for the regulation of apoptosis by phosphoinositide 3-OH kinase/protein kinase B in yeasts.

C. *CDC55*, *TPD3* (Protein Phosphatase 2A)

E4orf4 is a viral regulator protein which down-regulates the expression of genes activated by cAMP, affects alternative splicing of adenoviral mRNAs, and induces p53-independent apoptosis in transformed mammalian cells. Kornitzer et al. (2001) were able to show that E4orf4 induced an irreversible growth arrest at G2/M in *S. cerevisiae*, and that this is associated with the production of ROS. Genetic screens revealed that the killing activity of E4orf4 required the yeast protein phosphatase 2A components *CDC55* and *TPD3*. E4orf4 forms a complex with protein phosphatase 2A which, in turn, interacts with the anaphase-promoting complex/cyclosome, leading to its inactivation and concomitant cell cycle arrest/apoptosis.

D. Calcineurin and Calcium

Several lines of evidence suggest that Ca^{2+}/calmodulin/calcineurin signals affect the fungal death response. As discussed above, apoptosis induced by pheromone and salt was shown to be influenced by mutations in calmodulin or calcineurin (Huh et al. 2002; Severin and Hyman 2002). Furthermore, azole activity against *S. cerevisiae* can be reduced by the addition of Ca^{2+} and enhanced by the addition of EGTA (Edlind et al. 2002). Inhibitors of the Ca^{2+}-binding regulatory protein calmodulin (fluphenazine, calmidazolin and W-7), and inhibitors of a Ca^{2+}calmodulin-regulated phosphatase, calcineurin (cyclosporin and FK506), also enhance azole activity. Conversely, mutations constitutively activating calcineurin demonstrate reduced azole susceptibility. Disruption of *CRZ1* (a transcription factor regulated by calcineurin) also enhances azole sensitivity – indicating that the cell integrity pathway is important for the action of these drugs.

Sanglard et al. (2003) reported that FK506 treatment of fluconazole-treated cells induced a fungicidal, rather than a fungistatic response, which could have very important ramifications for future drug therapy regimens. Deletion of *CYP1* (cyclophilin) prevented the fungicidal activity, showing that this component of the signalling pathway was essential for fluconzaole toxicity. Although the mode of death of the cells has not yet been ascertained, it might be speculated that the inhibition of the calcium–calmodulin–calcineurin signalling pathway could induce apoptosis in the face of external stresses.

E. Sphingolipids, Ceramides and Protein Kinase C

The multiallelic *HetC* locus, which is involved with vegetative incompatibility in *Podospora anserina* (Saupe et al. 1994), is a member of a family of glycolipid transfer proteins. One member of this family, *ACD1* from *Arabidopsis thaliana*, has been found to induce an *accelerated cell death* phenotype (akin to apoptosis) during the plant hypersensitivity response (Brodersen et al. 2002). *ACD1* is involved specifically with the translocation of sphingosine (but not glycosylsphingolipids) across the plasma membrane, and may therefore have a direct role in death signalling. Whilst there is no obvious homolog of this gene in yeast, other lipid translocators have been identified, such as *CDR1* and *CDR2*, which could play a role in lipid-associated death signalling. Indeed, Smrti et al. (2002) have been able to show that *CDR1/2/3* ABC transporters in *C. albicans* can flip phosphatidylserine across the plasma membrane which is, of course, a conserved element in all apoptotic death responses.

Antifungal defensins produced by plants induce a number of changes in fungal membranes, including an increased efflux of K^+ ions, increased Ca^{2+} uptake, membrane potential changes, and membrane permeabilization (Thevissen et al. 1996, 1999). Using a genetic complementation approach, *IPT1* was identified as a gene affecting sensitivity towards the defensin DmAMP1 in *S. cerevisiae* (Thevissen et al. 2000), whilst *GCS* was found to reduce sensitivity to RsAFP2 in *Pichia pastoris* (Thomma et al. 2002). *IPT1* encodes an enzyme required for the biosynthesis of the sphingolipid mannosyldiinositolphosphorylceramide, $M(IP)_2C$ (Dickson et al. 1997), whilst *GCS* encodes a glucosyltransferase required for the biogenesis of glucosylceramides. Screening of *N. crassa* mutants by Ferket et al. (2003) also revealed several mutants with altered sphingolipid compositions which were resistant to defensins. Whether such effects can be linked to the well-described role of sphingolipids as signals of death, or if the effects were structural, affecting binding of the defensin to the fungal cells, is not clear.

The addition of D-*erythro*-sphingosine to lovastatin-treated *M. racemosus* cells accelerated apoptosis-like cell death (Roze and Linz 1998) – a point not made by the authors. D-*erythro*-sphingosine is an inhibitor of protein kinase C, and so there could be a role for this pathway in fungal cell death, particularly in view of its function in the cell-wall integrity pathway of yeast. Phosphorylated long-chain base sphingolipids accumulate in *S. cerevisiae ysr2 dpl1* nulls (Kim et al. 2000), and the effect is lethal. Lethality is lost when LCB4 (a long-chain base kinase) is also deleted, showing that the effect is specific to the phosphorylated sphingoid bases. The dying cells show many features of apoptosis (S. Kim, personal communication).

Mutation of the two genes required for the terminal step of triacylglycerol bisosynthesis (*DGA1* or *PLH1*) in *S. pombe* results in an accumulation of diacylglycerol which is associated with a loss of viability upon entry into stationary phase (Zhang et al. 2003). Death is accompanied by the production of ROS, nuclear fragmentation, dsDNA breakage and phosphatidylserine exposure. Exposure of wild-type cells to diacylglycerol is sufficient to induce apoptosis and expression of a bacterial diacylglycerol kinase prevents apoptosis, suggesting that the effect is specific to the build-up of diacylglycerol. Construction of a caspase (*pca1*) null, or treatment of the cells with a caspase inhibitor, failed to block the diacylglycerol-induced death response. Finding the molecular targets of the death-related lipid secondary messengers is clearly an important challenge in unravelling their pro-apoptotic functions.

F. MAP Kinase Cascades

Osmotin activates the mating, invasive growth and pseudohyphal growth MAP kinase cascades of *S. cerevisiae* independently of both the pheromone receptor and the associated G protein coupled alpha subunit encoded by *GPA1* (Yun et al. 1998). The pathway is activated very rapidly; as seen by the phosphorylation of Ste7, and precedes any of the phenotypic changes associated with PCD. Since exposure to osmotin at lethal or sub-lethal doses is unable to stimulate schmoo formation or to activate a *FUS1-lacZ* reporter, osmotin signalling through the MAP kinase cascade must target a set of death-related genes distinct from those involved with mating or morphogenesis.

V. Commitment, Effectors and Suppressors

Identifying the commitment points and effectors of the programmed death response in fungi remains the key to unlocking its therapeutic potential. Whilst this review aims to bring together what we do know, our knowledge in the field is strongly biased by hypotheses driven by our understanding of animal PCD. Hence, we now have candidate caspases and BCL-2-family proteins. In the future, we may do well to look for more fungal-specific pathways.

A. Reactive Oxygen Species

ROS production and the activation of antioxidant defence systems is thought to be involved with a number of differentiation events in fungi, including germination, yeast–hypha transitions and conidiation (Hansberg and Aguirre 1990; Hansberg et al. 1993; Toledo et al. 1995). Significantly, the majority of the studies outlined in this review have implicated ROS in the development of a fungal apoptotic phenotype.

The fact that intracellular ROS levels increase during apoptosis even in the absence of an external oxidative stress has been used to support a direct role for ROS in the generation of the apoptotic phenotype (Madeo et al. 1999; Frohlich and Madeo 2000). ROS production has been seen in every study which has looked for it, and typically it is an early event preceding the appearance of other apoptotic markers. ROS must play more than a secondary role in death since, in the majority of cases, blocking ROS accumulation prevents the appearance of apoptotic markers and enhancing ROS production stimulates apoptosis.

The interaction between stress responses and apoptotic outcomes is particularly important to consider in this context. Increased levels of ROS could originate from de novo sources such as an uncoupled electron transport chain or the induction of pro-oxidant enzymes such as laccase. Alternatively, they could result from a tempering of oxidative stress responses. Whilst there is considerable evidence in favour of the former hypothesis (e.g. Longo et al. 1996), little has been done to investigate whether the stress response can be inactivated during PCD. Contreras et al. (2002) found in a microarray study of Bax expression in yeast that at early time points, death is characterized by a repression of the major components of the antioxidant defence systems, including *CTA1*, *DDR48*, *GRE2*, *GRX1*, *HSP12*, *TRX1* and *TTR1*. This finding fits in well with the earlier report that the Ras–cAMP–protein kinase A pathway is stimulated during cell death and functions in part to inhibit STRE-mediated stress responses. Further array studies will need to be performed under a wider range of conditions to test the validity of this important proposition.

ROS production is commonly linked to the apoptotic response in animals. Whilst Bcl-2 interacts with Bax to negate its pro-apoptotic functions, Bcl-2 may also have a direct antioxidant role to play. In the short-term, it has been found that Bcl-2 expression prevents the death of both *sod1* and *sod1 sod2* yeast mutants entering into stationary phase (Longo et al. 1997). Such a result is consistent with an antioxidant role for Bcl-2, acting on basic elements which must be present in all eukaryotes. Surprisingly, in contrast to all other mutations known to suppress the growth defects of *sod1*- and *sod1 sod2*-deleted strains, Bcl-2 does not have any effect on the accumulation of Mn^{2+} or Cu^{2+} ions. Expression of Bcl-2 is, however, accompanied by an increase in catalase activity and an increase in GSH/GSSG ratios. Both changes suggest a direct role for Bcl-2 in promoting antioxidant defence mechanisms.

The nature of the active ROS themselves is not certain. Deletion of *TSA1*, a thiol peroxidase, increases the sensitivity of yeasts to osmotin (Narasimham et al. 2001). Since *TSA1* is believed to specifically detoxify hydrogen peroxide (Lee et al. 1999), it may be surmised that this is one of the intracellular mediators of the death response. Priault et al. (2002) have also reported that whereas a modulation of global ROS did not significantly affect Bax-induced killing, the direct modulation of lipid oxidation had a strong effect. Superoxide anions are thought to be a major component of the active ROS signal in animals (Cai and Jones 1998), but this has yet to be demonstrated in fungi.

Currently, we do not know enough to ascertain whether ROS only operate as signals for downstream apoptotic effectors and/or if ROS can act directly as effectors. Many studies have shown that ROS are damaging to DNA and lipids (Halliwell and Gutteridge 1989), and so it might be hypothesized that aspects of PCD, such as the accumulation of dsDNA breaks, could be directly attributable to this action. However, the production of ROS in the absence of measurable DNA damage has been seen after treatment of *C. albicans* with [Cu2(phen)3(mal)].2H_2O (Coyle et al. 2004), which could be taken to imply that ROS play a primary signalling role during apoptosis and do not act directly as DNA damaging effectors.

Two cases have been described in which ROS do not play a central role in the decision to die. The first involves the induction of apoptosis in *S. cerevisiae* with aspirin (Balzan et al. 2004), and the second is the induction of apoptosis by phytosphingosine in *A. nidulans* (Cheng et al. 2003). It appears that growth inhibition by aspirin cannot be reversed by free-radical scavengers such as N-acetyl-cysteine or vitamin E. ROS production is detectable, but it only occurs very late on, once apoptotic markers have already appeared. In this situation it would appear that in the early stages of death, aspirin is itself acting as an antioxidant, and that cells become committed to die in a ROS-independent fashion. At a later stage, a secondary accumulation of ROS occurs which swamps the intrinsic antioxidant properties of aspirin. During phytosphingosine-induced apoptosis of *A. nidulans*, ROS production precedes nuclear condensation, but the inhibition of ROS does not prevent cell death. Since death was also independent of the metacaspase *casA*, it appears that a second apoptotic pathway may exist in fungi which is also independent of ROS production.

The fact that ROS production lies at the core of most apoptotic responses may be attributable to the finding that in yeast, ROS, as a by-product of respiratory activities, are under the control of an ultradian clock (Lloyd et al. 2003). Senescence, which is associated with apoptosis in yeast, is therefore expected to ensue after a finite number of cycles of the 'ticking' of the ultradian clock which regulates ROS production. Since the clock is driven by inherent oscillations of central metabolic processes such as glycolysis and the electron transport chain, it will be a feature of all organisms with oxidative metabolism.

B. Cytochrome *c* and Mitochondrial Function

The intrinsic apoptosis pathway involves release of Cyt *c* from the mitochondrion, which leads to the formation of an apoptogenic complex (the mitochondrial apoptosome) which is able to activate caspases in the cytosol. When Cyt *c* is released from the inner mitochondrial membrane, normal electron flow in complex III of the respiratory chain is interrupted, leading to the generation of superoxide radicals (Cai and Jones 1998).

Yamaki et al. (2001), studying yeast apoptosis in an *ASF1/CIA1* mutant, found that cell death was accompanied by a decrease in the mitochondrial membrane potential, dysfunction of mitochondrial ATPases, and the release of Cyt *c*. Bax expression also leads to cytochrome *c* release in dying yeast cells (Manon et al. 1997; Kluck et al. 1999).

Ludovico et al. (2002) studied Cyt *c* release in stationary-phase cells (when mitochondria have accumulated) and observed that pro-apoptotic doses of acetic acid decreased the amount of Cyt *c* in mitochondria two-to threefold, and that this was offset by an increase in cytosolic Cyt *c*. Other mitochondrial components such as Cox II and V could not be detected in the cytosol, arguing against a non-specific lytic leakage of proteins. Staining of mitochondria with Mitotracker Red CM-H_2-XRos revealed the production of ROS in both the mitochondria and cytoplasm. Polarigraphic measurements of oxygen consumption in isolated mitochondria revealed a 75% decrease in the activity of NADH oxidase. Levels of a number of respiratory components in the mitochondrion were also reduced compared to untreated controls, suggesting significant changes in the organisation of the mitochondrial respiratory chain. Oligomycin did not inhibit yeast apoptosis, arguing that oxidative phosphorylation per se is not important for yeast apoptosis, but *ATP10*, ρ^0 and *cyc3* mutant cells did not undergo acetic acid-induced apoptosis, implicating mitochondria in the death response.

Linked to Cyt *c* release is a reduction in the level of cytochrome *c* oxidase which facilitates the uncoupling of the respiratory chain. Manon et al. (1997) linked the decrease in levels of Cyt *c* oxidase to the activation of Yme1, an AAA-family protease, which may be responsible for degrading components such as COX II. Under aerobic conditions, in a *yme1*-deleted strain, Bax-induced lethality was only slightly suppressed, indicating that the reduction in Cyt *c* oxidase in itself did not contribute significantly to the death response (Manon et al. 2001). Since *ATP10* deletion, but not oligomycin treatment, was able to inhibit yeast apoptosis, it may be surmised that a functional F_0F_1 ATPase complex is required for the release of Cyt *c*, but that there is no absolute requirement for mitochondrial ATP production per se.

Bax expression stimulates release of Cyt *c* from mitochondria and decreases cytochrome oxidase (Manon et al. 1997). However, Cyt *c* release per se was not essential in this case because a Cyt *c*–GFP construct, which remains stuck in the mitochondrial membrane, does not prevent Bax-induced killing (Roucou et al. 2000).

Whilst release of Cyt *c* has been demonstrated, its ability to act as a pro-apoptotic signal remains less certain. Yeast Cyt *c*, in contrast to Cyt *c* from many other organisms, has been found to be ineffective in inducing apoptotic changes in a mammalian nuclear apoptosis assay (Kluck et al. 1997).

The permeability transition pore is a complex upon which many apoptotic signals converge. It is composed of a voltage-dependent anion channel (VDAC) and a peripheral benzodiazepine receptor in the outer mitochondrial membrane. Several reports also suggest that Bax itself is also a part of the outer membrane complex (Marzo et al. 1998; Shimizu et al. 1999). The inner mitochondrial membrane components contain an adenine nucleotide translocator (ANT-1) and the chaperone cyclophilin D. Disintegration of the outer mitochondrial membrane and the release of pro-apoptotic factors such as Smac/Diablo and Cyt *c* is considered the point of no return in the implementation of a PCD response.

Whilst a number of early studies implicated the ADP/ATP carrier (*AAC1*, *AAC2*, *AAC3*) proteins and a requirement for respiratory function in Bax-induced cell death, Kissova et al. (2000) found that the cytotoxic effects of Bax expression in yeast do not require functional oxidative phosphorylation, respiration or ADP/ATP carriers. Such a major discrepancy could be attributable to the repression of the *GAL* promoter of Bax constructs in cells with dysfunctional mitochondria (Kissova et al. 2000). The view that Bax effects are independent of mitochondrial functions is, however, further supported by the studies made with Bax under the regulation of a tetracycline regulatable promoter (Priault et al. 1999a).

Using GFP-tagged mitochondrial proteins, Bax was shown to interfere with the mitochondrial protein import pathway which is specific for the proteins of the mitochondrial carrier family (Kissova et al. 2000). Bax becomes localised at the outer mitochondrial membrane (Priault et al. 1999a; Pavlov et al. 2001) and induces cell death which is accompanied by the release of Cyt *c* (Manon et al. 1997). This does not require any transition permeability of the inner mitochondrial membrane, or the VDAC proteins, *Por1* and *Por2* (Priault et al. 1999b; Gross et al. 2000; Harris et al. 2000; Polcic and Forte 2003) or the adenine nucleotide carriers (Priault et al. 1999a; Kissova et al. 2000).

Bax is believed to act upon mitochondria by creating a giant channel, independent of VDAC, in the outer mitochondrial membrane (Pavlov et al. 2001).

C. Caspases

Using nested, iterative PSI-BLAST searches, Uren et al. (2000) identified a family of caspases, metacaspases and paracaspases in plants, animals and fungi. In fungi, several metacaspases have subsequently been found. *S. cerevisiae* and *S. pombe* have a single member, *YCA1*/*MCA1* and *PCA1* respectively. *C. albicans* also has a single homolog (Ramsdale, unpublished data) whilst *A. fumigatus*, *A. nidulans* and *N. crassa* may have two each (Cheng et al. 2003; Mousavi and Robson 2004; Thrane et al. 2004).

Madeo et al. (2002b) was able to show that disruption of *YCA1* abrogated hydrogen peroxide-induced apoptosis and that Yca1 was cleaved in a manner typical of mammalian caspases. Yca1 also displayed a caspase-like proteolytic activity which was stimulated during apoptosis. Overexpression of *YCA1* increased caspase-like activity and increased the susceptibility of yeast to hydrogen peroxide-induced killing.

Yca1 is a 52-kDa protein with a 12-kDa cleavage product which may correspond to the small catalytic subunit found in true caspases. Alanine substitution of cysteine (A297C) prevents the formation of the small cleavage product and detectable caspase activity. Overexpression of *YCA1* and subsequent exposure to hydrogen peroxide revealed a caspase-like activity which was able to cleave the fluorescent substrates VEID-AMC and IETD-AMC but not DEVD-AMC. Significantly, this proteolytic activity could be abolished with the pan-caspase inhibitor zVAD-fmk. Such a substrate preference is indicative of animal initiator caspases, such as caspase-8.

Kang et al. (1999) showed that human caspase-8, when expressed in yeast, could be activated and that it was the only heterologous caspase with cytotoxic properties. Pre-processing of the procaspase per se was not associated with a significant increase in enzyme activity, suggesting perhaps a requirement for an additional *trans*-activating factor. The search for a factor such as Apaf-1 is ongoing. However, several proteins have been shown to interact with Yca1, including Sro77 and Wwm1 (Szallies et al. 2002; Wadskog et al. 2004). *WWM1* was found in an overexpression screen to be a very potent inhibitor of cell proliferation, causing cells to arrest in G1 (Stevenson et al. 2001). Significantly, *WWM1* overexpression phenotypes are suppressed by *YCA1* (Szallies et al. 2002).

In mammals, a mitochondrial mediator of apoptosis, DAP-3, which may act upstream (Miyazaki and Reed 2001) or downstream (Kissil et al. 1999) of caspase-8, has been identified. Berger et al. (2000) found that DAP-3 was highly conserved across many different groups, leading Madeo et al. (2002b) to show that the yeast DAP-3 homolog (*RSM23* – a component of the mitochondrial small subunit ribosome complex) could completely prevent induction of apoptosis by Yca1 overexpression, placing it downstream of caspase activation.

Characterization of *NMA111*, an HtrA2-like protein, in *S. cerevisiae* (Fahrenkrog et al. 2004) reveals that it shares a similar pro-apoptotic function to that of its mammalian homolog. Yeasts cells lacking Nma111 are able to survive better than wild-type at 50 °C and following exposure to pro-apoptotic doses of hydrogen peroxide. Furthermore, the *nma111* mutant also fails to exhibit hallmark features of apoptosis such as chromatin condensation, nuclear fragmentation or ROS accumulation. Critically, overexpression of Nma111 can directly stimulate apoptosis-like cell death. Under stress conditions, Nma111 aggregates in the nucleus and its pro-apoptotic function in yeast is dependent upon its serine-protease activity.

In mammals, HtrA2 stimulates cell suicide by reducing the levels of XIAP, an inhibitor of apoptosis protein (IAP). An IAP-like protein, Bir1, has also been described in yeast (Uren et al. 1999; Yoon and Carbon 1999; Li et al. 2000). Bir1 contains a baculovirus inhibitor repeat which is conserved amongst IAP proteins and is required for the interaction of the proteins with caspases. This protein has not so far been demonstrated to have a link to yeast apoptosis, but has been implicated in the regulation of cell division and arrest.

The first fungal homologs of poly-ADP ribose polymerase, a downstream target of effector caspases, have recently been discovered in *A. nidulans* and *N. crassa* (Thrane et al. 2004). During conidiation of *A. nidulans*, a time-dependent loss of poly-ADP ribose polymerase can be observed, which coincides with the programmed autolysis of the underlying mycelium. A caspase-3 inhibitor delayed, but did not abolish the degradation of poly-ADP ribose polymerase by fungal extracts, with endogenously high levels of caspase-like activity. The 81-kDa fungal poly-ADP ribose polymerase does not contain classical caspase-3 (DEVD) or caspase-8 (IETD) cleavage sites, but does contain a KVVDK site at a location which would give the 60-kDa fragment observed when it is exposed to extracts with high endogenous fungal caspase activity.

D. Nucleases

DNA fragmentation, with laddering, is most clearly associated with the PCD response of *Mucor racemosus* (Roze and Linz 1998). Incubation of lovastatin-treated cells in YPG at pH 4.5 prevented the activation of the DNA fragmentation response whilst incubation in a variety of buffers and media at pH 7.45 was permissive. It may therefore be that the endonuclease is pH-dependent, with an optimal neutral pH. Addition of EDTA, but not EGTA, to treated cells was also inhibitory for DNA fragmentation, perhaps indicating a requirement for Ca^{2+} and Mg^{2+} ions. DNA fragmentation could not be inhibited by cycloheximide, suggesting that the enzyme is constitutively produced.

DNA laddering per se has not so far been detected in yeasts undergoing apoptosis. It is now widely recognized that laddering is not a reliable feature of apoptosis in mammalian cells (e.g. Oberhammer et al. 1993), where it is often restricted to specific cell lines. Madeo et al. (2002a) have proposed that a lack of laddering may be due to basic differences in the architecture of yeast and mammalian chromatin (Lowary and Widom 1989). However, MNase treatment of isolated yeast chromatin does give a strong laddering effect (Kunoh et al. 2000), indicating that the architecture may not be too dissimilar. Differences in the apoptotic programme itself may therefore be more crucial.

During the sugar-induced apoptotic response of yeast, fragments of DNA about 30 kb in size can be detected by pulse field gel electrophoresis (Granot et al. 2003). In *S. pombe* (Ink et al. 1997), Bak expression has been shown to induce chromatin degradation with an accumulation of higher-order structures around 800 kb in size. In both cases, these fragments may be equivalent to the DFF40 hypersensitive sites in HeLa cells or the 30–50 kb DNA fragments which have often been reported in apoptotic animal cells (Brown et al. 1993; Oberhammer et al. 1993) prior to the formation of oligomeric ladders.

Although relatively little information is available concerning RNA degradation during apoptosis, specific cleavage of 28S, but not 18S, rRNA has been observed in several mammalian cell lines (Houge et al. 1995; Houge and Doskeland 1996). Whilst Granot et al. (2003) observed equal amounts of degradation in *S. cerevisiae*, dying *C. albicans* treated with either acetic acid or hydrogen peroxide show a preferential degradation of 28S rRNA (Ramsdale, personal observation) which is accentuated during necrosis. RNA processing activities are commonly found to be amongst the most significantly altered biological functions in proteome and array studies of dying mammalian cells (Brockstedt et al. 1998; Thiede et al. 2001). Although there are few comparable studies on fungi, similar changes are seen at the transcriptome level in the raw data of a number of studies (Ramsdale, unpublished data; Contreras et al. 2002; Enjalbert et al. 2003).

E. Endogenous BCL-2-Family Proteins

Using rabbit polyclonal antibodies, Bcl-2 and Bax-like proteins were detected in *Mucor racemosus* on Western blots, with sizes of 22 and 16–18 kDa respectively (though several other proteins of higher molecular weight were also detected in each analysis, so the evidence is not conclusive; Roze and Linz 1998). Bioinformatic studies have failed to reveal convincing homologues of the BCL-2-family proteins, but Komatsu et al. (2000) have shown that *S. pombe* Rad9 contains a BH3 domain which is required for its interaction with heterologously expressed Bcl-2 proteins. Screens for endogenous SpRAD9 interactants in yeasts could be useful in identifying components of an ancestral AIF.

VI. Global Responses of Dying Cells

Microarray technologies have allowed an in-depth exploration of the responses of fungi to a wide range of physiological and stress conditions (see Wilhelm and Bähler, Chap. 6, this volume). Several investigators have shown that a common signature subset of genes responds to many different stresses; components of the so-called environmental stress response (ESR; Gasch et al. 2000) or common environmental stress response (Causton et al. 2001; Chen et al. 2003). Typically during any given stress response, 30%–50% of the RNA encoded by the genome can alter in abundance (due either to transcriptional regulation or altered mRNA stability). The signature stress response accounts for a significant part of this, involving about 10%–15% of all genes.

Genes induced during the ESR are involved in glycolysis, antioxidant defence, protein folding and proteolysis, and vacuolar or mitochondrial functions. Repressed genes are responsible for many aspects of energy consumption and growth-related processes – most notably, ribosome biogenesis. The response is transient and is proportional to the intensity of the stress (Gasch and Werner-Washburne 2002). When the profiles of *S. cerevisiae* and *S. pombe* are compared, a significant degree of overlap is observed, indicating that the core responses are highly conserved.

The interrelationships between stress responses and death responses warrant closer inspection. In particular, it is important to know the extent to which responses overlap and to identify the major differences. Surprisingly, there are no published proteome or transcript profiling studies of stress-induced PCD in *S. cerevisiae* or *S. pombe*. A number of studies have investigated the global responses of fungi to antifungal treatments (Bammert and Fostel 2000; de Backer et al. 2001; Zhang et al. 2002a, b; Agarwal et al. 2003) or fungicides (Kitagawa et al. 2003). However, in these cases the agents investigated are either fungistatic or, if they are fungicidal, the array analyses have been performed on RNA collected from cells which have been treated with inhibitory, but nonetheless sub-lethal doses (e.g. 0.5× MIC). Unfortunately, this makes it impossible to demarcate death-specific functions.

Biella et al. (2002) reported that the response of *Cryphonectria parasitica* to infection with the dsRNA hypovirulence factor resembles PCD. Using microarrays to look for genes which were differentially regulated when strains became infected with the virus, 295 unique sequences (out of 2200) were found with changed abundance (Allen et al. 2003). Using differential display, Chen et al. (1996) observed that 65% of the global changes initiated by virus infection could be reproduced by manipulating G-protein and cAMP signalling pathways – further strong evidence linking the death responses to Ras-protein kinase A activity. Many of the other biological processes identified as significantly up-regulated during infection (e.g. SAM metabolism, polyamine biosynthesis and glutathione metabolism) have also been detected in an array study of acetic acid-induced apoptosis in *C. albicans* (Ramsdale, unpublished data) Furthermore, such changes have been linked to apoptosis through classical studies of both plants and animals (Kawalleck et al. 1992; Tschopp et al.1998).

In a pilot study of cell death in *C. albicans* which has integrated proteomics with transcriptomics (Ramsdale, unpublished data), cells undergoing apoptosis induced by acetic acid have been compared to control cells, stressed cells and necrotic cells. Significant numbers of *C. albicans* proteins and mRNAs appear to change in abundance during the stress and death responses. Analysis of the proteome reveals that >200 protein spots alter specifically in response to stress, 29 spots alter specifically during apoptosis and 50 spots change under necrotic conditions. Furthermore, 29 protein spots can be identified which are specific to the death response, but are not affected by stress-inducing conditions. In all, 144 proteins which change in a significant fashion in response to the stress and death stimuli have been identified by MALDI-ToF mass spectrometry. Whilst many of these proteins have no known function, others play key roles in cell rescue, core metabolism or de novo protein synthesis and proteasome-mediated degradation.

A similar picture is emerging from the parallel transcript profiling analysis of the same samples. *C. albicans* mRNAs have been identified which respond specifically to stress (145 transcripts), apoptosis (19 transcripts) or necrosis (108 transcripts). Many are also observed which change in response to both apoptosis and necrosis, but not stress (109 transcripts). These transcripts encode components of the proteasome and ubiquitin-mediated proteolysis pathways, proteases, glycolytic and TCA cycle enzymes, ergosterol biosynthetic enzymes and ribosomal proteins as well as a putative *C. albicans* metacaspase. One of the most striking findings is that the mRNA abundance of ribosomal proteins remains steady during both apoptotic and necrotic death responses but declines significantly during the stress response, indicating that dying cells are maintaining their protein biosynthesis machinery. The data indicate that necrosis, at least in the context of acetic acid-induced killing, is probably the endpoint of prior apoptotic events.

Both proteomics and transcript profiling can highlight cellular components which might be involved in the regulation or execution of the PCD response. Moreover, such studies emphasize that death is an active process, distinct in many ways from the stress response, which requires the full participation of a cell molecular repertoire.

VII. Fungi as Models for Apoptosis

The early view that yeast did not undergo apoptosis was seized upon by many as an opportunity to study mammalian and worm apoptosis genes in a 'clean' genetic environment. We now know that this is not the case, but nonetheless structure–function relationships and genetic screens revealing proteins which interact with components of the plant or animal cell death machinery have usefully been sought in the 'sterile' environment of the yeast genetics laboratory.

A. Genetic Screens

Overexpression of many genes in fungi can produce lethal effects. Often, death is not due to the increased activity of a dedicated pro-death protein but rather to an imbalance in some critical process. Nevertheless, such screens could provide a useful starting point to look for pro-death genes. If the screen is performed under conditions which stimulate PCD, then antagonists of this process might also be identified with anti-apoptotic properties. Many of the 'essential' genes described in fungi may also have anti-apoptotic roles. It has not yet been ascertained in a systematic fashion which of these induce killing (and, importantly, by what route) and which bring about cell cycle arrest. Future studies in this area might be particularly fruitful in revealing genes involved with yeast apoptosis.

1. Overexpression-Induced Lethality

Lethal outcomes have been observed after overexpressing genes encoding components of the cytoskeleton (Rose and Fink 1987; Burke et al. 1989; Berlin et al. 1990; Magdolen et al. 1993) and signal transduction cascades (Russell and Nurse 1987; Osmani et al. 1988; Whiteway et al. 1990; Millar et al. 1992; Ottilie et al. 1992). In genome-wide library screens using cDNA or genomic clones, lethal effects have been observed for *ABP1*, *ACT1*, *ARF2*, *ATE1*, *AUA1*, *BIK1*, *BNI1*, *BOI1*, *ERG6*, *GCL17*, *HSF1*, *KAR1*, *MCM1*, *NHP6*A, *NHP6*B, *NPS1*, *NSR1*, *NTH1*, *PRK1*, *PSP1*, *RBP1*, *RHO1*, *STE4*, *STE11*, *STE12*, *SAC7*, *SEC17*, *SIR1*, *SNU114*, *SRP40*, *TPK1*, *TPK3*, *TUB1* and *URA2* (Liu et al. 1992; Ramer et al. 1992, Espinet et al. 1995; Akada et al. 1997). Intriguingly, *BNI1* (Gin2) and *BOI1* (Gin7) produce cells with multiple DAPI-staining bodies – perhaps evidence of nuclear fragmentation and apoptosis (Akada et al. 1997).

This list provides further strong evidence in favour of a central role for Ras/protein kinase A signalling in the death responses of yeast. *TPK1*, *TPK3* and *NTH1* are obviously associated with Ras-protein kinase A signalling, whilst *RPB1* expression has been linked to Ras activity (Howard et al. 2002). Elevated expression of *RAS2* produces defects similar to those of cells expressing mutants of *RPB1* with a truncated CTD interacting domain. Ras2 activation is also synthetically lethal with many components of RNA pol II including components of the CTD mediator (*sin4*, *gal11*, *med6*, *srb4*) or TFIIH (*kin28*, *ccl1*, *rig2*, *tfb1*) complexes. Such a lethal effect cannot be attributed to a general defect in translation, since in the *sin4* mutant background global mRNA accumulation was not affected by Ras2 activation.

2. Proteasome-Dependent Lethality

Ligr et al. (2001) developed a screen to search for proteins whose degradation by the ubiquitin-proteasome system was required for viability. Overexpression of such a protein in a normal cell should be permissive for growth, but overexpression in a strain defective in proteolysis should be lethal. A screen was performed on a 2 μ-based *GAL*-regulated cDNA library in a *pre1-1 pre4-1 cyh2* mutant strain containing a *PRE1 CYH2* helper plasmid, pML1 (making the strain phenotypically wild-type with respect to growth, proteasome-dependent proteolysis and cycloheximide sensitivity). Growth on glucose and glucose cycloheximde media (which selected for loss of pML1), alongside replica plating of the original plates containing clones with a *pre1-1 pre1-4* background on galactose, revealed 125 library clones which caused growth arrest in *pre1-1 pre4-1* mutants but not in a *PRE1 pre4-1* background. After sequencing, 62 individual ORFs were identified as high-expression lethality genes (*HEL*). GFP-annexin and TUNEL staining, along with other apoptosis indicators such as nuclear condensation and fragmentation, were observed in six of these clones when overexpressed in a *pre1-1 pre4-1* background (*NSR1*, *PPA1*, *SAR1*, *STM1*, YNL208w-HEL10 and YOR309c-HEL13). ROS production could not be detected in any of the clones tested. Significantly, TUNEL and annexin staining were not detected in any of the other 62 clones. *NSR1* caused significant killing in the *PRE1 PRE4* strain, but the killing was exacerbated in the *pre1-1 pre4-1* background. Hence, it was included in the list of potential pro-apoptotic proteasome-regulated genes. Using *GAL1* promoter shut-off of an *STM1*:IRS reporter construct, stabilization of the Stm1 protein in *pre-1 pre4-1* cells was observed, demonstrating that Stm1 is a natural substrate of the active proteasome.

Disruption of *UBP10*, a gene which encodes a de-ubiquitinating enzyme in *S. cerevisiae*, results in slow growth and impairment of silencing at telomeres and at HM loci. In addition, mutant cells show an accumulation of ROS, DNA fragmentation and phosphatidylserine exposure – all hallmark features of apoptosis (Orlandi et al. 2004). A genome-wide analysis of a *ubp10* null strain revealed an alteration in the expression of many subtelomeric genes, as expected along with transcriptional changes indicative of an oxidative stress response. Construction of a *ubp10 sir4* double null strain prevented the large-scale re-modelling of the transcriptional response and abrogated the apoptotic phenotypes.

Down-regulated genes identified encoded a variety of transporters (tyrosine, glutamate, isoleucine, ammonium, biotin and peptide carriers), ribosomal proteins and translation factors (consistent with many stress-inducing conditions – see above discussion), and enzymes involved with the biosynthesis of arginine, lysine, tryptophan and various nucelotides. Up-regulated genes included enzymes involved with carbohydrate transport, metabolism and energy production, heat-shock proteins and a number of transcription factors (notably, *HAP2*, *HAP4*, *ROX1* and *YAP4*). Overall, such changes cluster with many other transcriptional profiles which are associated with oxidative stresses and the general stress response mediated by Msn2/4.

The array study indicates that correct de-ubiquitination is required for normal transcriptional balances, and that disruption of this can lead to apoptosis associated with the production of ROS and an associated oxidative stress response. In mammals, levels of the tumour suppressor gene p53 are in part controlled by the de-ubiquitination activities of the cell (de-ubiquitination stabilizes p53). Altered patterns of ubiquitination of pro- or anti-apoptotic proteins in yeast, in particular those involved with chromatin modelling, might therefore account for the death phenotypes observed when *UBP10* is disrupted.

3. Bax Suppressors

The pro-apoptotic members of the BCL-2 family of proteins, Bax and Bak, have been shown to kill yeast cells, and cell death is typically associated with an apoptotic phenotype (Ink et al. 1997; Jurgensmeier et al. 1997; Ligr et al. 1998). Not all fungi respond in the same way to these pro-apoptotic proteins. Expression of murine Bax in *Pichia pastoris*, for example, leads to growth arrest accompanied by the condensation of chromatin, and the accumulation of autophagic bodies (Martinet et al. 1999), but no other apoptotic features.

Yeasts have been used to examine structure–function relationships of mammalian pro-apoptotic proteins. Matsuyama et al. (1998a, 1999) found, using both mammalian and yeast cells expressing Bax, that it has two mechanisms of killing: putative pore-forming domain-dependent cell death (in both yeast and mammals) and dimerization-dependent cell death (only in mammalian cells). Minn et al. (1999) revealed that Bax could induce hyperpolarization of mitochondria in yeast, verifying a hypothesis for the mode of action of Bax in mammalian cell death responses.

Mutagenization of yeast with N-methyl-N′-nitro-N-nitrosoguanidine allowed the identification of mutants which were resistant to expression of Bax (Matsuyama et al. 1998b). Out of eight clones identified, only one, *ATP4*, was able to fully restore Bax-induced lethality when replaced with a wild-type copy. *ATP4* (subunit 4 of the F_0F_1-ATPase) is a component of a proton pump located on the inner membrane of the mitochondrion. Pharmacological inhibition of the proton pump with oligomycin reproduced the effects of deleting ATP4 on Bax-induced lethality. Furthermore, by comparing Bax-induced killing in the petites ρ^- and ATPδ knockout strains, the authors were able to conclude that respiration per se was not required for Bax-induced killing, but that an intact proton pump system was.

Expression of a soybean cDNA library in a Bax-expressing yeast strain has been used to screen for plant genes which prevented Bax-induced killing (Moon et al. 2002). From a total of five separate Bax-inhibiting clones, isolated ascorbate peroxidase (sAPX) was studied further. The overexpression of sAPX greatly suppressed the generation of Bax-associated ROS which is thought to be responsible for its protective effect.

Kuwana et al. (2002) have suggested that cardiolipin is required for Bax-mediated pore formation in liposome models, but Iverson et al. (2004) found that Cyt *c* release mediated by Bax in yeast was not affected in a *crd1* (cardiolipin-deficient) strain. Spontaneous release of Cyt *c* could be detected from mitochondria isolated from the *crd1* deficient strain (release was not detectable in WT mitochondria), which argues against Cyt *c* release being sufficient to induce apoptosis, as *crd1* mutants have normal growth. However, since Cyt *c* release could be observed only under a specific set of ionic conditions which promoted swelling of the CL deficient mitochondrial preparations, this conclusion must be viewed cautiously.

A genetic screen in yeast has been able to link Bax expression to both apoptotic and autophagic death. Camougrand et al. (2003) obtained a yeast mutant which was resistant to Bax-induced cell death by virtue of a decrease in the level of *UTH1*. Null *uth1* strains were also found to be resistant to Bax which could not be linked to a change in the localisation of Bax, its insertion into the mitochondrial membrane, or the release of Cyt *c*. *UTH1* disruption confers resistance to rapamycin under respiratory but not fermentative conditions, linking the gene and Bax to an autophagic pathway involving mitochondria.

A yeast two-hybrid screen revealed that Cnx1 (calnexin), an ER integral membrane protein with a cytosolic C-terminus, interacted with Bak in *S. pombe*. Moreover, expression of a truncated version of calnexin which lacked the Bak-binding C-terminus, in a *cnx1* knockout strain, could prevent Bak-induced cell death (Torgler et al. 1997). This study corroborates findings in several assessments of mammalian cells where Bak and other BCL-2-family proteins have been shown to co-localise with calnexin in the ER.

4. Caspase-Family Proteins

Mammalian and worm caspases have been expressed in *S. cerevisiae* and *S. pombe*. Ced-3 is lethal when expressed in *S. cerevisiae* (Tao et al. 1999), and lethality can be blocked by co-expression of ced-9. Ced-3, but not ced-4, toxicity was antagonized by co-expression of the caspase inhibitors CrmA and p35, showing that native apoptotic interactions can be accurately modelled in yeast.

Caspase-3 expression in yeast delays growth but does not induce killing (Wright et al. 1999), and expression of active forms of caspases 2,6-9 has no effect on growth, suggesting that there are no essential endogenous targets for these proteases in yeast. In *S. pombe*, both caspase-1 and caspase-3 induce growth arrest which is dependent upon proteolytic capacity (Ryser et al. 1999). Caspase-mediated death can be rescued by co-expression of the baculovirus protein p35 but not Bcl-2. Significantly, the yeast studies revealed that Bcl-2 can be cleaved by caspase-1 and -3.

Cascades of caspase activation have been recreated in yeast (Kang et al. 1999). Co-expression of caspase-8/-3 resulted in extensive cell lysis and a punctate DAPI staining of the cytoplasm which could be correlated with DNA fragmentation detected on agarose gels. Although oligosomal-sized fragments were not detected, a substantial amount of DNA about 2 kb in size could be detected. In addition, RNA was not seen in the cells expressing caspase-8/-3, suggesting some activation of RNAse activities. Endogenous yeast targets of the expressed caspases must contribute to this terminal phenotype, and so it will be of interest to determine their nature. It is often thought that caspase-8 mostly contributes to the downstream processing of other caspases, e.g. caspase-3, but cytotoxicity in yeast could imply that it has other targets or that it is actually activating the endogenous yeast metacaspase.

In order to identify novel caspases and their regulators, Hawkins et al. (1999) designed a reporter system for caspase activity in yeast. Co-expression of a caspase alongside a transcriptional activator (LexA-B42) fused to a transmembrane protein via a linker which contains multiple (but distinct) caspase cleavage sites releases the transcription factor from its membrane targeted support, allowing the activation of a *lacZ* reporter driven by a *lexA* UAS. This approach was successfully exploited to identify DIAP1, an inhibitor of the *Drosophila* caspase DCP-1, although there is no reason why the screen cannot be used to identify other novel caspases or small molecule inhibitors of caspases.

5. Other Non-Yeast Death Proteins

Mutations in the gene encoding p53 have been found to be the most common genetic alterations in human cancer. p53 is thought to exert its tumor suppression functions through an inhibition of cell proliferation and induction of apoptosis in response to DNA damage. Overexpression of wild-type human p53 has been found to block growth of *S. pombe* and *S. cerevisiae* (Bischoff et al. 1992; Nigro et al. 1992). In *S. pombe*, the p53 polypeptide localised to the nucleus and became phosphorylated at both the cdc2 site and the casein kinase II site (Bischoff et al. 1992).

In an attempt to identify p53-like proteins in yeast, Koerte et al. (1995) isolated a mutant which required wild-type p53 for its viability. The mutant, *rft1-1*, was defective in cell cycle progression and arrested at G1/S when p53 was depleted. Genetic and biochemical studies revealed that p53 suppressed the *rft1-1* mutation by forming a protein–protein complex with the Rft1 protein.

A core activity of p53 is an ability to stimulate transcription of genes which prevent cells with damaged DNA from entering S phase. p53 was able to strongly stimulate a reporter containing multiple repeats of the p53 recognition sequence RRRC(A/T)(T/A)GYYY in wild-type yeast, but could only weakly stimulate expression in a *trr1* null. The ectopic expression of the thioredoxin reductase gene, *TRR1*, could restore p53-dependent reporter gene activity to high levels. Such a study in yeast led to the proposition that p53 disulphides in p53 must be reduced in order for the protein to function as a transcription factor (Pearson and Merrill 1998).

Granzyme B and its pro-enzyme form have both been successfully expressed in the yeast *Pichia pastoris* (Pham et al. 1998), with no reported effects on yeast viability. The constructs described were fused in frame with yeast alpha-factor cDNA and were then released into the supernatant by the KEX2 signal peptidase, which may explain the observed lack of cytotoxicity.

VIII. Evolutionary Considerations

Regulated cell death has been described in all major phyla – perhaps with the exception of the archaebacteria (see introductory references). One overriding question yet to be answered by this review is why unicellular organisms such as bacteria and yeasts should have a PCD response at all. Of course, it is not possible to look back into the history of a yeast such as *S. cerevisiae* and ask what was the defining event which led to the acquisition of the ancestral death programme. However, we may be sure that it was either a serendipitous event or that its acquisition conferred some adaptive value.

Taking serendipity as a starting point, it is clear that some of the key players of the death response of 'higher' organisms such as mammals have their evolutionary roots in prokaryotic proteins, e.g. HtrA and Bax. The intrinsic cell suicide programme in particular may have its evolutionary origins at the dawn of the eukaryotic mode of life. We can only imagine the interplay which occurred between ancestral prokaryotes as they formed new alliances but, if the machinery for PCD was already present in prokaryotes, then it may have been retained during the subsequent emergence of the eukarya. This would then help to explain why PCD has been described in all eukaryotic lineages to date – it is simply built in. Alternatively, PCD in yeasts may be a hangover from a more recent evolutionary transition. We do not know, for example, if the ancestral relatives of the euascomycetous yeasts were multicellular. If they were, then they may have possessed and carried over PCD machinery for the same reasons which are advocated in extant multicellular organisms.

Since unicellular organisms are often most closely associated with their clonal relatives, explanations for the implementation of a PCD response based upon altruism also become plausible. Suicide of a single cell which is damaged, or infected within a clonal colony, can both release nutrients for use by the colony and prevent the spread of infectious agents (viruses, selfish genetic elements) or damaged hereditary material throughout a population. Finally, many microorgansims form intimate commensal, mutualistic and parasitic associations with other organisms which may require fine control over population size. It is not inconceivable, therefore, that PCD can be used to regulate population density, with the payoff that scarce spatial and nutritional resources are utilized more effectively.

IX. Conclusions

The biology of cell death in fungi is a large, relatively unexplored area. This is surprising, given the obvious need to control fungal infections in plants, animals and man. Whilst much effort has been invested in understanding the growth of fungi, and studies of their proliferation are at the forefront of current scientific research, studies of their death responses per se appear to have been sidelined in favour of pragmatic screens which aim to directly identify antifungal agents.

The failure to focus on death itself has led to the identification of many agents which are merely fungistatic, rather than fungicidal, and this has had a number of important consequences for the success or failure of therapeutic programmes. Mode-of-action studies often appear to follow almost as an after-thought, and these in themselves often do not explain why fungal cells die.

This chapter has examined what little we do know about the cell biology of fungal death responses, and shows how a combination of cell biological approaches, functional genetic analyses, genetic screens and global profiling technologies is just beginning to unravel this important, but neglected, aspect of fungal growth and development. Genomic or proteomic studies have not been widely used to investigate the nature of PCD in fungi to date. However, it is hoped that this review will stimulate such studies, and allow us to highlight both the similarities and differences between the events which regulate mammalian PCD and fungal PCD. I hope it is already apparent that a number of discrete endogenous cell suicide pathways exist in fungi, which might be usefully exploited in the search for novel therapies against fungal infections.

Acknowledgements. I would like to thank Andrew Phillips, Jon Crowe and Alistair Brown for their many helpful discussions. Support for this work has been provided by a Lloyd's Tercentenary Foundation Fellowship, with additional funds from the Scottish Hospitals Research Trust, the MRC and the BBSRC.

References

Agarwal AK, Rogers PD, Baerson SR, Jacob MR, Barker KS, Cleary JD, Walker LA, Nagle DG, Clark AM (2003) Genome-wide expression profiling of the response to polyene, pyrimidine, azole, and echinocandin antifungal agents in *Saccharomyces cerevisiae*. J Biol Chem 278:34998–35015

Ainsworth AM, Rayner ADM, Broxholme SJ, Beeching JR (1990) Occurrence of unilateral genetic transfer and genomic replacement between strains of *Stereum hirsutum* from non-outcrossing and outcrossing populations. New Phytol 115:119–128

Akada R, Yamamoto J, Yamashita I (1997) Screening and identification of yeast sequences that cause growth inhibition when overexpressed. Mol Gen Genet 254:267–274

Ali GS, Reddy ASN (2000) Inhibition of fungal and bacterial plant pathogens by synthetic peptides: in vitro growth inhibition, interaction between peptides and inhibition of disease progression. Mol Plant-Microbe Interact 13:847–859

Allen TD, Dawe AL, Nuss DL (2003) Use of cDNA microarrays to monitor transcriptional responses of the chestnut blight fungus Cryphonectria parasitica to infection by virulence-attenuating hypoviruses. Eukaryot Cell 2:1253–1265

Ameisen JC (1996) The origin of programmed cell death. Science 272:1278–1279

Anagnostakis SL, Day PR (1979) Hypovirulence conversion in Endothia parasitica. Phytopathology 69:1226–1229

Anglade P, Vyas S, Javoy-Agid F, Herrero MT, Michel PP, Marquez J, Mouatt-Prigent A, Ruberg M, Hirsch EC, Agid Y (1997) Apoptosis and autophagy in nigral neurons of patients with Parkinson's disease. Histol Histopathol 12:25–31

Aylmore RC, Todd NK (1986) Cytology of non-self hyphal fusions and somatic incompatibility in Phanerochaete velutina. J Gen Microbiol 132:581–591

Baek Y-U, Kim Y-R, Yim H-S, Kang S-O (2004) Disruption of γ-glutamylcysteine synthetase reults in absolute glutathione auxotrophy and apoptosis in *Candida albicans*. FEBS Lett 556:47–52

Balzan R, Sapienza K, Galea DR, Vassallo N, Frey H, Bannister WH (2004) Aspirin commits yeast cells to apoptosis depending on carbon source. Microbiology 150:109–115

Bammert GF, Fostel JM (2000) Genome-wide expression patterns in *Saccharomyces cerevisiae*: comparison of drug treatments and genetic alterations affecting biosynthesis of ergosterol. Antimicrob Agents Chemother 44:1255–1265

Barker MG, Brimage LJ, Smart KA (1999) Effect of Cu, Zn superoxide dismutase disruption mutation on replicative senescence in *Saccharomyces cerevisiae*. FEMS Microbiol Lett 177:199–204

Baudat F, Manova K, Yuen JP, Jasin M, Keeney S (2000) Chromosome synapsis defects and sexually dimorphic meiotic progression in mice lacking Spo11. Mol Cell 6:989–998

Berger T, Brigl M, Herrmann JM, Vielhauer V, Luckow B, Schlondorff D, Kretzler M (2000) The apoptosis mediator mDAP-3 is a novel member of a conserved family of mitochondrial proteins. J Cell Sci 113:3603–3612

Berlin V, Styles CA, Fink GR (1990) BIK1, a protein required for microtubule function during mating and mitosis in *Saccharomyces cerevisiae*, colocalizes with tubulin. J Cell Biol 111:2573–2586

Biella S, Smith ML, Aist JR, Cortesi P, Milgroom MG (2002) Programmed cell death correlates with virus transmission in a filamentous fungus. Proc R Soc Lond B 269:2269–2276

Bischoff JR, Casso D, Beach D (1992) Human p53 inhibits growth in *Schizosaccharomyces pombe*. Mol Cell Biol 12:1405–1411

Blanchard F, Rusiniak ME, Sharma K, Sun X, Todorov I, Castellano MM, Gutierrez, Baumann C, Burhans WC (2002) Targeted destruction of DNA replication protein cdc6 by cell death pathways in mammals and yeast. Mol Biol Cell 13:1536–1549

Brockstedt E, Rickers A, Kostka S, Laubersheimer A, Dorken B, Wittmann-Liebold B, Bommert K, Otto A (1998) Identification of apoptosis-associated proteins in a human Burkitt lymphoma cell line. J Biol Chem 273:28057–28064

Brodersen P, Petersen M, Pike HM, Olszak B, Skov S, Odum N, Jorgensen LB, Brown RE, Mundy J (2002) Knockout of Arabidopsis accelerated-cell-death11 encoding a sphingosine transfer protein causes activation of programmed cell death and defense. Genes Dev 16:490–502

Brown DG, Sun XM, Cohen GM (1993) Dexamethasone induced apoptosis involves cleavage of DNA to large fragments prior to internucleosomal fragmentation. J Biol Chem 268:3037–3039

Budovskaya YV, Stephan JS, Reggiori F, Klionsky DJ, Herman PK (2004) The Ras/cAMP-dependent protein kinase signaling pathway regulates an early step of the autophagy process in *Saccharomyces cerevisiae*. J Biol Chem 279:20663–20671

Burhans WC, Weinberger M, Marchetti MA, Ramachandran L, D'Urso G, Huberman JA (2003) Apoptosis-like yeast cell death in response to DNA damage and replication defects. Mutat Res 532:227–243

Burke D, Gasdaska P, Hartwell L (1989) Dominant effects of tubulin overexpression in *Saccharomyces cerevisiae*. Mol Cell Biol 9:1049–1059

Cai J, Jones DP (1998) Superoxide in apoptosis. Mitochondrial generation triggered by cytochrome *c* loss. J Biol Chem 273:11401–11404

Camougrand N, Grelaud-Coq A, Marza E, Priault M, Bessoule JJ, Manon S (2003) The product of the UTH1 gene, required for Bax-induced cell death in yeast, is involved in the response to rapamycin. Mol Microbiol 47:495–506

Castedo M, Perfettini JL, Roumier T, Kroemer G (2002) Cyclin-dependent kinase-1: linking apoptosis to cell cycle and mitotic catastrophe. Cell Death Differ 9:1287–1293

Cataldo AM, Barnett JL, Berman SA, Li J, Quarless S, Bursztajn S, Lippa C, Nixon RA (1995) Gene expression and cellular content of cathepsin D in Alzheimer's Disease brain: evidence for early up-regulation of the endosomal-lysosomal system. Neuron 14:671–680

Causton HC, Ren B, Koh SS, Harbison CT, Kanin E, Jennings EG, Lee TI, True HL, Lander ES, Young RA (2001) Remodeling of yeast genome expression in response to environmental changes. Mol Biol Cell 12:323–337

Celerin M, Merino ST, Stone JE, Menzle AM, Zolan ME (2000) Multiple roles of Spo11 in meiotic chromosome behavior. EMBO J 19:2739–2750

Chen B, Gao S, Choi GH, Nuss DL (1996) Extensive alteration of fungal gene transcript accumulation and elevation of G-protein-regulated cAMP levels by a virulence-attenuating hypovirus. Proc Natl Acad Sci USA 93:7996–8000

Chen D, Toone WM, Mata J, Lyne R, Burns G, Kivinen K, Brazma A, Jone N, Bahler J (2003) Global transcriptional responses of fission yeast to environmental stress. Mol Biol Cell 14:214–229

Cheng J, Park TS, Chio LC, Fischl AS, Ye XS (2003) Induction of apoptosis by sphingoid long-chain bases in *Aspergillus nidulans*. Mol Cell Biol 23:163–177

Chi S, Kitanaka C, Noguchi K, Mochizuki T, Nagashima Y, Shirouzu M, Fujita H, Yoshida M, Chen W, Asai A et al. (1999) Oncogenic Ras triggers cell suicide through the activation of a caspase-independent cell death program in human cancer cells. Oncogene 18:2281–2290

Contreras RH, Eberhardt I, Luyten WHML, Reekman RJ (2002) Bax-responsive genes for drug target identification in yeast and fungi. Patent International Publication Number WO 02/064766 A2

Cooke RC, Rayner ADM (1984) Ecology of saprophytic fungi. Longman, London, UK

Cornillon S, Foa C, Davoust J, Buonavista N, Gross JD, Golstein P (1994) Programmed cell death in Dictyostelium. J Cell Sci 107:2691–2704

Cortesi P, McCulloch CE, Song H, Lin H, Milgroom MG (2001) Genetic control of horizontal virus transmission in the chestnut blight fungus, Cryphonectria parasitica. Genetics 159:107–118

Coyle B, Kinsella P, McCann M, Devereux M, O'Connor R, Clynes M, Kavanagh K (2004) Induction of apoptosis in yeast and mammalian cells by exposure to 1,10-phenanthroline metal complexes. Toxicol In Vitro 18:63–70

Cutler NS, Pan X, Heitman J, Cardenas ME (2001) The TOR signal transduction cascade controls cellular differentiation in response to nutrients. Mol Biol Cell 12:4103–4113

Cyert MS, Thorner J (1992) Regulatory subunit (CNB1 gene product) of yeast Ca^{2+}/calmodulin-dependent phosphoprotein phosphatases is required for adaptation to pheromone. Mol Cell Biol 12:3460–3469

De Backer MD, Ilyina T, Ma X-J, Vandoninck S, Luyten WHML, Vanden Bossche H (2001) Genomic profiling of the response of *Candida albicans* to itraconazole treatment using a DNA microarray. Antimicrob Agents Chemother 45:1660–1670

Debets F, Yang X, Griffiths AJF (1994) Vegetative incompatibility in Neurospora: its effect on horizontal transfer of mitochondrial plasmids and senescence in natural populations. Curr Genet 26:113–119

Del Carratore R, Della Croce C, Simili M, Taccini E, Scavuzzo M, Sbrana S (2002) Cell cycle and morphological alterations as indicative of apoptosis promoted by UV irradiation in *S. cerevisiae*. Mutat Res 513:183–191

Dementhon K, Paoletti M, Pinan-Lucarre B, Loubradou-Bourges N, Sabourin M, Saupe SJ, Clave C (2003) Rapamycin mimics the incompatibility reaction in the fungus *Podospora anserina*. Eukaryot Cell 2:238–246

Dickson RC, Nagiec EE, Wells GB, Nagiec MM, Lester RL (1997) Synthesis of mannose-(inositol-P)2-ceramide, the major sphingolipid in *Saccharomyces cerevisiae*, requires the IPT1 (YDR072c) gene. J Biol Chem 272:29620–29625

Drees BL, Sundin B, Brazeau E, Caviston JP, Chen GC, Guo W, Kozminski KG, Lau MW, Moskow JJ, Tong A et al. (2001) A protein interaction map for cell polarity development. J Cell Biol 154:549–571

D'Souza CA, Heitman J (2001) Conserved cAMP signaling cascades regulate fungal development and virulence. FEMS Microbiol Rev 25:349–364

Du C, Fang M, Li Y, Li L, Wang X (2000) Smac, a mitochondrial protein that permits cytochrome *c*-dependent caspase activation by eliminating IAP inhibition. Cell 102:33–42

Edlind T, Smith L, Henry K, Katiyar S, Nickels J (2002) Antifungal activity in *Saccharomyces cerevisiae* is modulated by calcium signalling. Mol Microbiol 46:257–268

Elkabetz Y, Shapira I, Rabinovich E, Bar-Nun S (2004) Distinct steps in dislocation of luminal endoplasmic reticulum-associated degradation substrates: roles of endoplasmic reticulum-bound p97/Cdc48p and proteasome. J Biol Chem 279:3980–3989

Engelberg-Kulka H, Glaser G (1999) Addiction modules and programmed cell death and antideath in bacterial cultures. Annu Rev Microbiol 53:43–70

Enjalbert B, Nantel A, Whiteway M (2003) Stress-induced gene expression in *Candida albicans*: absence of a general stress response. Mol Biol Cell 14:1460–1467

Epple P, Apel K, Bohlmann H (1997) Overexpression of an endogenous thionin enhances resistance of Arabidopsis against Fusarium oxysporum. Plant Cell 9:509–520

Espinet C, de la Torre MA, Aldea M, Herrero E (1995) An efficient method to isolate yeast genes causing overexpression-mediated growth arrest. Yeast 11:25–32

Esser K (2006) Heterogenic incompatibility. In: Esser K (ed) The Mycota, vol I, 2nd edn. Kües U, Fischer R (eds) Growth, differentiation and sexuality. Springer, Berlin Heidelberg New York (in press)

Esser K, Blaich R (1994) Heterogenic incompatibility in fungi. In: Wessels JGH, Meinhardt F (eds) The Mycota, vol I. Growth, differentiation and sexuality. Springer, Berlin Heidelberg New York, pp 211–228

Fabrizio P, Battistella L, Vardava R, Gattazzo C, Liou L-L, Diaspro A, Dossen JW, Gralla EB, Longo VD (2004) Superoxide is a mediator of an altruistic aging program in *Saccharomyces cerevisiae*. J Cell Biol 166:1055–1067

Fahrenkrog B, Sauder U, Aebi U (2004) The *S. cerevisiae* HtrA-like protein Nma111p is a nuclear serine protease that mediates yeast apoptosis. J Cell Sci 117:115–126

Ferket KK, Levery SB, Park C, Cammue BP, Thevissen K (2003) Isolation and characterization of Neurospora crassa mutants resistant to antifungal plant defensins. Fungal Genet Biol 40:176–85

Fernandes L, Corte-Real M, Loureiro V, Loureiro-Dias MC, Leao C (1997) Glucose respiration and fermentation in *Zygosaccharomyces bailii* and *Saccharomyces cerevisiae* express different sensitivity patterns to ethanol and acetic acid. Lett Appl Microbiol 25:249–253

Fraser A, James C (1998) Fermenting debate: do yeast undergo apoptosis? Trends Cell Biol 8:219–221

Frohlich KU, Madeo F (2000) Apoptosis in yeast-a monocellular organism exhibits altruistic behaviour. FEBS Lett 473:6–9

Frohlich KU, Madeo F (2001) Apoptosis in yeast: a new model for aging research. Exp Gerontol 37:27–31

Garcia-Olmedo F, Molina A, Alamillo JM, Roderiguez-Palenzuela P (1998) Plant defense peptides. Biopolymers 47:479–491

Gasch AP, Werner-Washburne M (2002) The genomics of yeast responses to environmental stress and starvation. Funct Integr Genomics 2:181–192

Gasch AP, Spellman PT, Kao CM, Carmel-Harel O, Eisen MB, Storz G, Botstein D, Brown PO (2000) Genomic expression programs in the response of yeast cells to environmental changes. Mol Biol Cell 11:4241–4257

Glass NL, Jacobson DJ, Shiu PKT (2000) The genetics of hyphal fusion and vegetative incompatibility in filamentous ascomycete fungi. Annu Rev Genet 34:165–186

Golstein P, Aubry L, Levraud JP (2003) Cell-death alternative model organisms: why and which? Nat Rev Mol Cell Biol 4:798–807

Granot D, Levine A, Dor-Hefetz E (2003) Sugar-induced apoptosis in yeast cells. FEMS Yeast Res 4:7–13

Gross A, Pilcher K, Blachly-Dyson E, Basso E, Jockel J, Bassik MC, Korsmeyer SJ, Forte M (2000) Biochemical and genetic analysis of the mitochondrial response of yeast to BAX and BCL-X(L). Mol Cell Biol 20:3125–3136

Halliwell B, Gutteridge JMC (1989) Free radicals in biology and medicine. Oxford University Press, Oxford Hansberg W, Aguirre J (1990) Hyperoxidant states cause microbial cell differentiation by cell isolation from dioxygen. J Theor Biol 142:201–221

Hansberg W, de Groot H, Sies H (1993) Reactive oxygen species associated with cell differentiation in *Neurospora crassa*. Free Radic Biol Med 14:287–93

Harman D (1962) Role of free radicals in mutation, cancer, aging and maintenance of life. Radic Res 16:752–763

Harris MH, Vander Heiden MG, Kron SJ, Thompson CB (2000) Role of oxidative phosphorylation in Bax toxicity. Mol Cell Biol 20:3590–3596

Hartl DL, Dempster ER, Brown SW (1975) Adaptive significance of vegetative incompatibility in *Neurospora crassa*. Genetics 81:553–569

Hawkins CJ, Wang SL, Hay BA (1999) A cloning method to identify caspases and their regulators in yeast: Identification of Drosophila IAP1 as an inhibitor of the Drosophila caspase DCP-1. Proc Natl Acad Sci USA 96:2885–2890

Helmerhorst EJ, Troxler RF, Oppenheim FG (2001) The human salivary peptide histatin 5 exerts its antifungal activity through the formation of reactive oxygen species Proc Natl Acad Sci USA 98:14637–14642

Herker E, Jungwirth H, Lehmann KA, Maldener C, Frohlich KU, Wissing S, Buttner S, Fehr M, Sigrist S, Madeo F (2004) Chronological aging leads to apoptosis in yeast. J Cell Biol 164:501–507

Hiramoto F, Nomura N, Furumai T, Oki T, Igarashi Y (2003) Apoptosis-like cell death of *Saccharomyces cerevisiae* induced by a mannose-binding antifungal antibiotic, pradimicin. J Antibiot (Tokyo) 56:768–772

Hochman A (1997) Programmed cell death in prokaryotes. Crit Rev Microbiol 23:207–214

Holstege FCP, Jennings EG, Wyrick JJ, Lee TI, Hengartner CJ, Green MR, Golub TR, Lander ES, Young RA et al. (1998) Dissecting the regulatory circuitry of a eukaryotic genome. Cell 95:717–728

Houge G, Doskeland SO (1996) Divergence towards a dead end? Cleavage of the divergent domains of ribosomal RNA in apoptosis. Experientia 52:963–967

Houge G, Robaye B, Eikhom TS, Golstein J, Mellgren G, Gjertsen BT, Lanotte M, Doskeland SO (1995) Fine mapping of 28S rRNA sites specifically cleaved in cells undergoing apoptosis. Mol Cell Biol 15:2051–2062

Howard SC, Budovskaya YV, Chang YW, Herman PK (2002) The C-terminal domain of the largest subunit of RNA polymerase II is required for stationary phase entry and functionally interacts with the Ras/PKA signaling pathway. J Biol Chem 277:19488–19497

Huh GH, Damsz B, Matsumoto TK, Reddy MP, Rus AM, Ibeas JI, Narasimhan ML, Bressan RA, Hasegawa PM (2002) Salt causes ion disequilibrium-induced programmed cell death in yeast and plants. Plant J 29:649–659

Ink B, Zornig M, Baum B, Hajibagheri N, James C, Chittenden T, Evan G (1997) Human Bak induces cell death in *Schizosaccharomyces pombe* with morphological changes similar to those with apoptosis in mammalian cells. Mol Cell Biol 17:2468–2474

Iverson SL, Enoksson M, Gogvadze V, Ott M, Orrenius S (2004) Cardiolipin is not required for Bax-mediated cytochrome *c* release from yeast mitochondria. J Biol Chem 279:1100–1107

Jacobson DJ, Beurkens K, Klomparens KL (1998) Microscopic and ultrastructural examination of vegetative incompatibility in partial diploids heterozygous at het loci in Neurospora crasssa. Fungal Genet Biol 23:45–56

Jazwinski SM (1993) The genetics of aging in the yeast *Saccharomyces cerevisiae*. Genetica 91:35–51

Jeon BW, Kim KT, Chang SI, Kim HY (2002) Phosphoinositide 3-OH kinase/protein kinase B inhibits apoptotic cell death induced by reactive oxygen species in *Saccharomyces cerevisiae*. J Biochem 131:693–699

Jurgensmeier JM, Krajewski S, Armstrong RC, Wilson GM, Oltersdorf T, Fritz LC, Reed JC, Ottilie S (1997) Bax- and Bak-induced cell death in the fission yeast *Schizosaccharomyces pombe*. Mol Biol Cell 8:325–339

Kamada Y, Funakoshi T, Shintani T, Nagano K, Ohsumi M, Ohsumi Y (2000) Tor-mediated induction of autophagy via an Apgl protein kinase complex. J Cell Biol 150:1507–1513

Kang JJ, Schaber MD, Srinivasula SM, Alnemri ES, Litwack G, Hall DJ, Bjornsti MA (1999) Cascades of mammalian caspase activation in the yeast *Saccharomyces cerevisiae*. J Biol Chem 274:3189–3198

Kawalleck P, Plesch G, Hahlbrock K, Somssich IE (1992) Induction by fungal elicitor of S-adenosyl-L-methionine synthetase and S-adenosyl-Lhomocysteine hydrolase mRNAs in cultured cells and leaves of Petroselinum crispum. Proc Natl Acad Sci USA 89:4713–4717

Kim S, Fyrst H, Saba J (2000) Accumulation of phosphorylated sphingoid long chain bases results in cell growth inhibition in *Saccharomyces cerevisiae*. Genetics 156:1519–1529

Kissil JL, Cohen O, Raveh T, Kimchi A (1999) Structure-function analysis of an evolutionary conserved protein, DAP3, which mediates TNF-alpha- and Fas-induced cell death. EMBO J 18:353–362

Kissova I, Polcic P, Kempna P, Zeman I, Sabova L, Kolarov J (2000) The cytotoxic action of Bax on yeast cells does not require mitochondrial ADP/ATP carrier but may be related to its import to the mitochondria. FEBS Lett. 471:113–118

Kitagawa E, Momose Y, Iwahashi H (2003) Correlation of the structure of agricultural fungicides to gene expression in *Saccharomyces cerevisiae* upon exposure to toxic doses. Environ Sci Technol 37:2788–2793

Klionsky DJ, Emr SD (2000) Autophagy as a regulated pathway of cellular degradation. Science 290:1717–1721

Klionsky DJ, Ohsumi Y (1999) Vacuolar import of proteins and organelles from the cytoplasm. Annu Rev Cell Dev Biol 15:1–32

Kluck RM, Martin SJ, Hoffman BM, Zhou JS, Green DR, Newmeyer DD (1997) Cytochrome c activation of

CPP32-like proteolysis plays a critical role in a Xenopus cell-free apoptosis system. EMBO J 16:4639–4649
Kluck RM, Esposti MD, Perkins G, Renken C, Kuwana T, Bossy-Wetzel E, Goldberg M, Allen T, Barber MJ, Green DR et al. (1999) The pro-apoptotic proteins, Bid and Bax, cause a limited permeabilization of the mitochondrial outer membrane that is enhanced by cytosol. J Cell Biol 147:809–822
Koerte A, Chong T, Li X, Wahane K, Cai M (1995) Suppression of the yeast mutation rft1-1 by human p53. J Biol Chem 270:22556–22564
Komatsu K, Hopkins KM, Lieberman HB, Wang H (2000) *Schizosaccharomyces pombe* Rad9 contains a BH3-like region and interacts with the anti-apoptotic protein Bcl-2. FEBS Lett 481:122–126
Kontoyiannis DP, Murray PJ (2003) Fluconazole toxicity is independent of oxidative stress and apoptotic effector mechanisms in *Saccharomyces cerevisiae*. Mycoses 46:183–186
Kornitzer D, Sharf R, Kleinberger T (2001) Adenovirus E4orf4 protein induces PP2A-dependent growth arrest in *Saccharomyces cerevisiae* and interacts with the anaphase-promoting complex/cyclosome. J Cell Biol 154:331–344
Kunoh T, Sakuno T, Furukawa T, Kaneko Y, Harashima S (2000) Genetic characterization of rbt mutants that enhance basal transcription from core promoters in *Saccharomyces cerevisiae*. J Biochem (Tokyo) 128:575–584
Kurjan J (1993) The pheromone response pathway in *Saccharomyces cerevisiae*. Annu Rev Genet 27:147–179
Kuwana T, Mackey MR, Perkins G, Ellisman MH, Latterich M, Schneiter R, Green DR, Newmeyer DD (2002) Bid, Bax, and lipids cooperate to form supramolecular openings in the outer mitochondrial membrane. Cell 111:331–342
Lai CY, Jaruga E, Borghouts C, Jazwinski SM (2002) A mutation in the ATP2 gene abrogates the age asymmetry between mother and daughter cells of the yeast *Saccharomyces cerevisiae*. Genetics 162:73–87
Lam E (2004) Controlled cell death, plant survival and development. Nat Rev Mol Cell Biol 5:305–315
Laun P, Pichova A, Madeo F, Fuchs J, Ellinger A, Kohlwein S, Dawes I, Frohlich KU, Breitenbach M (2001) Aged mother cells of *Saccharomyces cerevisiae* show markers of oxidative stress and apoptosis. Mol Microbiol 39:1166–1173
Lee J, Spector D, Godon C, Labarre J, Toledano M (1999) A new antioxidant with alkyl hydroperoxide defense properties in yeast. J Biol Chem 274:4537–4544
Leslie JF, Zeller KA (1996) Heterokaryon incompatibility in fungi: more than just another way to die. J Genet 75:415–424
Lewis K (2000) Programmed death in bacteria. Microbiol Mol Biol Rev 64:503–514
Li P, Nijhawan D, Budihardjo I, Srinivasula SM, Ahmad M, Alnemri ES, Wang X (1997) Cytochrome c and dATP-dependent formation of Apaf-1/caspase-9 complex initiates an apoptotic protease cascade. Cell 91:479–489
Li F, Flanary PL, Altieri DC, Dohlman HG (2000) Cell division regulation by BIR1, a member of the inhibitor of apoptosis family in yeast. J Biol Chem 275:6707–6711
Liao RS, Rennie RP, Talbot JA (1999) Assessment of the effect of amphotericin B on the vitality of *Candida albicans*. Antimicrob Agents Chemother 43:1034–1041
Ligr M, Madeo F, Frohlich E, Hilt W, Frohlich KU, Wolf DH (1998) Mammalian Bax triggers apoptotic changes in yeast. FEBS Lett 438:61–65
Ligr M, Velten I, Frohlich E, Madeo F, Ledig M, Frohlich KU, Wolf DH, Hilt W (2001) The proteasomal substrate Stm1 participates in apoptosis-like cell death in yeast. Mol Biol Cell 12:2422–2432
Liu H, Krizek J, Bretscher A (1992) Construction of a GAL1-regulated yeast cDNA expression library and its application to the identification of genes whose overexpression causes lethality in yeast. Genetics 132:665–673
Liu X, Kim CN, Yang J, Jemmerson R, Wang X (1996) Induction of apoptotic program in cell-free extracts: requirement for dATP and cytochrome *c*. Cell 86:147–157
Lloyd D, Lemar KM, Salgado LE, Gould TM, Murray DB (2003) Respiratory oscillations in yeast: mitochondrial reactive oxygen species, apoptosis and time; a hypothesis. FEMS Yeast Res 3:333–339
Lockshin RA, Zakeri Z (2004) Caspase-independent cell death? Oncogene 23:2766–2773
Longo VD, Gralla EB, Valentine JS (1996) Superoxide dismutase activity is essential for stationary phase survival in *Saccharomyces cerevisiae*. J Biol Chem 271:12275–12280
Longo VD, Ellerby LM, Bredesen DE, Valentine JS, Gralla EB (1997) Human Bcl-2 reverses survival defects in yeast lacking superoxide dismutase and delays death of wild-type yeast. J Cell Biol 137:1581–1588
Lowary PT, Widom J (1989) Higher-order structure of *Saccharomyces cerevisiae* chromatin. Proc Natl Acad Sci USA 86:8266–8270
Lu BC (1991) Cell degeneration and gill remodelling during basidiocarp development in the fungus *Coprinus cinereus*. Can J Bot 69:1161–1169
Lu BC, Gallo N, Kues U (2003) White-cap mutanats and meiotic apoptosis in the basidiomycete *Coprinus cinereus*. Fungal Genet Biol 39:82–93
Ludovico P, Sousa MJ, Silva MT, Leao C, Corte-Real M (2001) *Saccharomyces cerevisiae* commits to a programmed cell death process in response to acetic acid. Microbiology 147:2409–2415
Ludovico P, Rodrigues F, Almeida A, Silva MT, Barrientos A, Corte-Real M (2002) Cytochrome c release and mitochondria involvement in programmed cell death induced by acetic acid in *Saccharomyces cerevisiae*. Mol Biol Cell 13:2598–2606
Ludovico P, Sansonetty F, Silva MT, Corte-Real M (2003) Acetic acid induces a programmed cell death process in the food spoilage yeast *Zygosaccharomyces bailii*. FEMS Yeast Res 3:91–96
Madeo F, Frohlich E, Frohlich KU (1997) A yeast mutant showing diagnostic markers of early and late apoptosis. J Cell Biol 139:729–734
Madeo F, Frohlich E, Ligr M, Grey M, Sigrist SJ, Wolf DH, Frohlich KU (1999) Oxygen stress: a regulator of apoptosis in yeast. J Cell Biol 145:757–767
Madeo F, Engelhardt S, Herker E, Lehmann N, Maldener C, Proksch A, Wissing S, Frohlich KU (2002a) Apoptosis in yeast: a new model system with applications in cell biology and medicine. Curr Genet 41:208–216
Madeo F, Herker E, Maldener C, Wissing S, Lachelt S, Herlan M, Fehr M, Lauber K, Sigrist SJ, Wesselborg S et al. (2002b) A caspase-related protease regulates apoptosis in yeast. Mol Cell 9:911–917

Magdolen V, Drubin DG, Mages G, Bandlow W (1993) High levels of profilin suppress the lethality caused by overproduction of actin in yeast cells. FEBS Lett 316:41–47

Manon S, Chaudhuri B, Guerin M (1997) Release of cytochrome *c* and decrease of cytochrome *c* oxidase in Bax-expressing yeast cells, and prevention of these effects by coexpression of Bcl-xL. FEBS Lett 415:29–32

Manon S, Priault M, Camougrand N (2001) Mitochondrial AAA-type protease Yme1p is involved in Bax effects on cytochrome *c* oxidase. Biochem Biophys Res Commun 289:1314–1319

Marek SM, Wu J, Louise Glass N, Gilchrist DG, Bostock RM (2003) Nuclear DNA degradation during heterokaryon incompatibility in *Neurospora crassa*. Fungal Genet Biol 40:126–137

Martinet W, Van den Plas D, Raes H, Reekmans R, Contreras R (1999) Bax-induced cell death in Pichia pastoris. Biotechnol Lett 21:821–829

Marzo I, Brenner C, Zamzami N, Jurgensmeier JM, Susin SA, Vieira HL, Prevost MC, Xie Z, Matsuyama S, Reed JC et al. (1998) Bax and adenine nucleotide translocator cooperate in the mitochondrial control of apoptosis. Science 281:2027–2031

Mathieu AL, Gonin S, Leverrier Y, Blanquier B, Thomas J, Dantin C, Martin G, Baverel G, Marvel J (2001) Activation of the phosphatidylinositol 3-kinase/Akt pathway protects against interleukin-3 starvation but not DNA damage-induced apoptosis. J Biol Chem 276:10935–10942

Matsuyama S, Schendel SL, Xie Z, Reed JC (1998a) Cytoprotection by Bcl-2 requires the pore-forming alpha5 and alpha6 helices. J Biol Chem 273:30995–31001

Matsuyama S, Xu Q, Velours J, Reed JC (1998b) The mitochondrial F_0F_1-ATPase proton pump is required for function of the proapoptotic protein Bax in yeast and mammalian cells. Mol Cell 1:327–336

Matsuyama S, Nouraini S, Reed JC (1999) Yeast as a tool for apoptosis research. Curr Opin Microbiol 2:618–623

Mazzoni C, Mancini P, Verdone L, Madeo F, Serafini A, Herker E, Falcone C (2003) A truncated form of KlLsm4p and the absence of factors involved in mRNA decapping trigger apoptosis in yeast. Mol Biol Cell 14:721–729

McCann M, Geraghty M, Devereux M, O'Shea D, Mason J, O'Sullivan L (2000) Insights into the mode of action of the anti-Candida activity of 1,10-phenanthroline and its metal chelates. Metal-Based Drugs 7:185–193

Mendoza I, Rubio F, Rodriguez-Navarro A, Pardo JM (1994) The protein phosphatase calcineurin is essential for NaCl tolerance of *Saccharomyces cerevisiae*. J Biol Chem 271:8792–8796

Metzstein MM, Stanfield GM, Horvitz HR (1998) Genetics of programmed cell death in *C. elegans*: past, present and future. Trends Genet 14:410–416

Migheli A, Piva R, Wei J, Attanasio A, Casolino S, Hodes ME, Dlouhy SR, Bayer SA, Ghetti B (1997) Diverse cell death pathways result from a single missense mutation in weaver mouse. Am J Pathol 151:1629–1638

Millar JB, Russell P, Dixon JE, Guan KL (1992) Negative regulation of mitosis by two functionally overlapping PTPases in fission yeast. EMBO J 11:4943–4952

Minn AJ, Kettlun CS, Liang H, Kelekar A, Vander Heiden MG, Chang BS, Fesik SW, Fill M, Thompson CB (1999) Bcl-xL regulates apoptosis by heterodimerization-dependent and -independent mechanisms. EMBO J 18:632–643

Miyazaki T, Reed JC (2001) A GTP-binding adapter protein couples TRAIL receptors to apoptosis-inducing proteins. Nat Immunol 2:493–500

Money NP (2003) Suicidal mushrooms. Nature 423:26

Moon H, Baek D, Lee B, Prasad DT, Lee SY, Cho MJ, Lim CO, Choi MS, Bahk J, Kim MO et al. (2002) Soybean ascorbate peroxidase suppresses Bax-induced apoptosis in yeast by inhibiting oxygen radical generation. Biochem Biophys Res Commun 290:457–462

Moore EJ (1965) Fungal gel tissue ontogenesis. Am J Bot 52:389–395

Moore D (2004) Programmed cell death alive and well in fungi. Mycol Res 107:1251–1252

Moser MJ, Geiser JR, Davis TN (1996) Ca^{2+}-calmodulin promotes survival of pheromone-induced growth arrest by activation of calcineurin and Ca^{2+}-calmodulin-dependent protein kinase. Mol Cell Biol 16:4824–4831

Mousavi SA, Robson GD (2003) Entry into the stationary phase is associated with a rapid loss of viabiliy and an apoptotic-like phenotype in the opportunistic pathogen Aspergillus fumigatau. Fungal Genet Biol 39:221–229

Mousavi SA, Robson GD (2004) Oxidative and amphotericin B-mediated cell death in the opportunistic pathogen *Aspergillus fumigatus* is associated with an apoptotic-like phenotype. Microbiology 150:1937–1945

Narasimhan ML, Damsz B, Coca MA, Ibeas JI, Yun DJ, Pardo JM, Hasegawa PM, Bressan RA (2001) A plant defense response effector induces microbial apoptosis. Mol Cell 8:921–930

Nestelbacher R, Laun P, Vondrakova D, Pichova A, Schuller C, Breitenbach M (2000) The influence of oxygen toxicity on yeast mother cell-specific aging. Exp Gerontol 35:63–70

Nigro JM, Sikorski R, Reed SI, Vogelstein B (1992) Human p53 and CDC2Hs genes combine to inhibit the proliferation of *Saccharomyces cerevisiae*. Mol Cell Biol 12:1357–1365

Oberhammer F, Wilson JW, Dive C, Morris ID, Hickman JA, Wakeling AE, Walker PR, Sikorska M (1993) Apoptotic death in epithelial cells: cleavage of DNA to 300 and/or 50 kb fragments prior to or in the absence of internucleosomal fragmentation. EMBO J 12:3679–3684

Orlandi I, Bettiga M, Alberghina L, Vai M (2004) Transcriptional profiling of ubp10 null mutant reveals altered subtelomeric gene expression and insurgence of oxidative stress response. J Biol Chem 279:6414–6425

Osiewacz HD (2004) Aging and mitochondrial dysfunction in the filamentous fungus *Podospora anserina*. In: Nystrom T, Osiewacz HD (eds) Model systems in ageing. Springer, Berlin Heidelberg New York, pp 17–38

Osiewacz HD, Hamann A (2006) Senescence and longevity. In: Esser K (ed) The Mycota, vol I, 2nd edn. Kües U, Fischer R (eds) Growth, differentiation and sexuality. Springer, Berlin Heidelberg New York (in press) Osmani SA, Pu RT, Morris NR (1988) Mitotic induction and maintenance by overexpression of a G2-specfic gene that encodes a potential protein kinase. Cell 53:237–244

Ottilie S, Chernoff J, Hannig G, Hoffman CS, Erikson RL (1992) The fission yeast genes pyp^{1+} and pyp^{2+} encode

protein tyrosine phosphatases that negatively regulate mitosis. Mol Cell Biol 12:5571–5580
Pavlov EV, Priault M, Pietkiewicz D, Cheng EH, Antonsson B, Manon S, Korsmeyer SJ, Mannella CA, Kinnally KW (2001) A novel, high conductance channel of mitochondria linked to apoptosis in mammalian cells and Bax expression in yeast. J Cell Biol 155:725–731
Pearson GD, Merrill GF (1998) Deletion of the *Saccharomyces cerevisiae* TRR1 gene encoding thioredoxin reductase inhibits p53-dependent reporter gene expression. J Biol Chem 273:5431–5434
Pham CT, Thomas DA, Mercer JD, Ley TJ (1998) Production of fully active recombinant murine granzyme B in yeast. J Biol Chem 273:1629–1633
Phillips AJ, Sudbery I, Ramsdale M (2003) Apoptosis induced by environmental stresses and amphotericin B in *Candida albicans*. Proc Natl Acad Sci USA 100:14327–14332
Pinan-Lucarre B, Paoletti M, Dementhon K, Coulary-Salin B, Clave C (2003) Autophagy is induced during cell death by incompatibility and is essential for differentiation in the filamentous fungus *Podospora anserina*. Mol Microbiol 47:321–333
Polcic P, Forte M (2003) Response of yeast to the regulated expression of proteins in the Bcl-2 family. Biochem J 37:393–402
Priault M, Camougrand N, Chaudhuri B, Schaeffer J, Manon S (1999a) Comparison of the effects of bax-expression in yeast under fermentative and respiratory conditions: investigation of the role of adenine nucleotides carrier and cytochrome *c*. FEBS Lett 456:232–238
Priault M, Chaudhuri B, Clow A, Camougrand N, Manon S (1999b) Investigation of bax-induced release of cytochrome *c* from yeast mitochondria permeability of mitochondrial membranes, role of VDAC and ATP requirement. Eur J Biochem 260:684–691
Priault M, Bessoule JJ, Grelaud-Coq A, Camougrand N, Manon S (2002) Bax-induced cell death in yeast depends on mitochondrial lipid oxidation. Eur J Biochem 269:5440–5450
Qi H, Li TK, Kuo D, Nur-E-Kamal A, Liu LF (2003) Inactivation of Cdc13p triggers MEC1-dependent apoptotic signals in yeast. J Biol Chem 278:15136–15141
Raju NB, Perkins DD (2000) Programmed ascospore death in the homothallic ascomycete *Coniochaete tetraspora*. Fungal Genet Biol 30:213–221
Ramer SW, Elledge SJ, Davis RW (1992) Dominant genetics using a yeast genomic library under the control of a strong inducible promoter. Proc Natl Acad Sci USA 89:11589–11593
Ramsdale M (1998) Genomic conflict in fungal mycelia: a subcellular population biology. In: Worrall JJ (ed) The structure and dynamics of fungal populations. Kluwer, Dordrecht, pp 139–174
Ramsdale M, Rayner ADM (1996) Phenotype-genotype relations in allopatrically-derived heterokaryons of *Heterobasidion annosum* (Fr.) Bref. New Phytol 133:303–319
Rayner ADM (1991) The challenge of the individualistic mycelium. Mycologia 83:48–71
Roberts P, Moshitch-Moshkovitz S, Kvam E, O'Toole E, Winey M, Goldfarb DS (2003) Piecemeal microautophagy of nucleus in *Saccharomyces cerevisiae*. Mol Biol Cell 14:129–141
Rose MD, Fink GR (1987) KAR1, a gene required for function of both intranuclear and extranuclear microtubules in yeast. Cell 48:1047–1060
Roucou X, Prescott M, Devenish RJ, Nagley P (2000) A cytochrome *c*-GFP fusion is not released from mitochondria into the cytoplasm upon expression of Bax in yeast cells. FEBS Lett 471:235–239
Roze LV, Linz JE (1998) Lovastatin triggers an apoptosis-like cell death process in the fungus *Mucor racemosus*. Fungal Genet Biol 25:119–133
Russell P, Nurse P (1987) Negative regulation of mitosis by wee^{1+}, a gene encoding a protein kinase homolog. Cell 49:559–567
Ryser S, Vial E, Magnenat E, Schlegel W, Maundrell K (1999) Reconstitution of caspase-mediated cell death signaling in *Schizosaccharomyces pombe*. Curr Genet 36:21–28
Sámi L, Emri T, Pócsi I (2001) Autolysis and ageing of *Penicillium chrysogenum* cultures under carbon starvation: glutathione metabolism and formation of reactive oxygen species. Mycol Res 105:1246–1250
Sanglard D, Ischer F, Marchetti O, Entenza J, Bille J (2003) Calcineurin A of *Candida albicans*: involvement in antifungal tolerance, cell morphogenesis and virulence. Mol Microbiol 48:959–976
Saupe S, Descamps C, Turcq B, Bégueret J (1994) Inactivation of the *Podospora anserina* vegetative incompatibility locus *het-c*, whose product resembles a glycolipid transfer protein, drastically impairs ascospore production. Proc Natl Acad Sci USA 91:5927–5931
Sekiguchi T, Nakashima T, Hayashida T, Kuraoka A, Hashimoto S, Tsuchida N, Shibata Y, Hunter T, Nishimoto T et al. (1995) Apoptosis is induced in BHK cells by the tsBN462/13 mutation in the CCG1/TAFII250 subunit of the TFIID basal transcription factor. Exp Cell Res 218:490–498
Severin FF, Hyman AA (2002) Pheromone induces programmed cell death in *S. cerevisiae*. Curr Biol 12:R233–R235
Shimizu S, Narita M, Tsujimoto Y (1999) Bcl-2 family proteins regulate the release of apoptogenic cytochrome *c* by the mitochondrial channel VDAC. Nature 399:483–487
Shirogane T, Fukada T, Muller JM, Shima DT, Hibi M, Hirano T (1999) Synergistic roles for Pim-1 and c-Myc in STAT3-mediated cell cycle progression and antiapoptosis. Immunity 11:709–719
Sinclair DA, Guarente L (1997) Extrachromosomal rDNA circles – a cause of aging in yeast. Cell 91:1033–1042
Smriti, Krishnamurthy S, Dixit BL, Gupta CM, Milewski S, Prasad R (2002) ABC transporters Cdr1p, Cdr2p and Cdr3p of a human pathogen *Candida albicans* are general phospholipid translocators. Yeast 19:303–318
Soloca R, Silva R, Ludovico P, Sansonetti F, Martinez-Peinado J, Corte-Real M (2003) High sugar concentrations trigger *Saccharomyces cerevisiae* into a programmed cell death process. In: Abstr Vol 21st Int Yeast Genetics and Molecular Biology Meet, 7–12 July 2003, Goteborg, Sweden, Abstr 18-50. Yeast 20:S306
Stanhill A, Schick N, Engelberg D (1999) The yeast Ras/cyclic AMP pathway induces invasive growth by

suppressing the cellular stress response. Mol Cell Biol 19:7529–7538
Stevenson LF, Kennedy BK, Harlow E (2001) A large-scale overexpression screen in *Saccharomyces cerevisiae* identifies previously uncharacterized cell cycle genes. Proc Natl Acad Sci USA 98:3946–3951
Szallies A, Kubata BK, Duszenko M (2002) A metacaspase of Trypanosoma brucei causes loss of respiration competence and clonal death in the yeast *Saccharomyces cerevisiae*. FEBS Lett 517:144–150
Tao W, Walke DW, Morgan JI (1999) Oligomerized Ced-4 kills budding yeast through a caspase-independent mechanism. Biochem Biophys Res Commun 260:799–805
Telford WG, King LE, Fraker J (1992) Comparative evaluation of several DNA binding dyes in the detection of apoptosis associated chromatin degradation by flow cytometry. Cytometry 13:137–143
Tessier C, Prigent-Tessier A, Ferguson-Gottschall S, Gu Y, Gibori G (2001) PRL antiapoptotic effect in the rat decidua involves the PI3K/ protein kinase B-mediated inhibition of caspase-3 activity. Endocrinology 142:4086–4094
Thevissen K, Ghazi A, De Samblanx GW, Brownlee C, Osborn RW, Broekaert WF (1996) Fungal membrane responses induced by plant defensins and thionins. J Biol Chem 271:15018–15025
Thevissen K, Terras FR, Broekaert WF (1999) Permeabilization of fungal membranes by plant defensins inhibits fungal growth. Appl Environ Microbiol 65:5451–5458
Thevissen K, Cammue BP, Lemaire K, Winderickx J, Dickson RC, Lester RL, Ferket KK, Van Even F, Parret AH, Broekaert WF (2000) A gene encoding a sphingolipid biosynthesis enzyme determines the sensitivity of *Saccharomyces cerevisiae* to an antifungal plant defensin from dahlia (Dahlia merckii). Proc Natl Acad Sci USA 97:9531–9536
Thiede B, Dimmler C, Siejak F, Rudel T (2001) Predominant identification of RNA-binding proteins in Fas-induced apoptosis by proteome analysis. J Biol Chem 276:26044–26050
Thomma BPHJ, Cammue BPA, Thevissen K (2002) Plant defensins. Planta 216:193–202
Thrane C, Kaufmann U, Stummann BM, Olsson S (2004) Activation of caspase-like activity and poly (ADP-ribose) polymerase degradation during sporulation in *Aspergillus nidulans*. Fungal Genet Biol 41:361–368
Toledo I, Rangel P, Hansberg W (1995) Redox imbalance at the start of each morphogenetic step of Neurospora crassa conidiation. Arch Biochem Biophys 319:519–524
Torgler CN, de Tiani M, Raven T, Aubry JP, Brown R, Meldrum E (1997) Expression of Bak in *S. pombe* results in a lethality mediated through interaction with the calnexin homologue Cnx1. Cell Death Differ 4:263–271
Tschopp J, Thome M, Hofmann K, Meinl E (1998) The fight of viruses against apoptosis. Curr Opin Genet Dev 8:82–87
Uchiyama Y (2001) Autophagic cell death and its execution by lysosomal cathepsins. Arch Histol Cytol 64:233–246
Uetz P, Giot L, Cagney G, Mansfield TA, Judson RS, Knight JR, Lockshon D, Narayan V, Srinivasan M, Pochart P et al. (2000) A comprehensive analysis of protein–protein interactions in *Saccharomyces cerevisiae*. Nature 403:623–627
Umar MH, van Griensven LJLD (1997) Morphogenetic cell death in developing primordia of *Agaricus bisporus*. Mycologia 89:274–277
Umar MH, van Griensven LJLD (1998) The role of morphogenetic cell death in the histogenesis of the mycelial cord of *Agaricus bisporus* and in the development of macrofungi. Mycol Res 102:719–735
Uren AG, Beilharz T, O'Connell MJ, Bugg SJ, van Driel R, Vaux DL, Lithgow T (1999) Role for yeast inhibitor of apoptosis (IAP)-like proteins in cell division. Proc Natl Acad Sci USA 96:10170–10175
Uren AG, O'Rourke K, Aravind LA, Pisabarro MT, Seshagiri S, Koonin EV, Dixit VM (2000) Identification of paracaspases and metacaspases: two ancient families of caspase-like proteins, one of which plays a key role in MALT lymphoma. Mol Cell 6:961–967
Van Diepeningen AD, Debets AJM, Hoekstra RF (1997) Heterokaryon incompatibility blocks virus transfer among natural isolates of black Aspergilli. Curr Genet 32:209–217
Verhagen AM, Ekert PG, Pakusch M, Silke J, Connolly LM, Reid GE, Moritz RL, Simpson RJ, Vaux DL (2000) Identification of DABLO, a mammalian protein that promotes apoptosis by binding to and antagonizing IAP proteins. Cell 102:43–53
von Gise A, Lorenz P, Wellbrock C, Hemmings B, Berberich-Siebelt F, Rapp UR, Troppmair J (2001) Apoptosis suppression by Raf-1 and MEK1 requires MEK- and phosphatidylinositol 3-kinase-dependent signals. Mol Cell Biol 21:2324–2336
Wadskog I, Maldener C, Proksch A, Madeo F, Adler L (2004) Yeast lacking the SRO7/SOP1-encoded tumor suppressor homologue show increased susceptibility to apoptosis-like cell death on exposure to NaCl stress. Mol Biol Cell 15:1436–1444
Watanabe K, Morishita J, Umezu K, Shirahige K, Maki H (2002) Involvement of RAD9-dependent damage checkpoint control in arrest of cell cycle, induction of cell death, and chromosome instability caused by defects in origin recognition complex in *Saccharomyces cerevisiae*. Eukaryot Cell 1200–1212
Wawryn J, Krzepilko A, Myszka A, Bilinski T (1999) Deficiency in superoxide dismutases shortens life span of yeast cells. Acta Biochim Pol 46:249–253
Weinberger M, Trabold PA, Lu M, Sharma K, Huberman JA, Burhans WC (1999) Induction by adozelesin and hydroxyurea of origin recognition complex-dependent DNA damage and DNA replication checkpoints in *Saccharomyces cerevisiae*. J Biol Chem 274:35975–35984
Whiteway M, Hougan L, Thomas DY (1990) Overexpression of the STE4 gene leads to mating response in haploid *Saccharomyces cerevisiae*. Mol Cell Biol 10:217–222
Wissing S, Ludovico P, Herker E, Buttner S, Engelhardt SM, Decker T, Link A, Proksch A, Rodrigues F, Corte-Real M et al. (2004) An AIF orthologue regulates apoptosis in yeast. J Cell Biol 166:969–974
Wright ME, Han DK, Carter L, Fields S, Schwartz SM, Hockenbery DM (1999) Caspase-3 inhibits growth in *Saccharomyces cerevisiae* without causing cell death. FEBS Lett 446:9–14
Wu D, Chen P-J, Chen S, Hu Y, Nunez G, Ellis RE (1999) *C. elegans* MAC-1, an essential member of the AAA family of ATPases, can bind CED-4 and prevent cell death. Development 126:2021–2031

Wunder D, Dong J, Baev D, Edgerton M (2004) Human salivary histatin 5 fungicidal action does not induce programmed cell death pathways in *Candida albicans*. Antimicrob Agents Chemother 48:110–115

Yamaki M, Umehara T, Chimura T, Horikoshi M (2001) Cell death with predominant apoptotic features in *Saccharomyces cerevisiae* mediated by deletion of the histone chaperone ASF1/CIA1. Genes Cells 6:1043–1054

Yarmolinsky MB (1995) Programmed cell death in bacterial populations. Science 267:836–837

Yoon HJ, Carbon J (1999) Participation of Bir1p, a member of the inhibitor of apoptosis family, in yeast chromosome segregation events. Proc Natl Acad Sci USA 96:13208–13213

Yun DJ, Ibeas JI, Lee H, Coca MA, Narasimhan ML, Uesono Y, Hasegawa PM, Pardo JM, Bressan RA (1998) Osmotin, a plant antifungal protein, subverts signal transduction to enhance fungal cell susceptibility. Mol Cell 1:807–817

Zhang L, Zhang Y, Zhou Y, An S, Zhou Y, Cheng J (2002a) Response of gene expression in *Saccharomyces cerevisiae* to amphotericin B and nystatin measured by microarrays. J Antimicrob Chemother 49:905–915

Zhang L, Zhang Y, Zhou Y, Zhao Y, Zhou Y, Cheng J (2002b) Expression profiling of the response of *Saccharomyces cerevisiae* to 5-fluorocytosine using a DNA microarray. Int J Antimicrob Agents 20:444–450

Zhang Q, Chieu HK, Low CP, Zhang S, Heng CK, Yang H (2003) *Schizosaccharomyces pombe* cells deficient in triacylglycerol synthesis undergo apoptosis upon entry into the stationary phase. J Biol Chem 278:47145–47155

Zou H, Henzel WJ, Liu X, Lutschg A, Wang X (1997) Apaf-1, a human protein homolog to *C. elegans* CED-4, participates in cytochrome *c*-dependent activation of caspase-3. Cell 90:405–413

8 Genomic Analysis of Cellular Morphology in *Candida albicans*

M. Whiteway[1], A. Nantel[1]

CONTENTS

Abbreviations: DNA, deoxyribonucleic acid; cDNA, copy DNA; mRNA, messenger ribonucleic acid; PCR, polymerase chain reaction; cy3, cyanine 3; cy5, cyanine 5; ORF, open reading frame; YPD, yeast extract peptone dextrose; cAMP, cyclic adenosine monophosphate; UAU, uracil–arginine–uracil cassette

I. Introduction

Genomics can be defined as the scientific approach to biological questions that is based in some way on the knowledge of the complete sequence of the DNA of an organism. Many of the approaches that make up the science of genomics are not new; they are simply made more efficient through the application of information resulting from this complete sequence. Comprehensivity is gained from this knowledge – one can look at the expression of every gene in the organism through microarray analysis, one can build collections of simple cells that each lack a specific gene, and make that collection encompass all the genes, one can build constructs to investigate the cellular location of every protein in a cell, and one can compare and contrast the genetic constitution of closely and distantly related organisms and see the processes of evolution at work. This ability to address scientific questions within the framework of the entire genome sequence of an organism is revolutionizing the approach to the investigation of all living systems.

Genomic technologies are increasingly being applied to the analyses of fungi. The ascomycete yeast *Saccharomyces cerevisiae* was the first eukaryotic cell to have its genome completely sequenced. This sequencing effort involved an international effort encompassing many laboratories, each responsible for a part of the genome. However, the development of high-throughput, industrialized sequencing centers has established conditions where a comprehensive, several-fold redundant sequence of a fungal genome can be generated in a few months at a reasonable cost. This technical capacity is putting the sequences of most experimentally or economically important fungi into publicly accessible databases. These sequences underlie the ability to address questions of cellular evolution and genetic function on a comprehensive, genome-wide level.

The genome of *Candida albicans* represented the first sequence of a human pathogenic fungus to be accessible in a public database. This tour de force was accomplished by the Stanford Genome Technology Center (http://www-sequence.stanford.edu/group/candida; Jones et al. 2004). Subsequent efforts from the Candida community, spearheaded by the Biotechnology Research Center in Montreal, have led to the development of an accurately annotated version of this sequence (http://candida.bri.nrc.ca/), and ongoing efforts at the Montreal Institute and the Magee group at the University of Minnesota are creating a final assembly of the genome. Ultimately, all the genomic information available about *C. albicans* will be collected and distributed through the Candida Genome Database (CGD; http://genome-www.stanford.edu/fungi/Candida/) developed a-

[1] Biotechnology Research Institute, National Research Council of Canada, 6100 Royalmount Ave., Montreal, Quebec H4P 2R2, Canada

The Mycota XIII
Fungal Genomics
Alistair J.P. Brown (Ed.)

round the model of the Saccharomyces Genome Database (SGD; http://www.yeastgenome.org). While annotation and assembly provide the genome in a more readily accessible form, the genomic era of *C. albicans* started with the release of the initial sequence information from Stanford.

II. Genomic Technologies Applied to *C. albicans*

Candida albicans is a particularly attractive organism for the application of genome-based experimental strategies, because these approaches can circumvent some of the technical problems associated with studies of this fungus. *C. albicans* is diploid, and prior to the genome sequencing effort it was also defined as an asexual fungus. Recently, however, the recognition that the genome contained sequences homologous to the mating-type loci of *S. cerevisiae* permitted researchers to uncover a complex mating regulatory circuitry, and to construct strains that exhibit a mating capacity (Hull et al. 2000; Magee and Magee 2000). Still, even if a sexual cycle were to be defined completely, the diploid nature of *C. albicans* precludes the identification of the recessive mutants in cellular processes that have made *S. cerevisiae* one of the pre-eminent model organisms in the analysis of cellular function. The genome sequence, however, provides access to gene function through alternate routes. For example, transcriptional profiling can identify genes whose expression is regulated in a defined manner, and such profiles can be used to group genes into functional classes in the same way that mutations are able to group genes based on their phenotypic consequences. The ability to scan the genome for genes with important predicted functions, or to identify open reading frames that exhibit interesting evolutionary relationships (such as being found only in mammals as well as *Candida*, but not found in other fungi, or being unique to *C. albicans*) also allow researchers to focus their efforts on genes with potentially unique and important characteristics.

Transcriptional profiling makes use of DNA microarrays. Microarrays consist of nucleotide sequences affixed to a solid support, which can be used to probe for the presence, and to a certain extent for the quantity, of nucleotides in a given target. In general, there are three major platforms for these arrays; DNAs attached to a membrane surface, the two-color array pioneered by the Brown group at Stanford (DeRisi et al. 1997), and the multiple oligonucleotide array technology of Affymetrix. The Affymetrix technology was the first to be widely exploited. In this approach, which borrows heavily from the concepts of the semiconductor industry, short oligonucleotides of specific sequences are synthesized directly at a defined location on a solid surface. Each gene to be probed can be represented by several oligonucleotides with a perfect match, as well as several nucleotides with single base mismatches. Thus, the Affymetrix array consists of an enormous number of short oligonucleotides representing perfect or near-perfect matches to each of the genes of the organism to be profiled. When these arrays are used to determine the transcription profile of a particular cell type under a particular condition, mRNA is isolated, turned into labeled cDNA and then hybridized to the array. The strength of the signals representing the perfect matches are compared to the strengths of the signals coming from the mismatched oligonucleotides, and computer analysis is used to determine the strength and quality of each measured interaction. Repetition of the process is used to establish the statistical quality of the determined signals.

Affymetrix arrays require a sophisticated production capacity that is beyond the scope of most research laboratories. Therefore, most of the published transcriptional profiles were produced through the use of either membranes or the two-color array technology developed at Stanford. In general, these arrays are constructed by physically spotting the probe nucleotides at specific locations on a solid surface such as a membrane or a coated glass microscope slide. The spotting strategy can range from the use of pins to ink-jet printers (Hughes et al. 2001), and probes can be PCR products, cDNAs or oligonucleotides. If the latter, the oligonucleotides are typically in the range of 70 bases long, and therefore are considerably longer than the probes used for the Affymetrix technology. The membrane system involves spotting probes on surfaces such as nitrocellulose, and probing typically with radioactively labeled samples that are scored absolutely, not comparatively. Much of the early transcription profiling work for *C. albicans* used membrane-based arrays (Lane et al. 2001; Murad et al. 2001a). Unlike the membrane array system, the two-color system is directly comparative. To use this system to investigate a transcription profile, mRNA is extracted from cells under two different conditions, such as a mutant compared to a wild-type strain, or a given strain grown under two different culture conditions. The RNA from one sample is copied into cDNA and labeled with a fluorescent dye, typically Cy3 or Cy5, and the RNA from the other sample is labeled with the other dye. The two samples are combined, and hybridized to the array. If the two RNAs for a specific probe are equally abundant, then the laser reads the combined Cy3 and Cy5 signals as a yellow signal at that point on the array; if one or the other sample contains a higher level of message, then either the Cy3 (green) or the Cy5 (red) signal predominates. Normalization of signals, signal quality and signal strength are monitored and, as with the Affymetrix technique, experimental repetitions are used to provide statistical robustness to the data. This approach was used initially to investigate the expression of a subset of the *C. albicans* genome to look at azole sensitive and resistant isolates (Cowen et al. 2002; Rogers and Barker 2002), while the first large scale uses of this technology to investigate transcriptional regulation of *C. albicans* were studies on the expression profiles of *Candida* cells growing in the presence of the antifungal drug itraconazole (DeBacker et al. 2001) and on cells undergoing the yeast-to-hyphal transition (Nantel et al. 2002).

Microarrays are not simply useful for the determination of transcription profiles. Because the probes will associate with any homologous, labeled DNA, an array can be used to determine the presence or absence of a gene in a genome. This strategy can be used in strain comparisons for taxonomic purposes. It can also be useful in the assembly of a genome. Purified chromosomes or chromosome fragments can be hybridized to comprehensive gene chips to test the quality of contig assemblies. All the genes on a specific contig should hybridize to a given chromosome or chromosome fragment; this provides a quality control for the assembly of the contigs, and also connects the contigs to a specific chromosome. This approach has proven useful in the assembly of the *C. albicans* genome. Pulse field gel electrophoresis was used to separate all the chromosomes of *C. albicans* strain SC5314, and these separated chromosomes were purified, labeled with fluorescent dyes, and hybridized to arrays consisting of over 6000 open reading frames. The data for these experiments are available online at http://candida.bri.nrc.ca/candida/index.cfm?page=CaChrom. Not all chromosomes could be completely purified, but substantial enrichment was accomplished in each case. This permitted each ORF to be attributed to a chromosome with little ambiguity, and when this assignment was mapped to the currently established contigs, the majority of contigs were uniquely connected to a chromosome. However, on some occasions the contigs could map to two chromosomes, and this permits the identification of misassembled contigs and defines the point at which the misconnection was made.

A second genomic technology involves the construction of libraries of mutant strains. The diploid nature of *C. albicans* precludes the simple production of collections of recessive mutants. However, the tremendous utility of the systematic collection of *S. cerevisiae* ORF disruptions has proven the power of comprehensive null mutant collections. The effort to create the disruption set in *S. cerevisiae* involved a substantial international undertaking, and only recently, with the construction of a set of regulated promoter shut-offs for essential genes (Mnaimneh et al. 2004), has the collection truly approached being comprehensive. However, projects are currently underway to apply the systematic disruption approach to *C. albicans* (Bruno and Mitchell 2004). The annotated genome is the starting point for such studies. A fungal drug discovery company, Elitra Canada, has developed the most advanced disruption technology. Researchers at this company have currently replaced one allele of about 80% of the *C. albicans* genes with a sequence coded selectable marker, and for about half these genes have replaced the endogenous promoter of the other allele with a regulated promoter. This approach, dubbed the GRACE strategy (for gene replacement and controlled expression), has created the ability to shut off genes representing a major fraction of the entire *C. albicans* genome (Roemer et al. 2003). Alternative approaches have also been developed to provide comprehensive mutant collections. One strategy has been to examine the phenotypes of cells that are missing a single copy of a given gene. Such an approach depends on haplo-insufficiency of the disrupted gene causing a detectable phenotype (Uhl et al. 2003). In addition, a clever strategy to select homozygosis of an insertion has been developed that obviates the need to perform double disruptions of each gene to be inactivated (Davis et al. 2002). This approach makes use of a disruption cassette that includes a selectable marker embedded in, and inactivating a second selectable marker. Excision of the first marker (*ARG4*) through recombination will result in the activation of the previously inactivated marker (*URA3*). This UAU cassette is used to disrupt a single copy of a chromosomal gene, and the disruptant is able to proliferate. If the chromosomal disruptant is not in an essential gene, then the population arising from the initial event will include mitotic recombinants that have homozygosed the disruptant allele, creating a rare cell that has two copies of the disruptant allele, and no wild-type gene copy. This rare cell has the unique ability to excise the *ARG4* marker from one copy to create a $URA3^+$ allele, while maintaining the $ARG4^+$ copy at the other allele, and thus can give rise to ARG^+ URA^+ events. Therefore, the selection of the ARG^+ URA^+ cells will identify cells that lack a functional copy of the targeted gene. Systematic application of this cassette can create a knockout collection of the non-essential genes (Bruno and Mitchell 2004).

III. *Candida albicans* Biology: Morphogenesis and Virulence

C. albicans is an important human pathogen because it can opportunistically exploit weaknesses in the human defenses against infection. One of the defining characteristics of *C. albicans* is the morphological plasticity of the organism. It can grow as physically separated yeast cells though a budding process, as chains of attached, elongated cells called pseudohyphae created by a directed budding process, or as true hyphae, where a parallel-sided tube is extended at the apical tip, and separate cellular compartments are created by septation (Sudbery et al. 2004; Whiteway and Oberholzer 2004). In addition, cells can take either a white or an opaque form; white cells have a smooth, rounded shape, while opaque cells are much more elongated and have a pimpled surface (Soll 1992). These different forms are under genetic control, because mutations can dramatically perturb the regulation of morphology, but they are also under environmental control because changing conditions can efficiently direct the cells toward one morphological form or another. Intriguingly, changes in the shape of the individual cells are reflected in changes in the shape of colonies created by these cells. White cells form smooth, domed colonies on agar surfaces, while opaque cells form colonies with a characteristic flattened appearance. Colonies formed from cells with a high proportion of hyphal and pseudohyphal cells tend to be more wrinkled than are colonies formed from yeast cells. The ability to undergo mating is also linked to cell shape. Cells with the genetic constitution to be mating competent can only mate when they are in the opaque state; white cells, regardless of the genotype, are unable to mate (Miller and Johnson 2002). The mating process itself involves changes in cellular morphology; in response to mating pheromones, the cells send out mating projections that serve to link the two mating cells (Soll 2004). Thus, cellular morphology plays an important role in the overall biology of *C. albicans*.

These cellular and colony morphologies are important components of the virulence of this pathogen. The infective properties of white and opaque cells are different; white cells are more virulent in systemic infections, while opaque cells can be more virulent in cutaneous infections (Kvaal et al. 1997). The ability to shift from a yeast form to a hyphal cell also appears critical for virulence. Systemic *Candida* infections typically contain yeast, hyphal and pseudohyphal cells. Mutants trapped in either the yeast form or the pseudohyphal form are less virulent than wild type when tested in mouse models of systemic infection, suggesting that the ability to change the cellular form plays a fundamental role in the infection process (Braun and Johnson 1997; Lo et al. 1997; Rocha et al. 2001). Aspects of colony formation are also implicated in virulence. Biofilms represent a specific colony form capable of strong attachment to a surface (Kumamoto 2002). Such cellular assemblies can provide a serious problem in hospital settings; biofilm formations on, for example, catheters can be difficult to dislodge, and the cells within biofilms can exhibit heightened resistance to antifungal drugs (Ramage et al. 2001). Therefore, an understanding of the control of cellular morphology is an important goal in experimental analysis of *C. albicans*, both in terms of our overall knowledge of basic biology of the organism as well as for our ability to develop new therapies against disease caused by *Candida* infections.

Because cellular morphology is under intrinsic genetic control, as well as under control of the environment, research efforts have been directed at determining the signaling processes responsible for the internal control as well as transmission of the external information to the intracellular machinery controlling cell shape. Genomic approaches are now providing critical information during these studies. The following sections provide an overview of the genomic analyses of several distinct morphological states.

A. Yeast-to-Hyphae Transition

Wild-type *C. albicans* cells can undergo the yeast-to-hyphal transition readily in response to external conditions, and this process is the most extensively studied of the variations in shape exhibited by this fungus. Classical inducing conditions include transferring yeast cells to YPD culture media containing serum and increasing the growth temperature to 37 °C. These conditions can trigger an almost quantitative shift of the yeast cells to hyphal cells. However, many other conditions can also generate an efficient shift from the yeast growth pattern to the hyphal mode. These include growth in Lee's medium at 37 °C, addition of N-acetylglucosamine, and changes in the pH of the growth medium. Many genes have been identified that are characteristically expressed as part of the transition. These include genes encoding a number of cell wall proteins, such as Ece1p and Hwp1p, as well as enzymes such as the secreted aspartyl proteases encoded by the *SAP4, 5* and *6* genes. The use of DNA microarrays has permitted the comprehensive analysis of the transcription profile of cells undergoing the yeast-to-hyphal transition in response to serum (Nantel et al. 2002). This work, and additional hybridizations performed in our laboratory over the last 2 years, has increased to approximately 400 the numbers of genes whose transcripts show a statistically significant change in abundance during this transition. The most significant of these modulated transcripts are shown in Table 8.1, and a complete list is available upon request. As evidenced in Table 8.2, there are significant differences between the functions of the gene products whose transcript are either upregulated or downregulated during the transition. We have noted that the list of upregulated transcript shows a much greater reproducibility between experimenters. One possible explanation may be that many of the gene products in the downregulated list are implicated in growth rate and metabolism.

Classical genetic studies have allow us to identify genes that are functionally required for the transition from the yeast to the hyphal cell. Several of these genes encode transcription factors, while others encode components of signal transduction networks. Genomic analyses of these transcription factors and signaling network members is providing insight into how external conditions serve to regulate the yeast-to-hyphal transition.

The observation that pseudohyphal growth in *S. cerevisiae* involved transcription factors that had homologs in *C. albicans* led the Fink laboratory to investigate whether these *C. albicans* transcription factors controlled hyphal development. They constructed strains that lacked the Efg1p and Cph1p proteins, and showed that these strains were unable to enter hyphal development in the presence of serum and high temperature, two conditions that are usually highly efficient in inducing hyphal development. These mutants were also unable to escape macrophage engulfment, while the wild-type cells could trigger hyphal development and exit the macrophage (Lo et al. 1997). Transcription profiling was done to test the hypothesis that loss of the transcription factors would disrupt the normal expression of hyphal-specific genes. The *cph1 efg1* double mutant strain was unable to activate the normal pattern of gene expression in response to heat and serum; the profile that was detected suggested that the cells were able to recognize the temperature increase but were insensitive to the presence of serum. Very similar results were observed for the single *efg1* mutant, suggesting that the bulk of the signaling passes through Efg1p (Nantel et al. 2002). Thus, the morphological consequences of the transcription factor mutations are reflected in the changes in the transcription profiles of the cells.

The recently identified Efh1p transcription factor is in the same family as Efg1p (Doedt et al. 2004). Deletion of *EFH1* did not have a dramatic effect on cellular morphology on its own, but caused hyperfilamentation in embedded conditions when coupled to deletion of *EFG1*. Transcription profiling of the disruption strains were consistent with this observed relationship. Few genes were influenced by deletion of *EFH1* under normal yeast growth conditions, while many were either up- or downregulated by loss of *EFG1*. Many genes were identified that were misregulated in the double mutant but were properly controlled in either single mutant (Doedt et al. 2004). Transcription profiling has also been used to examine the consequences of Efg1p and Efh1p overexpression. Efg1p acted mainly as a repressor, downregulating over 50 genes, but did activate the expression of several genes, many of which are also activated during serum-induced hyphal development. Efh1p acted primarily as an ac-

Table. 8.1. The 100 most significantly modulated transcripts observed during the yeast-to-hyphae transition induced by serum and high temperature[a]

Array spot identifier	Gene	orf19 number	Fold change	*p*-value	Product
Upregulated transcripts					
Contig4-2242_0004	ECE1	orf19.3374	42.9	0.0000	Secreted cell elongation protein
Contig4-2245_0009		orf19.7304	3.2	0.0000	Hypothetical protein
Contig4-2413_0004		orf.6705	2.1	0.0001	sec7 domain unknown protein
Contig4-2830_0009	SAP5	orf19.5585	11.2	0.0001	Secreted aspartyl proteinase 5
Contig4-2436_0002	HWP1	orf.1321	9.2	0.0001	Hyphal wall protein
Contig4-2682_0003	SNZ1	orf.2947	2.1	0.0001	Stationary-phase protein
Contig4-3065_0005	SOD5	orf.2060	7.2	0.0003	Copper–zinc superoxide dismutase
Contig4-2056_0002	IHD1	orf.5760	5.2	0.0006	Induced in hyphal development; membrane protein
Contig4-2558_0011	SMA2	orf.5644	1.4	0.0007	Conserved protein; spore cell wall biogenesis
Contig4-2734_0005	SAP6	orf.5542	4.1	0.0007	Candidapepsin 6 precursor
Contig4-2876_0008	CHA2	orf.1996	3.1	0.0008	Catabolic serine/threonine dehydratase
Contig4-2999_0011		orf.4924	1.2	0.0011	Splicing factor 3b
Contig4-3048_0002		orf.7531	2.0	0.0011	Conserved hypothetical protein
Contig4-2584_0004	CTA2	orf.631	1.4	0.0012	Transcription factor
Contig4-2187_0001	PHR1	orf.3829	2.7	0.0017	pH-regulated GPI-anchored membrane protein that is required for morphogenesis
Contig4-2714_0002		orf.815	2.2	0.0017	DOCK180 protein
Contig4-3060_0001	PHR1	orf.3829	3.3	0.0017	pH-regulated GPI-anchored membrane protein that is required for morphogenesis
Contig4-2790_0004		orf.6604	1.4	0.0021	Hypothetical protein
Contig4-2038_0003	SAP4	orf.5716	6.6	0.0022	Candidapepsin 4 precursor
Contig4-2567_0004		orf.4182	1.3	0.0022	Weak similarity to 3-oxoacyl-[acyl-carrier-protein] reductase
Contig4-2819_0014		orf.4666	2.0	0.0023	Hypothetical protein
Contig4-1945_0002	CTA2	orf.631	1.3	0.0027	Transcription factor
Contig4-3059_0006	HGT2	orf.3668	4.6	0.0027	Hexose transporter
Contig4-2831_0003	GRP4	orf.3150	1.8	0.0029	Similar to plant dihydroflavonol 4-reductase
Contig4-3108_0039		orf.7602	1.3	0.0031	Activator of HSP90 ATPase
Contig4-1814_0003	PFY1	orf.5076	1.5	0.0033	Profilin
Contig4-2034_0002	ALR1	orf.1607	1.6	0.0033	Putative divalent cation transporter
Contig4-2810_0012	IHD2	orf.6021	1.8	0.0033	Induced in hyphal development
Contig4-2461_0005	CBR1	orf.1801	1.3	0.0040	Cytochrome-b5 reductase
Contig4-2578_0004	IRO1	orf.1715	1.8	0.0040	Transcription factor
Contig4-2876_0017	PRY4	orf.6202	2.9	0.0044	Pathogenesis-related protein, repressed by TUP1 protein 4
Contig4-2859_0007	MCM6	orf.2611	1.2	0.0045	Involved in replication
Contig4-2768_0006	RBT12	orf.3384	3.5	0.0050	Repressed by TUP1 protein 1
Contig4-2775_0009		orf.3679	1.3	0.0050	Conserved hypothetical protein
Contig4-2948_0008	RIS1	orf.5675	1.2	0.0050	SNF2 family DNA-dependent ATPase
Contig4-2956_0007	CTA2	orf.631	1.4	0.0050	Transcription factor
Contig4-2779_0003	SPT14	orf.487	1.2	0.0053	N-acetylglucosaminyl-phosphatidylinositol (GPI) biosynthetic protein
Contig4-2911_0012	PRC2	orf.4135	1.6	0.0053	Carboxypeptidase Y precursor
Contig4-2972_0005	TUB2	orf.6034	1.6	0.0053	beta-tubulin
Contig4-2696_0012	EXG1	orf.2952	1.5	0.0053	exo-1,3-beta-glucanase
Contig4-2815_0003		orf.1105.2	1.8	0.0053	Hypothetical protein
Contig4-2151_0007		orf.4749	4.0	0.0059	Hypothetical membrane protein
Contig4-2189_0002	APR1	orf.1891	1.8	0.0059	Vacuolar aspartic proteinase precursor
Contig4-3053_0003	MOB2	orf.6044	1.3	0.0059	Protein kinase activator involved in cell cycle and shape regulation
Contig4-2283_0006		orf.3889	1.4	0.0059	Component of the COPII-coated vesicles
Contig4-2801_0007	MSB2	orf.1490	1.5	0.0059	Integral membrane protein; cell surface flocculin
Contig4-2624_0009		orf.1782	1.3	0.0060	Conserved hypothetical protein
Contig4-2946_0009		orf.2903	1.6	0.0062	Conserved hypothetical protein
Contig4-3102_0012		NoHit	1.4	0.0062	No homolog in orf
Contig4-2898_0012	CTA2	orf.631	1.3	0.0065	Transcription factor
Contig4-3002_0007	RNH1	orf.5564	1.5	0.0065	Ribonuclease H, exon 2
Contig4-3058_0013	RHO2	orf.2204.2	1.5	0.0065	GTP-binding protein of the rho subfamily
Contig4-2963_0002		NoHit	1.7	0.0065	No homolog in orf

Table. 8.1. (continued)

Array spot identifier	Gene	orf19 number	Fold change	*p*-value	Product
Contig4-2537_0005	SEC17	orf.2518	1.2	0.0066	Transport vesicle fusion protein
Downregulated transcripts					
Contig4-2692_0012		orf.925	1.4	0.0000	Hypothetical protein
Contig4-2520_0008	HYP2	orf.3426	1.8	0.0000	Translation initiation factor eIF5A.1
Contig4-2843_0005	RCK2	orf.2268	1.4	0.000	Ca^{2+}/calmodulin-dependent serine/threonine protein kinase
Contig4-2262_0002	MRPS5	orf.989	1.3	0.0007	Mitochondrial ribosomal protein S5
Contig4-3020_0010	RPL13	orf.2994	1.6	0.0007	Ribosomal protein
Contig4-2882_0010	APE2	orf.5197	1.4	0.0008	Alanine/arginine aminopeptidase
Contig4-2864_0006	ATP5	orf.5419	1.3	0.0011	F_1F_0-ATPase subunit
Contig4-3035_0011	HMO1	orf.6645	1.6	0.0017	High-mobility group-like protein
Contig4-2154_0002	ADK1	orf.683	1.5	0.0017	Adenylate kinase
Contig4-2826_0005	CBF1	orf.2876	2.0	0.0021	Centromere-binding kinetochore protein
Contig4-2921_0012	ALD5	orf.5806	1.8	0.0021	Aldehyde dehydrogenase
Contig4-2469_0001	RHD1	orf.54	3.1	0.0022	Repressed in hyphal development; family of conserved protein of unknown function
Contig4-2944_0009	MRP20	orf.3350	1.7	0.0022	Mitochondrial ribosomal protein
Contig4-2970_0009		orf.2582	1.4	0.0023	Highly conserved hypothetical protein
Contig4-2812_0004	GDH3	orf.4716	1.9	0.0024	NADP-glutamate dehydrogenase
Contig4-1966_0003	ADK12	orf.3391	1.7	0.0029	Adenylate kinase
Contig4-3022_0004	STM1	orf.5943.1	1.7	0.0029	Purine motif triplex-binding protein; specific affinity for guanine-rich quadruplex nucleic acids
Contig4-2973_0009	RPS0A	orf.6975	1.5	0.0030	Ribosomal protein S0A
Contig4-3094_0036	COR1	orf.4016	1.4	0.0031	Part of ubiquinol cyt-c reductase complex
Contig4-3104_0059	CLC1	orf.4594	2.0	0.0031	Clathrin light chain
Contig4-2977_0010		orf.6062.3	1.4	0.0031	Conserved hypothetical protein
Contig4-3037_0004	STP3	orf.5917	1.6	0.0033	Zinc finger protein involved in pre-tRNA splicing
Contig4-2863_0003	VSP24	orf.2031	1.4	0.0037	Endosomal Vps protein complex subunit involved in secretion
Contig4-2426_0003	FCY1	orf.4195.1	1.7	0.0044	Cytosine deaminase
Contig4-2933_0007	CSP37	orf.2531	1.9	0.0044	Cell surface protein
Contig4-3006_0011		orf.6748	1.3	0.0044	Translation initiation factor
Contig4-3083_0027		orf.2668	2.3	0.0044	gag-pol polyprotein
Contig4-1953_0001	TOM22	orf.3696	1.6	0.0046	Mitochondrial import receptor protein
Contig4-3087_0001	RPL35	orf.5964.2	2.2	0.0048	Ribosomal protein L35A
Contig4-2863_0021	MSF1	orf.2039	1.2	0.0050	Phenylalanyl-tRNA synthetase alpha subunit, mitochondrial
Contig4-3099_0016		orf.897	1.2	0.0052	Conserved hypothetical protein; coiled coils
Contig4-2521_0007		orf.4517	1.3	0.0053	Conserved hypothetical membrane protein
Contig4-2748_0005	RPS23	orf.6253	1.4	0.0053	Ribosomal protein S23A
Contig4-2763_0003	RPL5	orf.6541	1.5	0.0053	Ribosomal protein L5
Contig4-2813_0019	CVB1	orf.1970	1.3	0.0053	Complements *S. cerevisiae* vacuole biogenesis mutant, unpublished data
Contig4-2929_0011		orf.3940	1.2	0.0053	Hypothetical protein
Contig4-2613_0002	COX20	orf.6461	1.3	0.0054	Cytochrome oxidase
Contig4-2997_0025	DYS1	orf.1626	1.3	0.0054	Deoxyhypusine synthase
Contig4-2748_0004		orf.6252	1.6	0.0059	Conserved hypothetical protein
Contig4-2732_0018	RPL24A	orf.3789	1.5	0.0062	Ribosomal protein L24A (rp29) (YL21) (L30A)
Contig4-2970_0018	OPT9	orf.2584	3.5	0.0062	Oligopeptide transporter specific for tetra- and pentapeptides; possible pseudogene
Contig4-2904_0001	NAM9	orf.498	1.3	0.0062	Mitochondrial S4 ribosomal protein
Contig4-3097_0058	MRP7	orf.7203	1.4	0.0065	Mitochondrial large ribosomal subunit
Contig4-2649_0003	MRPL32	orf.549	1.6	0.0067	Mitochondrial ribosomal large subunit protein
Contig4-2622_0004	SFU1	orf.4869	1.2	0.0068	GATA-type transcriptional activator of nitrogen-regulated genes
Contig4-3001_0005	ATP3	orf.3223	1.2	0.0068	ATP synthase gamma subunit

[a]These were identified from a Student t-test analysis (multiple testing correction: Benjamini and Hochberg false discovery rate) of the combined results of 14 independent RNA preparations and 20 microarray hybridizations prepared by Nantel et al. (2002), Lee et al. (2004) and Oberholzer (unpublished data). Transcripts are sorted by statistical significance

Table. 8.2. Selected GO terms associated with gene products whose transcripts are significantly modulated during the yeast-to-hyphae transition

GO term	# of genes
Upregulated transcripts (172 genes)	
C:COPII-coated vesicle	4
C:bud neck	4
C:bud tip	2
C:cytoplasmic microtubule	5
C:plasma membrane	18
C:polar microtubule	4
C:spindle pole body	3
F:1,3-beta-glucanosyltransferase	2
F:histone deacetylase	4
F:Rho small monomeric GTPase	3
F:aspartic-type endopeptidase	5
F:structural constituent of cytoskeleton	5
F:thioredoxin peroxidase	6
P:ER to Golgi transport	5
P:beta-1,6 glucan biosynthesis	4
P:bud growth	2
P:bud site selection	3
P:cell wall organization and biogenesis	14
P:establishment of cell polarity	6
P:homologous chromosome segregation	5
P:inactivation of MAPK	4
P:nuclear migration	4
P:regulation of redox homeostasis	8
P:small GTPase-mediated signal transduction	3
Downregulated transcripts (228 genes)	
C:19S proteasome regulatory particle	8
C:actin cortical patch	2
C:centromere	5
C:hydrogen-transporting ATP synthase, stator stalk	7
C:nucleus	16
C:ribosome	28
C:respiratory chain complex III	4
C:spindle pole body	2
F:RNA binding	5
F:cytoskeletal adaptor	2
F:general RNA polymerase II transcription factor	3
F:hydrogen-transporting two-sector ATPase	7
F:proteasome endopeptidase	8
F:protein transporter	7
F:structural constituent of ribosome	39
F:translation initiation factor	7
P:35S primary transcript processing	7
P:ATP synthesis coupled proton transport	7
P:UDP-N-acetylglucosamine biosynthesis	2
P:actin filament organization	2
P:aerobic respiration	8
P:establishment of cell polarity	3
P:mRNA splicing	5
P:mitochondrial translocation	7
P:protein biosynthesis	62
P:regulation of transcription from Pol II promoter	4
P:secretory pathway	4
P:ubiquitin-dependent protein catabolism	8
P:vesicle-mediated transport	3

tivator, upregulating over 50 genes, and downregulating less than 30. Only about 10% of the genes were co-regulated by overproduction of the two transcription modulators (Doedt et al. 2004).

Other transcription factors play a role in hyphal development, and the global effects of some of these genes have been investigated by transcriptional profiling. Nrg1p and Tup1p are transcriptional regulators that play a role in hyphal development. Deletion of either *TUP1* (Braun and Johnson 1997) or *NRG1* (Murad et al. 2001b) leads to activation of filamentous development and results in the constitutive production of pseudohyphae under standard yeast growth conditions. The two mutants do not, however, create identical phenotypes; in response to serum stimulation, the *nrg1* mutant is still able to form true hyphae, while the *tup1* mutant remains locked in the pseudohyphal growth pattern (Murad et al. 2001b). Transcriptional profiling has been used to assess the consequences of these deletions. Partial genome arrays spotted on filters were used to profile both the *nrg1* mutant and the *tup1* mutant (Murad et al. 2001a). Among the genes that were derepressed in both mutants were hyphal-specific genes such as *HYR1*, *HWP1* and *ECE1*. Overall, Tup1p and Nrg1p were needed for the proper expression of many genes, and together with Mig1p formed overlapping clusters of co-regulated genes. This profiling work is consistent with Nrg1p/Tup1p controlling the expression of genes involved in hyphal development, while Mig1p/Tup1p control aspects of carbon metabolism. The profiling data also suggest that Mig1p and Nrg1p do more than act as co-regulators of the Tup1p repressor; removal of either Nrg1p or Mig1p affects expression of genes that are not affected by Tup1p depletion (Murad et al. 2001a).

Other cellular components have been implicated in the transition to hyphal growth. In particular, the loss of components of the cAMP regulatory circuit affects hyphal development. Mutation of *CDC35*, which encodes adenylyl cyclase, totally abolishes the ability to trigger hyphal development under all conditions tested (Rocha et al. 2001). Deletion of *RAS1*, a regulator of adenylyl cyclase, blocks true hyphal development in response to serum and temperature induction, but the mutant cells still can form pseudohyphae (Feng et al. 1999; Leberer et al. 2001). Mutations in the cAMP-dependent protein kinase subunits also influence hyphal development (Bockmuhl et al. 2001; Cloutier et al. 2003). Transcription profiling has been applied to investigate the consequences of deletion of three of the elements of the cAMP regulon – *CDC35*, encoding the adenylyl cyclase (Harcus et al. 2004), as well as *RAS1* and *EFG1* (Nantel et al. 2002; Harcus et al. 2004). In addition, profiling has investigated the function of an Efg1p homolog, Efh1p (Doedt et al. 2004). The loss of either *CDC35* or *RAS1* has related consequences on gene expression in *C. albicans*. In addition to disrupting the transcriptional response to hyphal development triggered by serum and high temperature, the *cdc35* and *ras1* mutants affect the expression of many genes involved in general metabolism and stress responses. The loss of adenylyl cyclase function changes the expression of hundreds of genes; loss of *RAS1* influences expression levels of a subset of those genes, and little else. The transcription profiles are very different for cells deleted for the *EFG1* gene. The genes requiring *EFG1* overlap with those requiring *CDC35* and *RAS1* for only the hyphal-induced set; other cellular processes have the *EFG1* and *CDC35/RAS1* genes acting in distinct ways. These differences in gene expression profiles are reflected in the physiology of the cells; for example, *ras1* and *cdc35* mutants are more resistant to chemicals or enzymes that attack the cell wall, while *efg1* mutants are more sensitive. Finally, this work demonstrated that significant regulatory connections remain to be identified, especially those that regulate cAMP-dependent changes in gene expression that are independent of Efg1p (Harcus et al. 2004).

Another enzyme with an important role in hyphal signaling is the Sit4p phosphatase. Morphological and virulence studies of homozygous *sit4* mutants have indicated that this type 2A protein phosphatase is required for proliferation, virulence and hyphal growth. Although serum-treated *sit4* null cells have a pseudohyphal phenotype and fail to elongate into true hyphae, transcriptional profiling has further indicated that the effects of the *sit4* mutation are quite distinct from what was observed with other hyphal-impaired mutants (Lee et al. 2004). Most of the canonical hyphal-specific genes, such as *ECE1*, *RBT1*, *SAP4* and *HWP1*, show a normal increase in transcript abundance. The pseudohyphal phenotype might be attributed to problems in cell wall reorganization, as the *sit4* mutants fail to upregulate transcripts encoding the exo-beta-glucanase Xog1p and the endo-1-3- beta-glucanase Acf1p. Furthermore, as was observed with the *efg1* and *efg1/cph1* null mutants, *sit4* mutants fail to properly respond to the presence of both serum and high temperature, which resulted instead in a mild activation of their heat shock response (Lee et al. 2004).

In addition to transcription profiling, large-scale genomic screens have been applied to the question of hyphal development. A modification of the transposon mutagenesis strategy that has been very successful in *S. cerevisiae* has been used to create disruptions that occur on average once every 2.5 kb in the *C. albicans* genome (Uhl et al. 2003). Screening for insertions that cause a modification in the yeast–hyphal transition identified over 140 genes that affect the process when one copy of the gene is disrupted. These mutations affected a number of known genes implicated in hyphal development, such as *TUP1*, *NRG1*, *CBK1* and *CZF1*, and also identified a large number of genes that were unique to *C. albicans*. Therefore, both expression level measurements and functional studies are being used on a genome-wide level to provide insights into this medically important cellular transition in *C. albicans*.

B. Biofilms

A second morphological state that has been investigated with microarrays is that of the biofilm. Biofilms are structurally complex collections of cells of various morphologies that are associated with solid surfaces, and are significantly more resistant to antifungal drugs, thus providing a source for re-infections. García-Sánchez et al. (2004) used macro- and microarrays to study the transcriptional profiles in biofilms that arise under different culture conditions. The Gcn4p transcription factor was shown to be necessary for biofilm formation, as it is required for the activation of numerous genes involved in protein synthesis. One possible role for excess protein synthesis lies in the production of the extracellular matrix that provides the structural basis of the biofilm. Furthermore, it was suggested that activation of the genes involved in the sulfur amino acid biosynthesis/salvage pathway may be involved in cell wall rearrangement and the production of quorum-sensing molecules. Comparison with biofilm composed of yeast-only *efg1/cph1* mutants allowed Garcia-Sanchez et al. (2004) to discriminate between genes that are involved with biofilm formation and those implicated in morphological switching. Although transcripts for *CDR1*, *ERG25* and *ERG16* were more abundant under certain biofilm-inducing conditions, the correlations between these higher-order structures and antifungal resistance remains unclear. To further elucidate this important question, it might be interesting to use transcriptional profiling to determine whether cells in these various morphological states respond differently to azole treatment.

C. Mating Projections

One of the initial consequences of the sequencing of the *C. albicans* genome was the identification of the *MTL* loci. These regions encode proteins similar to the transcription factors that control the definition of cell type in *S. cerevisiae*. Further investigations established that the typical *C. albicans* strain was heterozygous for the *MTL* locus, a genotype that in *S. cerevisiae* would preclude mating (Soll 2004). Therefore, researchers tested whether strains manipulated to be homozygous for each of the *MTL* loci could be induced to mate with each other. Inefficient mating was detected (Hull et al. 2000; Magee and Magee 2000), and the subsequent recognition that this mating took place significantly only when the two mating partners were in the opaque state (see next section) allowed Miller and Johnson (2002) to make a detailed characterization of the mating process.

Analysis of the genome sequence allowed researches to identify a putative gene encoding the *MTL*α-specific mating pheromone (Bennett et al. 2003; Lockhart et al. 2003; Panwar et al. 2003). A peptide based on the amino acid sequence of the predicted pheromone has been synthesized, and several groups have examined the consequences of pheromone treatment on cell cycle progression and cellular morphology. Depending on the strain and the assay conditions, pheromone treatment can cause significant cell cycle arrest (Panwar et al. 2003). Perhaps the most striking phenotypic modulation is that of the formation of mating projections (Lockhart et al. 2003). Analysis of the expression of genes in response to mating pheromone confirms that the pheromone can modulate transcription patterns (Bennett et al. 2003). Over 60 genes are induced by the mating pheromone. This induction is slow relative to the induction of pheromone-responsive genes in *S. cerevisiae*, although several of these genes represent the homologs of induced yeast genes. However, there are many pheromone-induced genes that are not found in *S. cerevisiae*, and there is also a group of transcripts that are induced by both mating pheromone and the yeast- to-hyphal transition.

Among the pheromone-induced gene products that are also hyphal specific is the cell wall protein Hwp1p. Hwp1p is a substrate for mammalian transglutaminase, and this characteristic of the protein is exploited by *C. albicans* cells to facilitate their adherence to epithelial cells. Intriguingly, in the mating process Hwp1p is localized to the conjugation tips of the **a**-mating cells responding to α-factor (Daniels et al. 2003). This establishes that the mating projection is a discrete cellular state; it is not just an elongated bud because Hwp1p is not expressed in buds. *HWP1* is expressed only in mating-type homozygous cells responding to α-factor, a direct test of the mating factor from **a** cells is not possible because this compound has not been identified.

Overall then, the mating process creates novel cellular morphologies. Cells responding to mating pheromones can initiate a cellular elongation that is superficially like a germ tube, but functionally distinct. Transcription profiling has identified a series of genes that are induced specifically during this process, and the involvement of these proteins in the morphological transition is now open to analysis.

D. White–Opaque Transition

The white–opaque transition leads to dramatic changes in both the shape of the individual cells and the shape of the colony. Initial work defining the characteristics of the white and opaque cells identified genes that were expressed uniquely in one cell type or the other. For example, *SAP1* (White and Agabian 1995), *CDR3* (Balan et al. 1997) and *OP4* (Morrow et al. 1993) were found to be opaque specific, while *WHI1* and *EFG1* were white specific (Sonneborn et al. 1999b). Recent analysis of the two cell types using transcription profiling has made possible the identification of the entire spectrum of white- and opaque-specific genes (Lan et al. 2002). These profiles establish that over 150 genes are more expressed in the white state, and over 200 genes are more expressed in the opaque state. There are major differences in the expression of metabolic genes in the two cell types, consistent with the white cells exhibiting a fermentative profile, and opaque cells an oxidative metabolic state. These distinctions, as well as the fact the fatty acid β-oxidation pathway is upregulated in opaque cells, can help explain the observation that the opaque cells are more effective pathogens on skin, while the white cells are more pathogenic during systemic infections.

A second cellular process that is different between the white and opaque cells is mating. Mating is exclusive to opaque cells, and genes such as those encoding the pheromone receptor *STE3* and the mating factor α-factor are upregulated in opaque cells, relative to white cells (Lan et al. 2002). The ability to switch to the opaque phase is limited to cells that are homozygous for the mating type; this observation clarified why only a minority of strains are capable of undergoing the white–opaque switch. This implies that the mating-type heterozygous state represses the opaque morphological phase (Miller and Johnson 2002), and therefore cells require two infrequent events, first a switch to mating-type homozygosity and then a switch to the opaque state, to create a mating competent state. Because the formation of a single mating competent cell appears infrequent, the chance of generating two cells of opposite mating competencies in the same vicinity is extremely small. This would make mating a highly improbable event in nature, and as such would make it apparently unreasonable for *C. albicans* to maintain this capacity.

Other regulatory circuits may interact with the white–opaque switching machinery to promote mating competence at a greater frequency. Inactivation of Efg1p transcription factor creates cells that are morphologically similar to opaque cells, and *EFG1* is one of the genes that is considered white specific. In addition, the metabolic state of opaque cells and *efg1* mutant cells are both shifted to a non-fermentative, oxidative mode (Lan et al. 2002; Doedt et al. 2004). Therefore, it is possible that some combination of environmental cues may facilitate the creation of a mating competent state at frequencies higher than that observed in laboratory conditions. Recent evidence shows that *HBR1*, a hemoglobin response gene, is implicated in the mating process (Pendrak et al. 2004). Hbr1p functions as a repressor of white–opaque switching; haplo-insufficiency at *HBR1* results in white–opaque switching without the need to create homozygosity at *MTL*. Therefore, environmental regulation of gene expression may promote the formation of mating competent cells that have not undergone the genomic rearrangements characteristic of experimentally controlled mating.

E. Chlamydospore Formation

Chlamydospores are a unique cellular form of *C. albicans* that consist of large, thick-walled cells that form on the ends of suspensor cells. These suspensor cells are typically attached to hyphae. This cell type is induced under oxygen-limited, nutrient-poor conditions; growth under glass cover slips on cornmeal agar, or growth directly embedded in agar can be used to induce chlamydospore formation. Both Efg1p and Hog1p are required for chlamydospore formation, but not for hyphal formation, in embedded conditions (Sonneborn et al. 1999a; Alonso-Monge et al. 2003).

A collection of over 200 defined homozygosed insertions was created by the UAU insertion cassette strategy (Davis et al. 2002), and screened for defects in chlamydospore formation. Three genes (*IWS2*, *SCH9* and *SUV3*) are essential for chlamydospore formation, while loss of three other genes (*RIM101*, *RIM13* and *MDS3*) delays their formation (Nobile et al. 2003). This large-scale approach to functional analysis provides a connection between the pH regulatory pathway that influences hyphal development, and chlamydospore formation. This work shows how systematic analysis of disruption constructs can provide new insights into morphological plasticity in *C. albicans*.

IV. Conclusions

After only a few years, genomic technologies have yielded an abundance of novel concepts and may already be changing the ways by which researchers are classifying morphological states. In addition to the simple identification of modulated genes, the detailed transcriptional profiles that can be obtained with DNA microarrays facilitate a more precise classification of the various types of morphologies that are observed in response to environmental and genetic cues. This is especially useful, as cells with a similar morphology may in fact arise from completely different mechanisms. For example, the elongated cell phenotype that is observed when *efg1* or *sit4* mutants are treated with serum and high temperature may appear similar under a microscope but have, in fact, very different transcriptional profiles, especially in relation to the canonical hyphal-specific genes. Nevertheless, correlations between these profiles and the responses of *Candida albicans* to environmental stresses, such as heat shock, indicate that both factor are involved in coordinating signals from two external stimuli (serum and high temperature) into a single morphological response. Similar correlations between different datasets are becoming more common and help us to understand the interconnections between the various signaling cascades that modulate morphogenesis. Furthermore, profiling of the components of a signal transduction chain will help us determine whether novel components remain to be discovered, as was the case with our recent study of cAMP signaling (Harcus et al. 2004). Finally, novel screening strategies are taking advantage of our knowledge of the *Candida* genome to identify novel modulators. The genes encoding these modulators can then be mutated and the consequences of these mutations on transcriptional profiles compared to available data. Thus, the increasing amount and quality of these large datasets, coupled with advances in the quality of the *Candida* genome annotation, is bound to produce additional discoveries and, hopefully, unexpected but informative correlations.

Acknowledgements. This work was supported by the NRC Genomics and Health Research Initiative, as well as by CIHR grant number HOP-67260 (to A. Nantel) and CIHR grant MOP42516 (to M. Whiteway). We would like to thank all members of the BRI Genetics group for discussions. This is NRC publication # 46234.

References

Alonso-Monge R, Navarro-Garcia F, Roman E, Negredo AI, Eisman B, Nombela C, Pla J (2003) The Hog1 mitogen-activated protein kinase is essential in the oxidative stress response and chlamydospore formation in *Candida albicans*. Eukaryot Cell 2:351–361

Balan I, Alarco AM, Raymond M (1997) The *Candida albicans* CDR3 gene codes for an opaque-phase ABC transporter. J Bacteriol 179:7210–7218

Bennett RJ, Uhl MA, Miller MG, Johnson AD (2003) Identification and characterization of a *Candida albicans* mating pheromone. Mol Cell Biol 23:8189–8201

Bockmuhl DP, Krishnamurthy S, Gerads M, Sonneborn A, Ernst JF (2001) Distinct and redundant roles of the two protein kinase A isoforms Tpk1p and Tpk2p in morphogenesis and growth of *Candida albicans*. Mol Microbiol 42:1243–1257

Braun BR, Johnson AD (1997) Control of filament formation in *Candida albicans* by the transcriptional repressor TUP1. Science 277:105–109

Bruno VM, Mitchell AP (2004) Large-scale gene function analysis in *Candida albicans*. Trends Microbiol 12:157–161

Cloutier M, Castilla R, Bolduc N, Zelada A, Martineau P, Bouillon M, Magee BB, Passeron S, Giasson L, Cantore ML (2003) The two isoforms of the cAMP-dependent protein kinase catalytic subunit are involved in the control of dimorphism in the human fungal pathogen *Candida albicans*. Fungal Genet Biol 38:133–141

Cowen LE, Nantel A, Whiteway MS, Thomas DY, Tessier DC, Kohn LM, Anderson JB (2002) Population genomics of drug resistance in *Candida albicans*. Proc Natl Acad Sci USA 99:9284–9289

Daniels KJ, Lockhart SR, Staab JF, Sundstrom P, Soll DR (2003) The adhesin Hwp1 and the first daughter cell localize to the a/a portion of the conjugation bridge during *Candida albicans* mating. Mol Biol Cell 14:4920–4930

Davis DA, Bruno VM, Loza L, Filler SG, Mitchell AP (2002) *Candida albicans* Mds3p, a conserved regulator of pH responses and virulence identified through insertional mutagenesis. Genetics 162:1573–1581

De Backer MD, Ilyina T, Ma XJ, Vandoninck S, Luyten WH, Vanden Bossche H (2001) Genomic profiling of the response of *Candida albicans* to itraconazole treatment using a DNA microarray. Antimicrob Agents Chemother 45:1660–1670

DeRisi JL, Iyer VR, Brown PO (1997) Exploring the metabolic and genetic control of gene expression on a genomic scale. Science 278:680–686

Doedt T, Krishnamurthy S, Bockmuhl DP, Tebarth B, Stempel C, Russell CL, Brown AJ, Ernst JF (2004) APSES proteins regulate morphogenesis and metabolism in *Candida albicans*. Mol Biol Cell 15:3167–3180

Feng Q, Summers E, Guo B, Fink G (1999) Ras signaling is required for serum-induced hyphal differentiation in *Candida albicans*. J Bacteriol 181:6339–6346

Garcia-Sanchez S, Aubert S, Iraqui I, Janbon G, Ghigo JM, d'Enfert C (2004) *Candida albicans* biofilms: a developmental state associated with specific and stable gene expression patterns. Eukaryot Cell 3:536–545

Harcus D, Nantel A, Marcil A, Rigby T, Whiteway M (2004) Transcription Profiling of cyclic AMP signaling in *Candida albicans*. Mol Biol Cell 15:4490–4499

Hughes TR, Mao M, Jones AR, Burchard J, Marton MJ, Shannon KW, Lefkowitz SM, Ziman M, Schelter JM, Meyer MR et al. (2001) Expression profiling using microarrays fabricated by an ink-jet oligonucleotide synthesizer. Nat Biotechnol 19:342–347

Hull CM, Raisner RM, Johnson AD (2000) Evidence for mating of the "asexual" yeast *Candida albicans* in a mammalian host. Science 289:307–310

Jones T, Federspiel NA, Chibana H, Dungan J, Kalman S, Magee BB, Newport G, Thorstenson YR, Agabian N, Magee PT et al. (2004) The diploid genome sequence of *Candida albicans*. Proc Natl Acad Sci USA 101:7329–7334

Kumamoto CA (2002) Candida biofilms. Curr Opin Microbiol 5:608–611

Kvaal CA, Srikantha T, Soll DR (1997) Misexpression of the white-phase-specific gene WH11 in the opaque phase of *Candida albicans* affects switching and virulence. Infect Immun 65:4468–4475

Lan CY, Newport G, Murillo LA, Jones T, Scherer S, Davis RW, Agabian N (2002) Metabolic specialization associated with phenotypic switching in *Candida albicans*. Proc Natl Acad Sci USA 99:14907–14912

Lane S, Birse C, Zhou S, Matson R, Liu H (2001) DNA array studies demonstrate convergent regulation of vir-

ulence factors by Cph1, Cph2, and Efg1 in *Candida albicans*. J Biol Chem 276:48988–48996
Leberer E, Harcus D, Dignard D, Ushinsky S, Thomas DY, Schroppel K (2001) Ras links cellular morphogenesis to virulence by regulation of the MAP kinase and cAMP signalling pathways in the pathogenic fungus *Candida albicans*. Mol Microbiol 42:673–687
Lee CM, Nantel A, Jiang L, Whiteway M, Shen SH (2004) The serine/threonine protein phosphatase SIT4 modulates yeast-to-hypha morphogenesis and virulence in *Candida albicans*. Mol Microbiol 51:691–709
Lo HJ, Kohler JR, DiDomenico B, Loebenberg D, Cacciapuoti A, Fink GR (1997) Nonfilamentous *C. albicans* mutants are avirulent. Cell 90:939–949
Lockhart SR, Zhao R, Daniels KJ, Soll DR (2003) Alpha-pheromone-induced "shmooing" and gene regulation require white-opaque switching during *Candida albicans* mating. Eukaryot Cell 2:847–855
Magee BB, Magee PT (2000) Induction of mating in *Candida albicans* by construction of MTLa and MTLalpha strains. Science 289:310–313
Miller MG, Johnson AD (2002) White-opaque switching in *Candida albicans* is controlled by mating-type locus homeodomain proteins and allows efficient mating. Cell 110:293–302
Mnaimneh S, Davierwala AP, Haynes J, Moffat J, Peng WT, Zhang W, Yang X, Pootoolal J, Chua G, Lopez A et al. (2004) Exploration of essential gene functions via titratable promoter alleles. Cell 118:31–44
Morrow B, Srikantha T, Anderson J, Soll DR (1993) Coordinate regulation of two opaque-phase-specific genes during white-opaque switching in *Candida albicans*. Infect Immun 61:1823–1828
Murad AM, d'Enfert C, Gaillardin C, Tournu H, Tekaia F, Talibi D, Marechal D, Marchais V, Cottin J, Brown AJ (2001a) Transcript profiling in *Candida albicans* reveals new cellular functions for the transcriptional repressors CaTup1, CaMig1 and CaNrg1. Mol Microbiol 42:981–993
Murad AM, Leng P, Straffon M, Wishart J, Macaskill S, MacCallum D, Schnell N, Talibi D, Marechal D, Tekaia F et al. (2001b) NRG1 represses yeast-hypha morphogenesis and hypha-specific gene expression in *Candida albicans*. EMBO J 20:4742–4752
Nantel A, Dignard D, Bachewich C, Harcus D, Marcil A, Bouin A-P, Sensen CW, Hogues H, Van het Hoog M, Gordon P et al. (2002) Transcription profiling of *Candida albicans* cells undergoing the yeast to hyphal transition. Mol Biol Cell 13:3452–3465
Nobile CJ, Bruno VM, Richard ML, Davis DA, Mitchell AP (2003) Genetic control of chlamydospore formation in *Candida albicans*. Microbiology 149:3629–3637
Panwar SL, Legrand M, Dignard D, Whiteway M, Magee PT (2003) MFalpha1, the gene encoding the alpha mating pheromone of *Candida albicans*. Eukaryot Cell 2:1350–1360
Pendrak ML, Yan SS, Roberts DD (2004) Hemoglobin regulates expression of an activator of mating-type locus alpha genes in *Candida albicans*. Eukaryot Cell 3:764–775
Ramage G, Wickes BL, Lopez-Ribot JL (2001) Biofilms of *Candida albicans* and their associated resistance to antifungal agents. Am Clin Lab 20:42–44
Rocha CRC, Schroppel K, Harcus D, Marcil A, Dignard D, Taylor BN, Thomas DY, Whiteway M, Leberer E (2001) Signalling through adenylyl cyclase is essential for hyphal growth and virulence in the pathogenic fungus *Candida albicans*. Mol Biol Cell 12:3631–3643
Roemer T, Jiang B, Davison J, Ketela T, Veillette K, Breton A, Tandia F, Linteau A, Sillaots S, Marta C et al. (2003) Large-scale essential gene identification in *Candida albicans* and applications to antifungal drug discovery. Mol Microbiol 50:167–181
Rogers PD, Barker KS (2002) Evaluation of differential gene expression in fluconazole-susceptible and - resistant isolates of *Candida albicans* by cDNA microarray analysis. Antimicrob Agents Chemother 46:3412–3417
Soll DR (1992) High-frequency switching in *Candida albicans*. Clin Microbiol Rev 5:183–203
Soll DR (2004) Mating-type locus homozygosis, phenotypic switching and mating: a unique sequence of dependencies in *Candida albicans*. Bioessays 26:10–20
Sonneborn A, Bockmuhl DP, Ernst JF (1999a) Chlamydospore formation in *Candida albicans* requires the Efg1p morphogenetic regulator. Infect Immun 67:5514–5517
Sonneborn A, Tebarth B, Ernst JF (1999b) Control of white-opaque phenotypic switching in *Candida albicans* by the Efg1p morphogenetic regulator. Infect Immun 67:4655–4660
Sudbery P, Gow N, Berman J (2004) The distinct morphogenic states of *Candida albicans*. Trends Microbiol 12:317–324
Uhl MA, Biery M, Craig N, Johnson AD (2003) Haploinsufficiency-based large-scale forward genetic analysis of filamentous growth in the diploid human fungal pathogen *C. albicans*. EMBO J 22:2668–2678
White TC, Agabian N (1995) *Candida albicans* secreted aspartyl proteinases: isoenzyme pattern is determined by cell type, and levels are determined by environmental factors. J Bacteriol 177:5215–5221
Whiteway M, Oberholzer U (2004) Candida morphogenesis and host-pathogen interactions. Curr Opin Microbiol 7:350–357

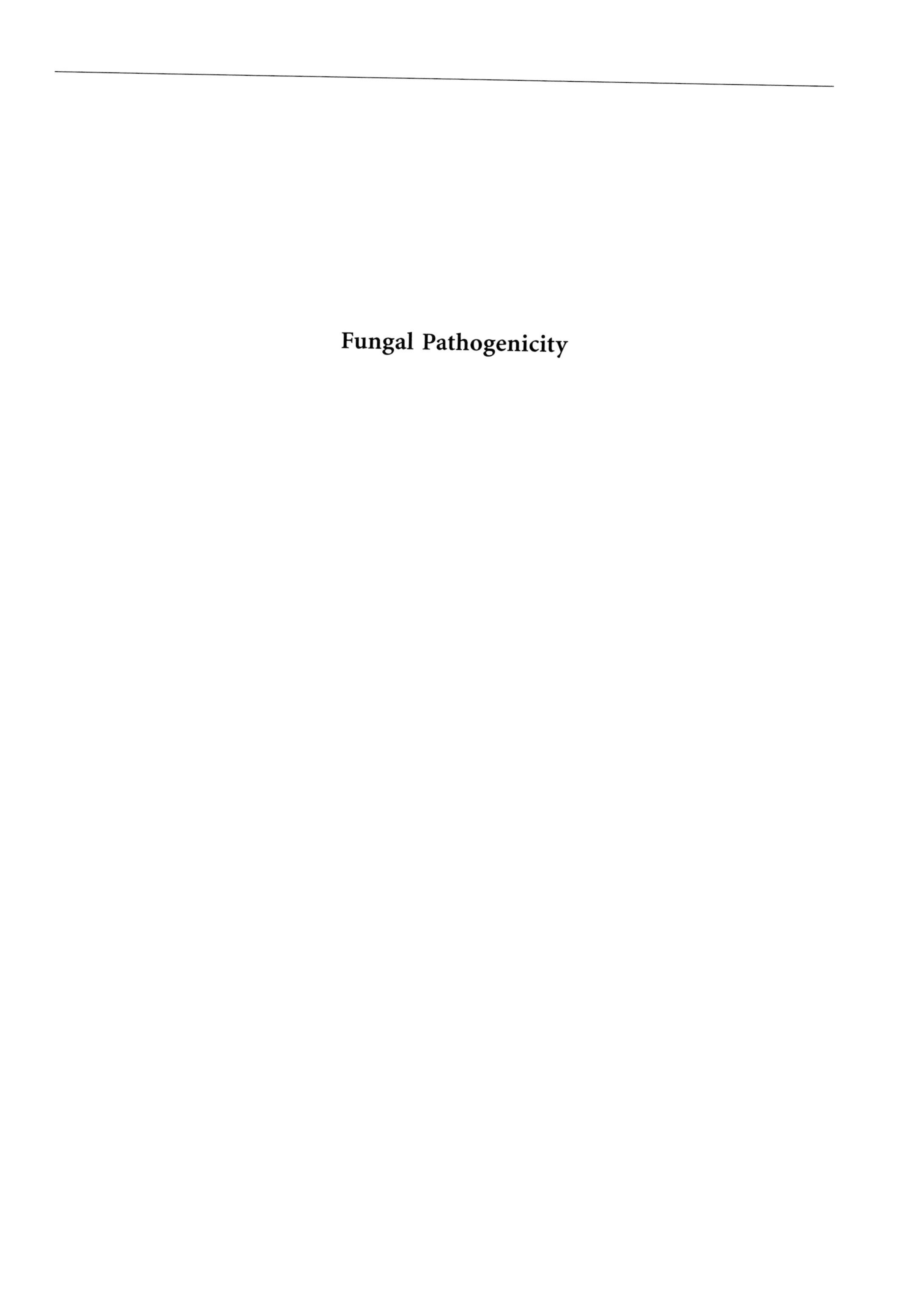

Fungal Pathogenicity

9 Postgenomic Approaches to Analyse *Candida albicans* Pathogenicity

C.A. Munro[1], C. Fradin[2], O. Bader[2], B. Hube[2]

CONTENTS

Abbreviations: ALS, agglutinin-like sequence; GPI, glycosylphosphatidylinositol; ORF, open reading frame; RHE, reconstituted human epithelium; *SAP*, secreted aspartyl proteinase genes; TCA, tricarboxylic acid

I. Introduction

The yeast *Candida albicans* can act as both a harmless commensal in healthy individuals and an aggressive pathogen in immunocompromised patients, for example, when the normal microbial flora is removed or unbalanced by extensive antibacterial treatment or when physical barriers of the human body are damaged due to injury or surgical procedures (Odds 1988; Calderone 2002).

C. albicans has probably co-evolved with human beings for millions of years and, through permanent competition with the bacterial flora and continuous interactions with human cells, has become highly adapted to survival on epithelial surfaces (Odds et al. 2004). As a pathogen, *C. albicans* can not only colonise a multiplicity of body sites but can also invade tissue, survive in the bloodstream and cause life-threatening systemic infections when host defence mechanisms are weakened (De Repentigny 2004). Therefore, *C. albicans* is able to exist and proliferate in radically changing environments with strong changes in oxygen and carbon dioxide levels, pH, osmolarity, availability of nutrients, and temperature (Hube 2004). Furthermore, the fungus needs to counteract the attack of immune cells, amongst others by phagocytosis and the secretion of antimicrobial agents. Such a high degree of flexibility requires the expression of special sets of attributes during commensal growth and at each stage of infection in response to the specific demands of the many microenvironments which the fungus encounters. In this context, it is rather surprising that *C. albicans*, although highly flexible in source of nutrients and robust to relatively harsh conditions, is rarely found outside the body in the global environment (e.g. soil), in contrast to most other human pathogenic fungi such as *Histoplasma capsulatum*, *Cryptococcus neoformans* and *Aspergillus fumigatus* (Hube 2004).

Growth, morphology, metabolism, responses to the environment and expression of attributes especially required for pathogenicity during the different stages of infection are ultimately regulated by genes and their products, and adaptation to changing environment occurs predominately on the transcriptional level. With the complete genome sequence of *C. albicans* available, we are now in a position to analyse the pathogenesis of infections with

[1] School of Medical Sciences, Institute of Medical Sciences, Foresterhill, University of Aberdeen, Aberdeen AB25 2ZD, UK
[2] Robert Koch Institut, NG4, Nordufer 20, 13353 Berlin, Germany

The Mycota XIII
Fungal Genomics
Alistair J.P. Brown (Ed.)

this fungus not only on a traditional gene-by-gene basis but also by using much broader approaches. By means of in silico methods, such as when comparing the genome of *C. albicans* with the genomes of less virulent related species or when comparing the genomes of strains of *C. albicans* differing in pathogenic potential, we may identify genes or genetic properties which are crucial for infection, which exist only in pathogenic fungi or which are even unique to this fungus. By searching for sequences homologous to genes known to be associated with disease in other pathogenic microorganisms, we may identify as yet unknown putative virulence factors of *C. albicans*. The discovery of new gene families may hint at functions crucial for survival on or within a human host. Global experimental approaches based on the availability of complete genome sequences will identify genes directly associated with the infection process. This includes the use of microarrays representing all identified open reading frames of *C. albicans*, which will identify not only infection-associated genes but also global transcription patterns, the interplay of gene networks, and regulators of these networks. Finally, we can apply postgenomic approaches to analyse the interactions between *C. albicans* and its host in vivo or in models which mimic the in vivo situation. Thus, postgenomic approaches will enable us to further shed light on the complex process of the pathogenesis of *C. albicans* infections.

II. The Hunt for Pathogenicity-Related Genes Using Comparative Genomics and Predictive Algorithms

The multitude of sequencing projects completed or in progress has brought with them a powerful opportunity to search for genes which may be involved in pathogenicity and virulence – comparative genomics. Here, the full genome sequences of two or more species are compared to one another, making it possible to analyse them for similarities as well as distinct differences between species or groups. A recent example is the study of yeast genome evolution by the Genolevures project (Dujon et al. 2004). Here, the authors were able to describe the major evolutionary events which led to the specification of an entire phylum, by analysing the genomes of five yeasts (*Saccharomcyes cerevisiae*, *Kluveromyces lactis*, *Candida glabrata*, *Debaryomyces hansenii* and *Yarrowia lipolytica*) for major rearrangements, genome duplication events and gene losses. The diversity of changes taken by each branch reflects the long evolutionary time span separating these organisms – indeed, this is as long as that for the phylum of chordates (a fact which should be kept in mind when we discuss the experiments described below).

The most straightforward way to find attributes unique to pathogenic species is to generate two groups of genomes, one containing non-pathogenic and the other pathogenic species, and then to simply extract all those genes and features which are unique to or possibly even conserved within the latter. This approach has been used successfully in the analysis of bacterial pathogenicity (reviewed in Brosch et al. 2001; Whittam and Bumbaugh 2002).

If the aim is to identify genes unique to a pathogenic organism, it is of utmost importance to use well-annotated open reading frame (ORF) sets with high coding probability. Comparison of several closely related genomes can yield an improved set of protein-coding open reading frames. As shown by Kellis et al. (2003), the comparison of four *Saccharomyces* species (*S. cerevisiae*, *S. paradoxus*, *S. mikatae* and *S. bayanus*) led to the elimination of 500 ORFs with very low coding probability, the inclusion of 43 new genes, and the correction of several splice sites, yielding an improved catalogue of 5538 protein-encoding ORFs of a size larger than 100 amino acids. For *C. albicans*, genome sequences of such closely related species are not yet available, but the ongoing sequencing of the *Candida dubliniensis* genome, the closest known but less virulent relative, and the genome of the closely related apathogenic yeast *Debaryomyces hansenii* (teleomorph of *Candida famata*), commonly used for cheese production, will provide a first step in this direction.

A. Comparing the *C. albicans* Translated ORF Set to the Outside World

Strain SC5314 was chosen for the *C. albicans* sequencing project (Table 9.1). This strain is a clinical isolate belonging to the predominant clade of closely related *C. albicans* strains which represents almost 40% of all isolates worldwide (Odds et al. 2004). SC5314 and its derivatives are now the most commonly utilised laboratory strains. *C. albicans* lacks a known haploid or homozygous stage and unfortunately, SC5314 displays, as all other known

Table. 9.1. Websites of *C. albicans* genome and annotation projects

Project	Website
Candida genome sequencing at Stanford Genome Technology Center	www.sequence.stanford.edu/group/candida/
Candida Genome Database	candidagenome.org
CandidaDB at the Insitut Pasteur	genolist.pasteur.fr/CandidaDB
Candida information at the University of Minnesota	alces.med.umn.edu/Candida.html
CanoDB at the University of San Francisco	agabian.ucsf.edu/canoDB/
Candida albicans at national research council of Canada	candida.bri.nrc.ca/candida/index.cfm

C. albicans strains, a high degree of heterozygosity throughout its diploid genome which made assembling the entire sequence a major challenge (Jones et al. 2004).

With the race to create DNA microarrays for genome-wide transcriptional profiling, annotation of the *C. albicans* genome sequence was carried out in parallel to the assembly and completion of the genome sequencing project by several different groups (Table 9.1). This has led to the development of several different annotation databases whose datasets are presently being incorporated into the Candida Genome Database (Table 9.1). Currently, the only human pathogenic yeast sequenced completely in addition to *C. albicans* is *C. glabrata*, a fungus more closely related to *S. cerevisiae* than to *C. albicans*. Many other fungal sequencing projects are underway (an up-to-date list can be found at the Genomes OnLine Database; Bernal et al. 2001), also for several fungal pathogens of humans and plants, from different phylae. Simply assembling evolutionary distant organisms into pathogen/non-pathogen groups might yield a distinct set of "universal" pathogenicity genes. This approach is discussed by Ahmad et al. (Chap. 3, this volume). However, most of these organisms have evolved into specific niches, and have probably developed different approaches towards survival in those environments. Hence, a comparison on this level would most likely yield a reliable and conserved set of housekeeping genes, and a multitude of overlapping subgroups which would reflect different nutritional and ecological adaptations as well as phylogeny.

In the original paper describing the *C. albicans* genome sequence (Jones et al. 2004), the authors presented a BLASTp-based classification of the predicted protein sequences from 6419 ORFs of the reduced haploid genome of *C. albicans*, against *S. cerevisiae*, *Schizosaccharomyces pombe* and the human databases. They found that almost half of the *C. albicans* ORFs (3027, the "housekeeping genes") had putative orthologues in all three species, 1119 had a match in *S. cerevisiae* and in either one of the other species, 613 had matches only in *S. cerevisiae*, and 1426 had no match at all. The remaining 234 ORFs were not found in *S. cerevisiae* but either only in *S. pombe* (91) or in human (83) or in both (60). The high number with matches only in the human genome is very intriguing and raises the question of their origin. Other yeast species have genes with close homologs found only in bacteria – *Y. lipolytica* contains eight such genes, *K. lactis* five and *D. hansenii* one (Dujon et al. 2004). One possible explanation is that these genes may have been acquired from other distinct species (such as humans or bacteria). However, horizontal gene transfer, which is very important in prokaryotes, has not yet been demonstrated in fungi. The evolution of these genes in *C. albicans* requires further investigation.

The functional classification of the genes unique to *C. albicans* within this comparison (Jones et al. 2004) shows numerous distinct differences in oxidative metabolism. *C. albicans* possesses several additional components for respiration, such as an extra complex I of the electron transport chain, a pyruvate dehydrogenase kinase for possible flow regulation from glycolysis to the TCA cycle, a complete family of secreted lipases (Hube et al. 2000; see Sect. IV) and other enzymes in fatty acid metabolism and, last but not least, additional amino acid catabolic pathways, including one for cysteine. This difference in sulphur metabolism is also reflected by an additional direct cysteine biosynthesis pathway, which may indicate an increased significance for glutathione metabolism in *C. albicans*.

Finally, *C. albicans* has a number of genes related to the pH regulatory genes of *Aspergillus* (Davis et al. 2000), possibly reflecting an adaptation to the environment of the digestive tract. Also encoded are a small family of chloride channels, with members resembling types expressed in mammals (Jones et al. 2004).

In CandidaDB (d'Enfert et al. 2005), the haploid genome of *C. albicans* was found to contain 5917 protein-coding ORFs, including some shorter than 100 amino acids. A comparison of these predicted protein sequences to all publically available sequences revealed 548 proteins specific to *C. albicans* and *D. hansenii*, and around 600 sequences unique to *C. albicans* (C. d'Enfert and F. Tekaia, personal communication). These latter sequences presumably contribute to pathogenicity and other traits which separate these two otherwise closely related organisms.

Assuming that genes responsible for pathogenicity are not found in closely related but rather in apathogenic species or strains, there is another commonly used method for comparative genomics using cross-hybridisation of genomic DNA to a DNA microarray of the species of interest (Murray et al. 2001; Daran-Lapujade et al. 2003; Moran et al. 2004). Here the spots of interest, representing a gene not present or very divergent at the nucleotide level, will give only a weak signal or no signal at all. A drawback of this method is that one misses all of those genes absent from the array but present in the organism used for comparison.

Within the framework of the various approaches presented above, one important issue remains to be discussed. Genes which are present in a pathogen and absent in non-pathogens clearly have a higher probability of being involved in pathogenicity, this probability increasing the smaller the genetic distance becomes between the organisms compared. A good example of this is the family of secreted aspartic proteinases, which are unique to *C. albicans* (Naglik et al. 2003a, 2004; see Sect. IV). Originating from several gene duplications, the members of this family have subsequently diversified with respect to their expression patterns and their functions in pathogenesis. However, the inference that proteins present in both organisms cannot be involved in pathogenesis is not always correct. For example, the *PLD1* gene in *S. cerevisiae* is essential for sporulation, while *PLD1* in *C. albicans* (which has a deduced amino acid sequence very similar to that of Pld1p in *S. cerevisiae*) plays a role in dimorphism and invasion, attributes involved in pathogenesis (Hube et al. 2001). Similarly, agglutinin-like sequences play a role in mating of *S. cerevisiae*, but have adapted to promote adhesion to host cells in *C. albicans* (see Sect. IV).

Kellis et al. (2003) found evidence for rapidly evolving proteins, mainly due to changes in the copy number of trinucleotide repeats. Such a feature may well be missed in microarray studies if it lies outside the region selected for probe design, but it may be very significant in terms of function. The same is true for (point) mutations which change substrate specificity or regulatory features of a protein and, of course, the regulatory elements of the promoter region.

B. Genes Encoding Proteins with Unknown Function

Researchers are faced with another problem – currently, about 25% of predicted proteins in *C. albicans* have no known function and contain no identifiable motif signatures. Some of these have similarities with proteins of unknown function in other organisms. Most proteins of at least partially known function are involved in processes which have been extensively studied, such as metabolic enzymes and structural or regulatory proteins. Within groups of genes identified in studies of host/pathogen interaction in *C. albicans*, the proportion of proteins of unknown function can be much larger, up to 35% (Fradin et al. 2005), which reflects the current status of detailed biological understanding of pathogenesis. Here, the so-called guilt by association principle can be of great help – genes involved in one and the same cellular process are likely to be co-regulated (Allocco et al. 2004). Hence, comparative genomics offers another promising approach, namely, the identification of *cis*-regulatory elements in closely related species by promoter sequence comparison. It is feasible that *cis*-acting elements are more conserved than are the rest of the upstream or downstream regulatory region of a protein-coding ORF. Thus, it should be possible to identify motifs which are conserved across orthologous genes. Following this strategy, Kellis et al. (2003) have identified 36 known *cis*-regulatory motifs in *S. cerevisiae*, which corresponds to 85% of the full set with highly conserved patterns. In addition, they found 42 new motifs of which 25 correlated well with functional classification based on gene ontology categories and expression patterns of the adjacent protein-coding regions. Taken together with the data produced by cluster analysis of transcriptional profiling experiments, this leads to the large-scale functional classification of gene products of no known function.

Sequence-based prediction algorithms are progressively becoming more sophisticated, gaining very high importance in whole-genome annotation projects. Today, the main focus lies on large-scale protein functional classification (e.g. InterProScan, www.ebi.ac.uk/InterProScan/) and prediction of subcellular localisation. In the latter, the prediction of amino acid motifs which target proteins to the secretory pathway (signal peptides) is the furthest developed, and several studies have aimed to computationally annotate the "secretome", or subsets thereof, from the *Candida* genome sequence (Lee et al. 2003; De Groot et al. 2003; Alberti-Segui et al. 2004; Eisenhaber et al. 2004; d'Enfert et al. 2005). For the analysis of the host–pathogen interface, the subset of these proteins localised to the cell surface is of special interest, as this enables them to directly and physically interact with the host. Research has mainly focused on the prediction of proteins which are attached to the cell wall via remnants of glycosylphosphatidylinositol (GPI) anchors (De Groot et al. 2003), revealing over 100 potential candidates in *C. albicans*, at least one-third more than in all of the other species (all non-pathogenic) analysed. *C. albicans* GPI-proteins not found in *S. cerevisiae* include the Als, Hyr1, Csa1/Rbt5 and Hwp1/Rbt1 protein families (De Groot et al. 2003). Compared to *S. cerevisiae*, the dataset in *C. albicans* also has a higher number of hypothetical proteins of unknown function and sugar hydrolases (Eisenhaber et al. 2004). *C. albicans* uniquely contains GPI-anchored superoxide dismutases Sod4p (Pga2p), Sod5p (Pga3p) and Sod6p (Pga9p), as reported by De Groot et al. (2003), Martchenko et al. (2004) and Fradin et al. (2005).

In summary, computer-based methodologies are already vital in identifying and analysing genes possibly involved in microbial pathogenesis in postgenomic analysis, and the impact of their contributions will certainly increase in the next few years. Nevertheless, the definite elucidation of the function and role of identified genes will always require experimental proof. This may be done on a gene-by-gene basis. However, promising global experimental approaches to elucidate gene functions and to identify virulence genes have already been used (see Sect. VIII) and will, in combination with computer-based methodologies, boost our knowledge about the process of infection.

III. Genes with Possible Virulence Functions

Different microbial pathogens of humans face similar conditions during colonisation and proliferation on or within a host. Microbial cells attach to human cells and surfaces, use the nutrients available, adapt to changing pH and oxygen levels, develop stress tolerance, develop mechanisms to evade the immune system, and damage host tissue in order to promote invasion. Thus, pathogenic microorganisms share a similar repertoire of strategies for growth and adaptation in association with their hosts and pathogenesis. Orthologues of known fitness and virulence genes in bacteria or other fungi have already been identified in the sequenced genome of *C. albicans*, and it is likely that additional genes will be found. Therefore, we can identify fitness and virulence genes which are novel to *C. albicans* by searching for sequences with similarity to genes known to be associated with these strategies in other pathogenic microorganisms. For example, in order to survive phagocytosis by immune cells, genes involved in an antioxidative response must be expressed to avoid or survive the oxidative burst. This is a reaction displayed by numerous pathogenic bacteria (Ruckdeschel 2002; Samuel et al. 2003; Ng et al. 2004), and similar genes have been discovered in *C. albicans* (Fradin et al. 2003; Rubin-Bejerano et al. 2003; Martchenko et al. 2004; Fradin et al. 2005; see Sect. X). Other bacterial gene products are essential for growth in carbohydrate-poor conditions of the phagosome (McKinney et al. 2000), and also seem to play a critical role for survival of *C. albicans* in macrophages (Lorenz and Fink 2001; see Sect. V).

Extracellular hydrolytic enzymes contribute to virulence of pathogenic microorganisms, and genes encoding secreted proteolytic and lipolytic enzymes with known or putative virulence functions have been identified by pre- and postgenomic approaches (Ghannoum 2000; Hube et al. 2000; Naglik et al. 2003a). In particular, *C. albicans* genes encoding extracellular lipases have been identified by a combination of screening a genomic fosmid library (Magee and Scherer 1998) and the first sequence data released by the Stanford *C. albicans* genome project (Hube et al. 2000).

One essential component for growth which has limited availability within the human host is iron. Therefore, pathogenic microorganisms have

Table. 9.2. Genes involved in iron metabolism identified in the genome of *C. albicans*

Function	Gene names	References
Siderophore transporter	*ARN1/SIT1*, CA3740	Ardon et al. (2001), Heymann et al. (2002), Hu et al. (2002), CandidaDB (see Table 9.1)
Ferric permeases	*FTR1*, *FTR2*, CA5345, CA5354	Ramanan and Wang (2000), CandidaDB
Iron transporter	*FTH1* (CA1521), *FTH2* (CA1493)	CandidaDB
Reductases	*CFL1*, *CFL95*, CA3460, CA1701, CA2397, CA3461, CA0947	Hammacott et al. (2000), Knight et al. (2002), CandidaDB
Ferroxidases	*FET3*, CA2922, CA2923, CA2924, CA1431, CA5401, CA2920	Eck et al. (1999), CandidaDB

gained numerous ways for iron uptake and maintenance of intracellular iron homeostasis (Crosa 1997; Rodriguez and Smith 2003; Perkins-Balding et al. 2004). This includes siderophore production and uptake of siderophores, transferrin or lactoferrin receptors, enzymes which reduce and oxidise iron, iron permeases and iron transporters. Iron uptake mechanisms are also crucial for growth and survival of environmental microorganisms in competition with other microorganisms. Not surprisingly, a large number of genes involved in iron metabolism have been identified in the genome of *C. albicans* by screening for sequences similar to those of *S. cerevisiae* genes encoding siderophore transporters (Ardon et al. 2001; Heymann et al. 2002; Hu et al. 2002), iron reductases (Hammacott et al. 2000; Knight et al. 2002), ferroxidases (Eck et al. 1999), or ferric permeases (Ramanan and Wang 2000; Lan et al. 2004; Table 9.2). It should be pointed out that it is very likely that *C. albicans* has genes necessary not only for adaptation to and interaction with host cells but also for survival in competition with other members of the normal microbial flora. The search for genes involved in iron metabolism in *C. albicans* led to the identification of large gene families (see Sect. IV), reflecting the importance of iron as a vital nutrient for growth and survival of *C. albicans* as both a commensal and a pathogen.

IV. Gene Families

Gene families in ascomycetes seem to be relatively frequent (Dujon et al. 2004). Some families with orthologous proteins in distinct species show either relatively low sequence identities (25%–50%), probably representing ancient gene duplications, or highly similar proteins (90%–100%), reflecting recent duplications and/or sequence homogenisation by gene conversion (Dujon et al. 2004). In *C. albicans*, several gene families have been identified whose functions may be associated with fitness, adaptation to the host environment, and virulence (Jones et al. 2004). These include genes encoding proteinases (Naglik et al. 2003a), lipases (Hube et al. 2000), adhesins (Hoyer 2001), superoxide dismutases (Martchenko et al. 2004; Fradin et al. 2005), oligopeptide transporters (Lubkowitz et al. 1997), iron transporter and permeases (Ramanan and Wang 2000; CandidaDB; Table 9.2), multicopper oxidases (Eck et al. 1999; CandidaDB; Table 9.1), iron reductases (Hammacott et al. 2000; Knight et al. 2002; CandidaDB; Table 9.2), mannosyltransferases (Hobson et al. 2004, Munro et al. 2005), peroxisomal carnitine acetyl transferases (Prigneau et al. 2004), and putative permeases with transmembrane regions (CandidaDB). A few of these families, such as secreted aspartic proteinases (*SAPs*; Naglik et al. 2003a, 2004), agglutinin-like adhesins (*ALSs*; Hoyer 2001) and extracellular lipases (*LIPs*; Hube et al. 2000), which all have at least nine members, have been investigated in more detail (Fradin and Hube 2003).

The extracellular proteinase activity of *C. albicans* is due to the secretion of aspartic proteinases encoded by ten *SAP* genes (reviewed by Naglik et al. 2003a). Although genes encoding aspartic proteinases exist in *S. cerevisiae* (e.g. *PRA1*, *BAR1*, and the yapsines *YPS1*, *3*, *5*, *6*, *7*), the majority of the *SAP* gene family seem to be unique to *C. albicans*. Even the most related *Candida* species, *C. dubliniensis*, does not contain such a variety of genes encoding these enzymes (Moran et al. 2004). Saps seem to be involved in a number of processes relevant to the pathogenesis of *C. albicans*. For example, Saps may contribute to cell damage and tissue invasion by the hydrolysis of host proteins such as keratin and collagen. Almost all investigated proteins of the human immune system, including immunoglobulins IgG, IgA and sIgA, alpha-macroglobulin, proteins associated with leukocytes and salivary proteins, are hydrolysed by purified Sap2p proteinase. Since adhesion of *C. albicans* to epithelial and endothelial cells was shown to be inhibited by the classical aspartic proteinase inhibitor pepstatin A, Saps may also function in attachment processes, possibly by modifying surface proteins.

The *ALS* gene family was identified by the similarity of their products to the *S. cerevisiae* mating-associated adhesion glycoprotein α-agglutinin (Sag1) encoded by AGα1 (*SAG1*; Hoyer 2001). Despite a relatively high sequence similarity between Als1p and Sag1p, Als1p is not believed to be involved in *C. albicans* mating (Magee et al. 2002). Als1p has been demonstrated to mediate the adherence of *C. albicans* to epithelial and endothelial cells via the amino terminal domains of these proteins (Gaur and Klotz 1997; Fu et al. 1998; Gaur et al. 1999), and Als5p was shown to bind to host extracellular matrix proteins (Gaur and Klotz 1997). Sheppard et al. (2004) have extended these studies and demonstrated that different members of the Als family have different adhesion profiles to a variety of host substrates, this specificity being mediated through the N-terminus. Furthermore, immunohistological analysis has shown that Als antigens are produced in murine disseminated candidosis (Hoyer et al. 1999).

Extracellular lipase activity encoded by ten members of the *LIP* gene family has no counterpart in *S. cerevisiae* or any other ascomycetes (Fu et al. 1997; Hube et al. 2000; Stehr et al. 2004). The most closely related gene identified in databases was from *Mycobacterium tuberculosis*, which probably encodes a lipase (Hube et al. 2000). The open reading frames of all ten lipase genes encode highly similar proteins, with up to 80% identical amino acid sequences (Hube et al. 2000). These lipases may have evolved to adapt to the permanent association of *C. albicans* with the human body, and may have important functions in colonisation and infection processes. One obvious role of lipases may be the utilisation of lipid substrates on human tissue, such as the skin or the intestinal tract. The high number of *LIP* genes may provide an adaptive advantage to persist on these surfaces, in the absence of carbohydrates, and may assist *C. albicans* in competing with the normal microbial flora. Furthermore, release of fatty acids due to lipolytic activity may modify the surrounding pH of fungal cells, thereby optimising the activity of other proteins such as the secreted proteinases. As has been shown for other microbial lipases, *Candida* lipases may also directly affect the host immune system by inhibiting cell-mediated chemotaxis, by damaging phagocytic cells or by triggering local inflammatory responses.

Fradin and Hube (2003) have recently speculated why *C. albicans* (and perhaps other eukaryotic pathogenic microorganisms) may possess such large gene families. They concluded that the possession of a variety of gene families by pathogenic microorganisms may confer at least five possible advantages.

1. Gene families may have evolved to facilitate the coordinated regulation of members of this family, together with other virulence attributes.
2. The different gene products may have adapted and co-evolved to function in different tissues and/or environments during colonisation and infection.
3. Members of a gene family may be functionally redundant and act by providing a second protein when another member of the family fails, is removed or is otherwise lost.
4. They may encode proteins with similar activities but distinct and different functions.
5. Finally, the concomitant expression of a number of similar, but functionally distinct, members of a gene family may result in synergistic effects to promote colonisation or infection. Thus, several gene products may act in unison to carry out a series of tasks to possibly provide the microorganism with a biological advantage.

V. Role of Metabolism During Infection

Some of the most notable differences which exist between the genomes of the pathogen *C. albicans* and the model yeast *S. cerevisiae* are concerned with metabolism. *C. albicans* appears to have the ability to utilise a larger range of nutrients for growth, and this may be one of the attributes which has enabled *C. albicans* to exist as a commensal organism in healthy individuals and to evolve into a successful pathogen in the immunocompromised. In order to colonise and cause disease in a wide range of host sites, *C. albicans* must be able to survive and grow in vastly different environments. As *C. albicans* spreads from mucosal surfaces to deep-seated organs via the bloodstream, it must cope with changes in pH, oxygen availability, and variations in carbon and nitrogen sources. Even as a commensal it can adapt to colonise the oral cavity, vaginal tract and gastrointestinal tract, which are different environments where the fungus must compete with other microbes which make up the natural flora in these niches.

During an infection, *C. albicans* will enounter cells of the immune response. When engulfed by macrophages, *C. albicans* yeast cells switch to hyphal growth. Not only does *C. albicans* survive the hostile environment within these phagocytes, but also hyphal growth enables the fungus to grow out and escape these host defences. In a key paper, Lorenz and Fink (2001) described for the first time changes in the transcriptional profile of a fungus (in this case, *S. cerevisiae*) in response to phagocytosis by mammalian macrophages. They showed that *S. cerevisiae* responded to conditions within the phagolysosome by up-regulating genes of the glyoxylate cycle. The glyoxylate cycle enables simple two-carbon compounds to be assimilated by the TCA cycle. The authors inferred from their results that the phagosome was low in complex carbon compounds causing nutrient deprivation, a major stress to the fungal cells. No conventional stress responses were observed. Lorenz and Fink (2001) went on to show the importance of the glyoxylate cycle in the virulence of *C. albicans*, by generating a mutant in the isocitrate lyase gene (*ICL1*) which was avirulent when tested in a murine model of systemic candidosis. So, here we have an example of the genome-wide response of a non-pathogen (*S. cerevisiae*) providing valuable leads to factors which contribute to virulence in a pathogen (*C. albicans*).

Prigneau et al. (2003) identified seven genes in *C. albicans* induced by macrophage phagocytosis, using differential display coupled to reverse transcription (DDRT)-PCR. Four of the seven genes were predicted to encode peroxisomal proteins involved in β-oxidation (acyl CoA oxidase, Pxp2p, carnitine acetyl transferase, Cat3p) and the glyoxylate cycle (isocitrate lyase, Icl1p, malate permease, Mae1p). A fifth gene encoded cytosolic formate dehydrogenase, which is also associated with the glyoxylate cycle because formate is an indirect product. The remaining two genes encoded putative plasma membrane sensors for extracellular nutrients: Snf3p, a glucose sensor/transporter, and Gpr1p, a G-protein coupled receptor. These findings again suggest that the major stress associated with phagocytosis is nutrient starvation, and that *C. albicans* is induced to break down lipids as an alternative carbon source.

Like many other pathogens, *C. albicans* has the capability to switch phenotypes. In some species, for example, the African trypanosome, this involves altering their surface proteins, and thus they can evade the host's immune system (Barry and McCulloch 2001). This phenomenon, termed antigenic variation, is reversible and stochastic. In *C. albicans*, phenotypic switching results in altered physiology, morphology and pathogenicity in murine models. Many clinical isolates exhibit a high frequency of switching (Hellstein et al. 1993). The best characterised phenotypic switching in *C. albicans* involves the transition from white to opaque cells in strains homozygous for the *MTL* (mating type-like) locus. Accompanying the change from white to opaque colonies are alterations in cell size and shape, adhesion, ability to form hyphae, hydrophobicity, drug susceptibility, sensitivity to neutrophils and pathogenicity (Soll 2002). White cells are more virulent in a systemic murine model, and opaque cells are more successful in a cutaneous infection experimental model (Kvaal et al. 1999). Therefore, both switch phenotypes appear to have an advantage at different host sites. In addition, the white-to-opaque transition is critical for *C. albicans* mating, with opaque cells mating at a million-fold higher frequency than do white cells (Miller and Johnson 2002). A genome-wide comparison of the expression profile of white and opaque cells of the classical switching strain WO-1 revealed the differential expression of 373 genes (Lan et al. 2002). Approximately one-third of these genes were involved in metabolism. The transcriptional profile of white cells reflected fermentative metabolism and, in contrast opaque cells, gave patterns consistent with oxidative metabolism. White cells had higher levels of three glucose transporters (*HXT3*, *HXT4* and *HXT7*) and genes encoding glycolytic enzymes (hexokinase, phosphofructokinase and pyruvate kinase). By contrast, opaque cells had higher levels of genes encoding TCA cycle and β-oxidation enzymes. The abundance of other genes involved in amino acid metabolism, amino acid, phosphate and sulphate transport also differed between the two switch types. These differences in metabolic status between the two switch types may provide the advantage to survive and colonise vastly different host sites. The up-regulation of the β-oxidation pathway in the opaque phase correlates with its success in the cutaneous model, the skin being rich in lipids but relatively poor in sugars. Therefore, we can add metabolic specialisation to the list of properties which are modified as a result of phenotypic switching, and which may contribute to *C. albicans* success as a pathogen with the ability to colonise a diverse range of anatomical sites. This issue is discussed further by Brown (Chap. 10, this volume).

VI. Role of Morphology

One of the most intriguing properties of *C. albicans* is the ability to switch between yeast and hyphal forms of growth (dimorphism). The signalling pathways regulating this transition are discussed by Whiteway and Nantel (Chap. 8, this volume). In this chapter, we will focus on particular aspects of the role of dimorphism in pathogenesis in the postgenomic era.

It has long been postulated that the transition to filamentous growth is associated with invasive growth of *C. albicans*. However, during infection both morphological forms are found, and both forms may play their individual roles during infection. Furthermore, many pathogenic fungi such as *H. capsulatum*, *Penicillium marneffei* and *Paracoccidioides brasiliensis* are also dimorphic, but here the hyphal form is the environmental growth form which switches to a yeast form in the human body. Thus, the ability to produce filaments is not generally associated with fungal virulence (Gow et al 2002).

Clearly, dimorphism in *C. albicans* is regulated by a complex transcriptional programme which not only modulates the morphology but also includes the expression of a number of hyphae-associated genes, encoding surface proteins (*HWP1*, *HYR1*, *ALS3/8*), secreted proteinases (*SAP4-6*), and detoxification enzymes, such as superoxide dismutases *SOD5* (Hube et al. 1994; Brown and Gow 1999; Nantel et al. 2002). During systemic infections, the morphological transition and expression of hyphal-associated factors may benefit *C. albicans* in several ways (Hube 2004; Fradin et al. 2005).

1. Hyphal cells secrete factors which inhibit killing by neutrophils (Smail et al. 1992).
2. *C. albicans* cells phagocytosed by macrophages produce hyphae which penetrate the surrounding membranes, causing the death of the macrophage (Borg-von Zeppelin et al. 1998).
3. Hyphal cells, but not yeast cells, have been shown to be endocytosed by endothelial cells, a mechanism which is discussed as a potential strategy of *C. albicans* to escape from the bloodstream (Zink et al. 1996; Phan et al. 2000).
4. Hyphal cells have stronger adherence properties (e.g. due to the expression of the Als adhesins; Hoyer 2001), which may help the fungus to adhere to endothelial cells.
5. The higher expression level of superoxide dismutases may counteract the oxidative burst of phagocytic cells (Nantel et al. 2002; Fradin et al. 2005).
6. Finally, hyphal cells are known to have greater invasive properties in tissue, which assists the fungus in penetrating into the surrounding, deeper tissue of blood vessels (Calderone and Fonzi 2001; Gow et al. 2003).

Although it has long been proposed that the yeast form may be the preferred cell shape for haematogenous dissemination (Rooney and Klein 2002), *C. albicans* normally switches to hyphal growth when exposed to blood plasma. However, hyphal morphology per se is not sufficient for pathogenesis. Instead, the entire transcriptional programme associated with the transition (Odds et al. 2003) is likely to account for any hyphae-associated pathogenesis. It should be noted that cells which are locked in the yeast phase still have the ability to cross the endothelial barrier of blood vessels, but clearly have reduced virulence potential (Saville et al. 2003). Furthermore, as mentioned above, the yeast form and genes associated with the yeast form are likely to play important roles during other types of infection at different body sites.

VII. Role of the Cell Surface During Host/Pathogen Interactions

The cell wall of *C. albicans* is in intimate contact with host cells and tissues during an infection. Exposed on the *C. albicans* outer surface are highly glycosylated mannoproteins which act as adhesins (Hoyer 2001; Sundstrom 2002), antigenicity factors and immunomodulators (Suzuki 2002; Lopez-Ribot et al. 2004), and contribute to biofilm formation (Douglas 2003; Garcia-Sanchez et al. 2004). These properties involve both the protein and carbohydrate moieties of mannoproteins (Buurman et al. 1998; Sheppard et al. 2004; Munro et al. 2005). *C. albicans* can adhere to and colonise a wide range of niches in the host, reflecting a versatile cell wall composition. Indeed, experiments in vitro have shown that the fungal cell wall is highly dynamic, and alters its structure and architecture in response to changing environments, different stresses and in response to defects in cell wall components (reviewed by Klis et al. 2002). Many of the environmental conditions which lead to cell wall modifications would be encountered during an infection as the fungus colonises new sites in the body, for example, changes in pH, oxygen and nutrient availability. The ability of the cell to respond to a wide variety of environmental conditions reflects the complex and intricate network regulating cell wall biogenesis. The cell wall proteome in particular seems adaptable, responding to fluctuating pH and oxygen levels, for example (Klis et al. 2002). Two classes of proteins are covalently attached to the *C. albicans* cell wall: (1) the Pir proteins (proteins with internal repeats), attached to β(1,3) glucan by an alkali-sensitive bond and (2) GPI-anchored proteins, attached via a GPI-anchor remnant to β(1,6) glucan. The GPI-cell wall proteins share a conserved structure which includes a serine/threonine-rich domain. This class includes some of the best studied cell wall proteins of *C. albicans*, including the Als family of adhesins (Hoyer 2001), the pH-regulated glycosyltransferases Phr1p and Phr2p (Saporito-Irwin et al. 1995; Muhlschlegel and Fonzi 1997; Fonzi 1999) and the hyphal-specific proteins Hyr1p (Bailey et al. 1996) and Hwp1p, a substrate for mammalian transglutaminase (Staab et al. 1999).

The availability of an annotated genome sequence has greatly facilitated the analysis of the cell wall proteome, as discussed in Sect. II. Cell wall proteins can be fractionated depending on their mechanism of attachment to the cell wall, reducing the enormous complexity of cell wall analysis. A comparison has been made by two-dimensional gel electrophoresis of cell wall proteins in yeast and hyphal cells (Pitarch et al. 2002), revealing a heterogeneous set of proteins associated with the cell wall. Along with the known hyphal-specific proteins, a large number of cell surface-associated proteins were up-regulated under hyphal-inducing conditions. More recently, De Groot et al. (2004) have developed a method to analyse covalently linked cell wall proteins using tandem liquid chromatography and mass spectrometry. Exponential growth phase yeast cells, grown under laboratory conditions, expressed 14 proteins which were covalently attached to the cell wall. These included the adhesins Als1p, Als4p and flocculin-like Pga24p, the superoxide dismutase Sod4p, carbohydrate active enzymes Cht2p, Crh11p, Pga4p, Phr1p, Scw1p, and others with no known function, Ecm33.3p, Pir1p, Pga29p, Rbt5p and Ssr1p. The next challenge will be to examine the cell wall proteome when cells are grown under conditions which mimic in vivo environments.

The cell wall proteome of *C. albicans* has greatly diverged from that of *S. cerevisiae*, including the addition of the Als family of adhesins. Comparison of the genomes of the closely related *C. albicans* and *C. dubliniensis* has also revealed significant sequence divergence amongst the GPI-proteins, and the absence of several members of the Als family (Moran et al. 2004). The structure and number of cell surface proteins appear to be evolutionary less constrained than is the case for metabolic enzymes – for example, enabling the cell wall proteome to become diversified between species, leading to variations in the complement of adhesins giving specificity for a variety of host tissues. Many of these cell wall proteins are antigenic and elicit an antibody response in vivo (reviewed by Lopez-Ripot et al. 2004). Such proteins may be suitable candidates for the development of serodiagnosis of systemic candidiasis or as vaccines for novel therapies.

Analysis of the *C. albicans* genome sequence has revealed that many of the cell wall synthetic genes in *C. albicans* are similar to their *S. cerevisiae* counterparts. There are, however, some notable differences. For example, *C. albicans* has an additional chitin synthase gene *CHS8*, giving it two Class I enzymes homologous to *ScCHS1* (Munro and Gow 2001; Munro et al. 2003). Also, *S. cerevisiae* has two β(1,3) glucan synthase subunits *FKS1* and *FKS2* whereas *C. albicans* has only one essential enzyme Gsl21p (Mio et al. 1997). There are a number of examples where gene families of functionally redundant proteins have been expanded in *C. albicans*, compared to *S. cerevisiae*. These include *MNT*-like, *MNN1*-like, *MNN2*-like and *MNN4*-like gene families (Hobson et al. 2004; Bates, Munro and Gow, personal communication). The significance of this is not yet known. The presence of gene families complicates gene function analysis, as the true homologues of the *S. cerevisiae* genes are hard to predict by amino acid sequence alone (see Sect. II). Mutants lacking single members of the *MNN1*-like and *MNN2*-like families have no distinguishable phenotype (Bates, Hughes, Munro and Gow, personal communication). Therefore, further analysis will require multiple gene knockouts, which are difficult in *C. albicans*, or complementation tests with *S. cerevisiae* mutants. It will be interesting to determine whether members of these gene families exhibit differential expression patterns which may give clues to the reasons why these families have become expanded.

The fungal cell wall is under complex regulation at the transcriptional and post-transcriptional level. A large cohort of cell wall-related genes are differentially regulated during the dimorphic transition, and some are hyphal specific (*HYR1*, *ALS3*, *HWP1*). Therefore, the pathways which regulate cell wall biosynthesis will also play an important role in pathogenesis. Using cell wall-specific microarrays comprising 117 probes, Efg1p, a basic helix-loop-helix transcription factor, has been identified as a major regulator of cell wall biosynthesis controlling both hyphal-specific genes such as *HWP1* and yeast-specific genes (Sohn et al. 2003). *EFG1* had already been implicated in regulating morphogenesis (Stoldt et al. 1997). Using the same cell wall arrays, Lotz et al. (2004) have examined the role of Rim101p, the transcription factor involved in pH-dependent gene regulation. Rim101p was already known to activate *PHR1* and *PRA1* in response to alkaline pH, and *PHR2* in response to acid pH. Out of the 117 genes on the array, 32 were differentially regulated by Rim101p, with nine activated including the hyphal-specific genes *HWP1*, *ALS1*, *ALS5* and *RBT1*. *RIM101* negatively regulated 23 genes, including *RBR1* which encodes a GPI-protein. The disruption of the *RBR1* gene resulted in defective pH-induced hyphal growth. Expression of *RBR1* was dependent on the transcriptional repressor Nrg1p, and Rim101p was shown to repress Nrg1p. Therefore, a complex regulatory circuit exists which links morphogenesis, pH responses and cell wall dynamics, allowing *C. albicans* to respond to external cues.

VIII. Global Approaches to Identify Virulence Genes

The availability of the genome sequence has been invaluable in providing the *C. albicans* research community with easily accessible gene sequence data. However, the analysis of gene function in *C. albicans* has been hampered by its diploid nature. Current methods which rely on homologous recombination to create null mutants are time-consuming and laborious. Recent advances using PCR-based methods, flipper cassettes and a positive selectable marker have improved this process (Morschhäuser et al. 1999; Wilson et al. 2000; Gola et al. 2003; Shen et al. 2004; Reuß et al. 2004). Over the last 3 years, several strategies have been employed to develop methods for genome-wide or large-scale analysis. Although some have been pioneered by companies in drug discovery programmes aimed to identify essential genes, they may provide useful insights also into genes involved in pathogenicity and growth in vivo. In 2001, De Backer et al. described an approach that used antisense RNA and promoter interference to analyse gene function. Transformation with an antisense cDNA library yielded over 2000 transformants – 87 genes were identified which were required for growth in vivo, and 45 of these were of unknown function. The role of these genes in in vivo growth has not been published. Roemer et al. (2003) developed the GRACE (Gene Replacement And Conditional Expression) method primarily to screen for novel antifungal drug targets. In this approach, a single copy of each gene is disrupted with a PCR-derived cassette comprising a *HIS3* selectable marker and unique bar codes for tagging. The second copy of each gene is placed under the control of a tetracycline regulatable promoter, with the *SAT-1* gene conferring resistance to noureseothricin used as a dominant selectable marker. This results in the systematic generation of a bank of conditional mutants with expression controlled by the absence or presence of the repressor tetracycline.

A total of 567 *C. albicans* genes were found to be essential for growth in vitro using this technique, with only 61% of *S. cerevisiae* essential genes being also essential in *C. albicans*. This highlights the important differences between these two species. GRACE technology can also be applied to look at gene function in vivo. An important part of drug target validation is that the gene selected plays a role in pathogenesis and growth in vivo. A subset of 140 GRACE strains have been tested in vivo using a murine model of systemic candidosis, feeding mice tetracycline or doxycycline to switch off the expression of the target gene. Tetracycline may be administered prior to, during or after challenge with *C. albicans* (Kauffman et al. 2004). Data for six genes which were identified as essential for growth in vitro, and which included two conditional mutants for *BRX1* and YJL109C, showed 80% or greater survival in the presence of doxycycline. Upon removal of the repressor, mice infected with the YJL109C strain still had a high fungal burden in the kidneys. The *BRX1* strain was completely cleared from the animals, even in the absence of doxycycline. The remaining four genes, *TRP5*, *PDC1*, *MRPL6* and *EXG1*, were not essential for growth in vivo, emphasising the need for more detailed, large-scale analysis of the requirements for the growth of *C. albicans* in vivo.

Large-scale gene analysis performed in academic laboratories has involved the generation of mutant libraries by transposon mutagenesis. Davis et al. (2002) modified a Tn7 transposon with a *UAU1* marker cassette comprised of the *ARG4* gene flanked by sections of the *URA3* gene which can recombine to excise *ARG4*. Addition of the *UAU1* cassette to the transposon allows one to select for homozygous insertion mutations. The *Tn7–UAU1* was transposed into a *C. albicans* genomic library, and 253 insertion cassettes were isolated and transformed into *C. albicans* to generate heterozygous insertion mutants. The *UAU1* cassette was then employed to identify homozygous mutants which had arisen through mitotic recombination or gene conversion, and which now expressed both *ARG4* and *URA3* markers. A total of 217 homozygous mutants were isolated, with insertions in 197 ORFs. These homozygous mutants were screened for defects in alkaline pH-induced hyphal growth, and insertions in three genes gave the desired phenotype – *SLA2*, *RIM13* and *MSD3*. This was the first time that *MSD3* had been implicated in pH response and filamentous growth.

Uhl et al. (2003) introduced a modified Tn7 transposon into genomic DNA fragments to create a library which was then transformed back into *C. albicans*. A library of 18,000 *C. albicans* strains was generated which was estimated to have a transposon insertion every 2.5 kb of haploid sequence. Each strain harboured one heterozygous insertion mutation, and was screened for phenotypes arising from loss of function in one copy of the target gene, a phenomenon termed haploinsufficiency. Such a gene dosage effect is commonly observed in *C. albicans* (Braun and Johnson 1997; Munro et al. 1998; Gale et al. 1998). The mutant library was first screened for transformants with modifications in filamentous growth under two inducing conditions, Spider medium and in the presence of 1% foetal calf serum. Approximately 2% (340) of the strains tested had reproducibly modified hyphal formation, as judged by alterations in colony morphology. Identification of the transposition point in each strain narrowed down the number of unique genes affected in hyphal formation to a total of 146. Among these genes were several already implicated in filamentous growth (*TUP1*, *RFG1*, *PDE2*, *CZF1*, *NRG1*, *NOT1*), but 27% lacked significant similarity to any genes of *S. cerevisiae* or in the databases.

Genetic screens provide valuable insights into cellular processes in *C. albicans*, and highlight important differences between this pathogen and the model yeast *S. cerevisiae*. As the ability to switch between yeast and hyphae is required for virulence, the genes identified in this screen may have a potential role in pathogenicity. These approaches may not be so amenable for in vivo studies, due to the large number of animals required to screen such libraries. However, preliminary screens could be performed in vitro using a number of model systems, and only those strains exhibiting phenotypes where defects in pathogenicity may be inferred would then be tested in vivo.

IX. Transcriptional Profiling During Infection

A number of technologies such as reverse transcriptase-PCR (RT-PCR) and in vivo expression technology (IVET) have been used to study the expression of selected genes from *C. albicans* during infection (reviewed in Hube 2004). Much broader, genome-wide approaches including differential display (Prigneau et al. 2003), cDNA subtractive hybridisation (Fradin et al. 2003), antibody-based strategies (Cheng et al. 2003), and DNA microarrays (Rubin-Bejerano et al. 2003; Fradin et al. 2003, 2005) have been used to identify infection-associated genes. Microarray analysis provides a fascinating tool to unravel the complex genetic processes underlying the interaction between microorganisms and the host, and will prove invaluable to our understanding of these diseases (Bryant et al. 2004). The ultimate goal of whole-genome expression studies of pathogenic microorganisms is the identification of microbial genes which are differentially regulated in the host (Schoolnik 2002). However, there are a number of major technical challenges for in vivo transcriptional profiling, including (1) contamination with host tissue/RNA, (2) mRNA half-life and instability, (3) small tissue samples, necessitating amplification of RNA/cDNA probes, (4) heterogeneous population of cells which colonise or invade tissue, (5) heterogeneity of biological samples, and (6) heterogeneity of microbial strains in patient samples (Hube 2004). These technical problems may be reduced by simply using culture conditions which mimic certain features of the host (Enjalbert et al. 2003). Expression profiles obtained from microorganisms grown in media simulating host microenvironments may yield a portrait of interacting metabolic pathways and multistage developmental programmes, and disclose regulatory networks (Schoolnik 2002). More sophisticated ex vivo or in vitro infection models may mimic mucosal (Schaller et al. 1998) or systemic infections (Fradin et al. 2003, 2005), or may focus on the interaction with a particular type of host cell (Lorenz and Fink 2001; Rubin-Bejerano et al. 2003).

X. Gene Expression During Host/Pathogen Interaction: Examples

The balance between commensalism and pathogenicity is determined by both the host and *C. albicans*. To determine how this balance is regulated, it is crucial to analyse in more detail the host–fungus interface. The availability of the entire human and *C. albicans* genome sequences facilitates the high-throughput analysis of human and fungal genes regulated in response to different infections or infection-like situations.

Several studies have been performed in order to answer different questions which are essential to understand the mechanisms of pathogenicity of *C. albicans*. The following points need to be clarified.

1. Which virulence factors are important for which type of infection?
2. Which environments represent specific host cells or sites for *C. albicans*?
3. Which host cells can efficiently prevent *C. albicans* infection, and how is this achieved?
4. Which host responses can be modulated by *C. albicans*?

Incubation of *C. albicans* with different mammalian cell lineages representing different cell types has been widely used as a simple model to study *C. albicans*-host interaction. This type of model provides some important information about the behaviour of *C. albicans* in the presence of host cells, as well as the host response against this fungus, but it is very important to complement these studies with models more representative of the complexity of the different host environments (Fig. 9.1).

A. *C. albicans* Facing Epithelial and Endothelial Barriers

Adhesion of *C. albicans* to epithelial cells is the first important step of colonisation. Molecular methods have been used to identify genes involved in adhesion to epithelial cells. Non-adherent *S. cerevisiae* were transformed with a *C. albicans* genomic library, and adherent transformants were found to contain *AAF1*, the first *C. albicans* adherence gene described (Barki et al. 1993), *ALS5* (formerly named *ALA1*; Gaur and Klotz 1997), which is a member of the *ALS* gene family mentioned above encoding cell wall glycoproteins, and *EAP1*, which encodes another putative GPI-anchored protein (Li and Palecek 2003). These proteins were further characterised and shown to promote adhesion of *C. albicans* to oral epithelial or kidney cells. In order to mimic more closely the complexity of mucosal infections, models based on reconstituted human oral, oesophageal or vaginal epithelia (RHE) have been used as in vitro models of mucosal candidosis (Schaller et al. 1998, 2003; Li et al. 2002; Green et al. 2004). The use of these models has brought new insights into the colonisation and invasion mechanisms of *C. albicans*. For example, several members of the *ALS* gene family were shown to be expressed in the RHE model, and are possibly involved in the formation of a pseudomembranous structure on top of the epithelium, resembling a biofilm (Green et al. 2004). The ability of *C. albicans* to grow in close associations with surfaces under biofilm-like conditions may have several benefits for the fungus (Baillie and Douglas 2000). Therefore, Garcia-Sanchez et al. (2004) studied *C. albicans* during biofilm formation by analysing the genome-wide transcriptional profile. In this study, *ALS1* was found to be expressed, supporting the view that the *ALS* gene family is associated with biofilm formation. Expression of two other genes (*CHK1* and *CSSK1*), encoding a putative two-component histidine kinase and a response regulator, was gradually induced by *C. albicans* during adherence and colonisation to epithelial tissue (Li et al. 2002).

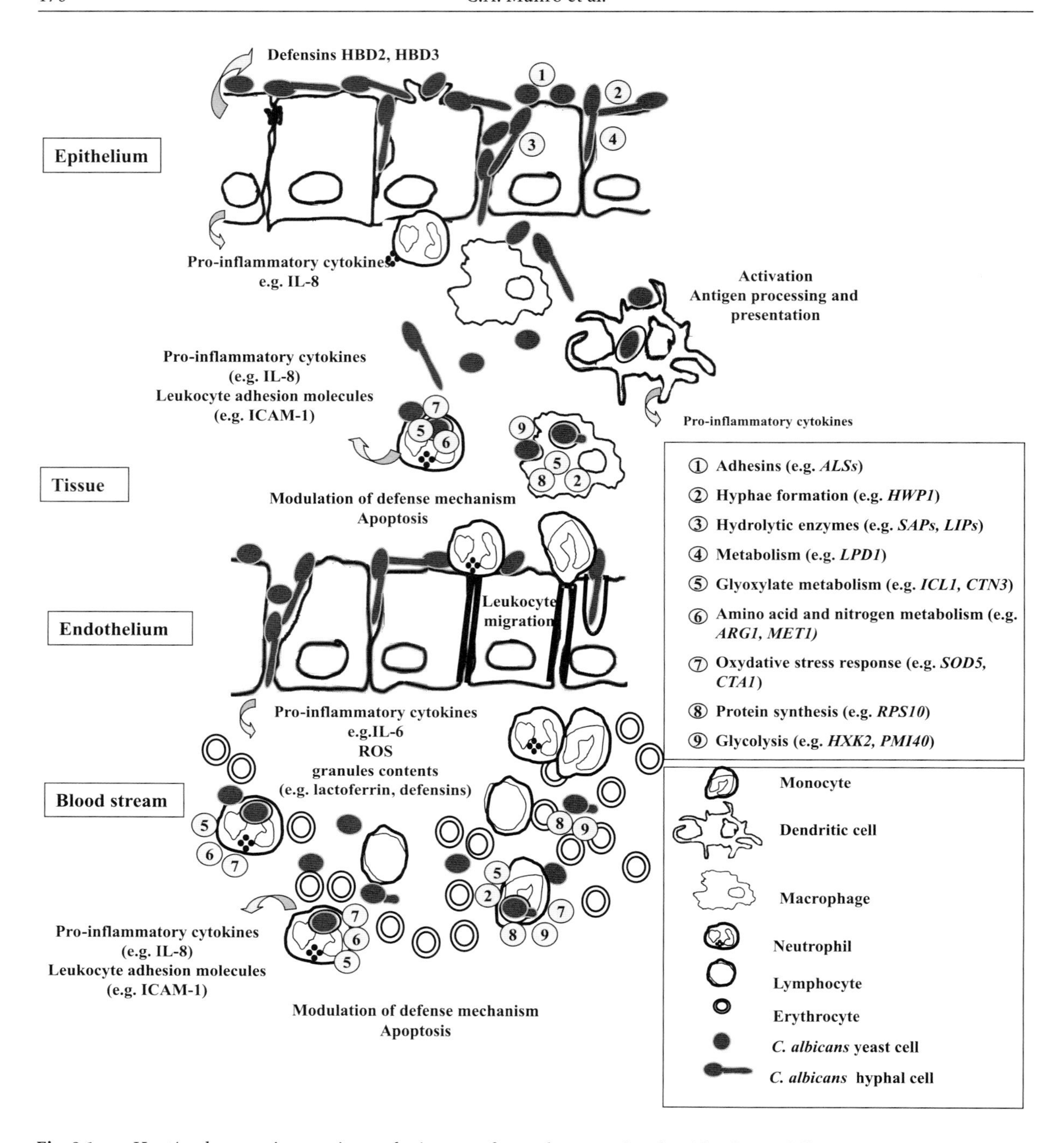

Fig. 9.1. Host/pathogen interactions during surface infections (epithelium), deep-seated infections (tissue) and bloodstream infections (endothelium and bloodstream). Shown are (1) *C. albicans* genes which are known to be associated with these different stages of infection (numbered and listed on the *right-hand side*) and (2) factors and processes characteristic of the host response to *C. albicans* infections

The identification of genes which were induced in vivo during oral thrush confirmed the importance of genes involved in diverse functions in this process. These included the regulation of yeast-to-hyphal morphogenesis (*HWP1*), adhesion to host cells, nutrient uptake, phospholipid biosynthesis and amino acid catabolism (*LPD1*) during the interaction of *C. albicans* with the epithelial layers (Cheng et al. 2003).

Genes encoding hydrolytic enzymes were also expressed during colonisation and penetration of mucosal models (Hube and Naglik 2002). The *SAP* genes have been the most studied gene family in *C. albicans* (see Sect. IV), and the order of their temporal expression was dependent upon the type of RHE infection used (Schaller et al. 1998, 2000, 2003). In addition, the expression of these genes was analysed in vivo in samples from patients suffering from oral or vaginal infections (Naglik et al. 2003b), and compared to the expression pattern of samples from *Candida* carriers (Naglik et al. 1999). These studies showed that *SAP1* and *SAP3* were more commonly expressed in patients than in carriers, suggesting that these genes may be involved in mucosal infections but not in commensal growth on mucosal surfaces. Transcripts from lipase and phospholipase genes were also detected during RHE infection, during experimental mucosal infection of mice, and in patient samples (Naglik et al. 2003b; Schofield et al. 2003; Stehr et al. 2004).

As *C. albicans* is a commensal microorganism, epithelial cells should have attributes to maintain the homeostasis between resident *C. albicans* and the epithelial barrier. Antimicrobial agents such as lactoferrin or IgA, which are present in the saliva, as well as other components secreted by epithelial cells, like the defensins, are probably essential to control the microbial colonisation of the epithelium. RT-PCR data showed that the β-defensins *HBD2* and *HBD3* are induced in oral tissues infected with *C. albicans* (Dunsche et al. 2001, 2002). Interestingly, hyphal cells seem to be the main growth form able to induce this epithelial response. The yeast-to-hyphal transition is commonly associated with better adhesion and penetration properties (see Sect. VI), which might explain why epithelial cells respond more strongly and efficiently to hyphae than do yeast cells. Furthermore, a non-specific defence mechanism, represented by epidermal cell proliferation, was enhanced during an acute experimental *C. albicans* infection (Korting et al. 1998; Bowers et al. 1999).

In vitro infection models of mucosal surfaces lack non-epithelial cell factors such as dendritic cells, macrophages and neutrophils, which might have a dramatic influence on the outcome of *C. albicans* infections in vivo. Phagocytes are usually found at the site of mucosal infection, where they are probably recruited by chemoattractants secreted by epithelial cells. Keratinocytes can indeed be stimulated by *C. albicans* to produce pro-inflammatory cytokines (Wollina et al. 2004), and neutrophils were attracted by physiologically active *Candida* cells after RHE infection (Schaller et al. 2004). This pro-inflammatory response can also induce stimulation of T-cells, which are crucial for host defence against *C. albicans*.

After penetration through the epithelium, *C. albicans* may reach the endothelial barrier and again must display a programme to combat weakened host defences. As for epithelial cells, the interactions between *C. albicans* and endothelial cells have not yet been widely studied using functional genomics. Cultured endothelial cells harvested from human umbilical vein with collagenase were used to study genes expressed by these cells after stimulation with *C. albicans*. Endothelial cells responded to in vitro *C. albicans* infection by expressing genes involved in the pro-inflammatory response (Orozco et al. 2000). This response is important, as it can determine the success of the infection by (1) initiating acute-phase response, (2) recruiting leukocytes from the bloodstream to the site of infection, (3) initiating the adaptive immune response and (4) increasing specific immune responses. The recruitment of leukocytes, especially neutrophils, by endothelial cells was confirmed by the expression of genes encoding leukocyte adhesion molecules such as E-selectin, and ICAM-1 by endothelial cells stimulated with *C. albicans* (Filler et al. 1996). Taken together, these data show that endothelial cells have the capacity to enhance the host response against *C. albicans* in vivo.

B. *C. albicans* Facing Phagocytes

Neutrophils play a key role in the host defences against *C. albicans*, as neutropenic patients are more susceptible to disseminated candidiasis. These cells can also play an important role during superficial infections, as they rapidly migrate from the bloodstream to the site of infection. Their recruitment is mediated by chemotactic factors which are produced by epithelial cells, such as Interleukin-8 (Wollina et al. 2004). Ex vivo, after contact with neutrophils, the majority of *C. albicans* cells are phagocytosed (Peltroche-Llacsahuanga et al. 2000). Phagocytosis and killing of an ingested microorganism by neutrophils is a complex mechanism involving oxidative and non-oxidative agents acting in concert or independently of each other. The phagosome of neutrophils constitutes an extremely hostile environment with an acidic or alkaline pH, antimicrobial peptides, reactive oxidative species, and relatively poor nutrient conditions (Hampton et al. 1998; Faurschou and Borregaard 2003). Neutrophils efficiently inhibit the yeast-to-hyphal transition of *C. albicans*, either by phagocytosis or via soluble factors. For example, antimicrobial peptides such as lactoferrin, which is contained in the neutrophil's granules and released in the phagosome and extracellular compartment, are very toxic to *C. albicans* and were shown to inhibit *Candida* hypha formation (Okutomi et al. 1997; Wakabayashi et al. 1998). By contrast, *C. albicans* cells which were phagocytosed by monocyte-derived macrophages were shown to have the potential to produce hyphal cells, which may assist escaping and killing of the macrophages. Transcript profiling of *C. albicans* incubated with purified polymorphonuclear cells, consisting mainly of neutrophils, showed that genes induced in response to neutrophils reflect the aggressive environment *C. albicans* is exposed to (Rubin-Bejerano et al. 2003; Fradin et al. 2005). The fungus must overcome the nutrient deficiency of the phagosome, and it does this by inducing methionine and arginine biosynthetic pathways (genes such as *MET1* and *ARG1*; Rubin-Bejerano et al. 2003). The neutrophil phagosome seems to be a poor source of nitrogen and carbohydrates, as genes involved in other amino acid pathways, nitrogen metabolism and the glyoxylate cycle were also induced by neutrophils (Fradin et al. 2005). Nitrogen deprivation seemed reduced or completely absent in monocyte phagosomes. By contrast, *C. albicans* cells also appear to be deprived of sources of carbohydrate in macrophages, as the genes involved in glyoxylate metabolism were induced after phagocytosis of *C. albicans* by macrophages, and genes such as *ICL1* were crucial for survival of macrophages (see Sect. V). Furthermore, transcript profiling showed that *C. albicans* cells were less physiologically active after incubation with neutrophils than were cells phagocytosed by macrophages (Fradin et al. 2005). Several genes encoding components of the protein synthesis machinery (*RPS10*) and glycolytic enzymes (*HXK2*, *PMI40*) were down-regulated in the presence of these phagocytes. The transcriptome also revealed that *C. albicans* faced a strong oxidative environment in the presence of neutrophils. The NADPH oxidase of neutrophils initiated the so-called oxidative burst, and *C. albicans* responded by expressing a large set of genes in order to survive this stress (Fradin et al. 2005). Effectively, genes encoding enzymes involved in the dismutation of the superoxide anion in H_2O_2, as well as those encoding enzymes involved in the transformation of H_2O_2 (*CTA1*) into water, were induced. Interestingly, the expression of the *SOD5* gene, which is normally hyphae regulated, was strongly expressed in yeast cells, even though neutrophils block the yeast-to-hyphal-transition (Fradin et al. 2005). The extracellular enzyme Sod5p seems particularly important for the fungus when facing extracellular reactive oxygen species. Consequently, a mutant lacking this gene was more susceptible to killing by neutrophils (Fradin et al. 2005).

Clearly, neutrophils dominate the host response against *C. albicans* in blood. However, the effect of neutrophils on the transcript profile, growth and survival of *C. albicans* in blood is attenuated when compared with the response of *C. albicans* to isolated neutrophils (Fradin et al. 2005). One obvious explanation is that *C. albicans* can interact with other cells in the blood such as monocytes, allowing the fungus to escape and survive. This suggests that it is important to work with a model which is as close as possible to the in vivo situation. Even with such a model, it can be difficult to discriminate between the respective roles of each component in *C. albicans* behaviour, and between the different patterns of fungal gene expression in response to the different components.

As neutrophils are efficient phagocytic cells which do not have a real role in the initiation of a specific immune response, it would be of interest to determine whether these cells display a transcriptional response upon incubation with microorganisms, including *C. albicans*. Incubation of *C. albicans* with cells of a granulocytoid lineage, HL60, showed that a number of neutrophil genes were in fact regulated in response to interaction with the fungus. As expected, several genes involved in the pro-inflammatory response were induced by *C. albicans* in a dose-dependent manner (Mullick et al. 2004). Interestingly, some genes involved in innate defence, such as the defensin gene *HNP1*, were repressed, showing a modulation of the neutrophils' response by *C. albicans*. *C. albicans* also induced the neutrophils to undergo apoptosis. Evidently, *C. albicans* not only expresses different factors to overcome the neutrophil defence mechanism but the fungus can also modulate neutrophil functions. However, it should be noted that neutrophils are known to be apoptotic cells and it has to be clarified whether the apoptosis is due to *C. albicans* itself blocking the neutrophil defence mechanism or whether the apoptosis is a normal neutrophil programme to stem inflammation at the site of infection.

Monocyte-derived macrophages and dendritic cells have a definitive role in the initiation of the adaptive immune response. *C. albicans*, but not *S. cerevisiae*, has been shown to induce apoptosis of macrophages, reducing their defence activities against the fungus (Ibata-Ombetta et al. 2003). Microarray analysis of the response of dendritic cells to *C. albicans* revealed induction of a different set of genes (Huang et al. 2001). Genes involved in the innate response (e.g. phagocytosis and pathogen recognition) were rapidly down-regulated, suggesting that a more specific immune programme was induced during the incubation time. Factors contributing to the recruitment of immune cells to the site of infection, as well as mediators of shape change and migratory behaviour of activated dendritic cells, were induced by *C. albicans*. Furthermore, several antigen processing and presentation genes were highly expressed, suggesting preparation for a more specific immune response against *C. albicans*. Nevertheless, it should be noted that overall, dendritic cells are less stimulated by *C. albicans* than by *Escherichia coli*. Also, *C. albicans* modulates only a subset of the genes regulated by *E. coli*. These observations might highlight the specific modulation of dendritic cell functions by *C. albicans*.

XI. Conclusions

The enormous influx of information from genome sequencing projects is revolutionising the science of microbial pathogenesis. This ranges from understanding the most basic aspects of gene content and pathogen genome organisation, to elucidating the regulatory networks of virulence gene expression, and the investigation of the global patterns of host response to infection (Whittam and Bumbough 2002). Since the release of the complete *C. albicans* genome sequence in 2000, significant progress has been made in unravelling the intriguing biology of this medically important fungal pathogen of humans and its mechanisms of virulence (Berman and Sudbery 2002). Gaps in our knowledge will be filled by a combination of comparative and functional genomics, including techniques such as proteomics, bioinformatics, structural biology and microarrays (Brosch et al. 2001).

Comparative genomics will uncover novel virulence determinants and hidden aspects of pathogenesis (Whittam and Bumbaugh 2002). The role of metabolic and biosynthetic pathways which have been neglected in the past as elements of fungal response to host environments have now come into focus (see Brown, Chap. 10, this volume). The role of morphology, a more traditional field in analysing virulence attributes of *C. albicans*, will be further elucidated and pathways with novel key regulatory factors will be identified. More molecules associated with the cell surface and involved in direct host/fungus interactions will be discovered and their function(s) analysed. New global approaches to identify virulence genes will be developed. Finally, genome-wide transcriptional profiling will continue to be used to investigate the response of *C. albicans* cells to their natural environments using infection models. Together with the knowledge of the genome of the human host and the use of host microarrays, these approaches will allow us to look simultaneously at both the host and the fungus, providing fascinating insights into the complex process of pathogenesis, and the transition from commensalism to parasitism of *C. albicans*. This knowledge will ultimately enable us to develop new treatments or strategies in the fight against fungal infections.

Acknowledgements. Donna MacCallum, Ricardo Almeida, Sascha Thewes and Antje Albrecht for help in preparing the manuscript. Our own investigations were supported by the Wellcome Trust (063204), the Robert Koch-Institut, the Deutsche Forschungsgemeinschaft (Hu 528/7/8/10), and the European Commission (Galar Fungail Consortium QLK-2000-00795 and MRTN-CT-2003-504148).

References

Alberti-Segui C, Morales AJ, Xing H, Kessler MM, Willins DA, Weinstock KG, Cottarel G, Fechtel K, Rogers B (2004) Identification of potential cell-surface proteins in *Candida albicans* and investigation of the role of a putative cell-surface glycosidase in adhesion and virulence. Yeast 21:285–302

Allocco DJ, Kohane IS, Butte AJ (2004) Quantifying the relationship between co-expression, co-regulation and gene function. BMC Bioinformatics 5:18

Ardon O, Bussey H, Philpott C, Ward DM, Davis-Kaplan S, Verroneau S, Jiang B, Kaplan J (2001) Identification of a *Candida albicans* ferrichrome transporter and its characterization by expression in *Saccharomyces cerevisiae*. J Biol Chem 276:43049–43055

Bailey DA, Feldmann PJF, Bovey M, Gow NAR, Brown AJP (1996) The *Candida albicans* HYR1 gene, which is activated in response to hyphal development, belongs to a gene family encoding yeast cell wall proteins. J Bacteriol 178:5353–5360

Baillie GS, Douglas LJ (2000) Matrix polymers of Candida biofilms and their possible role in biofilm resistance to antifungal agents. J Antimicrob Chemother 46:397–403

Barki M, Koltin Y, Yanko M, Tamarkin A, Rosenberg M (1993) Isolation of a *Candida albicans* DNA sequence conferring adhesion and aggregation on *Saccharomyces cerevisiae*. J Bacteriol 175:5683–5689

Barry JD, McCulloch R (2001) Antigenic variation in trypanosomes: enhanced phenotypic variation in a eukaryotic parasite. Adv Parasitol 49:1–70

Berman J, Sudbery PE (2002) *Candida albicans*: a molecular revolution built on lessons from budding yeast. Nat Rev Genet 3:918–930

Bernal A, Ear U, Kyrpides N (2001) Genomes OnLine Database (GOLD): a monitor of genome projects world-wide. Nucleic Acids Res 29:126–127

Borg-von Zepelin M, Beggah S, Boggian K, Sanglard D, Monod M (1998) The expression of the secreted aspartyl proteinases Sap4 to Sap6 from *Candida albicans* in murine macrophages. Mol Microbiol 28:543–554

Bowers W, Blaha M, Alkhyyat A, Sankovich J, Kohl J, Wong G, Patterson D (1999) Artificial human skin: cytokine, prostaglandin, Hsp70 and histological responses to heat exposure. J Dermatol Sci 20:172–182

Braun BR, Johnson AD (1997) Control of filament formation in *Candida albicans* by the transcriptional repressor *TUP1*. Science 277:105–109

Brosch R, Pym AS, Gordon SV, Cole ST (2001) The evolution of mycobacterial pathogenicity: clues from comparative genomics. Trends Microbiol 9:452–458

Brown AJP, Gow NAR (1999) Regulatory networks controlling *Candida albicans* morphogenesis. Trends Microbiol 7:333–338

Bryant PA, Venter D, Robins-Browne R, Curtis N (2004) Chips with everything: DNA microarrays in infectious diseases. Lancet Infect Dis 4:100–111

Buurman ET, Westwater C, Hube B, Brown AJP, Odds FC, Gow NAR (1998) Molecular analysis of CaMnt1p, a mannosyl transferase important for adhesion and virulence of *Candida albicans*. Proc Natl Acad Sci USA 95:7670–7675

Calderone RA (2002) Introduction and historical perspective. In: Calderone RA (ed) *Candida* and candidiasis. ASM Press, Washington, DC, pp 3–13

Calderone RA, Fonzi WA (2001) Virulence factors of *Candida albicans*. Trends Microbiol 9:327–335

Cheng S, Clancy CJ, Checkley MA, Handfield M, Hillman JD, Progulske-Fox A, Lewin AS, Fidel PL, Nguyen MH (2003) Identification of *Candida albicans* genes induced during thrush offers insight into pathogenesis. Mol Microbiol 48:1275–1288

Crosa JH (1997) Signal transduction and transcriptional and posttranscriptional control of iron-regulated genes in bacteria. Microbiol Mol Biol Rev 61:319–336

Daran-Lapujade P, Daran JM, Kotter P, Petit T, Piper MD, Pronk JT (2003) Comparative genotyping of the *Saccharomyces cerevisiae* laboratory strains S288C and CEN.PK113-7D using oligonucleotide microarrays. FEMS Yeast Res 4:259–269

Davis D, Wilson RB, Mitchell AP (2000) RIM101-dependent and-independent pathways govern pH responses in *Candida albicans*. Mol Cell Biol 20:971–978

Davis DA, Bruno VM, Loza L, Filler SG, Mitchell AP (2002) *Candida albicans* Mds3p, a conserved regulator of pH responses and virulence identified through insertional mutagenesis. Genetics 162:1573–1581

De Backer MD, Nelissen B, Logghe M, Viaene J, Loonen I, Vandoninck S, de Hoogt R, Dewaele S, Simons FA, Verhasselt P et al. (2001) An antisense-based functional genomics approach for identification of genes critical for growth of *Candida albicans*. Nat Biotechnol 19:235–241

De Groot PW, Hellingwerf KJ, Klis FM (2003) Genome-wide identification of fungal GPI proteins. Yeast 20:781–796

De Groot PWJ, De Boer AD, Cunningham J, Dekker HL, De Jong L, Hellingwerf KJ, De Koster C, Klis F (2004) Proteomic analysis of *Candida albicans* cell wall reveals covalently bound carbohydrate-active enzymes and adhesins. Eukaryot Cell 3:955–965

d'Enfert C, Goyard S, Rodriguez-Arnaveilhe S, Frangeul L, Jones L, Tekaia F, Bader O, Albrecht A, Castillo L, Dominguez A et al. (2005) CandidaDB: a genome database for *Candida albicans* pathogenomics. Nucleic Acids Res 33:D353–D357

De Repentigny L (2004) Animal models in the analysis of *Candida* host–pathogen interactions. Curr Opin Microbiol 7:24–329

Douglas LJ (2003) *Candida* biofilms and their role in infection. Trends Microbiol 11(1):30–36

Dujon B, Sherman D, Fischer G, Durrens P, Casaregola S, Lafontaine I, De Montigny J, Marck C, Neuveglise C, Talla E et al. (2004) Genome evolution in yeasts. Nature 430:35–44

Dunsche A, Acil Y, Siebert R, Harder J, Schroder JM, Jepsen S (2001) Expression profile of human defensins and antimicrobial proteins in oral tissues. J Oral Pathol Med 30:154–158

Dunsche A, Acil Y, Dommisch H, Siebert R, Schroder JM, Jepsen S (2002) The novel human beta-defensin-3 is widely expressed in oral tissues. Eur J Oral Sci 110:121–124

Eck R, Hundt S, Hartl A, Roemer E, Kunkel W (1999) A multicopper oxidase gene from *Candida albicans*: cloning, characterization and disruption. Microbiology 145:2415–2422

Eisenhaber B, Schneider G, Wildpaner M, Eisenhaber F (2004) A sensitive predictor for potential GPI lipid modification sites in fungal protein sequences and its application to genome-wide studies for *Aspergillus nidulans*, *Candida albicans*, *Neurospora crassa*, *Saccharomyces cerevisiae* and *Schizosaccharomyces pombe*. J Mol Biol 337:243–253

Enjalbert B, Nantel A, Whiteway M (2003) Stress-induced gene expression in *Candida albicans*: absence of a general stress response. Mol Biol Cell 14:1460–1467

Faurschou M, Borregaard N (2003) Neutrophil granules and secretory vesicles in inflammation. Microbes Infect 5:1317–1327

Filler SG, Pfunder AS, Spellberg BJ, Spellberg JP, Edwards JE Jr (1996) *Candida albicans* stimulates cytokine production and leukocyte adhesion molecule expression by endothelial cells. Infect Immun 64:2609–2617

Fonzi WA (1999) *PHR1* and *PHR2* of *Candida albicans* encode putative glycosidases required for proper cross-linking of beta-1,3- and beta-1,6-glucans. J Bacteriol 181:7070–7079

Fradin C, Hube B (2003) Tissue infection and site-specific gene expression in *Candida albicans*. Adv Appl Microbiol 53:271–290

Fradin C, Kretschmar M, Nichterlein T, Gaillardin C, d'Enfert C, Hube B (2003) Stage-specific gene expression of *Candida albicans* in human blood. Mol Microbiol 47:1523–1543

Fradin C, De Groot P, MacCallum D, Schaller M, Klis F, Odds FC, Hube B (2005) Granulocytes govern the transcriptional response, morphology and proliferation of *Candida albicans* in human blood. Mol Microbiol 56(2):397–415

Fu Y, Ibrahim AS, Fonzi W, Zhou X, Ramos CF, Ghannoum MA (1997) Cloning and characterization of a gene (LIP1) which encodes a lipase from the pathogenic yeast *Candida albicans*. Microbiology 143:331–340

Fu Y, Rieg G, Fonzi WA, Belanger PH, Edwards JE Jr, Filler SG (1998) Expression of the *Candida albicans* gene *ALS1* in *Saccharomyces cerevisiae* induces adherence to endothelial and epithelial cells. Infect Immun 66:1783–1786

Gale CA, Bendel CM, McClellan M, Hauser M, Becker JM, Berman J, Hostetter MK (1998) Linkage of adhesion, filamentous growth, and virulence in *Candida albicans* to a single gene, *INT1*. Science 279:1355–1358

Garcia-Sanchez S, Aubert S, Iraqui I, Janbon G, Ghigo JM, d'Enfert C (2004) *Candida albicans* biofilms: a developmental state associated with specific and stable gene expression patterns. Eukaryot Cell 3:536–545

Gaur NK, Klotz SA (1997) Expression, cloning, and characterization of a *Candida albicans* gene, *ALA1*, that confers adherence properties upon *Saccharomyces cerevisiae* for extracellular matrix proteins. Infect Immun 65:5289–5294

Gaur NK, Klotz SA, Henderson RL (1999) Overexpression of the *Candida albicans ALA1* gene in *Saccharomyces cerevisiae* results in aggregation following attachment of yeast cells to extracellular matrix proteins, adherence properties similar to those of *Candida albicans*. Infect Immun 67:6040–6047

Ghannoum MA (2000) Potential role of phospholipases in virulence and fungal pathogenesis. Clin Microbiol Rev 13:122–143

Gola S, Martin R, Walther A, Dunkler A, Wendland J (2003) New modules for PCR-based gene targeting in *Candida albicans*: rapid and efficient gene targeting using 100 bp of flanking homology region. Yeast 20:1339–1347

Gow NA, Brown AJ, Odds FC (2002) Fungal morphogenesis and host invasion. Curr Opin Microbiol 5:366–371

Gow NA, Knox Y, Munro CA, Thompson WD (2003) Infection of chick chorioallantoic membrane (CAM) as a model for invasive hyphal growth and pathogenesis of *Candida albicans*. Med Mycol 41:331–338

Green CB, Cheng G, Chandra J, Mukherjee P, Ghannoum MA, Hoyer LL (2004) RT-PCR detection of *Candida albicans ALS* gene expression in the reconstituted human epithelium (RHE) model of oral candidiasis and in model biofilms. Microbiology 150:267–275

Hammacott JE, Williams PH, Cashmore AM (2000) *Candida albicans CFL1* encodes a functional ferric reductase activity that can rescue a *Saccharomyces cerevisiae fre1* mutant. Microbiology 146:869–876

Hampton MB, Kettle AJ, Winterbourn CC (1998) Inside the neutrophil phagosome: oxidants, myeloperoxidase, and bacterial killing. Blood 92:3007–3017

Hellstein J, Vawter-Hugart H, Fotos P, Schmid J, Soll DR (1993) Genetic similarity and phenotypic diversity of commensal and pathogenic strains of *Candida albicans* isolated from the oral cavity. J Clin Microbiol 31:3190–3199

Heymann P, Gerads M, Schaller M, Dromer F, Winkelmann G, Ernst JF (2002) The siderophore iron transporter of *Candida albicans* (Sit1p/Arn1p) mediates uptake of ferrichrome-type siderophores and is required for epithelial invasion. Infect Immun 70:5246–5255

Hobson RP, Munro CA, Bates S, MacCallum DM, Cutler JE, Heinsbroek SE, Brown GD, Odds FC, Gow NA (2004) Loss of cell wall mannosylphosphate in *Candida albicans* does not influence macrophage recognition. J Biol Chem 279:39628–39635

Hoyer LL (2001) The *ALS* gene family of *Candida albicans*. Trends Microbiol 9:176–180

Hoyer LL, Clevenger J, Hecht JE, Ehrhart EJ, Poulet FM (1999) Detection of Als proteins on the cell wall of *Candida albicans* in murine tissues. Infect Immun 67:4251–4255

Hu CJ, Bai C, Zheng XD, Wang YM, Wang Y (2002) Characterization and functional analysis of the siderophore-iron transporter CaArn1p in *Candida albicans*. J Biol Chem 277:30598–30605

Huang Q, Liu D, Majewski P, Schulte LC, Korn JM, Young RA, Lander ES, Hacohen N (2001) The plasticity of dendritic cell responses to pathogens and their components. Science 294:870–875

Hube B (2004) From commensal to pathogen: stage- and tissue-specific gene expression of *Candida albicans*. Curr Opin Microbiol 7:336–341

Hube B, Naglik J (2002) Extracellular Hydrolases In: Calderone RA (ed) *Candida* and candidiasis. ASM Press, Washington, DC, pp 3–13

Hube B, Monod M, Schofield DA, Brown AJP, Gow NAR (1994) Expression of seven members of the gene family encoding secretory aspartyl proteinases in *Candida albicans*. Mol Microbiol 14:87–99

Hube B, Stehr F, Bossenz M, Mazur A, Kretschmar M, Schafer W (2000) Secreted lipases of *Candida albicans*: cloning, characterisation and expression analysis of a new gene family with at least ten members. Arch Microbiol 174:362–374

Hube B, Hess D, Baker CA, Schaller M, Schafer W, Dolan JW (2001) The role and relevance of phospholipase D1 during growth and dimorphism of *Candida albicans*. Microbiology 147:879–889

Ibata-Ombetta S, Idziorek T, Trinel PA, Poulain D, Jouault T (2003) *Candida albicans* phospholipomannan promotes survival of phagocytozed yeasts through modulation of Bad phosphorylation and macrophage apoptosis. J Biol Chem 278(15):13086–13093

Jones T, Federspiel NA, Chibana H, Dungan J, Kalman S, Magee BB, Newport G, Thorstenson YR, Agabian N,

Magee PT et al. (2004) The diploid genome sequence of *Candida albicans*. Proc Natl Acad Sci USA 101:7329–7334

Kauffman S, Roemer T, Breton A, Veillette K, Sherrill TP, Becker JM (2004) Mouse model utilizing conditional gene expression in *Candida albicans*. In: Abstr Vol 7th American Society for Microbiology Conf Candida and Candidiasis, 18–22 March 2004, Austin, TX, Abstr S7:2

Kellis M, Patterson N, Endrizzi M, Birren B, Lander ES (2003) Sequencing and comparison of yeast species to identify genes and regulatory elements. Nature 423:241–254

Klis FM, Mol P, Hellingwerf K, Brul S (2002) Dynamics of cell wall structure in *Saccharomyces cerevisiae*. FEMS Microbiol Rev 26:239–256

Knight SA, Lesuisse E, Stearman R, Klausner RD, Dancis A (2002) Reductive iron uptake by *Candida albicans*: role of copper, iron and the *TUP1* regulator. Microbiology 148:29–40

Korting HC, Patzak U, Schaller M, Maibach HI (1998) A model of human cutaneous candidosis based on reconstructed human epidermis for the light and electron microscopic study of pathogenesis and treatment. J Infect Dis 36:259–267

Kvaal C, Lachke SA, Srikantha T, Daniels K, McCoy J, Soll DR (1999) Misexpression of the opaque-phase-specific gene *PEP1* (*SAP1*) in the white phase of *Candida albicans* confers increased virulence in a mouse model of cutaneous infection. Infect Immun 67:6652–6662

Lan CY, Newport G, Murillo LA, Jones T, Scherer S, Davis RW, Agabian N (2002) Metabolic specialization associated with phenotypic switching in *Candida albicans*. Proc Natl Acad Sci USA 99:14907–14912

Lan CY, Rodarte G, Murillo LA, Jones T, Davis RW, Dungan J, Newport G, Agabian N (2004) Regulatory networks affected by iron availability in *Candida albicans*. Mol Microbiol 53:1451–1469

Lee SA, Wormsley S, Kamoun S, Lee AF, Joiner K, Wong B (2003) An analysis of the *Candida albicans* genome database for soluble secreted proteins using computer-based prediction algorithms. Yeast 20:595–610

Li F, Palecek SP (2003) *EAP1*, a *Candida albicans* gene involved in binding human epithelial cells. Eukaryot Cell 2:1266–1273

Li D, Bernhardt J, Calderone R (2002) Temporal expression of the *Candida albicans* genes *CHK1* and *CSSK1*, adherence, and morphogenesis in a model of reconstituted human esophageal epithelial candidiasis. Infect Immun 70:1558–1565

Lopez-Ribot JL, Casanova M, Murgui A, Martinez JP (2004) Antibody response to *Candida albicans* cell wall antigens. FEMS Immunol Med Microbiol 41:187–196

Lorenz MC, Fink GR (2001) The glyoxylate cycle is required for fungal virulence. Nature 412:83–86

Lotz H, Sohn K, Brunner H, Muhlschlegel FA, Rupp S (2004) *RBR1*, a novel pH-regulated cell wall gene of *Candida albicans*, is repressed by *RIM101* and activated by *NRG1*. Eukaryot Cell 3:776–784

Lubkowitz MA, Hauser L, Breslav M, Naider F, Becker JM (1997) An oligopeptide transport gene from *Candida albicans*. Microbiology 143:387–396

Magee PT, Scherer S (1998) Genome mapping and gene discovery in *Candida albicans*. ASM News 64:505–511

Magee BB, Legrand M, Alarco AM, Raymond M, Magee PT (2002) Many of the genes required for mating in *Saccharomyces cerevisiae* are also required for mating in *Candida albicans*. Mol Microbiol 46:1345–1351

Martchenko M, Alarco AM, Harcus D, Whiteway M (2004) Superoxide dismutases in *Candida albicans*: transcriptional regulation and functional characterization of the hyphal-induced *SOD5* gene. Mol Biol Cell 15:456–467

McKinney JD, Honer zu BK, Munoz-Elias EJ, Miczak A, Chen B, Chan WT, Swenson D, Sacchettini JC, Jacobs WR Jr, Russell DG (2000) Persistence of *Mycobacterium tuberculosis* in macrophages and mice requires the glyoxylate shunt enzyme isocitrate lyase. Nature 406:735–738

Miller MG, Johnson AD (2002) White-opaque switching in *Candida albicans* is controlled by mating-type locus homeodomain proteins and allows efficient mating. Cell 110:293–302

Mio T, AdachiShimizu M, Tachibana Y, Tabuchi H, Inoue SB, Yabe T, YamadaOkabe T, Arisawa M, Watanabe T, YamadaOkabe H (1997) Cloning of the *Candida albicans* homolog of *Saccharomyces cerevisiae GSC1/FKS1* and its involvement in beta-1,3-glucan synthesis. J Bacteriol 179:4096–4105

Moran G, Stokes C, Thewes S, Hube B, Coleman D, Sullivan D (2004) Comparative genomics using *Candida albicans* DNA microarrays reveals absence and divergence of virulence associated genes in *Candida dubliniensis*. Microbiology 150:3363–3382

Morschhäuser J, Michel S, Staib P (1999) Sequential gene disruption in *Candida albicans* by FLP-mediated site-specific recombination. Mol Microbiol 32:547–556

Muhlschlegel FA, Fonzi WA (1997) *PHR2* of *Candida albicans* encodes a functional homolog of the pH-regulated gene *PHR1* with an inverted pattern of pH-dependent expression. Mol Cell Biol 17:5960–5967

Mullick A, Elias M, Harakidas P, Marcil A, Whiteway M, Ge B, Hudson TJ, Caron AW, Bourget L, Picard S et al. (2004) Gene expression in HL60 granulocytoids and human polymorphonuclear leukocytes exposed to *Candida albicans*. Infect Immun 72:414–429

Munro CA, Gow NAR (2001) Chitin synthesis in human pathogenic fungi. Med Mycol 39 suppl 1:41–53

Munro CA, Schofield DA, Gooday GW, Gow NAR (1998) Regulation of chitin synthesis during dimorphic growth of *Candida albicans*. Microbiology 144:391–401

Munro CA, Whitton RK, Hughes HB, Rella M, Selvaggini S, Gow NAR (2003) *CHS8*-a fourth chitin synthase gene of *Candida albicans* contributes to in vitro chitin synthase activity, but is dispensable for growth. Fungal Genet Biol 40:146–158

Munro CA, Bates S, Buurman ET, Hughes HB, MacCallum DM, Bertram G, Atrih A, Ferguson MA, Bain JM, Brand A et al. (2005) Mnt1p and Mnt2p of *Candida albicans* are partially redundant α-1,2-mannosyltransferases that participate in O-linked mannosylation and are required for adhesion and virulence. J Biol Chem 280:1051–1060

Murray AE, Lies D, Li G, Nealson K, Zhou J, Tiedje JM (2001) DNA/DNA hybridization to microarrays reveals gene-specific differences between closely related microbial genomes. Proc Natl Acad Sci USA 98:9853–9858

Naglik JR, Newport G, White TC, Fernandes-Naglik LL, Greenspan JS, Greenspan D, Sweet SP, Challacombe SJ, Agabian N (1999) In vivo analysis of secreted aspartyl proteinase expression in human oral candidiasis. Infect Immun 67:2482–2490

Naglik JR, Challacombe SJ, Hube B (2003a) *Candida albicans* secreted aspartyl proteinases in virulence and pathogenesis. Microbiol Mol Biol Rev 67:400–428

Naglik JR, Rodgers CA, Shirlaw PJ, Dobbie JL, Fernandes-Naglik LL, Greenspan D, Agabian N, Challacombe SJ (2003b) Differential expression of *Candida albicans* secreted aspartyl proteinase and phospholipase B genes in humans correlates with active oral and vaginal infections. J Infect Dis 188:469–479

Naglik J, Albrecht A, Bader O, Hube B (2004) *Candida* proteinases and virulence. Cell Microbiol 6:915–926

Nantel A, Dignard D, Bachewich C, Harcus D, Marcil A, Bouin AP, Sensen CW, Hogues H, van het Hoog M, Gordon P et al. (2002) Transcription profiling of *Candida albicans* cells undergoing the yeast-to-hyphal transition. Mol Biol Cell 13:3452–3465

Ng VH, Cox JS, Sousa AO, MacMicking JD, McKinney JD (2004) Role of *KatG* catalase-peroxidase in mycobacterial pathogenesis: countering the phagocyte oxidative burst. Mol Microbiol 52:1291–1302

Odds FC (1988) Candida and candidosis, 2nd edn. Bailliere Tindal, London Odds FC, Calderone RA, Hube B, Nombela C (2003) Virulence in *Candida albicans*: views and suggestions from a peer-group workshop. ASM News 69:54–55

Odds FC, Brown AJ, Gow NA (2004) *Candida albicans* genome sequence: a platform for genomics in the absence of genetics. Genome Biol 5:230–232

Okutomi T, Abe S, Tansho S, Wakabayashi H, Kawase K, Yamaguchi H (1997) Augmented inhibition of growth of *Candida albicans* by neutrophils in the presence of lactoferrin. FEMS Immunol Med Microbiol 18:105–112

Orozco AS, Zhou X, Filler SG (2000) Mechanisms of the proinflammatory response of endothelial cells to *Candida albicans* infection. Infect Immun 68:1134–1141

Peltroche-Llacsahuanga H, Schnitzler N, Schmidt S, Tintelnot K, Lutticken R, Haase G (2000) Phagocytosis, oxidative burst, and killing of *Candida dubliniensis* and *Candida albicans* by human neutrophils. FEMS Microbiol Lett 191:151–155

Perkins-Balding D, Ratliff-Griffin M, Stojiljkovic I (2004) Iron transport systems in Neisseria meningitidis. Microbiol Mol Biol Rev 68:154–171

Phan QT, Belanger PH, Filler SG (2000) Role of hyphal formation in interactions of *Candida albicans* with endothelial cells. Infect Immun 68:3485–3490

Pitarch A, Sanchez M, Nombela C, Gil C (2002) Sequential fractionation and two-dimensional gel analysis unravels the complexity of the dimorphic fungus *Candida albicans* cell wall proteome. Mol Cell Proteomics 1:967–982

Prigneau O, Porta A, Poudrier JA, Colonna-Romano S, Noel T, Maresca B (2003) Genes involved in beta-oxidation, energy metabolism and glyoxylate cycle are induced by *Candida albicans* during macrophage infection. Yeast 20:723–730

Prigneau O, Porta A, Maresca B (2004) *Candida albicans CTN* gene family is induced during macrophage infection: homology, disruption and phenotypic analysis of *CTN3* gene. Fungal Genet Biol 41:783–793

Ramanan N, Wang Y (2000) A high-affinity iron permease essential for *Candida albicans* virulence. Science 288:1062–1064

Reuß O, Vik Å, Kolter R, Morschhäuser J (2004) The *SAT1* flipper, an optimized tool for gene disruption in *Candida albicans*. In: Abstr Vol 7th American Society for Microbiology Conf Candida and Candidiasis, 18–22 March 2004, Austin, TX Rodriguez GM, Smith I (2003) Mechanisms of iron regulation in mycobacteria: role in physiology and virulence. Mol Microbiol 47:1485–1494

Roemer T, Jiang B, Davison J, Ketela T, Veillette K, Breton A, Tandia F, Linteau A, Sillaots S, Marta C et al. (2003) Large-scale essential gene identification in *Candida albicans* and applications to antifungal drug discovery. Mol Microbiol 50:167–181

Rooney PJ, Klein BS (2002) Linking fungal morphogenesis with virulence. Cell Microbiol 4:127–137

Rubin-Bejerano I, Fraser I, Grisafi P, Fink GR (2003) Phagocytosis by neutrophils induces an amino acid deprivation response in *Saccharomyces cerevisiae* and *Candida albicans*. Proc Natl Acad Sci USA 100:11007–11012

Ruckdeschel K (2002) Immunomodulation of macrophages by pathogenic *Yersinia* species. Arch Immunol Ther Exp (Warszawa) 50:131–137

Samuel JE, Kiss K, Varghees S (2003) Molecular Pathogenesis of Coxiella burnetii in a Genomics Era. Ann N Y Acad Sci 990:653–663

Saporito-Irwin SM, Birse CE, Sypherd PS, Fonzi WA (1995) *PHR1*, a pH regulated gene of *Candida albicans*, is required for morphogenesis. Mol Cell Biol 15:601–613

Saville SP, Lazzell AL, Monteagudo C, Lopez-Ribot JL (2003) Engineered control of cell morphology in vivo reveals distinct roles for yeast and filamentous forms of *Candida albicans* during infection. Eukaryot Cell 2:1053–1060

Schaller M, Schafer W, Korting HC, Hube B (1998) Differential expression of secreted aspartyl proteinases in a model of human oral candidosis and in patient samples from the oral cavity. Mol Microbiol 29:605–615

Schaller M, Schackert C, Korting HC, Januschke E, Hube B (2000) Invasion of *Candida albicans* correlates with expression of secreted aspartic proteinases during experimental infection of human epidermis. J Invest Dermatol 114:712–717

Schaller M, Bein M, Korting HC, Baur S, Hamm G, Monod M, Beinhauer S, Hube B (2003) The secreted aspartyl proteinases Sap1 and Sap2 cause tissue damage in an in vitro model of vaginal candidiasis based on reconstituted human vaginal epithelium. Infect Immun 71:3227–3234

Schaller M, Boeld U, Oberbauer S, Hamm G, Hube B, Korting HC (2004) Polymorphonuclear leukocytes (PMNs) induce protective Th1-type cytokine epithelial responses in an in vitro model of oral candidosis. Microbiology 150:2807–2813

Schofield DA, Westwater C, Warner T, Nicholas PJ, Paulling EE, Balish E (2003) Hydrolytic gene expression during oroesophageal and gastric candidiasis in immunocompetent and immunodeficient gnotobiotic mice. J Infect Dis 188:591–599

Schoolnik GK (2002) Functional and comparative genomics of pathogenic bacteria. Curr Opin Microbiol. 5:20–26

Shen J, Guo W, Kohler JR (2004) A heterologous dominant selectable marker for *C. albicans*. In: Abstr Vol 7th American Society for Microbiology Conf Candida and Candidiasis, 18–22 March 2004, Austin, TX

Sheppard DC, Yeaman MR, Welch WH, Phan QT, Fu Y, Ibrahim AS, Filler SG, Zhang M, Waring AJ, Edwards JE Jr (2004) Functional and structural diversity in the Als protein family of *Candida albicans*. J Biol Chem 279:30480–30489

Smail EH, Cronstein BN, Meshulam T, Esposito AL, Ruggeri RW, Diamond RD (1992) In vitro, *Candida albicans* releases the immune modulator adenosine and a second, high-molecular weight agent that blocks neutrophil killing. J Immunol 148:3588–3595

Sohn K, Urban C, Brunner H, Rupp S (2003) *EFG1* is a major regulator of cell wall dynamics in *Candida albicans* as revealed by DNA microarrays. Mol Microbiol 47:89–102

Soll DR (2002) Phenotypic switching. In: Calderone RA (eds) *Candida* and candidosis. ASM Press, Washington DC, pp 123–142

Staab JF, Bradway SD, Fidel PL, Sundstrom P (1999) Adhesive and mammalian transglutaminase substrate properties of *Candida albicans* Hwp1. Science 283:1535–1538

Stehr F, Felk A, Gacser A, Kretschmar M, Mahnss B, Neuber K, Hube B, Schafer W (2004) Expression analysis of the *Candida albicans* lipase gene family during experimental infections and in patient samples. FEMS Yeast Res 4:401–408

Stoldt VR, Sonneborn A, Leuker CE, Ernst JF (1997) Efg1, an essential regulator of morphogenesis of the human pathogen *Candida albicans*, is a member of a conserved class of bHLH proteins regulating morphogenetic processes in fungi. EMBO J 16:1982–1991

Sundstrom P (2002) Adhesion in *Candida* spp. Cell Microbiol 4:461–469

Suzuki S (2002) Serological differences among pathogenic Candida spp. In: Calderone RA (eds) *Candida* and candidosis. ASM Press, Washington DC, pp 29–36

Uhl MA, Biery M, Craig N, Johnson AD (2003) Haploinsufficiency-based large-scale forward genetic analysis of filamentous growth in the diploid human fungal pathogen *C. albicans*. EMBO J 22:2668–2678

Wakabayashi H, Abe S, Teraguchi S, Hayasawa H, Yamaguchi H (1998) Inhibition of hyphal growth of azole-resistant strains of *Candida albicans* by triazole antifungal agents in the presence of lactoferrin-related compounds. Antimicrob Agents Chemother 42:1587–1591

Whittam TS, Bumbaugh AC (2002) Inferences from whole-genome sequences of bacterial pathogens. Curr Opin Genet Dev 12:719–725

Wilson RB, Davis D, Enloe BM, Mitchell AP (2000) A recyclable *Candida albicans URA3* cassette for PCR product-directed gene disruptions. Yeast 16:65–70

Wollina U, Kunkel W, Bulling L, Funfstuck C, Knoll B, Vennewald I, Hipler UC (2004) *Candida albicans*-induced inflammatory response in human keratinocytes. Mycoses 47:193–199

Zink S, Nass T, Rosen P, Ernst JF (1996) Migration of the fungal pathogen *Candida albicans* across endothelial monolayers. Infect Immun 64:5085–5091

10 Integration of Metabolism with Virulence in *Candida albicans*

A.J.P. Brown[1]

CONTENTS

Abbreviations: 3AT, 3-aminotriazole; bHLH, β-helix loop helix domain; bZIP, leucine zipper domain; GCN response, general amino acid control

I. Introduction

Candida albicans is the major systemic pathogen of humans (Odds 1988; Calderone 2002). This fungus exists as a relatively harmless commensal organism in the oral cavity and gastrointestinal tracts of at least half of all individuals. However, when the defences of the host become compromised, *C. albicans* can cause mucocutaneous infections such as oral or vaginal candidiasis. About 75% of women suffer at least one vaginal *Candida* infection in their lifetime, and about 5% of these infections are recurrent. In severely immunocompromised individuals, *C. albicans* can establish deep-seated systemic infections. For example, in patients undergoing chemotherapy or organ transplant surgery, the fungus is able to disseminate via the bloodstream and colonise internal organs. In some patient groups, one-third to one-half of these systemic infections are fatal. Hence, depending upon the immunocompetence of the host, *C. albicans* can infect numerous sites in the host, including the skin, mouth, gastrointestinal and urogenital tracts, bloodstream, kidneys, liver and other internal organs (Ruhnke 2002; Filler and Kullberg 2002; Kullberg and Filler 2002).

Risk factors which are thought to predispose patients to candidemia include protracted neutropenia (reflecting the importance of neutrophils in host defences), the use of broad-spectrum antibiotics (which eliminate the endogenous bacterial flora) and the application of indwelling intravenous devices such as catheters (Wey et al. 1989; Kullberg and Filler 2002). The significance of catheters relates to the ability of *C. albicans* to form biofilms on the surfaces of these devices (Kumamoto 2002; Douglas 2003), which can become a source of recurrent infection. Furthermore, *C. albicans* cells growing within biofilms are relatively resistant to antifungal agents (Baillie and Douglas 1999). Hence, biofilms represent another medically relevant microenvironment in which *C. albicans* can thrive. Of particular relevance to this review is the observation that the fungus adjusts its metabolic activities during growth in biofilms (Garcia-Sanchez et al. 2004).

Several factors are thought to promote the virulence of *C. albicans* (Odds 1994; Haynes 2001; Calderone and Fonzi 2001; van Burik and Magee 2001). Yeast-(pseudo)hyphal morphogenesis is thought to promote fungal dissemination and invasion (Gow et al. 2002; Saville et al. 2003). Adhesins are thought to promote adherence to host tissue and colonisation (Hoyer 2001; Sundstrum 2002). Secreted aspartyl proteinases and lipases may promote invasion, counteract host defences and provide nutrients (Hube and Naglik 2002). High-frequency switching between different phenotypic forms might help the fungus to evade host defences (Soll 2002). These virulence attributes are thought to be required to differing extents during disease establishment and progression (Odds 1994; Brown and Gow 1999). The

[1] Aberdeen Fungal Group, School of Medical Sciences, Institute of Medical Sciences, University of Aberdeen, Foresterhill, Aberdeen AB25 2ZD, UK

The Mycota XIII
Fungal Genomics
Alistair J.P. Brown (Ed.)

impact of genomics upon our understanding of *C. albicans* pathogenicity is discussed by Munro and colleagues in Chap. 9 (this volume).

Some years ago we predicted that the regulation of metabolism and virulence might be mechanistically linked in *C. albicans* (Brown and Gow 1999; Brown et al. 2000). We reasoned that these links might allow this pathogen to adjust its metabolic programme in parallel with its portfolio of virulence attributes in response to the new microenvironments it encounters during disease establishment and progression. In this chapter I review recent studies which address this prediction and the significant impact which *C. albicans* genomics has had in this field. Genomics has revealed unexpected links between metabolism and virulence in *C. albicans* which were not exposed by pre-genomic experimentation.

II. The Experimental Dissection of *Candida albicans*

C. albicans has been the focus of a large number of research studies because of its clinical importance, and the number of publications on *Candida* species has increased steadily over the previous two decades (Calderone 2002). This increase has been due, at least in part, to the development of an improved molecular toolbox for what used to be an experimentally intransigent fungus. The aim of this section is to provide a brief historical perspective on the development of this molecular toolbox, which has led to the establishment of *C. albicans* genomics.

C. albicans is a diploid fungus, and it was thought to be asexual (Magee 1998). However, *C. albicans* is indeed able to mate (Magee and Magee 2000; Hull et al. 2000; Miller and Johnson 2002), and the tetraploid products of these matings can be induced to shed chromosomes, such that diploid recombinants can be isolated (Bennett and Johnson 2003). Hence, a parasexual cycle has now been defined. However, meiosis has not been demonstrated in *C. albicans*. *C. albicans* genome sequencing has revealed homologues of many *Saccharomyces cerevisiae* genes involved in the sexual cycle, but *C. albicans* appears to lack critical factors which are required for meiosis in budding yeast (Tzung et al. 2001). Hence, no true sexual cycle has been established for *C. albicans*, and this has prevented the efficient application of classical genetic approaches in this pathogen. In the absence of genetic screens, the development of genomic screens has proved especially important for the dissection of *C. albicans* pathobiology.

A second problem hindered the molecular dissection of *C. albicans* until the mid-1990s. The CUG codon, which is normally decoded as leucine, is translated as serine in *C. albicans* (Santos et al. 1993). As a result, most standard reporter genes such as *Escherichia coli lacZ*, firefly luciferase or jellyfish GFP are not expressed in a functional form in *C. albicans*. This problem was circumvented by the development of specialised reporter genes for *C. albicans*, such as *Renilla reniformis* luciferase, codon-optimised GFP and *Streptococcus thermophilus* lacZ (Srikantha et al. 1996; Cormack et al. 1997; Uhl and Johnson 2001).

Other essential components of the molecular toolbox were developed in the 1990s (De Backer et al. 2000; Berman and Sudbery 2002). These included improved plasmid vectors, ectopic expression vectors and regulatable promoter systems (Bailey et al. 1996; Leuker et al. 1997; Care et al. 1999). Significantly, robust procedures were established for the generation of *C. albicans* mutants (Fonzi and Irwin 1993), which were improved further by the development of PCR-based gene disruption methods (Wilson et al. 1999, 2000; Enloe et al. 2000). These tools were exploited widely for gene analysis before the *C. albicans* genomics era, and they remain important for the functional analysis of genes identified by genome sequencing.

C. albicans genome sequencing began in the private sector with Incyte in the mid-1990s, but this was followed shortly thereafter by a public sequencing programme at the Stanford DNA Sequencing and Technology Center, which was funded by the US National Institute for Dental and Cranofacial Research and the Burroughs Wellcome Fund. The latest genome sequence assembly from the Stanford Group represents the diploid genome sequence of *C. albicans* (Assembly 19; Jones et al. 2004). Assembly 19 includes a haploid set of sequence contigs and a set of allelic contigs which describe the heterozygosities in the genome of *C. albicans* strain SC5314 (a clinical isolate from Bristol-Myers Squibb). In addition to the Stanford sequencing website (http://www-sequence.stanford.edu/group/candida/), two main *C. albicans* genome databases have been established. CandidaDB was first released in 2001 (http://genolist.pasteur.fr/CandidaDB; d'Enfert et al. 2005), and the *Candida* Genome Database was released in 2004 (http://genome-www.stanford.edu/fungi/Candida/).

The Stanford Group generously released their *C. albicans* genome sequence data at various stages of the sequencing process (http://www.sequence.stanford.edu/group/candida/). These preliminary sequence data provided a strong platform for many groups to initiate *C. albicans* transcript profiling, proteomics and large-scale gene deletion studies even before sequence Assembly 19 was released (Lane et al. 2001; Murad et al. 2001b; Cowen et al. 2002; Lan et al. 2002; Nantel et al. 2002; Pitarch et al. 2002; Bruneau et al. 2003; Chauhan et al. 2003; Fradin et al. 2003; Roemer et al. 2003; Rubin-Bejerano et al. 2003; Tsong et al. 2003; Garcia-Sanchez et al. 2004; Hernandez et al. 2004; Yin et al. 2004).

Initially, the experimental dissection of *C. albicans* was confined to biochemical and immunological approaches. The molecular era, which began in earnest in the 1990s, saw the isolation and characterisation of many genes involved in *C. albicans* pathobiology. However, in the absence of classical genetics, these genes were often targeted for study on the basis of their sequence relatedness to, or functional interactions with *S. cerevisiae* genes, and their analysis was mostly restricted to their predicted functions. As a result, unexpected roles for these genes often remained undetected. Arguably one of the greatest impacts of the *C. albicans* genomics era, which took off at the beginning of the new millennium, was the ability to examine the global roles of specific genes in a relatively unbiased fashion. As described below, this led to the finding that the regulation of metabolic and virulence functions appears to be integrated in this pathogen.

III. Overview of Metabolism in *Candida albicans*

Investigations of *C. albicans* metabolism began over half a century ago (e.g. Van Neil and Cohen 1942), and enzymes of central carbon metabolism were characterised as long ago as 1960 by Rao and co-workers. However, in general, studies of *C. albicans* metabolism have focused either directly or indirectly upon virulence attributes or the mode of action of antifungal drugs (Odds 1988). For example, considerable attention has been paid to the pathways involved in ergosterol biosynthesis and cell wall biogenesis, both of which are antifungal targets. Even now, *C. albicans* metabolism is not addressed directly in the latest definitive text on *Candida* and candidiasis (Calderone 2002). However, pathways of central and amino acid metabolism have been highlighted as potential antifungal targets by large-scale gene deletion studies in *C. albicans* (Roemer et al. 2003) and by transcript profiling studies (Lorenz and Fink 2001, 2002). With respect to antifungal therapy, the priority is to kill the fungus, rather than to inhibit its virulence. Hence, metabolic functions which are required for growth and survival in the host represent potential antifungal targets. Of course, the validity of such targets and the efficacy of drugs which hit these targets will depend upon the degree of cross-reactivity with the host. Nevertheless, an increased understanding of *C. albicans* metabolism and its regulation should facilitate the development of new antifungal therapies in the long term.

To date, *S. cerevisiae* has provided a reasonable metabolic paradigm for *C. albicans*. The pathways of central carbon metabolism are conserved in this fungus, including the glycolytic, gluconeogenic and pentose phosphate pathways, and the tricarboxylic and glyoxylate cycles (Odds 1988; Jones et al. 2004). Pathways for the generation of storage and cell wall carbohydrates are conserved. Furthermore, pathways of amino acid, lipid and nucleotide assimilation and anabolism appear to be conserved. However, metabolic differences do exist between *C. albicans* and *S. cerevisiae*, the most obvious of which relates to their patterns of sugar utilisation. Indeed, differences in the patterns of carbohydrate assimilation are used routinely to distinguish *C. albicans* from other microbes in the clinical setting (Wickerham 1951; Williamson et al. 1986).

With regard to the *regulation* of metabolism, the *S. cerevisiae* paradigm is less robust. This is hardly surprising, given that the pathogen *C. albicans* and the relatively benign *S. cerevisiae* have evolved in fundamentally different niches. *C. albicans* ferments glucose to ethanol and displays glucose repression. Also, at least some glucose-repressible functions appear to be regulated by the transcriptional repressor Mig1 in a fashion analogous to that of *S. cerevisiae* (Murad et al. 2001a, b). Furthermore, *C. albicans* contains a homologue of *S. cerevisiae Snf1* (Petter et al. 1997), which is required for the derepression of glucose-repressed genes in budding yeast. However, *C. albicans* Mig1 displays several differences vis-à-vis *S. cerevisiae* Mig1, including the lack of a putative *Snf1* phosphorylation site (Zaragoza et al. 2000). Also, *S. cerevisiae snf1* mutants are viable, whereas Snf1 appears to be essential for viability in *C. albicans* (Petter et al. 1997), suggesting that Snf1 might execute additional functions in this yeast. Therefore, significant differences do appear to exist between *C. albicans* and *S. cerevisiae* with respect to glucose repression mechanisms.

The differences extend to the regulation of hypoxic gene expression in *C. albicans* and *S. cerevisiae*. For example, the transcriptional regulator Rox1 controls hypoxic gene expression in *S. cerevisiae* in a haem-dependent fashion (Kastaniotis and Zitomer 2000). In contrast, its orthologue in *C. albicans*, Rfg1, regulates yeast-hypha morphogenesis, but not hypoxic gene expression (Kadosh and Johnson 2001).

Further significant differences between *C. albicans* and *S. cerevisiae* metabolism have been suggested by the genome sequence data. Jones et al. (2004) indicate that unlike *S. cerevisiae*, *C. albicans* possesses both mitochondrial- and nuclear-encoded subunits of electron transport complex I, a pyruvate dehydrogenase kinase, a large lipase gene family and additional enzymes involved in fatty acid and amino acid catabolism. These differences might reflect an increased emphasis on respiratory and oxidative metabolism in *C. albicans* (Jones et al. 2004). This view is consistent with the observation that *C. albicans* converts a lower proportion of glucose to ethanol in comparison to *S. cerevisiae* (Bertram et al. 1996).

To summarise, whilst many metabolic pathways are conserved between *S. cerevisiae* and *C. albicans*, these organisms display significant differences in metabolic regulation. Nevertheless, many investigators still extrapolate from *S. cerevisiae* to *C. albicans* when interpreting their experimental data.

IV. Conservation of a Metabolic Response in *Candida albicans*

To date, only one metabolic response has been compared directly in *C. albicans* and *S. cerevisiae* at the genomic level: General Amino Acid Control, or the GCN response. The environmental stimulus which triggers the GCN response is amino acid starvation, and this signal stimulates the induction of almost all amino acid biosynthetic pathways. This broad metabolic response can be activated by starvation for even a single amino acid.

The GCN response has been well characterised at the molecular level in *S. cerevisiae* (Hinnebusch 1988; Natarajan et al. 2001; Hinnebusch and Natarajan 2002). Briefly, amino acid starvation leads to the intracellular accumulation of uncharged tRNAs, which bind to the histidyl tRNA synthetase-like domain of Gcn2 (Wek et al. 1995). This activates the protein kinase activity of Gcn2, which then phosphorylates the α-subunit of the translation initiation factor eIF2 (Dever et al. 1992). eIF2a phosphorylation increases the affinity of eIF2 for eIF2B, its guanine nucleotide exchange factor, and inhibits the regeneration of recharged eIF2-GTP required for translation initiation (Krishnamoorthy et al. 2001). This leads to a decrease in the overall rate of translation initiation in the yeast cell, but to an increase in the translation of the *GCN4* mRNA.

The translation of the *GCN4* mRNA is regulated by four short upstream open reading frames (uORFs), which lie in its unusually long 5′-leader region (Mueller and Hinnebusch 1994). Essentially, this 5′-leader normally acts to repress the translation of the main Gcn4-encoding open reading frame on the *GCN4* mRNA under non-starvation conditions. However, in response to amino acid starvation, this repression is released when eIF2-GTP levels are reduced (Mueller and Hinnebusch 1994). The resultant increase in *GCN4* mRNA translation leads to an elevation in the abundance of the Gcn4 protein. Gcn4 is a bZIP transcription factor which binds as a dimer to GCRE elements (5′-TGA(C/G)TCA) located in the promoters of target genes, and activates their transcription (Oliphant et al. 1989; Ellenberger et al. 1992). These target genes include amino acid biosynthetic enzymes on all amino acid biosynthetic pathways, with the exception of the cysteine pathway (Natarajan et al. 2001). Therefore, amino acid biosynthesis is induced in response to amino acid starvation via a signalling pathway involving the key regulators Gcn2 and Gcn4.

Transcript profiling of amino acid starvation in *S. cerevisiae* has revealed that the GCN response is much broader than was initially expected (Natarajan et al. 2001). In these experiments the histidine analogue, 3-aminotriazole (3AT), was used to activate the GCN response. Exposure to 3AT, which causes histidine starvation, is a classic means of activating the GCN response (Hinnebusch 1988). The transcript profiling experiments showed that, in addition to inducing the expression of amino acid biosynthetic genes, Gcn4 induces the expression of aminoacyl tRNA synthetases, amino acid transporters, vitamin biosynthetic pathways and glycogen biosynthetic functions, and represses the expression of ribosomal proteins (Natarajan et al. 2001). The transcript profiling also confirmed the earlier observation that Gcn4 regulates purine biosynthetic functions (*ADE* genes; Rolfes and Hinnebusch 1993). Furthermore, the transcript profiling revealed that Gcn4 controls the expression of numerous transcription factors including Arg80, Leu3, Lys14, Met4, Met28, Bas1, Gln3, Rtg3, Pip2, Gat1, Uga3, Mal13, Cup9 and Rim101. This led Natarajan et al. (2001) to propose that Gcn4 is a master regulator of gene expression in *S. cerevisiae*.

Gcn4-like proteins are conserved in other fungi (Paluh et al. 1988; Wanke et al. 1997; Hoffmann et al. 2001; Tripathi et al. 2002; Tournu et al. 2005). Furthermore, the mRNAs encoding these proteins carry unusually long 5′-leader regions with multiple short uORFs. This suggested that these mRNAs might be regulated at the translational level, like the *S. cerevisiae GCN4* mRNA. *S. cerevisiae* Gcn4 levels are also regulated at the transcriptional level and via accelerated protein turnover (Kornitzer et al. 1994; Albrecht et al. 1998), but most control appears to be executed at the translational level (Hinnebusch and Natarajan 2002). In contrast, the expression of *GCN4*-like genes in *C. albicans* and *Aspergillus nidulans* appears to be regulated primarily at the transcriptional level (Hoffmann et al. 2001; Tournu et al., unpublished data). Therefore, subtle differences do exist between fungi with respect to their GCN responses. Hence, we have compared the global GCN responses in *S. cerevisiae* and *C. albicans* (Yin et al. 2004; Tournu et al., unpublished data). Our aims were firstly to test whether the transcriptional responses observed by Natarajan et al. (2001) are reflected in the *S. cerevisiae* proteome, and secondly to determine the extent of conservation of this GCN response in *C. albicans*.

Initially, the *S. cerevisiae* Gcn4 proteome was examined. Protein extracts were prepared from wild-type and *gcn4* cells exposed to 3AT or carrier alone. Of 956 spots identified reproducibly on the resultant two-dimensional gels, 107 were reproducibly elevated by 3AT, and 80 of these changes were Gcn4-dependent (Fig. 10.1; Yin et al. 2004). Therefore, a significant proportion (11%) of the *S. cerevisiae* proteome responded to 3AT, which was roughly analogous to the corresponding *S. cerevisiae* transcriptome (Natarajan et al. 2001). Forty-five percent of the 52 Gcn4-dependent changes identified corresponded to amino acid biosynthetic enzymes on the arginine, asparagine, cysteine, chorismate, glutamine, glycine, histidine, isoleucine–valine, leucine, lysine, methionine, serine, threonine and tryptophan pathways (Yin et al. 2004). Clearly, this proteomic study confirmed that histidine starvation (using 3AT) invokes a broad response across most (if not all) amino acid biosynthetic pathways. In addition, 9% of the Gcn4-dependent changes represented purine biosynthetic enzymes (Ade proteins), and 8% were involved in the biosynthesis of vitamins and cofactors. Proteins involved in carbon metabolism and energy generation were also elevated by 3AT in a Gcn4-dependent fashion. Hence, many of the changes observed by transcript profiling (Natarajan et al. 2001) were reflected in the proteome (Yin et al. 2004). Indeed, there was a positive correlation between the *S. cerevisiae* GCN transcriptome and proteome (Spearman rank correlation coefficient $= 0.59, p<1.89 \times 10^{-06}$).

An analogous proteomic experiment has been performed for *C. albicans* (Fig. 10.1; Yin et al. 2004). Of 942 spots identified reproducibly on the two-dimensional gels, 218 (23%) were reproducibly elevated by 3AT, and 147 of these changes were Gcn4-dependent. Thirty percent of the 31 Gcn4-dependent protein changes identified by peptide mass fingerprinting corresponded to amino acid biosynthetic enzymes on the arginine, chorismate, isoleucine–valine, leucine, lysine, methionine and serine pathways. In addition, a significant number of *C. albicans* proteins involved in carbon metabolism and energy generation were elevated by 3AT in a Gcn4-dependent fashion. To this extent, the *C. albicans* GCN proteome closely mirrored the *S. cerevisiae* GCN proteome. However, one interesting difference was observed between *C. albicans* and *S. cerevisiae*. All four Ade proteins identified in our *S. cerevisiae* two-dimensional map were induced by 3AT. In contrast, none of the seven Ade proteins identified in the *C. albicans* two-dimensional map were elevated in response to 3AT.

More recently, transcript profiling has been performed on the *C. albicans* GCN response (Tournu et al., unpublished data; http://www.pasteur.fr/recherche/unites/Galar_Fungail/). About 12% of the ca. 6200 *C. albicans* genes analysed were regulated by 3AT in a Gcn4-

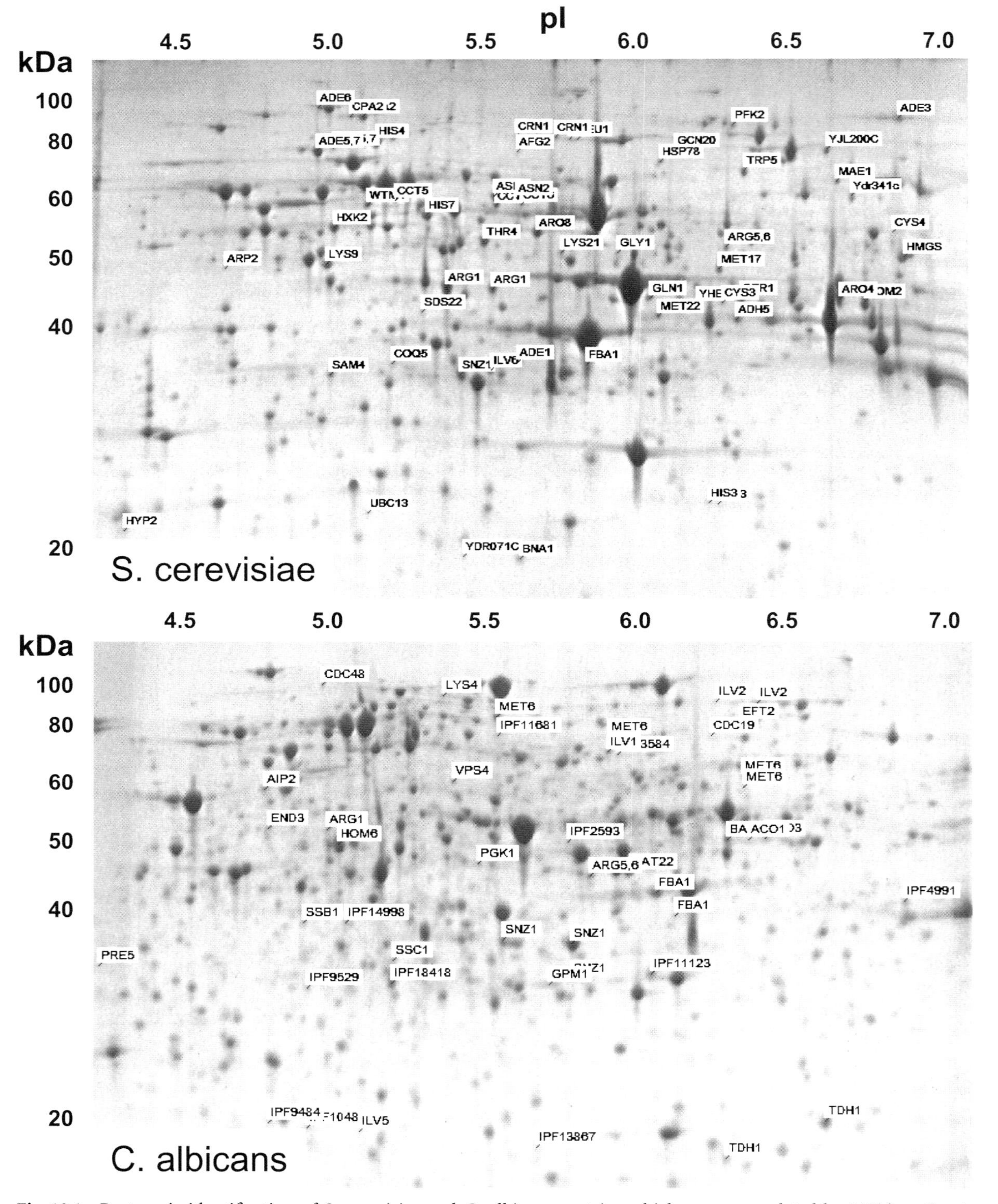

Fig. 10.1. Proteomic identification of *S. cerevisiae* and *C. albicans* proteins which are up-regulated by 3AT in a Gcn4-dependent fashion

dependent manner. This was significantly less than the 27% of *C. albicans* proteins which seemed to be regulated in a Gcn4-dependent manner (Yin et al. 2004). Indeed, there was no positive correlation between the *C. albicans* GCN transcriptome and proteome (Spearman rank correlation coefficient = 0.09). Hence, the levels of many proteins might be post-transcriptionally regulated during the GCN response in *C. albicans*. Consistent with this idea, transcript profiling revealed the induction of many putative peptidase genes by 3AT (*APE3, PRC1, PRC3, PRE3, PRE8, PUP2, RPN1, RPN2, RPN6, RPN7, RPN8, RPN10, RPT1, RPT4, RPT5, RPT6, IPF4866*), suggesting that protein turnover rates increase following amino acid starvation in *C. albicans*. Nevertheless, there were strong similarities between the *C. albicans* and *S. cerevisiae* transcriptomes with respect to the types of cellular function which were regulated by Gcn4. In both cases, amino acid biosynthetic genes were significantly elevated, and ribosomal protein genes were significantly repressed (Natarajan et al. 2001; Tournu et al., unpublished data). However, unlike *S. cerevisiae*, *C. albicans* ADE mRNAs were not induced by 3AT in a Gcn4-dependent fashion. Hence, our recent transcript profiling data are entirely consistent with the proteomic data with respect to two important conclusions. First, in general the global GCN response, as defined in *S. cerevisiae*, has been conserved in *C. albicans*. Second, in contrast to the situation in *S. cerevisiae*, purine biosynthetic genes are not regulated in response to amino acid starvation in *C. albicans*.

Molecular studies have revealed another significant difference between the GCN responses of *S. cerevisiae* and *C. albicans*. Amino acid starvation induces cellular morphogenesis in the pathogenic fungus, but does not do so in the benign yeast (Tripathi et al. 2002). The formation of pseudohyphae by *C. albicans* in response to amino acid starvation is dependent upon Gcn4. Furthermore, ectopic expression of Gcn4 in *C. albicans* stimulates pseudohyphal development in the absence of an environmental stimulus. Hence, Gcn4 co-ordinates morphogenetic and metabolic responses in this organism (Fig. 10.2A). Collectively, the data suggest subtle but potentially significant differences between the GCN responses in *S. cerevisiae* and *C. albicans*.

Does the GCN response contribute to the virulence of *C. albicans*? The inactivation of Gcn4 does not attenuate the virulence of this fungus in the mouse model of systemic candidiasis (Brand et al. 2004). In this model, *C. albicans* cells are injected directly into the bloodstream of the host. Hence, these experiments indicate that the GCN response is not required for *C. albicans* to establish a systemic infection once it has entered the bloodstream. However, Gcn4 is required for *C. albicans* to form nor-

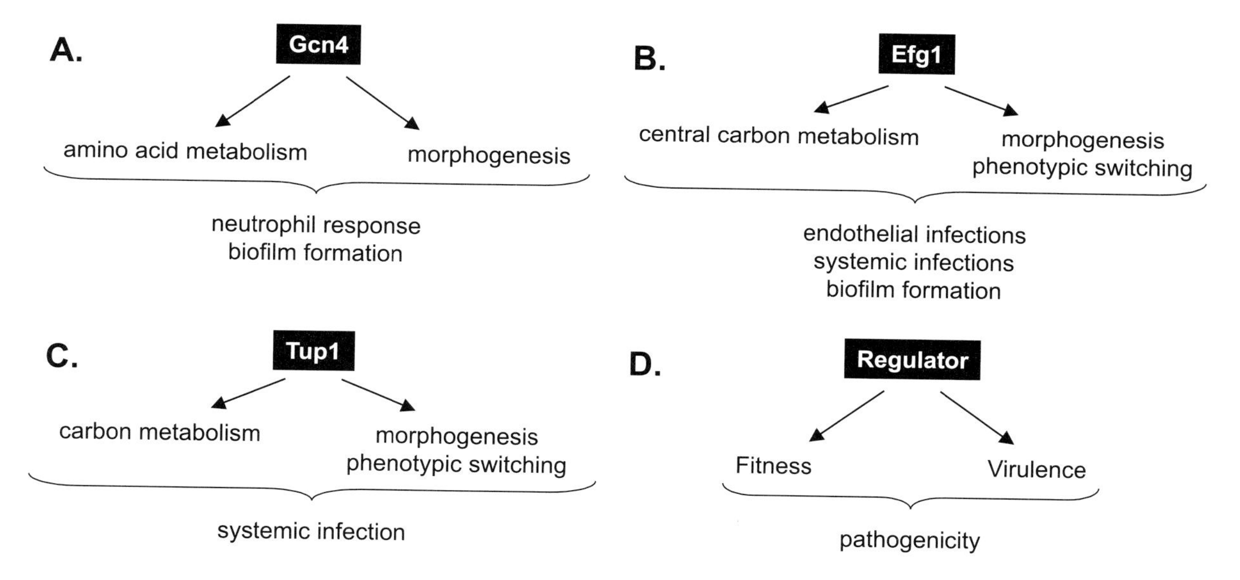

Fig. 10.2. Co-ordinated regulation of metabolism and virulence attributes in *C. albicans*. Genomic analyses have revealed the influence of specific regulators (*box*) upon metabolism (*left*) and virulence (*right*) and, in some cases, the relevance of these responses to specific medically relevant environments (*below*). See text for details. **A** Gcn4; **B** Efg1; **C** Tup1; **D** general model

mal biofilms (Garcia-Sanchez et al. 2004). Hence, the GCN response does appear to be required for survival in at least one medically important niche. It remains to be seen whether *C. albicans* depends upon the GCN response in other infection sites, for example, during oral candidiasis, vaginitis or during commensalism.

V. Integration of Metabolism with Virulence

A. Amino Acid Metabolism

As described above (Sect. IV), a detailed examination of the GCN response in *C. albicans* has revealed a well-defined molecular link between a morphogenetic and metabolic response. Amino acid starvation in *C. albicans* stimulates pseudohyphal development as well as the induction of amino acid biosynthetic genes and other metabolic functions. Both of these responses depend upon the transcription factor Gcn4 (Tripathi et al. 2002). Hence, through Gcn4, the regulation of a virulence attribute is integrated with the control of amino acid metabolism in *C. albicans* (Fig. 10.2A). This section will review further experimental observations which lend weight to the view that the control of virulence and metabolism is integrated in this pathogen.

The exact molecular mechanism by which Gcn4 stimulates morphogenesis in *C. albicans* is not known. However, Gcn4 is known to activate morphogenesis specifically via the Ras-cAMP signalling pathway (Tripathi et al. 2004). Mutations on the MAP kinase pathway do not prevent the ability of ectopically expressed Gcn4 to stimulate morphogenesis, but the inactivation of Efg1 blocks this response. Hence, Gcn4 appears to co-ordinate morphogenesis with amino acid metabolism by connecting specific signalling pathways.

A further mechanistic link between morphogenesis and amino acid metabolism has been described recently. Csy1 is thought to be the main amino acid sensor in *C. albicans*. The inactivation of Csy1 blocks the transcriptional induction of amino acid permease genes in response to the presence of amino acids in the growth medium (Brega et al. 2004). *C. albicans csy1* mutants display morphogenetic defects in response to serum and pH induction. However, the molecular mechanisms by which Csy1 links amino acid sensing with cellular morphogenesis remain to be defined. It is attractive to speculate that, like Gcn4, Csy1 might mediate its morphogenetic effects via the Ras-cAMP pathway.

Recent expression profiling of *C. albicans* cells growing in biofilms has revealed another unexpected link between amino acid metabolism and virulence (Garcia-Sanchez et al. 2004). In this study, cells growing in biofilms were compared with those in planktonic cultures under a variety of different growth conditions. The rationale was to identify sets of genes which were regulated specifically in response to growth in biofilms, rather than in response to the conditions used to generate the biofilms. Amino acid biosynthetic genes were amongst those which were consistently up-regulated in biofilms. These included *ARO*, *CYS*, *HIS*, *ILV*, *MET*, *SER* and *TRP* genes, suggesting that many amino acid biosynthetic pathways are induced during biofilm formation. Garcia-Sanchez et al. (2004) reasoned that this induction might be mediated by the GCN response (Tripathi et al. 2002; Yin et al. 2004), and hence they tested the effects of inactivating Gcn4 upon biofilm formation. Biofilm formation was significantly inhibited in *C. albicans gcn4* cells, compared with control cells in which the GCN4 gene had been reintroduced (Garcia-Sanchez et al. 2004). Therefore, the GCN response appears to be required for efficient biofilm formation.

As described above (Sect. I), *C. albicans* biofilms which form on the catheters of hospital patients appear to be a relatively frequent source of bloodstream infections in these patients. The regulation of amino acid metabolism still appears to be important when *C. albicans* cells enter the bloodstream. This view is supported by a global study of fungal gene expression during phagocytosis by human neutrophils (Rubin-Bejerano et al. 2003). Neutrophils represent an important weapon in the host's defences against disseminated candidiasis. Following phagocytosis by neutrophils, *C. albicans* cells remain in the yeast form and are killed, whereas the fungus can undergo morphogenesis inside cultured macrophages and escape from these cells (Lo et al. 1997; Rubin-Bejerano et al. 2003). The transcript profiling study of Rubin-Bejerano et al. (2003) focussed mainly upon *S. cerevisiae*. However, these authors did show that, to a great extent, *C. albicans* behaves analogously to *S. cerevisiae* following exposure to neutrophils. Genes on the methionine and arginine biosynthetic pathways were induced when *C. albicans* was exposed to human neutrophils.

The corresponding *S. cerevisiae* response required fresh serum, because it was lost when the serum was inactivated. Also, this metabolic response required direct contact between the fungal and human cells, because it was lost if a membrane separated the cells. Furthermore, the response was specific to neutrophils, and was not observed in monocytes. *C. albicans* genes on other amino acid biosynthetic pathways were generally not affected by exposure to neutrophils (Rubin-Bejerano et al. 2003), suggesting that the neutrophil response is distinct from the GCN response described above. This might explain why blocking the GCN response does not inhibit disseminated candidiasis (Brand et al. 2004). Instead, the neutrophil response seems more reminiscent of stress responses, which also induce the expression of genes on the methionine and arginine biosynthetic pathways (supplementary data in Enjalbert et al. 2003). Of course, this might reflect the potent cidal effect of neutrophils upon *C. albicans* cells.

Whilst *MET* and *ARG* genes were induced when *C. albicans* cells were exposed to neutrophils in RPMI medium containing fresh serum, the expression of *ADE* genes was decreased (Rubin-Bejerano et al. 2003). This observation reinforces the idea that purine and amino acid biosynthesis is regulated by distinct mechanisms in *C. albicans* (Yin et al. 2004). However, in the context of this review, the main conclusion which can be drawn from this study of Rubin-Bejerano et al. (2003) is that *C. albicans* regulates amino acid biosynthetic pathways as part of its response to attack by neutrophils.

A recent study has suggested that *C. albicans* responds slightly differently from *S. cerevisiae* during phagocytosis by mammalian macrophages (Lorenz et al. 2004). Under the conditions analysed in this study, the *C. albicans* cells survived phagocytosis, formed germ tubes and escaped from the macrophage. Transcript profiling of the fungal response suggested that major changes in carbon metabolism accompanied phagocytosis by the macrophages (discussed below), and that arginine biosynthesis was significantly induced. The response of *MET* genes was not highlighted in this study (Lorenz et al. 2004).

B. Carbon Metabolism

Carbon assimilation is essential for the generation of new biomass, i.e. growth. Therefore, the rapid growth of *C. albicans* in the immunocompromised host must depend upon the efficient assimilation of available carbon sources in vivo. Furthermore, as described above (Sect. V.A), the fungus must adjust its metabolic programme as it encounters new microenvironments in the host. A number of studies have indicated that the regulation of carbon metabolism, like amino acid metabolism, is intimately linked to the control of virulence in *C. albicans*. This section will focus mainly on the pathways of central carbon metabolism: glycolysis, gluconeogenesis and the tricarboxylic (TCA) and glyoxylate cycles.

Over 10 years ago it was reported that the levels of glycolytic mRNAs change during yeast-hypha morphogenesis in *C. albicans* (Swoboda et al. 1994). It was concluded that these changes, which were observed by Northern blotting of *PYK1*, *ADH1*, *PGK1* and *GPM1* mRNAs, reflected the underlying physiological changes which accompany morphogenesis, rather than bone fide morphogenetic regulation of these genes. A strict definition of morphogenetic regulation, which excluded genes which are regulated by medium composition alone, was helpful in terms of identifying hypha-specific genes (Bailey et al. 1996). However, this shifted attention away from potential links between *C. albicans* physiology and development. Attention has only returned to these potential links following the publication of several transcript profiling studies which have reinforced and extended this early observation.

Nantel et al. (2002) published a transcript profiling study of *C. albicans* morphogenesis in which the behaviour of wild-type and mutant cells was compared during serum-induced morphogenesis. Naturally, the discussion in this paper focussed to a large extent on the behaviour of hypha-specific and signalling genes and, as expected, hypha-specific genes such as *ECE1*, *HWP1*, *RBT1* and *SAP4, 5, 6* were induced during hyphal development. However, of particular relevance to this review is the fact that numerous genes encoding glycolytic enzymes were also shown to be regulated during hyphal development, including *ENO1*, *FBA1*, *PYK1* (*CDC19*), *TPI1* and *PGI1* (Nantel et al. 2002). Clearly, the regulation of these genes might be indirect, as implied in the early study by Swoboda et al. (1994), and these genes do appear to respond to physiological stimuli such as ambient growth temperature (Nantel et al. 2002). However, Nantel and co-workers also showed that these glycolytic genes respond to the key morphogenetic regulator Efg1, and this observation has been confirmed by an independent transcript profiling study (Doedt et al. 2004). Hence, the link between morphogenetic and glycolytic regulation might be more direct than was originally anticipated.

Ernst and co-workers were the first to identify the bHLH transcription factor, Efg1, as an important regulator of yeast-hypha morphogenesis in *C. albicans* (Stoldt et al. 1997). The inactivation or depletion of Efg1 severely attenuates hyphal development (Lo et al. 1997; Stoldt et al. 1997), and hence Efg1 has been described as an activator of morphogenesis. Efg1 has now been shown to control chlamydospore formation as well as hyphal development, and the inactivation of Efg1 also affects phenotypic switching (Stoldt et al. 1997; Sonneborn et al. 1999; Srikantha et al. 2000). Efg1 has also been shown to regulate the expression of adhesins such as Als3 (formally known as Als8) and Hwp1 (Sharkey et al. 1999; Lane et al. 2001; Leng et al. 2001; Nantel et al. 2002). Hence, Efg1 is a key developmental regulator which controls the expression of several virulence attributes in *C. albicans* including yeast-hypha morphogenesis, phenotypic switching and adhesins. Not surprisingly, the inactivation of Efg1 (albeit in combination with a *cph1* mutation) attenuates the virulence of *C. albicans* in the mouse model of systemic infection, reduces invasion in epithelial infection models, and inhibits biofilm formation (Lo et al. 1997; Gow et al. 2003; Korting et al. 2003; Garcia-Sanchez et al. 2004).

Efg1 acts in concert with the structurally and functionally related bHLH factor, Efh1 (Doedt et al. 2004). Whereas Efh1 is a transcriptional activator, Efg1 appears to act mechanistically as a transcriptional repressor in *C. albicans* (Tebarth et al. 2003; Doedt et al. 2004). Recently, Ernst's group published a detailed analysis of the Efg1 and Efh1 regulons in *C. albicans* (Doedt et al. 2004). In this experiment, transformants which overexpressed Efg1 or Efh1 were included alongside *efg1* and *efh1* mutants. A clear picture emerged in which the central metabolic pathways are regulated by Efg1 and, to a lesser extent, by Efh1. The glycolytic genes *GLK1*, *HXK2*, *PGI1*, *PFK1*,2, *FBA1*, *TPI1*, *GAP1*, *PGK1*, *GPM1* and *ENO1* were all up-regulated by Efg1. In contrast, gluconeogenic (*FBP1*), TCA cycle (*CIT1*, *MDH1*, *FUM12*, *SDH12*, *KGD1*) and respiratory functions (*ATP1*, *ATP17*, *PET9*) were down- regulated. Therefore, Efg1 stimulates fermentative metabolism and represses respiratory metabolism. It is not clear whether Efg1 regulates these metabolic genes by acting directly at their promoters, or indirectly by controlling the activities of other regulatory factors. Although Efg1 has been shown to bind DNA in a sequence-specific fashion in vitro (Leng et al. 2001), a consensus sequence for its normal binding site in vivo has not been defined unambiguously. Hence, this issue cannot be addressed by screening the promoters of these metabolic genes for a consensus Efg1 binding site. Nevertheless, this metabolic regulation by Efg1 is physiologically significant, because *efg1* cells are more sensitive to an inhibitor of respiration, antimycin A (Doedt et al. 2004). Hence, Efg1 regulates metabolism as well as virulence attributes (Figs. 10.2B and 10.3).

The link between central carbon metabolism and virulence attributes in *C. albicans* is further strengthened by another transcript profiling study (Lan et al. 2002). The focus of this study was phenotypic switching in *C. albicans*. *C. albicans* strain WO1 switches at high frequency between white and opaque forms. White cells are slightly ellipsoidal and their cell walls have a relatively smooth surface when examined by scanning electron microscopy. In contrast, opaque cells are larger than are white cells, are more elongated, contain a large vacuole and display pimples on their cell surface (Soll 2002). Significantly, clinical isolates of *C. albicans* from diseased patients undergo phenotypic switching at higher rates than do commensal strains (Hellstein et al. 1993). Furthermore, white cells are more virulent than are opaque cells in the mouse model of systemic candidiasis, whereas opaque cells are more virulent than are white cells in a cutaneous mouse model (Kvaal et al. 1997, 1999). Therefore, phenotypic switching is closely associated with *C. albicans* virulence.

Lan et al. (2002) compared the global expression patterns of the two main switch phenotypes of *C. albicans* strain WO1 – white and opaque cells. Well-defined sets of genes were found to be up- or down-regulated in opaque cells, compared with white cells during growth on rich or defined media. Genes involved in adhesion, stress responses, drug resistance and signalling were found to be regulated. Interestingly, about one-third of the regulated genes encoded metabolic functions. Glycolytic genes were down-regulated in opaque cells (*HXT3*, *HXT4*, *HXK1*, *PFK2* and *PYK1* (*CDC19*)), whereas TCA genes were up-regulated (*IDP2*, *MDH1*, *MLS1*). Also, genes involved in fatty acid β-oxidation were up-regulated in opaque cells (*FAA2*, *POX1*, *ECI1*, *FOX2*, *FOX3*). This led Lan et al. (2002) to suggest that white and opaque cells express different metabolic programmes which might be associated with their preference for different anatomical niches in the host. White cells favour fermentative metabolism, whereas opaque cells favour respiratory metabolism (Fig. 10.3).

The findings of Lan et al. (2002) are entirely consistent with those of Doedt et al. (2004). The inactivation of Efg1 leads to the generation of opaque cells, and *EFG1* is expressed at low levels in opaque cells. As described above, fermentative metabolism is down-regulated and oxidative metabolism is up-regulated in *efg1* and opaque cells. The opposite is true in *EFG1* over-expressing cells and in white cells. Hence, Efg1 appears to link mechanistically central carbon metabolism with phenotypic switching, as well as with yeast-hypha morphogenesis (Fig. 10.2B). Efg1 activity is regulated by the Ras-cAMP signalling pathway (Bockmuhl and Ernst 2001). In *S. cerevisiae*, the Ras-cAMP pathway is known to regulate growth and carbon metabolism (Thevelein 1994; Rolland et al. 2001).

Global analyses of another regulator, Tup1, have reinforced the view that the regulation of carbon metabolism and virulence attributes are mechanistically linked in *C. albicans*. *C. albicans* Tup1 was first identified as a repressor of hyphal development (Braun and Johnson 1997). *C. albicans tup1* mutants grow constitutively in an elongated pseudohyphal form, even in the absence of a morphogenetic stimulus. Ectopic Tup1 expression affects white–opaque phenotypic switching, and *tup1* mutants display attenuated virulence (Murad et al. 2001a; Zhao et al. 2002). By analogy with its homologue in *S. cerevisiae* (Smith and Johnson 2000), *C. albicans* Tup1 is thought to operate as a global transcriptional repressor which is targeted to specific promoters through protein–protein interactions with specific DNA binding proteins. Nrg1 is thought to be one such protein. Like Tup1, Nrg1 is a repressor of hyphal development in *C. albicans* (Braun et al. 2001;

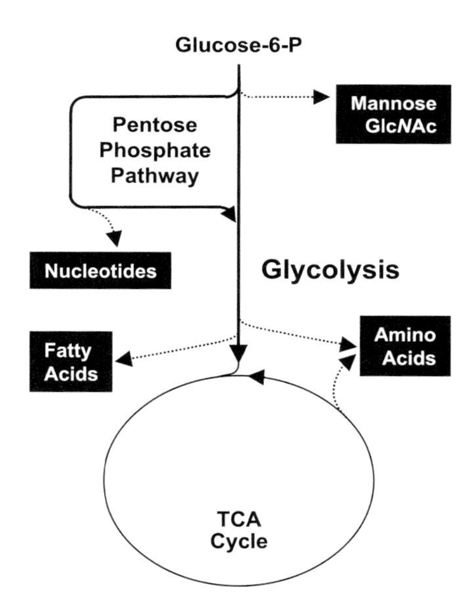

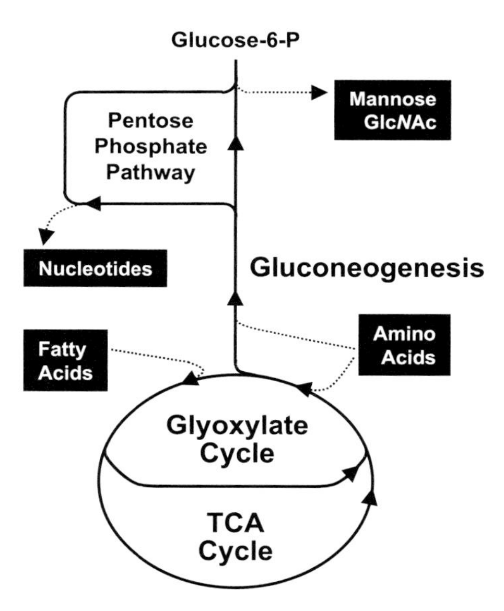

Up-regulated by Efg1
White cells
(Bloodstream; organs)

Down-regulated by Efg1
Down-regulated bu Tup1
Opaque cells
(Phagocytosis; skin infection)

Fig. 10.3. Regulation of central carbon metabolism in *C. albicans*. Fermentative metabolism (*left*) is up-regulated by Efg1, observed in white cells, and appears to be activated in the bloodstream and possibly in organs. Inright contrast, respiratory metabolism (*right*) is down-regulated by Efg1 and Tup1, observed in opaque cells, and appears to be activated in during phagocytosis and possibly skin infections (see text for details). Flux to or from peripheral pathways (nucleotide, fatty acid and amino acid metabolism) is inferred, but will depend in actuality upon the available nutrients in each microenvironment. Flux towards the cell wall precursors mannose and *N*-acetylglucosamine (GlcNAc) is inferred from the observation that the cell wall comprises about 30% of *C. albicans* biomass, and therefore that these precursors must be synthesised under all growth conditions

Murad et al. 2001a). Nrg1 is a sequence-specific DNA binding protein which interacts with Nrg1 response elements in the promoters of hyphal genes to repress their transcription. This repression is dependent upon Tup1, which is consistent with the idea that Nrg1 targets Tup1 to the promoters of hypha-specific genes to repress their expression.

Preliminary transcript profiling, using arrays containing about one-third of *C. albicans* genes, confirmed that Tup1 represses hypha-specific genes in concert with Nrg1 (Murad et al. 2001b). These genes included *ECE1* and the adhesin genes *ALS3* and *HWP1*. This has been further confirmed by more recent transcript profiling experiments with whole-genome arrays (Mavor 2004; http://www.pasteur.fr/recherche/unites/Galar_Fungail/). However, these experiments have also revealed that Tup1 represses numerous other functions in addition to morphogenesis, including pathways of central and peripheral carbon metabolism (http://www.pasteur.fr/recherche/unites/Galar_Fungail/). Gluconeogenic genes (*PCK1*, *FBP1*) and TCA and glyoxylate cycle genes (*IDP2*, *MDH11*, *ICL1*) were derepressed in *tup1* cells, indicating that Tup1 repressed non-fermentative metabolism. Consistent with this idea, genes required for the catabolism of non-fermentative carbon sources were derepressed in *tup1* cells (*GAL1,10*, *ADH3,4,5*, *POX4*). Therefore, like Efg1, Tup1 appears to provide another mechanistic link between carbon metabolism and morphogenesis and phenotypic switching (Figs. 2C and 3).

There is some evidence that the regulation of carbon metabolism by Tup1 might be relatively direct, operating via Tup1-targeting proteins such as Nrg1 and Mig1. Several glycolytic genes contain Nrg1 response elements in their promoters (*PGI1*, *FBA1*, *GPM1*), and recent proteomic data from our laboratory have indicated that Fba1 and Gpa1 protein levels are regulated by Nrg1 (Yin et al., unpublished data). Hence, these genes would appear to be regulated by Nrg1-Tup1. Mig1 and Tup1 have been shown by Northern blotting to regulate the gluconeogenic gene *PCK1*, and transcript profiling has revealed that Mig1 and Tup1 regulate additional genes involved in carbon metabolism (Murad et al. 2001a, b).

Collectively, these data reinforce the links between carbon metabolism and morphogenesis (and other virulence attributes). However, the relevance of these links to pathogenesis is only inferred. For example, the observations (1) that Efg1 and Tup1 control carbon metabolism and morphogenesis, and (2) that *efg1* and *tup1* mutants display attenuated virulence do not necessarily imply that the regulation of carbon metabolism is essential for pathogenicity. Is this the case? So far, no papers have been published which describe the global expression profiles of *C. albicans* cells isolated from diseased tissue. No doubt these types of experiment will be published in the near future, as soon as the technical challenges posed by the limited availability of fungal biomass and contamination with host tissue have been overcome. Meanwhile, evidence highlighting the importance of carbon metabolism for fungal pathogenicity is based on two main observations, both of which arose through the expression profiling of fungal cells using ex vivo infection models.

The first observation arose through the transcript profiling of *S. cerevisiae* cells following phagocytosis by macrophages (Lorenz and Fink 2001). Glyoxylate cycle genes (*CIT2*, *ICL1*, *MDH2*, *MLS1*) were shown to be amongst the most strongly induced genes in these phagocytosed yeast cells, and these glyoxylate cycle genes were induced much more strongly than were TCA cycle-specific genes (*FUM1*, *IDH1*, *KGD1*, *SDH1,2,3,4*). Northern blotting was then used to demonstrate that *ICL1* is induced in phagocytosed *C. albicans* cells (Lorenz and Fink 2001), and this has been confirmed by an independent DDRT-PCR-based screen for *C. albicans* genes which are induced following ingestion by macrophages (Prigneau et al. 2003). Lorenz and Fink (2001, 2002) noted that glyoxylate cycle genes are required for the virulence of other microbial pathogens, and hence they tested whether this was the case for *C. albicans*. Their prediction was confirmed by demonstrating that the inactivation of *ICL1* attenuates the virulence of *C. albicans*. This study suggests that *C. albicans* cells are deprived of fermentative carbon sources following ingestion by a macrophage (Fig. 10.3). Most significantly, these authors highlighted the importance of central metabolic pathways for the survival of this pathogenic fungus in vivo.

These observations have been reinforced by a recent analysis of the transcriptional response of *C. albicans* to phagocytosis by murine macrophages (Lorenz et al. 2004). This study confirmed that glycoxylate cycle genes are induced following phagocytosis, and revealed that lipid β-oxidation genes are also induced (*FOX2*, *POX1*, *POX2*). This led the authors to suggest that, following phagocytosis by macrophages, *C. albicans* reprograms its metabolism to generate glucose via lipid catabolism, the glycoxylate cycle and gluconeogenesis (Lorenz et al. 2004).

The second observation was generated by the transcript profiling of *C. albicans* cells following exposure to human blood (Fradin et al. 2003). This study, which has provided important clues about the regulation of numerous virulence attributes during the early stages of disseminated candidiasis, is discussed in some detail in the chapter by Munro and colleagues (Chap. 9, this volume). Hence, my discussion of their data will be restricted to observations relating to carbon metabolism. Both glycolytic genes (*PGI1, PFK2, FBA1, GAP1, PGK1, ENO1, PYK1* (*CDC19*)) and glyoxylate cycle genes (*ICL1, MLS1, MDH1, ACS1*) were induced in *C. albicans* cells within 20 min of exposure to human blood. Fradin et al. (2003) then used RT-PCR to examine the expression of a subset of these fungal genes following the intravenous injection of *C. albicans* cells into a murine host. This confirmed, for the first time, the induction of the glyoxylate cycle genes in vivo.

The simultaneous induction of fermentative and respiratory metabolism might seem counterintuitive (Fradin et al. 2003). However, the explanation is probably that the population of fungal cells examined contained a mixture of cells in different microenvironments. Presumably, glyoxylate cycle genes were induced in phagocytosed *C. albicans* cells, whereas glycolytic genes were induced in those fungal cells which remained in the plasma and, hence, retained access to glucose (Fig. 10.3; Fradin et al. 2003). This highlights the importance of developing parallel technologies allowing one to analyse individual fungal cells within complex populations which occupy different microenvironments in the mammalian host (Barelle et al. 2004). Nevertheless, in the context of this review, the work of Fradin et al. (2003) clearly confirms that *C. albicans* undergoes significant metabolic reprogramming when it comes in contact with the host.

VI. Co-Evolution of Virulence and Fitness

C. albicans is an opportunistic pathogen. It causes systemic infections relatively rarely, and the development of such infections is probably more dependent upon the immune status of the host than on the virulence status of the fungus. Hence, it could be argued that evolutionary pressure upon *C. albicans* might have been directed more towards its survival as a commensal or mucocutaneous pathogen, rather than as a systemic pathogen.

Many *C. albicans* genes, including those encoding metabolic enzymes, have been reported to be required for virulence (reviewed by Navarro-Garcia et al. 2001). This has sparked debate in the field about the definition of a virulence gene. "Virulence genes" have now been defined as those encoding factors which interact directly with host components, such as adhesins or secreted aspartyl proteinases (Odds et al. 2001, 2003). Factors such as Efg1 and Tup1 which control the expression of virulence genes should be termed "virulence regulators". Odds et al. (2003) also suggest that factors which are required for microbial function but which do not interact directly with the host should not be termed virulence attributes. Nevertheless, although position effects which influence the expression of the *URA3* marker potentially compromise the interpretation of some virulence studies (Brand et al. 2004), it is clear that some non-virulence attributes, such as metabolic genes, are required for virulence (Navarro-Garcia et al. 2001; Roemer et al. 2003). These include genes involved in lipid and nucleotide biosynthesis. Clearly, these attributes are required for the growth and survival of the fungus in the host. Hence, these functions will be referred to here as "fitness attributes". The inactivation of fitness attributes would be expected to attenuate the ability of the fungus to assimilate nutrients or combat environmental stress, for example, and hence would attenuate its ability to establish an infection (Fig. 10.2D).

There is mounting evidence for the evolution of fitness attributes in *C. albicans*. For example, this pathogen appears to have evolved specialised stress responses which differ significantly from stress responses in *S. cerevisiae* and *Schizosaccharomyces pombe* (Enjalbert et al. 2003; Nicholls et al. 2004; Smith et al. 2004). These specialised stress responses appear to reflect the evolutionary adaptation of *C. albicans* to its host. Unlike budding and fission yeast, *C. albicans* does not appear to recognise an ambient temperature of 37 °C as a stress, and this fungus is relatively resistant to the oxidative stresses it encounters following phagocytosis (Smith et al. 2004). The inactivation of Hog1, which is required to activate the oxidative stress responses, has been reported to attenuate the virulence of *C. albicans* (Alonso-Monge et al. 1999). Similarly, genome sequencing has provided evidence of metabolic specialisation in *C. albicans* (Jones et al. 2004), as discussed in Sect. III. Metabolic genes and pathways appear to have evolved to allow this fungus to assimilate those nutrients which are available to it within the various microenvironments it encounters during the initial colonisation of the host and subsequent disease progression. Hence, fitness attributes appear to have evolved alongside virulence attributes in *C. albicans*.

Not only have fitness and virulence attributes themselves been evolving, but also the mechanisms which regulate their activity appear to have been evolving in *C. albicans*. There is clear evidence for the co-ordinated regulation of virulence attributes in a niche-specific fashion (Lane et al. 2001; Murad et al. 2001a; Nantel et al. 2002). This presumably reflects the differential contributions of the various virulence attributes within different microenvironments in the host (Brown and Gow 1999; Brown 2002). Similarly, as described here (Sect. V), there is mounting evidence for differential metabolic regulation in a niche-specific fashion in *C. albicans* (Lorenz and Fink 2001; Fradin et al. 2003; Rubin-Bejerano et al. 2003). In addition, some data suggest that stress responses might be regulated in a niche-specific fashion in *C. albicans* (Fradin et al. 2003). Therefore, it is attractive to speculate that virulence and fitness attributes are regulated in such a manner as to ensure that both are expressed appropriately in particular microenvironments. This review has highlighted emerging evidence which suggests that metabolic reprogramming is mechanistically linked to the control of virulence through the regulators Efg1, Tup1 and Gcn4 (Sect. V; Lan et al. 2002; Tripathi et al. 2002; Doedt et al. 2004). Hence, *C. albicans* appears to have evolved specific signalling mechanisms which facilitate the co-ordinated regulation of the virulence and fitness attributes which it requires for survival in the host (Fig. 10.2D).

VII. Conclusions and Future Perspectives

In this chapter I have argued that metabolism is intimately linked with virulence in *C. albicans*. I have also highlighted the critical role of *C. albicans* genomics in revealing these links. Unexpected links between metabolism and virulence were first revealed by the global perspectives provided by transcript profiling. However, these findings have raised major new questions which need to be addressed.

How is metabolism integrated with virulence? As described above, initial studies indicate that Efg1, Tup1 and Gcn4 are involved, but how do they link metabolism and virulence mechanistically? Tup1 might achieve this by regulating metabolic and virulence genes via specific DNA binding proteins such as Nrg1, Mig1 and Rfg1 (Kadosh and Johnson 2001; Murad et al. 2001a, b). However, this prediction needs to be tested experimentally. Gcn4 appears to regulate morphogenesis via the Ras-cAMP pathway (Tripathi et al. 2002), but the specific link between Gcn4 and Ras-cAMP signalling has not been defined. Efg1 is known to regulate the expression of both metabolic and virulence genes (Doedt et al. 2004), but the mechanisms by which this is achieved remain obscure.

Other factors are probably involved in the co-ordination of virulence and fitness attributes, and these remain to be identified. However, the increasing availability of large mutant collections will facilitate the design of global screens for mutations which affect both virulence attributes and fitness. These include the doxycycline-conditional mutants in the GRACE collection (Roemer et al. 2003), and the collections of transposon insertion mutants generated by the Mitchell and Johnson groups (Nobile et al. 2003; Uhl et al. 2003). In addition, the potential of an antisense-based screen (de Backer et al. 2001) for the isolation of this type of pleiotropic factor should be tested.

How does *C. albicans* regulate its fitness and virulence attributes in response to specific host environments? The limited data addressing this important issue have been derived largely from elegant analyses of a limited number of genes or proteins in infection models, and in patient samples using RT-PCR and immunological techniques (Glee et al. 1995; Staab et al. 1996; Hoyer et al. 1999; Naglik et al. 1999). Whilst transcript profiling has been performed on ex vivo infection models (Fradin et al. 2003; Rubin-Bejerano et al. 2003), further technological developments are required before genomic technologies can be applied to in vivo samples, for example, in combination with laser capture dissection microscopy. Even with such developments, problems will remain. For example, most of these technologies average the responses of a fungal population. However, in vivo samples are non-homogenous and complex, with individual fungal cells being exposed to differing microenvironments. Hence, to gain a true understanding of fungal behaviour during disease progression, technologies such as Single Cell

Profiling must be used in parallel with transcript profiling and proteomics. Single cell profiling allows the molecular responses of individual fungal cells to be examined in vivo (Barelle et al. 2004).

Finally, what happens to metabolic reprogramming when fungal cells respond and adapt to antifungal drugs? Transcript profiling and proteomic studies of *C. albicans* cells as they acquire drug resistance have suggested that metabolism is affected (Cowen et al. 2002; Bruneau et al. 2003; Rogers and Barker 2003). These changes in metabolism do not appear to be restricted to the antifungal targets themselves. For example, they are not limited to ergosterol biosynthesis following exposure to an azole antifungal. Is this metabolic reprogramming required for drug adaptation, or is it an indirect by-product of exposure to the antifungal agent?

What relevance has the issue of *C. albicans* fitness and virulence to the patient suffering from a *Candida* infection? To date, the academic community has focussed most attention upon *C. albicans* virulence attributes and has paid relatively little attention to metabolism. Ironically, most virulence attributes are likely to be poor targets for antifungal therapy because most are polygenic and not essential for survival in vivo. In contrast, most successful antifungal drug screens have focussed on the fitness of *C. albicans*. Also, central metabolic functions do appear to be essential for fitness in vivo (Lorenz and Fink 2001; Roemer et al. 2003). Clearly, a complete understanding of *C. albicans* pathogenicity is dependent upon a characterisation of fitness as well as virulence attributes.

Acknowledgements. I am grateful to many colleagues for stimulating debates about *Candida* genomics, especially to Frank Odds, Neil Gow, other members of the Aberdeen Fungal Group, my partners in the European Galar Fungail Consortium, and my collaborators with the COGEME Consortium, particularly Ken Haynes and Jan Quinn. My work is supported by grants from the Biotechnology and Biological Sciences Research Council (1/P17124, 1/G188883, BBS/B/06679, BBS/B/13764, BB/C501176/1), the British Society for Antimicrobial Chemotherapy (PG/01), the British Council and National Research Council of Canada (00CRP04), the European Commission (MRTN-CT-2003-504148) and the Wellcome Trust (063204, 068143).

References

Albrecht G, Mosch H-U, Hoffman B, Reusser U, Braus GH (1998) Monitoring the Gcn4 protein-mediated response in the yeast Saccharomyces cerevisiae. J Biol Chem 273:12696–12702

Alonso-Monge R, Navarro-Garcia F, Molero G, Diez-Orejas R, Gustin M, Pla J, Sanchez M, Nombela C (1999) Role of the mitogen-activated protein kinase Hog1p in morphogenesis and virulence of Candida albicans. J Bacteriol 181:3058–3068

Bailey DA, Feldmann PJF, Bovey M, Gow NAR, Brown AJP (1996) The Candida albicans HYR1 gene, which is activated in response to hyphal development, belongs to a gene family encoding yeast cell wall proteins. J Bacteriol 178:5353–5360

Baillie GS, Douglas LJ (1999) Candida biofilms and their susceptibility to antifungal agents. Methods Enzymol 310:644–656

Barelle CJ, Manson C, MacCallum D, Odds FC, Gow NAR, Brown AJP (2004) GFP as a quantitative reporter of gene regulation in Candida albicans. Yeast 21:333–340

Bennett RJ, Johnson AD (2003) Completion of a parasexual cycle in Candida albicans by induced chromosome loss in tetraploid strains. EMBO J 22:2505–2515

Berman J, Sudbery PE (2002) Candida albicans: a molecular revolution built on lessons from budding yeast. Nat Rev Genet 3:918–930

Bertram G, Swoboda RK, Gow NAR, Gooday GW, Brown AJP (1996) Structure and regulation of the Candida albicans ADH1 gene encoding an immunogenic alcohol dehydrogenase. Yeast 12:115–128

Bockmuhl DP, Ernst JF (2001) A potential phosphorylation site for an A-type kinase in the Efg1 regulator protein contributes to hyphal morphogenesis of Candida albicans. Genetics 157:1523–1530

Brand A, MacCallum DM, Brown AJP, Gow NAR, Odds FC (2004) Ectopic expression of URA3 can influence the virulence phenotypes and proteome of Candida albicans but can be overcome by targeted reintegration of URA3 at the RPS10 locus. Eukaryot Cell 3:900–909

Braun BR, Johnson AD (1997) Control of filament formation in Candida albicans by the transcriptional repressor TUP1. Science 277:105–109

Braun BR, Kadosh D, Johnson AD (2001) NRG1, a repressor of filamentous growth in Candida albicans, is down-regulated during filament induction. EMBO J 20:4753–4761

Brega E, Zufferey R, Mamoun CB (2004) Candida albicans Csy1 is a nutrient sensor important for activation of amino acid uptake and hyphal morphogenesis. Eukaryot Cell 3:135–143

Brown AJP (2002) Morphogenetic signalling pathways in Candida albicans. In: Calderone R (ed) Candida and candidiasis. ASM Press, Washington, DC, pp 95-106

Brown AJP, Gow NAR (1999) Regulatory networks controlling Candida albicans morphogenesis. Trends Microbiol 7:333–338

Brown AJP, Barelle CJ, Budge S, Duncan J, Harris S, Lee PR, Leng P, Macaskill S, Murad AMA, Ramsdale M et al. (2000) Gene regulation during morphogenesis in Candida albicans. In: Ernst JF, Schmidt A (eds) Contributions to microbiology: dimorphism in human pathogenic and apathogenic yeasts. Karger, Basel, vol 5, pp 112–125

Bruneau JM, Maillet I, Tagat E, Legrand R, Supatto F, Fudali C, Caer JP, Labas V, Lecaque D, Hodgson J (2003) Drug induced proteome changes in Candida albicans: comparison of the effect of beta(1,3) glucan synthase inhibitors and two triazoles, fluconazole and itraconazole. Proteomics 3:325–336

Calderone R (2002) Candida and candidiasis. ASM Press, Washington, DC Calderone R, Fonzi WA (2001) Virulence factors of Candida albicans. Trends Microbiol 9:327–335

Care RS, Trevethick J, Binley KM, Sudbery PE (1999) The MET3 promoter: a new tool for Candida albicans molecular genetics. Mol Microbiol 34:792–798

Chauhan N, Inglis D, Roman E, Pla J, Li D, Calera JA, Calderone R (2003) Candida albicans response regulator gene SSK1 regulates a subset of genes whose functions are associated with cell wall biosynthesis and adaptation to oxidative stress. Eukaryot Cell 2:1018–1024

Cormack B, Bertram G, Egerton M, Gow NAR, Falkow S, Brown AJP (1997) Yeast Enhanced Green Fluorescent Protein (yEGFP): a reporter of gene expression in Candida albicans. Microbiology 143:303–311

Cowen LE, Nantel A, Whiteway MS, Thomas DY, Tessier DC, Kohn LM, Anderson JB (2002) Population genomics of drug resistance in Candida albicans. Proc Natl Acad Sci USA 99:9284–9289

De Backer MD, Magee PT, Pla J (2000) Recent developments in molecular genetics of Candida albicans. Annu Rev Microbiol 54:463–498

De Backer MD, Nelissen B, Logghe M, Viaene J, Loonen I, Vandoninck S, de Hoogt R, Dewaele S, Simons FA, Verhasselt P et al. (2001) An antisense-based functional genomics approach for identification of genes critical for growth of Candida albicans. Nat Biotechnol 19:235–241

d'Enfert C, Goyard S, Rodriguez-Arnaveilhe S, Frangeul L, Jones L, Tekaia F, Bader O, Albrecht A, Castillo L, Dominguez A et al. (2005) CandidaDB: a genome database for Candida albicans pathogenomics. Nucleic Acids Res 33:D353–D357

Dever TE, Feng L, Wek RC, Cigan AM, Donahue TD, Hinnebusch AG (1992) Phosphorylation of initiation factor 2α by protein kinase GCN2 mediates gene-specific translational control of GCN4 in yeast. Cell 68:585–596

Doedt T, Krishnamurthy S, Bockmühl DP, Tebarth B, Stempel C, Russell CL, Brown AJP, Ernst JF (2004) APSES proteins regulate morphogenesis and metabolism in Candida albicans. Mol Biol Cell 15:3167–3180

Douglas LJ (2003) Candida biofilms and their role in infection. Trends Microbiol 11:30–36

Ellenberger TE, Brandl CJ, Struhl K, Harrison SC (1992) The GCN4 basic region leucine zipper binds DNA as a dimer of uninterrupted a helices: crystal structure of the protein-DNA complex. Cell 71:1223–1237

Enjalbert B, Nantel A, Whiteway M (2003) Stress-induced gene expression in Candida albicans: absence of a general stress response. Mol Biol Cell 14:1460–1467

Enloe B, Diamond A, Mitchell AP (2000) A single-transformation gene function test in diploid Candida albicans. J Bacteriol 182:5730–5736

Filler, SG, Kullberg BJ (2002) Deep-seated candidal infections. In: Calderone R (ed) Candida and candidiasis. ASM Press, Washington, DC, pp 341–348

Fonzi WA, Irwin MY (1993) Isogenic strain construction and gene mapping in Candida albicans. Genetics 134:717–728

Fradin C, Kretschmar M, Nichterlein T, Gaillardin C, d'Enfert C, Hube B (2003) Stage-specific gene expression of Candida albicans in human blood. Mol Microbiol 472:1523–1543

Garcia-Sanchez S, Aubert S, Iraqui I, Janbon G, Ghigo J-M, d'Enfert C (2004) Candida albicans biofilms: a developmental state associated with specific and stable gene expression patterns. Eukaryot Cell 3:536–545

Glee PM, Sundstrom P, Hazen KC (1995) Expression of surface hydrophobic proteins by Candida albicans in vivo. Infect Immun 63:1373–1379

Gow NAR, Brown AJP, Odds FC (2002) Fungal morphogenesis and host invasion. Curr Opin Microbiol 5:366–371

Gow NAR, Knox Y, Munro CA, Thompson WD (2003) Infection of chick chorioallantoic membrane (CAM) as a model for invasive hyphal growth and pathogenesis of Candida albicans. Med Mycol 41:331–338

Haynes K (2001) Virulence in Candida species. Trends Microbiol 9:591–596

Hellstein J, Vawter-Hugart H, Fotos P, Schmid J, Soll DR (1993) Genetic similarity and phenotypic diversity of commensal and pathogenic strains of Candida albicans isolated from the oral cavity. J Clin Microbiol 31:3190–3199

Hernandez R, Nombela C, Diez-Orejas, Gil C (2004) Two-dimensional reference map of Candida albicans hyphal forms. Proteomics 4:374–382

Hinnebusch AG (1988) Mechanisms of gene regulation in the general control of amino acid biosynthesis in Saccharomyces cerevisiae. Microbiol Rev 52:248–273

Hinnebusch AG, Natarajan K (2002) Gcn4p, a master regulator of gene expression is controlled at multiple levels by diverse signals of starvation and stress. Eukaryot Cell 1:22–32

Hoffmann B, Valerius O, Andermann M, Braus GH (2001) Transcriptional autoregulation and inhibition of mRNA translation of amino acid regulator gene cpcA of filamentous fungus Aspergillus nidulans. Mol Biol Cell 12:2846–2857

Hoyer LL (2001) The ALS gene family of Candida albicans. Trends Microbiol 9:176–180

Hoyer LL, Clevenger J, Hecht JE, Ehrhart EJ, Poulet FM (1999) Detection of als proteins on the cell wall of Candida albicans in murine tissues. Infect Immun 67:4251–4255

Hube B, Naglik J (2002) Extracellular hydolases. In: Calderone R (ed) Candida and candidiasis. ASM Press, Washington, DC, pp 107–122

Hull CM, Raisner RM, Johnson AD (2000) Evidence for mating of the "asexual" yeast Candida albicans in a mammalian host. Science 289:307–310

Jones T, Federspiel NA, Chibana H, Dungan J, Kalman S, Magee BB, Newport G, Thorstenson YR, Agabian N, Magee PT et al. (2004) The diploid genome sequence of Candida albicans. Proc Natl Acad Sci USA 101:7329–7334

Kadosh D, Johnson AD (2001) Rfg1, a protein related to the S. cerevisiae hypoxic regulator Rox1, controls filamentous growth and virulence in C albicans. Mol Cell Biol 21:2496–2505

Kastaniotis AJ, Zitomer RS (2000) Rox1 mediated repression. Oxygen dependent repression in yeast. Adv Exp Med Biol 475:185–95

Kornitzer D, Raboy B, Kulka RG, Fink GR (1994) Regulated degradation of the transcription factor Gcn4. EMBO J 13:6021–6030
Korting HC, Hube B, Oberbauer S, Januschke E, Hamm G, Albrecht A, Borelli C, Schaller M (2003) Reduced expression of the hyphal-independent Candida albicans proteinase genes SAP1 and SAP3 in the efg1 mutant is associated with attenuated virulence during infection of oral epithelium. J Med Microbiol 52:623–632
Krishnamoorthy T, Pavitt GD, Zhang F, Dever TE, Hinnebusch AG (2001) Tight binding of the phosphorylated a subunit of initiation factor 2 (eIF2α) to the regulatory subunits of guanine nucleotide exchange factor eIF2B is required for inhibition of translation initiation. Mol Cell Biol 21:5018–5030
Kullberg BJ, Filler SG (2002) Candidemia. In: Calderone R (ed) Candida and candidiasis. ASM Press, Washington, DC, pp 327–340
Kumamoto CA (2002) Candida biofilms. Curr Opin Microbiol 5:608–611
Kvaal CA, Srikantha T, Soll DR (1997) Misexpression of the white-phase-specific gene WH11 in the opaque phase of Candida albicans affects switching and virulence. Infect Immun 65:4468–4475
Kvaal C, Lachke SA, Srikantha T, Daniels K, McCoy J, Soll DR (1999) Misexpression of the opaque-phase-specific gene PEP1 (SAP1) in the white phase of Candida albicans confers increased virulence in a mouse model of cutaneous infection. Infect Immun 67:6652–6662
Lan C-Y, Newport G, Murillo LA, Jones T, Scherer S, Davis RW, Agabian N (2002) Metabolic specialization associated with phenotypic switching in Candida albicans. Proc Natl Acad Sci USA 99:14907–14912
Lane S, Birse C, Zhou S, Matson R, Liu H (2001) DNA array studies demonstrate convergent regulation of virulence factors by Cph1, Cph2, and Efg1 in Candida albicans. J Biol Chem 276:48988–48996
Leng P, Lee PR, Wu H, Brown AJP (2001) Efg1, a morphogenetic regulator in Candida albicans, is a sequence-specific DNA binding protein. J Bacteriol 183:4090–4093
Leuker CE, Sonneborn A, Delbruck S, Ernst JF (1997) Sequence and promoter regulation of the PCK1 gene encoding phosphoenolpyruvate carboxykinase of the fungal pathogen Candida albicans. Gene 192:235–240
Lo HJ, Kohler JR, DiDomenico B, Loebenberg D, Cacciapuoti A, Fink GR (1997) Nonfilamentous *C. albicans* mutants are avirulent. Cell 90:939–949
Lorenz MC, Fink GR (2001) The glyoxylate cycle is required for fungal virulence. Nature 412:83–86
Lorenz MC, Fink GR (2002) Life and death in a macrophage: role of the glycoxylate cycle in virulence. Eukaryot Cell 1:657–662
Lorenz MC, Bender JA, Fink GR (2004) Transcriptional response of Candida albicans upon internalization by macrophages. Eukaryot Cell 3:1076–1087
Magee PT (1998) Analysis of the Candida albicans genome. Methods Microbiol 26:395–415
Magee BB, Magee PT (2000) Induction of mating in Candida albicans by construction of MTLa and MTLalpha strains. Science 289:310–313
Mavor A (2004) Transcriptional regulation of morphogenesis in Candida albicans. PhD Thesis, University of Aberdeen Miller MG, Johnson AD (2002) White-opaque switching in Candida albicans is controlled by mating-type locus homeodomain proteins and allows efficient mating. Cell 110:293–302
Mueller PP, Hinnebusch AG (1994) Multiple upstream AUG codons mediate translational control of GCN4. Cell 45:201–207
Murad AMA, Leng P, Straffon M, Wishart J, Macaskill S, MacCallum D, Schnell N, Talibi D, Marechal D, Tekaia F et al. (2001a) NRG1 represses yeast-hypha morphogenesis and hypha-specific gene expression in Candida albicans. EMBO J 20:4742–4752
Murad AMA, d'Enfert C, Gaillardin C, Tournu H, Tekaia F, Talibi D, Marechal D, Marchais V, Cottin J, Brown AJP (2001b) Transcript profiling in Candida albicans reveals new cellular functions for the transcriptional repressors, CaTup1, CaMig1 and CaNrg1. Mol Microbiol 42:981–993
Naglik JR, Newport G, White TC, Fernandes-Naglik LL, Greenspan JS, Greenspan D, Sweet SP, Challacombe SJ, Agabian N (1999) In vivo analysis of secreted aspartyl proteinase expression in human oral candidiasis. Infect Immun 67:2482–2490
Nantel A, Dignard D, Bachewich C, Harcus D, Marcil A, Bouin A-P, Sensen CW, Hogues H, van het Hoog M, Gordon P et al. (2002) Transcript profiling of Candida albicans cells undergoing the yeast-to-hyphal transition. Mol Biol Cell 13:3452–2365
Natarajan K, Meyer MR, Jackson BM, Slade D, Roberts C, Hinnebusch AG, Marton MJ (2001) Transcriptional profiling shows that Gcn4p is a master regulator of gene expression during amino acid starvation in yeast. Mol Cell Biol 21:4347–4368
Navarro-Garcia F, Sanchez M, Nombela C, Pla J (2001) Virulence genes in the pathogenic yeast Candida albicans. FEMS Microbiol Rev 25:245–268
Nicholls S, Straffon M, Enjalbert B, Nantel A, Macaskill S, Whiteway M, Brown AJP (2004) Msn2/4-like transcription factors play no obvious roles in the stress responses of the fungal pathogen, Candida albicans. Eukaryot Cell 3:1111–1123
Nobile CJ, Bruno VM, Richard ML, Davis DA, Mitchell AP (2003) Genetic control of chlamydospore formation in Candida albicans. Microbiology 149:3629–3637
Odds FC (1988) Candida and candidosis, 2nd edn. Bailliere Tindall, London Odds FC (1994) Candida species and virulence. ASM News 60:313–318
Odds FC, Gow NAR, Brown AJP (2001) Fungal virulence studies come of age. Genome Biol 2:1009.1–1009.4
Odds FC, Calderone RA, Hube B, Nombela C (2003) Virulence in Candida albicans: views and suggestions from a peer-group workshop. ASM News 69:54–55
Oliphant AR, Brandl CJ, Struhl K (1989) Defining the sequence specificity of DNA binding proteins by selecting binding sites from random sequence oligonucleotides: analysis of yeast GCN4 protein. Mol Cell Biol 9:2944–2949
Paluh JL, Orbach MJ, Legerton TL, Yanofsky C (1988) The cross-pathway control gene of Neurospora crassa, cpc-1, encodes a protein similar to GCN4 of yeast and the DNA-binding domain of the oncogene v-jun-encoded protein. Proc Natl Acad Sci USA 85:3728–3732
Petter R, Chang YC, Kwon-Chung KJ (1997) A gene homologues to Saccharomyces cerevisiae SNF1 appears

to be essential for the viability of Candida albicans. Infect Immun 65:4909–4917

Pitarch A, Sanchez M, Nombela C, Gil C (2002) Sequential fractionation and two-dimensional gel analysis unravels the complexity of the dimorphic fungus Candida albicans cell wall proteome. Mol Cell Proteomics 112:967–982

Prigneau O, Porta A, Poudrier JA, Colonna-Romano S, Noel T, Maresca B (2003) Genes involved in β-oxidation, energy metabolism and glyoxylate cycle are induced by Candida albicans during macrophage infection. Yeast 20:723–730

Rao GR, Ramakrishnan T, Sirsi M (1960) Enzymes in Candida albicans: I pathways of glucose dissimilation. J Bacteriol 80:654–658

Roemer T, Jiang B, Davison J, Ketela T, Veillette K, Breton A, Tandia F, Linteau A, Sillaots S, Marta C et al. (2003) Large-scale essential gene identification in Candida albicans and applications to antifungal drug discovery. Mol Microbiol 50:167–181

Rogers PD, Barker KS (2003) Genome-wide expression profile analysis reveals coordinately regulated genes associated with stepwise acquisition of azole resistance in Candida albicans clinical isolates. Antimicrob Agents Chemother 47:1220–1227

Rolfes RJ, Hinnebusch AG (1993) Translation of the yeast transcriptional activator GCN4 is stimulated by purine limitation: implications for activation of the protein kinase GCN2. Mol Cell Biol 13:5099–5111

Rolland F, Winderickx J, Thevelein JM (2001) Glucose-sensing mechanisms in eukaryotic cells. Trends Biochem Sci 26:310–317

Rubin-Bejerano I, Fraser I, Grisafi P, Fink GR (2003) Phagocytosis by neutrophils induces an amino acid deprivation response in Saccharomyces cerevisiae and Candida albicans. PNAS USA 100:11007–11012

Ruhnke M (2002) Skin and mucous membrane infections. In: Calderone R (ed) Candida and candidiasis. ASM Press, Washington, DC, pp 307–325

Santos MAS, Keith G, Tuite MF (1993) Non-standard translational events in Candida albicans mediated by an unusual seryl-tRNA with a 5′-CAG-3′ (leucine) anticodon. EMBO J 12:607–616

Saville SP, Lazzell AL, Monteagudo C, Lopez-Ribot JL (2003) Engineered control of cell morphology in vivo reveals distinct roles for yeast and filamentous forms of Candida albicans during infection. Eukaryot Cell 2:1053–1060

Sharkey LL, McNemar MD, Saporito-Irwin SM, Sypherd PS, Fonzi WA (1999) HWP1 functions in the morphological development of Candida albicans downstream of EFG1, TUP1 and RBF1. J Bacteriol 181:5273–5279

Smith RL, Johnson AD (2000) Turning off genes by Ssn6-Tup1: a conserved system of transcriptional repression in eukaryotes. Trends Biochem Sci 25:325–330

Smith DA, Nicholls S, Morgan BA, Brown AJP, Quinn J (2004) A conserved stress-activated protein kinase regulates a core stress response in the human pathogen Candida albicans. Mol Biol Cell 15:4179–4190

Soll DR (2002) Phenotypic switching. In: Calderone R (ed) Candida and candidiasis. ASM Press, Washington, DC, pp 123–142

Sonneborn A, Tebarth B, Ernst J (1999) Control of White-Opaque phenotypic switching in Candida albicans by the Efg1p morphogenetic regulator. Infect Immun 67:4655–4660

Srikantha T, Klapach A, Lorenz WW, Tsai LK, Laughlin LA, Gorman JA, Soll DR (1996) The sea pansy Renilla reniformis luciferase serves as a sensitive bioluminescent reporter for differential gene expression in Candida albicans. J Bacteriol 178:121–129

Srikantha T, Tsai LK, Daniels K, Soll DR (2000) EFG1 null mutants of Candida albicans switch but cannot express the complete phenotype of white-phase budding cells. J Bacteriol 182:1580–1591

Staab JF, Ferrer CA, Sundstrom P (1996) Developmental expression of a tandemly repeated, proline and glutamine-rich amino acid motif on hyphal surfaces of Candida albicans. J Biol Chem 271:6298–6305

Stoldt VR, Sonneborn A, Leuker CE, Ernst J (1997) Efg1p, an essential regulator of morphogenesis of the human pathogen Candida albicans, is a member of a conserved class of bHLH proteins regulating morphogenetic processes in fungi. EMBO J 16:1982–1991

Sundstrum P (2002) Adhesion in Candida spp. Cell Microbiol 4:461–469

Swoboda RK, Bertram G, Delbruck S, Ernst JF, Gow NAR, Gooday GW, Brown AJP (1994) Fluctuations in glycolytic mRNA levels during the yeast-to-hyphal transition in Candida albicans reflect underlying changes in growth and are not a response to cellular dimorphism. Mol Microbiol 13:663–672

Tebarth B, Doedt T, Krishnamurthy S, Weide M, Monterola F, Dominguez A, Ernst JF (2003) Adaptation of the Efg1p morphogenetic pathway in Candida albicans by negative autoregulation and PKA-dependent repression of the EFG1 gene. J Mol Biol 329:949–962

Thevelein JM (1994) Signal-transduction in yeast. Yeast 10:1753–1790

Tripathi G, Wiltshire C, Macaskill S, Tournu H, Budge S, Brown AJP (2002) CaGcn4 co-ordinates morphogenetic and metabolic responses to amino acid starvation in Candida albicans. EMBO J 21:5448–5456

Tsong AE, Miller MG, Raisner RM, Johnson AD (2003) Evolution of a combinatorial transcriptional circuit: a case study in yeasts. Cell 115:389–399

Tournu H, Tripathi G, Gertram G, Macaskill S, Mavor A, Walker L, Odds FC, Gow NAR, Brown AJP (2005) Blobal role of the protein kinase, Gcn2, in the human pathogen, Candida albicans. Eukaryotic Cell (in press)

Tzung K-W, Williams RM, Scherer S, Federspiel N, Jones T, Hansen N, Bivolarevic V, Huizar L, Komp C, Surzycki R et al. (2001) Genomic evidence for a complete sexual cycle in Candida albicans. Proc Natl Acad Sci USA 98:3249–3253

Uhl MA, Johnson AD (2001) Development of Streptococcus thermophilus lacZ as a reporter gene for Candida albicans. Microbiology 147:1189–1195

Uhl MA, Biery M, Craig N, Johnson AD (2003) Haploinsufficiency-based large-scale forward genetic analysis of filamentous growth in the diploid human fungal pathogen *C. albicans*. EMBO J 22:2668–2678

van Burik J-AH, Magee PT (2001) Aspects of fungal pathogenesis in humans. Annu Rev Microbiol 55:743–772

Van Neil CB, Cohen AL (1942) The metabolism of Candida albicans. J Cell Comp Physiol 20:95–112

Wanke C, Eckert S, Albrecht G, van Hartingsveldt W, Punt PJ, van den Hondel CA, Braus GH (1997) The Aspergillus niger GCN4 homologue, cpcA, is transcriptionally regulated and encodes an unusual leucine zipper. Mol Microbiol 23:23–33

Wek SA, Zhu S, Wek RC (1995) The histidyl-tRNA synthetase-related sequence in the eIF-2a protein kinase GCN2 interacts with tRNA and is required for activation in response to starvation for different amino acids. Mol Cell Biol 15:4497–4506

Wey SB, Mori M, Pfaller MA, Woolson RF, Wenzel P (1989) Risk factors for hospital-acquired candidemia. A matched case-control study. Arch Intern Med 149:2349–2353

Wickerham LJ (1951) Taxonomy of yeast. Dept Agric Tech Bull 1029:1–19

Williamson MI, Samaranayake LP, MacFarlane TW (1986) Biotypes of Candida albicans using the API 20C system. FEMS Microbiol Lett 37:27–29

Wilson RB, Davis D, Mitchell AP (1999) Rapid hypothesis testing with Candida albicans through gene disruption with short homology regions. J Bacteriol 181:1868–1874

Wilson RB, Davis D, Enloe BM, Mitchell AP (2000) A recyclable Candida albicans URA3 cassette for PCR product-directed gene disruptions. Yeast 16:65–70

Yin Z, Stead D, Selway L, Walker J, Riba-Garcia I, Mclnerney T, Gaskell S, Oliver SG, Cash P, Brown AJP (2004) Divergence between Candida albicans and Saccharomyces cerevisiae in their global responses to amino acid starvation. Proteomics 4:2425–2436

Zaragoza O, Rodriguez C, Gancedo C (2000) Isolation of the MIG1 gene from Candida albicans and effects of its disruption on catabolite repression. J Bacteriol 182:320–326

Zhao R, Lockhart SR, Daniels K, Soll DR (2002) Roles of TUP1 in switching, phase maintenance, and phase-specific gene expression in Candida albicans. Eukaryot Cell 1:353–365

11 Regulators of *Candida glabrata* Pathogenicity

K. Haynes[1]

CONTENTS

Abbreviations: FOA, 5-fluoro-orotic acid; GLA, genome-wide location analysis

I. Introduction

The incidence of life-threatening fungal infections increased substantially world-wide from the 1980s to the 1990s (Beck-Sague and Jarvis 1993; Chen et al. 1997). The current reported incidence rates for candidiasis range from 0.2 to 0.55 per 1000 hospital admissions (Asmundsdottir et al. 2002; Tortorano et al. 2004). *Candida* species now cause more bloodstream infections than do all individual Gram-negative bacterial species, including *Escherichia coli* (Wisplinghoff et al. 2003). Indeed, candidaemia now accounts for 10%–15% of all bloodstream infections (Jarvis 1995), one of the top 15 causes of mortality in the United States, resulting in over 200,000 deaths per year (Pittet et al. 1997). Given the continuing use of broad-spectrum antibiotics and indwelling intravenous catheters, two major risk factors for the development of candidiaemia, plus the inexorable rise of at-risk patients, it seems unlikely that this threat will diminish (Calderone 2002). This increased incidence of candidiasis is matched by high direct and attributable mortality rates (Gudlaugsson et al. 2003). Additionally, patients contracting *Candida* infection stay in hospital for up to 30 days longer than do similar uninfected individuals (Wey et al. 1988). In one Spanish study the increase in direct costs associated with *Candida* infection was estimated at € 16,000 (Olaechea et al. 2004). Thus, *Candida* species exert a significant clinical and economic impact.

Candida glabrata is now responsible for about 15%–20% of candidiasis cases (Pfaller et al. 2001) and its incidence is still increasing (Pfaller and Diekema 2002; Tortorano et al. 2004). This is a cause for concern, as this species may be more resistant to antifungal agents than are other *Candida* species, including *Candida albicans* (Pfaller et al. 2004). A recent European study suggests that *C. glabrata* is more often associated with infection in surgical and solid tumour patients, and this is even more pronounced in older individuals (Tortorano et al. 2004).

This clinical problem clearly needs addressing, and a better understanding of how *C. glabrata* causes disease is imperative. Unfortunately, the increase in fungal disease caused by *C. glabrata* has not been matched by a concomitant rise in our understanding of the basic biology of the pathogen. Here I will attempt to review what is known about the regulation of *C. glabrata* virulence, within the context of *C. albicans* virulence and protein function in *Saccharomyces cerevisiae*.

II. Regulators of *C. albicans* Virulence

A large number of attributes have been shown to be important in *C. albicans* virulence (Navarro-Garcia et al. 2001). These include the ability to undertake morphogenetic transitions, to respond correctly to changes in the host environment, to secrete hydrolytic enzymes and to adhere to host surfaces. For these processes to occur in an appropriate lo-

[1] Department of Infectious Diseases, Imperial College London, Du Cane Road, London W12 0NN, UK

The Mycota XIII
Fungal Genomics
Alistair J.P. Brown (Ed.)

cation at the correct time, they must be regulated. Transcription factors are likely to play a large role in this regulation and have thus received a high amount of interest as potential virulence regulators in *C. albicans* (Table 11.1). Other reviews in this volume will concentrate on many of these proteins. Here I will discuss only two pairs of *C. albicans* transcription factors: Fkh2 and Ace2 plus Mds3 and Rim101. These have been linked to processes which are probably involved in *C. glabrata* virulence.

A. *C. albicans* Fkh2 and Ace2

S. cerevisiae FKH2 encodes a DNA-binding protein which is similar to the forkhead transcription factor of *Drosophila melanogaster* (Weigel et al. 1989). Fkh2 regulates, in partnership with Fkh1, expression of the *CLB2* cluster, a group of 33 genes, the transcription of which peaks in early mitosis (Zhu et al. 2000). These include the transcription factor-encoding genes *SWI5* and *ACE2* whose products regulate separately, and in tandem, genes expressed at the M/G1 boundary of the cell cycle. In *C. albicans* there is only a single forkhead homologue, *FKH2* (Bensen et al. 2002). The ability of *C. albicans fkh2/fkh2* mutants to cause disease in vivo has not been explored, as these cells have a slow growth phenotype (Bensen et al. 2002). However, their ability to damage both oral epithelial and vascular endothelial cells was much reduced in comparison to wild type *C. albicans*, suggesting a role in virulence (Bensen et al. 2002). In addition, the

Table. 11.1. Known *Candida albicans* transcription factors

Regulator	Regulatory domain[a]	TF type[b]	Mutant attenuated[c]	Reference
Ace2	Cell separation, adherence, biofilms	Zinc finger	Yes	(Kelly et al. 2004)
Ash1	Filamentous growth	GATA	Yes	(Inglis and Johnson 2002)
Brf1	Transcription from Pol III promoters	RNA Pol III	n.d.[d]	(Khoo et al. 1994)
Cap1	Response to drug and oxidative stress	bZIP	n.d.	(Alarco and Raymond 1999)
Cbf1	Chromosome segregation, methionine biosynthesis	Helix-loop helix-zipper	n.d.	(Eck et al. 2001)
Cph1	Conjugation, filamentous growth	Homeodomain	Yes[e]	(Liu et al. 1994; Lo et al. 1997)
Cph2	Hyphal growth	bHLH family	n.d.	(Lane et al. 2001)
Cwt1	Cell wall	Zinc finger	n.d.	(Moreno et al. 2003)
Czf1	Filamentous growth	Zinc finger	n.d.	(Whiteway et al. 1992)
Efg1	Cell adhesion, filamentous growth	bHLH family	Yes[e]	(Stoldt et al. 1997)
Fcr1	Response to drug	Zinc finger	n.d.	(Talibi and Raymond 1999)
Fcr3	Response to drug	bZIP	n.d.	(Yang et al. 2001)
Fkh2	Hyphal growth	Forkhead	n.d.	(Bensen et al. 2002)
Gat1	Nitrogen utilization	Zinc finger	Yes	(Limjindaporn et al. 2003)
Gcn4	Filamentous growth	bZIP	No	(Tripathi et al. 2002; Brand et al. 2004)
Mds3	Chlamydospore formation, hyphal growth, response to pH	No homology	Yes	(Davis et al. 2002)
Mig1	Carbon utilization	Zinc finger	No	(Zaragoza et al. 2000)
Nrg1	Cellular morphogenesis, stress response	Zinc finger	Yes	(Murad et al. 2001; Saville et al. 2003)
Pdc2	Glucose metabolism	DNA binding[f]	n.d.	(Kaiser et al. 1999)
Rap1	Binds telomeres	RPG-box binding	n.d.	(Biswas et al. 2003)
Rbf1	Filamentous growth	RPG-box binding	n.d.	(Ishii et al. 1997)
Rfg1	Filamentous growth	HMG domain	Yes	(Kadosh and Johnson 2001; Khalaf and Zitomer 2001)
Rim101	Chlamydospore formation, filamentous growth, response to pH	Zinc finger	Yes	(Ramon et al. 1999; Davis et al. 2000)
Suc1	Maltose and sucrose metabolism	Zinc finger	n.d.	(Kelly and Kwon-Chung 1992)
Tec1	Hyphal growth	TEA/ATTS family	Yes	(Schweizer et al. 2000)
Tup1	Filamentous growth	WD repeat	Yes	(Braun and Johnson 1997)

[a] As defined in CGD (www.candidagenome.org) [b] Family of transcriptional regulator [c] In murine infection experiments [d] Not determined [e] *cph1/cph1 efg1/efg1* double mutants are attenuated [f] Centromere protein B-like DNA binding and DDE endonuclease domains

deletion of the Fkh2-regulated gene *ACE2* in *C. albicans* results in complete attenuation of virulence in a murine model of candidiasis (Kelly et al. 2004). In complete contrast, *C. glabrata ace2* cells are almost 200-fold increased in their ability to cause disease (Kamran et al. 2004).

Why does the inactivation of homologous transcription factors in two *Candida* species result in such different effects on virulence? The deletion of *FKH2* in *C. albicans* results in a severe cell separation defect (Bensen et al. 2002). The cells undergo cytokinesis forming intact septa, but mother and daughter cells remain joined, resulting in large clumps of cells. The use of a *C. albicans* partial microarray demonstrated that the expression of *CHT2*, one of at least three chitinase genes in the genome, is down-regulated in *fkh2/fkh2* cells, suggesting a possible mechanistic explanation for the cell separation defect (Bensen et al. 2002). Similarly, *C. albicans ace2/ace2* mutants also have a cell separation defect and they exhibit defective expression of genes encoding proteins required for cell separation, including the chitinase Cht3 (Kelly et al. 2004). Expression of *CHT2* is not changed in *C. albicans ace2/ace2* mutants, but the cells are defective in separation. These data tentatively suggest that both functional Cht2 and Cht3 are required for *C. albicans* cells to separate efficiently, and they support the observation that there is differential regulation of each chitinase gene (McCreath et al. 1995; Nantel et al. 2002). *C. glabrata* has only a single chitinase gene, *CTS1*, which is not expressed in the absence of Ace2 (Kamran et al. 2004). It is possible that this differential regulation of chitinase gene expression results in cell wall modifications which interact in different ways with the host. These possibilities will be discussed in more detail below.

B. *C. albicans* Mds3 and Rim101

The *S. cerevisiae* zinc finger transcription factor Rim101 was first described as a positive regulator of *IME1*, which encodes a protein required for the expression of many meiotic genes (Smith et al. 1990; Su and Mitchell 1993). Subsequently, Rim101 has been shown to function in a *S. cerevisiae* pathway analogous to the PacC pH response pathway of *Aspergillus nidulans* (Penalva and Arst 2004). In *C. albicans*, Rim101 is also involved in regulating the response to changes in environmental pH (Davis et al. 2000), although this is done in tandem with at least one other regulator, Mds3 (Davis et al. 2002). Mds3 has two homologues in *S. cerevisiae*, Mds3 and Pmd1, which function as inhibitors of meiosis and have no known role in pH adaptation (Benni and Neigeborn 1997). Double *C. albicans mds3/mds3 rim101/rim101* mutants have more extreme phenotypes (growth inhibition at pH 9 and sensitivity to lithium chloride) than does each single mutant (Davis et al. 2002). This suggests that the two pathways act in parallel, a hypothesis which is further supported by the fact that a constitutive *RIM101-405* allele, which encodes a gain-of-function Rim101 protein, cannot restore alkaline-induced filamentous growth to a *C. albicans mds3/mds3* strain (Davis et al. 2002). Additionally, the double mutants were more attenuated in a murine model of candidiasis than was each of the single mutants (Davis et al. 2000, 2002).

It is not surprising that the ability to adapt to changes in environmental pH plays a role in *Candida* virulence. *PHR1* and *PHR2* encode glycosidases and have a pH-dependent expression pattern. *PHR1* is expressed in neutral and alkaline pH environments and suppressed at acidic pH. *PHR2* has an inverse pattern of expression. The two proteins are functional homologues of each other, as forced expression of *PHR1* in a *phr2* mutant at acid pH, and of *PHR2* at alkaline pH in a *phr1* background rescues the conditional growth and morphogenetic defects of each mutant (Muhlschlegel and Fonzi 1997). In addition, *C. albicans phr1/phr1* mutants are completely avirulent in a murine model of systemic candidiasis; murine blood has a pH of about 7.3 (De Bernardis et al. 1998). They have wild-type virulence in a rat vaginal model; the pH of the rat vagina is about 4.5. Conversely, *C. albicans phr2/phr2* cells exhibit wild-type virulence in the systemic mouse model but are attenuated in the rat vaginal model (De Bernardis et al. 1998). pH-dependent gene regulation is also required for virulence of other fungal pathogens. For example, *Sclerotinia sclerotiorum pac1* (*pac*C homologue) loss-of-function mutants show aberrant sclerocial development and reduced virulence on *Arabidopsis thaliana* and tomatoes (Rollins 2003). Pal1 regulates the expression of the virulence factor-encoding pectate lyase gene *pel*B in *Colletotrichum gloeosporioides* (Drori et al. 2003). Also, in *Fusarium oxysporum*, loss of function *pac*C$^{-/-}$ mutants are more virulent than are wild-type cells in a tomato root infection assay, whereas gain-of-function *pac*C^{C} mutants are attenuated (Caracuel et al. 2003). Recently we have shown that elimination of PacC, prevention of PacC processing or blocking passage of the ambient pH signal to PacC via inactivation of the *pal* genes result in attenuation of *A. nidulans* virulence in a murine intranasal infection model (Bignell et al. 2005). In stark contrast, constitutive activation of PacC results in increased virulence, even when the pH signal is blocked.

Taken together, these observations suggest that the ability to successfully adapt to variations in ambient pH may be a "universal" virulence trait in fungi. Homologues of *RIM101* (CAGL0E3762g), *MDS3* (CAGL0D01782g) and *PMD1* (CAGL0K02101g) are present in the *C. glabrata* genome (cbi.labri.fr/Genolevures/elt/CAGL), and it is probable that the encoded proteins will play a significant role in virulence.

III. *C. glabrata* Transcriptional Regulators

The analysis of transcriptional regulation in *C. albicans* virulence is relatively well developed (see M. Ramsdale, M. Whiteway and A. Nantel, C. Munro et al. and A.J.P. Brown, Chaps. 7, 8, 9 and 10 respectively, this volume; Table 11.1). However, to date only four transcription factors have been analysed in *C. glabrata*. In this section I will deal with each of these in turn.

A. *C. glabrata* Amt1

Initial work on *C. glabrata* transcription factors was not focused on understanding virulence, but rather on the regulation of metallothionein genes (*MT-I*, *MT-IIa* and *MT-IIb*) which encode small proteins chelating copper (Kagi and Schaffer 1988; Zhou and Thiele 1991; Thorvaldsen et al. 1993). The metal-responsive transcription factor Amt1 facilitates de-toxification following exposure to high external levels of copper (Zhou and Thiele 1993). Such exposure results in rapid auto-activation of *AMT1* expression which is greatly facilitated by a homopolymeric (dA–dT) nucleosomal element (Zhu and Thiele 1996). This analysis revealed that such nucleosomal elements may generally function in eukaryotes to enable rapid access of transcription factors, and may provide a mechanism for rapid transcriptional activation in response to specific signals (Koch and Thiele 1999).

The role of Amt1 in virulence has not been analysed in murine models. However, *C. glabrata* switch phenotypes express differing levels of the *MT-II* metallothionein genes independently of external copper levels (Lachke et al. 2000). High-frequency switching in *C. albicans* regulates the expression of many phase-specific genes, resulting in a remarkable level of phenotypic plasticity, including several phenotypes which facilitate pathogenesis (White and Agabian 1995). In *C. glabrata*, phenotypic switching has been demonstrated in hosts, both orally and vaginally (Brockert et al. 2003), and it is possible that this ability plays a role in virulence, although this remains to be tested.

B. *C. glabrata* Rap1

The ability to adhere to host cells is an accepted virulence attribute of *C. albicans*. The fungus is able to express a large number of adhesins which can interact with disparate host cell ligands including the integrin Int1 (Gale et al. 1998), the host transglutaminase substrate Hwp1 (Staab et al. 1999) and the ALS (agglutinin-like sequence) family of proteins (Hoyer 2001). In a search for *C. glabrata* adhesins, Brendan Cormack and colleagues at Johns Hopkins compared the ability of 4800 signature tagged mutants (Hensel et al. 1995) to adhere to cultured human (HEp2) epithelial cells (Cormack et al. 1999). They identified 16 non-adherent mutants, 14 of which had disruptions in the promoter region upstream of the same large open reading frame, *EPA1* (epithelial adhesin 1). *C. glabrata epa1* null strains have 95% reduced adherence to epithelial cells, but they are not attenuated in animal models (Cormack and Falkow 1999). It is possible that Epa1 is a member of a family of proteins which can compensate for its loss. Indeed, *EPA1* is a member of a family of sub-telomerically located genes which encode related glycosylphosphatidylinositol (GPI) anchored cell wall proteins with a consistent domain structure (Frieman et al. 2002; De Las Penas et al. 2003). Of the *EPA* genes, only *EPA1* is expressed to any extent in vitro. However, inactivation of Sir3 or truncation of Rap1 (*rap1-21*) results in de-repression of *EPA2-5* expression (De Las Penas et al. 2003). *RAP1* cannot be deleted in *C. glabrata*, as it appears to be an essential gene (Haw et al. 2001). In *S. cerevisiae*, Sir3 is one of four proteins which maintain repressed chromatin structures at telomeres and silent mating-type loci (Rine and Herskowitz 1987; Aparicio et al. 1991). They do this not by binding directly to DNA, but via interactions with DNA-binding proteins such as the repressor activator protein Rap1 (Shore 1994). Rap1 binds to $[C^{(1--3)}A]^n$ repeats in telomeric DNA, and interacts with Sir2, Sir3 and Sir4 to facilitate gene silencing (Shore and Nasmyth 1987) and with Rif1 and Rif2 to maintain telomere ends (Shore 1997). It is therefore likely that the de-repression of *EPA2-5* expression in *C. glabrata sir3* and *rap1-21* cells is due to release of sub-telomeric silencing. This was shown to be the case in an elegant series of experiments by De Las Penas et al. (2003). Transposon mutagenesis was used to place the *C. glabrata URA3* gene into intergenic regions of the *EPA* clusters at various distances from the telomere. Gene silencing would result in a strain capable of growth in the presence of 5-fluro-orotic acid. Wild-type cell strains containing the *URA3* gene up to about 24 kb from the telomeres could still grow in the presence of FOA, demonstrating active gene silencing. In *C. glabrata sir3* and *rap1-21* cells, this growth was completely ablated, suggesting that gene silencing was not active (De Las Penas et al. 2003). What role does gene silencing have in virulence? The deletion of *EPA1-5* and *HYR1* resulted in a modest decline in *C. glabrata* virulence as measured by organ burden (De Las Penas et al. 2003). However, an analysis of the *C. glabrata* genome sequence (cbi.labri.fr/Genolevures/elt/CAGL) by the Cormack group suggests the existence of at least 11 other Epa-like proteins, of which six are subtelomeric (De Las Penas et al. 2003). These could compensate for loss of Epa1-5 and Hyr1. No reports of the virulence of *sir3* and *rap1-21* mutants have been described. However, it is intriguing that the usual synteny between *C. glabrata* and *S. cerevisiae* breaks down in sub-telomeric regions (see L.J. Montcalm and K.H. Wolfe, Chap. 2, this volume).

C. *C. glabrata* Ste12

Ste12 belongs to a large family of fungal transcription factors, regulating processes involved in mating, filamentation, substrate invasion, cell wall integrity and virulence (Liu et al. 1994; Singh et al. 1994; Lo et al. 1997; Wickes et al. 1997; Vallim et al. 2000; Young et al. 2000; Chang et al. 2000, 2001; Borneman et al. 2001). There are two major subclasses within the Ste12 family of proteins: those with both homeodomain and C_2H_2 zinc finger DNA-binding domains such as *Penicillium marneffei* SteA, and those which only have the homeodomain like, *C. albicans* Cph1. *C. glabrata* Ste12 does not have zinc fingers but, unusually for the homeodomain group of Ste12 proteins, it lacks an apparent consensus binding motif for the regulatory proteins Dig1 and Dig2 (Pi et al. 1997).

In haploid *S. cerevisiae* strains, Ste12 regulates the response to mating pheromone and the invasive growth phenotype, whereas in diploids it plays a role in filamentous growth (Gustin et al. 1998; Roberts et al. 2000; Gancedo 2001). The *C. albicans* homologue Cph1 also plays a role in filamentous growth and, in combination with another transcription factor Efg1, a crucial role in virulence (Lo et al. 1997). In *C. glabrata*, Ste12 plays an essential role in filamentation and also regulates the expression of genes encoding proteins with functions in cell wall maintenance and biosynthesis, including two structural components, Cis3 and Tip1 (Calcagno et al. 2003). Furthermore, the inactivation of *C. glabrata* Ste12 results in the partial attenuation of virulence (Calcagno et al. 2003). Median survival times for mice infected with *C. glabrata ste12* cells is 5 days, compared with 3 days for animals infected with wild-type cells ($p < 0.01$), suggesting that Ste12-regulated processes play some role in mediating virulence. How can these processes be identified? The genomic sequence of *C. glabrata* has recently been reported and, although whole-genome oligonucleotide arrays have been designed, no transcript profiling experiments have appeared in the literature so far.

C. glabrata Ste12 can complement both the mating and filamentation defects of haploid or diploid *S. cerevisiae ste12* mutants respectively. This demonstrates that *C. glabrata* Ste12 can activate transcription from native *S. cerevisiae* Ste12 responsive promoters. This suggests that the identification of all *S. cerevisiae* promoters which bind Ste12 would be useful in an analysis of *C. glabrata* Ste12 regulation. Genome-Wide Location Analysis (GLA) combines chromatin immunoprecipitation and DNA microarray technology to identify protein–DNA interactions on a genome-wide scale (Ren et al. 2000). Briefly, cells are fixed with formaldehyde and disrupted by sonication to yield DNA fragments of 500–1000 bp. Transcription factor-bound DNA sequences are enriched by immunoprecipitation, the cross-links reversed, and the enriched DNA labelled with Cy-5 using ligation-mediated PCR. Un-enriched DNA is labelled in the same way with Cy-3. The extracts are then combined and used to probe an intergenic array (Ren et al. 2000). Initially, a total of 29 pheromone-induced genes which were directly regulated by Ste12 were reported (Ren et al. 2000). Subsequently, 55 intergenic regions were shown to be bound by Ste12 when cells were cultured in YPD (Lee et al. 2002). Most recently, Ste12-GLA was performed with cells cultured in one of three conditions: YPD; YPD with 5 mg/ml α-factor for 30 min to induce mating; and YPD plus 1% (v/v) butanol to encourage filamentation (Harbison et al. 2004). This revealed binding to 87 intergenic regions, potentially regulating the expression of 107 protein-encoding genes. In total, 143 genes which are potentially regulated by Ste12 have been identified by GLA. Of these, five are dubious ORFs and a further four are hypothetical ORFs. The remaining 134 genes (Table 11.2) encode functions mapped by GO Mapper (db.yeastgenome.org/cgi-bin/GO/goTermMapper) to morphogenesis, cytoskeletal organisation and biogenesis, cell wall organisation and biogenesis, signal transduction, response to stress and cytokinesis. All of these processes have been implicated in *Candida* virulence (Navarro-Garcia et al. 2001).

Remarkably, only the promoters of seven *S. cerevisiae* genes bound Ste12 in all three GLA experiments: *CIK1*, *ERG24*, *FUS1*, *HYM1*, *PCL2*, *PRM1* and *STE12*. Cik1 has microtubule motor activity and is important for the establishment of the spindle and chromosome segregation (Manning et al. 1999). Erg24 is a C-14 sterol reductase which acts in ergosterol biosynthesis (Lorenz and Parks 1992). The function of Fus1 is unknown, but it is located at the shmoo tip in the membrane, and plays an integral role in cell fusion, perhaps as a substrate for Cdc28 (McCaffrey et al. 1987; Bagnat and Simons 2002). Hym1 is a transcriptional repressor which localizes to sites of polarized growth during budding, and acts to mediate cell separation and regulation of cell shape, as a component of the RAM network (Nelson et al. 2003). The G1 cyclin Pcl2 forms a complex with the cyclin-dependent kinase Pho85, which localizes to sites of polarized growth, regulates entry into the mitotic cell cycle, and is essential for morphogenesis (Wang et al. 2001). The membrane protein Prm1 localizes to the shmoo tip and plays a role in cell fusion (Heiman and Walter 2000). Ste12 also binds to its own promoter. Homologues of all seven genes are present in the *C. glabrata* genome. The disruption of these genes and subsequent phenotypic analysis of the *C. glabrata* mutants will allow their role in virulence to be determined.

D. *C. glabrata* Ace2

In *S. cerevisiae*, the zinc finger transcription factor Ace2 activates expression of early G1-specific genes (McBride et al. 1999; Simon et al. 2001). Of a total of 31 Ace2 cell cycle targets defined by Ace2-GLA, 17 also bound the paralogous transcription factor Swi5 (Simon et al. 2001). Subsequent GLA analysis demonstrated that promoter elements binding both Ace2 and Swi5 occur statistically more often ($p = 0.005$) than expected by chance (Harbison et al. 2004). The pair of genes encoding these transcription factors arose as a result of an ancient genome duplication event (Wolfe and Shields 1997). *C. glabrata* has both *ACE2* and *SWI5* homologues, but *C. albicans* contains only a single *ACE2* gene (Kamran et al. 2004; Kelly et al. 2004).

The inactivation of Ace2 in *S. cerevisiae* results in a severe cell separation phenotype, the basis for which is partially understood. Ace2 is localized to the daughter cell nucleus after cytokinesis, as a result of Cbk1-Mob2-dependent phosphorylation, where it regulates the expression of at least eight daughter cell-specific genes including *AMN1*, *CTS1*, *DSE1-4*, *PRY3* and *SCW11* (Colman-Lerner et al. 2001; Weiss et al. 2002). The chitinase Cts1 and the three glucan 1,3-β-glucosidases Dse2, Dse4 and Scw11 are required to break down the trilaminar septum which links the mother and daughter cells (Kuranda and Robbins 1991; Cappellaro et al. 1998). Amn1 binds directly to the GTPase Tem1, preventing it from interacting with its target kinase Cdc15 and blocking activation of the mitotic exit network (Wang et al. 2003), a process which is co-ordinately regulated with cell separation (Yeong et al. 2002). Dse1, Dse3 and Pry3 are cell wall proteins of unknown function.

Thus, it appears that one of the primary roles of Ace2 is to regulate cell separation. However, Ace2 also appears to be involved in other processes. Evidence for this hypothesis comes from a number of observations. Only 31 of the 119 genes, the promoters of which bind Ace2, are cell cycle-regulated. Others include genes which encode proteins involved in response to oxidative stress (Yap1, Snq2 and Mcr1) and cell wall organisation and biogenesis (Dse1, Exg1, Hsp150, Psa1, Rot2 and Wsc4). Furthermore, recent proteomics work in *C. glabrata* has demonstrated a role for Ace2 in the regulation of metabolism, protein synthesis, folding and targeting, and aspects of cell growth and polarization (Stead et al. 2005).

Table. 11.2. *Saccharomyces cerevisiae* promoters identified by genome location analysis which bind Ste12

ORF	Gene	Biological process	Molecular function
YFL039C	*ACT1*	Cell wall organisation and biogenesis[a]	Structural constituent of cytoskeleton[a]
YAR015W	*ADE1*	Purine nucleotide biosynthesis[a]	Phosphoribosylaminoimidazole-succinocarboxamide synthase activity
YDR085C	*AFR1*	Signal transduction during conjugation with cellular fusion[a]	Receptor signalling protein activity
YNR044W	*AGA1*	Agglutination during conjugation with cellular fusion	Cell adhesion molecule binding
YGL032C	*AGA2*	Agglutination during conjugation with cellular fusion	Cell adhesion molecule binding
YDL192W	*ARF1*	ER to Golgi transport[a]	GTPase activity
YPR122W	*AXL1*	Bud site selection[a]	Metalloendopeptidase activity
YIL015W	*BAR1*	Protein catabolism	Aspartic-type endopeptidase activity
YER155C	*BEM2*	Cell wall organisation and biogenesis[a]	Signal transducer activity[a]
YPR171W	*BSP1*	Actin cytoskeleton organisation and biogenesis[a]	Molecular function unknown
YAR014C	*BUD14*	Cellular morphogenesis during vegetative growth	Molecular function unknown
YLR353W	*BUD8*	Pseudohyphal growth[a]	Molecular function unknown
YPL048W	*CAM1*	Regulation of translational elongation	Translation elongation factor activity
YAL038W	*CDC19*	Glycolysis[a]	Pyruvate kinase activity
YNL192W	*CHS1*	Budding	Chitin synthase activity
YBR023C	*CHS3*	Response to osmotic stress[a]	Chitin synthase activity
YMR198W	*CIK1*	Meiosis[a]	Microtubule motor activity
YNL298W	*CLA4*	Protein amino acid phosphorylation[a]	Protein serine/threonine kinase activity
YPR119W	*CLB2*	G2/M transition of mitotic cell cycle[a]	Cyclin-dependent protein kinase regulator activity
YMR199W	*CLN1*	Regulation of cyclin-dependent protein kinase activity	Cyclin-dependent protein kinase regulator activity
YKL179C	*COY1*	Golgi vesicle transport	Molecular function unknown
YLR216C	*CPR6*	Protein folding	Unfolded protein binding[a]
YGR189C	*CRH1*	Biological process unknown	Molecular function unknown
YGR218W	*CRM1*	mRNA-nucleus export[a]	Protein carrier activity
YDR179C	*CSN9*	Adaptation to pheromone during conjugation with cellular fusion[a]	Molecular function unknown
YBR158W	*CST13*	Negative regulation of exit from mitosis[a]	Protein binding
YIL036W	*CST6*	Transcription initiation from Pol II promoter[a]	Specific RNA polymerase II transcription factor activity
YPL177C	*CUP9*	Transcription initiation from Pol II promoter[a]	Specific RNA polymerase II transcription factor activity
YOL082W	*CVT19*	Protein-vacuolar targeting	Protein binding
YCR017C	*CWH43*	Cell wall organisation and biogenesis[a]	Molecular function unknown
YKL096W	*CWP1*	Cell wall organisation and biogenesis	Structural constituent of cell wall
YAL039C	*CYC3*	Cytochrome c-heme linkage	Holocytochrome-c synthase activity
YMR135C	*DCR1*	Negative regulation of gluconeogenesis	Molecular function unknown
YPL049C	*DIG1*	Invasive growth (sensu *Saccharomyces*)	Transcription factor binding
YOR092W	*ECM3*	Cell wall organisation and biogenesis	ATPase activity
YNL280C	*ERG24*	Ergosterol biosynthesis	C-14 sterol reductase activity
YJL157C	*FAR1*	Signal transduction during conjugation with cellular fusion[a]	Cyclin-dependent protein kinase inhibitor activity
YBR040W	*FIG1*	Cellular morphogenesis during conjugation with cellular fusion[a]	Molecular function unknown
YCR089W	*FIG2*	Cellular morphogenesis during conjugation with cellular fusion[a]	Molecular function unknown
YMR306W	*FKS3*	Biological process unknown	1,3-beta-glucan synthase activity
YCL026C-A	*FRM2*	Negative regulation of fatty acid metabolism	Molecular function unknown
YCL027W	*FUS1*	Conjugation with cellular fusion	Molecular function unknown
YMR232W	*FUS2*	Plasma membrane fusion	Molecular function unknown
YBL016W	*FUS3*	Protein amino acid phosphorylation[a]	MAP kinase activity
YOL030W	*GAS5*	Biological process unknown	1,3-beta-glucanosyltransferase activity
YMR136W	*GAT2*	Transcription	Transcription factor activity

[a] Annotated to more than one biological process or molecular function in SGD, only the first is shown

Table. 11.2. (continued)

ORF	Gene	Biological process	Molecular function
YKL104C	*GFA1*	Cell wall chitin biosynthesis	Glutamine-fructose-6-phosphate transaminase (isomerizing) activity
YDR309C	*GIC2*	Establishment of cell polarity (sensu Fungi)[a]	Small GTPase regulatory/interacting protein activity
YHR005C	*GPA1*	Signal transduction during conjugation with cellular fusion	GTPase activity
YFL027C	*GYP8*	Vesicle-mediated transport	Rab GTPase activator activity
YPR005C	*HAL1*	Positive regulation of transcription from Pol II promoter[a]	Molecular function unknown
YKL189W	*HYM1*	Cytokinesis, completion of separation[a]	Transcriptional repressor activity
YBR157C	*ICS2*	Biological process unknown	Molecular function unknown
YNL106C	*INP52*	Cell wall organisation and biogenesis[a]	Inositol-polyphosphate 5-phosphatase activity
YER019W	*ISC1*	Response to salt stress[a]	Phospholipase C activity
YCL055W	*KAR4*	Meiosis[a]	Transcription regulator activity
YMR065W	*KAR5*	Karyogamy during conjugation with cellular fusion	Molecular function unknown
YHR082C	*KSP1*	Protein amino acid phosphorylation	Protein serine/threonine kinase activity
YKR061W	*KTR2*	N-linked glycosylation[a]	Mannosyltransferase activity
YCR019W	*MAK32*	Host–pathogen interaction	Molecular function unknown
YNL145W	*MFA2*	Signal transduction during conjugation with cellular fusion	Pheromone activity
YLR332W	*MID2*	Cell wall organisation and biogenesis[a]	Transmembrane receptor activity
YJL186W	*MNN5*	Protein amino acid glycosylation	Alpha-1,2-mannosyltransferase activity
YGR014W	*MSB2*	Establishment of cell polarity (sensu Fungi)[a]	Osmosensor activity
YNL053W	*MSG5*	Protein amino acid dephosphorylation[a]	Prenylated protein tyrosine phosphatase activity
YHR086W	*NAM8*	Nuclear mRNA splicing, via spliceosome[a]	RNA binding[a]
YPR155C	*NCA2*	Aerobic respiration[a]	Molecular function unknown
YGL067W	*NPY1*	NADH metabolism	NAD+diphosphatase activity
YBR066C	*NRG2*	Invasive growth (sensu *Saccharomyces*)	Transcriptional repressor activity
YBR060C	*ORC2*	DNA replication initiation[a]	DNA replication origin binding
YDL127W	*PCL2*	Cell cycle	Cyclin-dependent protein kinase regulator activity
YBL017C	*PEP1*	Protein-vacuolar targeting[a]	Signal sequence binding
YKL127W	*PGM1*	Glucose 1-phosphate utilization[a]	Phosphoglucomutase activity
YGR233C	*PHO81*	Phosphate metabolism	Cyclin-dependent protein kinase inhibitor activity
YNL279W	*PRM1*	Plasma membrane fusion	Molecular function unknown
YIL037C	*PRM2*	Conjugation with cellular fusion	Molecular function unknown
YPL192C	*PRM3*	Karyogamy	Molecular function unknown
YPL156C	*PRM4*	Conjugation with cellular fusion	Molecular function unknown
YML047C	*PRM6*	Conjugation with cellular fusion	Molecular function unknown
YML046W	*PRP39*	Nuclear mRNA splicing, via spliceosome	RNA binding
YGL062W	*PYC1*	Gluconeogenesis[a]	Pyruvate carboxylase activity
YDL135C	*RDI1*	Actin filament organisation[a]	Signal transducer activity[a]
YPR165W	*RHO1*	Cell wall organisation and biogenesis[a]	GTPase activity[a]
YNL180C	*RHO5*	Rho protein signal transduction	GTPase activity
YNL178W	*RPS3*	Protein biosynthesis[a]	Structural constituent of ribosome
YHR083W	*SAM35*	Mitochondrial outer membrane protein import	Protein binding
YHR205W	*SCH9*	Protein amino acid phosphorylation[a]	Protein serine/threonine kinase activity
YBR024W	*SCO2*	Copper ion transport	Molecular function unknown
YMR305C	*SCW10*	Conjugation with cellular fusion	Glucosidase activity
YGL028C	*SCW11*	Cytokinesis, completion of separation	Glucan 1,3-beta-glucosidase activity
YNL272C	*SEC2*	Exocytosis	Guanyl-nucleotide exchange factor activity
YLR105C	*SEN2*	tRNA splicing	tRNA-intron endonuclease activity
YPR198W	*SGE1*	Response to drug[a]	Xenobiotic-transporting ATPase activity

[a] Annotated to more than one biological process or molecular function in SGD, only the first is shown

Table. 11.2. (continued)

ORF	Gene	Biological process	Molecular function
YOL031C	*SIL1*	SRP-dependent cotranslational membrane targeting, translocation	Molecular function unknown
YIL123W	*SIM1*	Microtubule cytoskeleton organisation and biogenesis	Molecular function unknown
YBL061C	*SKT5*	Response to osmotic stress[a]	Enzyme activator activity
YBR156C	*SLI15*	Protein amino acid phosphorylation[a]	Protein kinase activator activity
YDR240C	*SNU56*	Nuclear mRNA splicing, via spliceosome	mRNA binding
YGR013W	*SNU71*	Nuclear mRNA splicing, via spliceosome	RNA binding
YER018C	*SPC25*	Chromosome segregation[a]	Structural constituent of cytoskeleton
YHR152W	*SPO12*	Regulation of exit from mitosis[a]	Molecular function unknown
YOR247W	*SRL1*	Nucleobase, nucleoside, nucleotide and nucleic acid metabolism	Molecular function unknown
YKR091W	*SRL3*	Nucleobase, nucleoside, nucleotide and nucleic acid metabolism	Molecular function unknown
YDR312W	*SSF2*	Ribosomal large subunit assembly and maintenance[a]	rRNA binding
YDR086C	*SSS1*	Cotranslational membrane targeting	Protein transporter activity
YLR452C	*SST2*	Signal transduction[a]	GTPase activator activity
YHR084W	*STE12*	Pseudohyphal growth[a]	Transcription factor activity
YFL026W	*STE2*	Response to pheromone during conjugation with cellular fusion[a]	Mating-type alpha-factor pheromone receptor activity
YKL209C	*STE6*	Peptide pheromone export	ATPase activity, coupled to transmembrane movement of substances
YDR310C	*SUM1*	Chromatin silencing at telomere[a]	Transcriptional repressor activity
YPL163C	*SVS1*	Response to chemical substance	Molecular function unknown
YJL187C	*SWE1*	G2/M transition of mitotic cell cycle[a]	Protein kinase activity
YNL087W	*TCB2*	Biological process unknown	Molecular function unknown
YBR083W	*TEC1*	Positive regulation of transcription from Pol II promoter[a]	Specific RNA polymerase II transcription factor activity
YDR311W	*TFB1*	Transcription initiation from Pol II promoter[a]	General RNA polymerase II transcription factor activity
YBR162C	*TOS1*	Biological process unknown	Molecular function unknown
YBR082C	*UBC4*	Response to stress[a]	Ubiquitin-conjugating enzyme activity
YKR042W	*UTH1*	Mitochondrion organisation and biogenesis	Molecular function unknown
YMR197C	*VTI1*	Intra-Golgi transport[a]	v-SNARE activity
YKL095W	*YJU2*	Nuclear mRNA splicing, via spliceosome	Molecular function unknown
YBL060W		Biological process unknown	Molecular function unknown
YCL056C		Biological process unknown	Molecular function unknown
YGR038C-A		Ty element transposition	RNA binding[a]
YGR038C-B		Ty element transposition	DNA-directed DNA polymerase activity
YIL083C		Coenzyme A biosynthesis	Phosphopantothenate-cysteine ligase activity
YJR054W		Biological process unknown	Molecular function unknown
YLR412C-A		Biological process unknown	Molecular function unknown
YLR413W		Biological process unknown	Molecular function unknown
YNL024C-A		Biological process unknown	Molecular function unknown
YOL155C		Cell wall organisation and biogenesis	Glucosidase activity
YOR129C		Response to drug[a]	Structural constituent of cytoskeleton
YOR246C		Biological process unknown	Oxidoreductase activity
YOR343W-A		Ty element transposition	RNA binding[a]
YOR343w-B		Ty element transposition	DNA-directed DNA polymerase activity

[a] Annotated to more than one biological process or molecular function in SGD, only the first is shown

Inactivation of the single Ace2/Swi5-encoding gene in *C. albicans* results in defective cell separation accompanied by pseudohyhal growth, even in the absence of specific inducers. In addition, *C. albicans ace2/ace2* cells have a reduced ability to adhere to plastic surfaces, cannot form confluent biofilms, and are less able to invade solid agar (Kelly et al. 2004). The same authors showed that *C. albicans* Ace2 is localized to the daughter cell nucleus in yeast phase cells and regulates the expression of *SCW11*, *DSE1* and one of the *C. albicans* chitinase-encoding genes *CHT3*. These data suggest that the function of *C. albicans* Ace2 resembles that of *S. cerevisiae* Ace2. *C. glabrata* Ace2 also regulates the expression of *CTS1*, *DSE1* and *SCW11* (unpublished data) and has a strong cell separation phenotype (Kamran et al. 2004). In addition, *C. glabrata ACE2* can functionally complement the cell separation defect of *S. cerevisiae ace2* cells (unpublished data). These data strongly suggest that the *C. glabrata ACE2* gene also encodes a functional homologue of *S. cerevisiae* Ace2.

The role of the functionally related Ace2 proteins in fungal virulence has been explored in both *C. albicans* and *C. glabrata* (Kamran et al. 2004; Kelly et al. 2004). *C. albicans ace2/ace2* mutants are completely attenuated for virulence. Tail vein inoculation of equivalent doses of wild-type or reconstituted *ace2/ACE2* cells into DBA/2 mice results in 100% mortality within 5 days. This is accompanied by significant tissue burden. Conversely, when *C. albicans ace2/ace2* cells are inoculated, all mice survive to the end of the experiment (28 days) and only about 40% have *C. albicans* recoverable from kidneys (Kelly et al. 2004).

In stark and unexpected contrast, inactivation of *ACE2* in *C. glabrata* results in increased virulence. *C. glabrata ace2* cells are about 200-fold more virulent than is the wild-type strain (Kamran et al. 2004). Infection is characterized by fungal escape from the vasculature, tissue penetration, proliferation in vivo and massive over-stimulation of the pro-inflammatory arm of the innate immune system. Systemic levels of IL-6 and TNF-α, 18 h post-infection, were 16- and 38-fold higher respectively in *ace2*-infected mice compared to mice infected with an *ACE2*-reconstituted strain (Kamran et al. 2004). The obvious question arises – why does inactivation of proteins with apparently similar functions result in such different virulence phenotypes? Unfortunately, the answer still remains elusive. However, *C. albicans* Ace2 regulates the expression of only one of three chitinase genes in this species whereas *C. glabrata* Ace2 regulates expression of its sole chitinase-encoding gene (Kamran et al. 2004; Kelly et al. 2004). It is possible that significant differences in the regulatory domains of Ace2 in the two species contribute to the observed virulence differences. Comparison of the Ace2 regulons in the two species has not yet been done, but a recent proteomics investigation identified reproducible and statistically significant alterations in the levels of 61 proteins in wild-type and *C. glabrata ace2* cells. *C. glabrata* Ace2 regulates proteins involved in protein synthesis/turnover (Cdc33, Cof1, Efb1, Mnp1, Rpn10, Rps12, Sec28, Tif34, Ykl056c), transport (Dss4, Nce103) and interactions with the cellular environment (Atp2, Atp16). Notably, homologues of a number of proteins which are down-regulated in *C. glabrata ace2* cells play roles in *S. cerevisiae* stress responses. Fes1 encodes a Hsp70 nucleotide exchange factor, and the protein kinase inhibitor Lsp1 acts along with Pil1 to down-regulate heat stress resistance (Zhang et al. 2004) whereas Ahp1 and Tsa1 have thioredoxin peroxidase activities.

Can these observations be related to the virulence phenotype of *C. glabrata ace2* cells? In *S. cerevisiae* the RAM network regulates Ace2 function (Nelson et al. 2003). At least two RAM components, Ssd1 and Kic1, have been annotated to cell wall organisation and biogenesis. *S. cerevisiae ssd1* mutants are more virulent than are wild-type fungi in DBA2 mice, and can induce over-stimulation of pro-inflammatory cytokines from macrophages (Wheeler et al. 2003). The striking similarity in the virulence phenotypes of *S. cerevisiae ssd1* and *C. glabrata ace2* cells suggests that they might share a common basis. Interestingly, *S. cerevisiae ssd1* cells have significant alterations in cell wall composition. The question therefore arises – do alterations in the *C. glabrata* proteome suggest a role for Ace2 in cell wall organisation and biogenesis? It is known that some proteins which were down-regulated in *ace2* cells (Stead et al. 2005) have roles in cell growth, polarity and morphogenesis, and cell wall remodelling (Act1, Ahp1, Cdc33, Cof1, Dss4, Fsa1, Sec28, Tub2). By analogy with *S. cerevisiae*, altered levels of many of these proteins would result in significant changes in the *C. glabrata* cell wall and secretome (Roberts et al. 1997). Ace2 also regulates the expression of the endochitinase-encoding gene *CTS1* in *C. glabrata*, which plays a major role in cell wall remodelling at the site of cell separation. Additionally, proteins identified in this study (YJL123c) have been localized to COPI-coated vesicles. Other Ace2-regulated proteins are involved in signalling (e.g. Bmh1). We have previously demonstrated that inactivation of transcription factors activated by well-defined signalling pathways can result in alterations to *C. glabrata* cell wall integrity (Calcagno et al. 2003). The 14-3-3 proteins Bmh1 and Bmh2 are essential for pseudohyphal production in *S. cerevisiae*, a morphological event which has major effects on cell wall structure (Roberts et al. 1997). Furthermore, the central carbon metabolic functions regulated by Ace2 are involved in the provision of sugars essential for the biosynthesis of cell wall carbohydrates and the post-translational modification of cell surface proteins.

Is there any other evidence to support the view that Ace2 regulates cell wall biogenesis? Systems analysis of *S. cerevisiae* Ace2 functional connections using the STRING tool (string.embl.de) suggests important links to a number of genes encoding functions which significantly affect cell wall biology. These include ubiquitin-dependent protein turnover (e.g. Ubp15, Rsp5), glycoprotein degradation (Mnl1), maintenance of the cell integrity pathway (Bck1), and transcriptional regulation (Fkh1, Fkh2). Taken together, these observations begin to suggest a mechanistic explanation for the increased virulence of *C. glabrata ace2* mutants. Analysis of the *C. albicans* Ace2 regulon and proteome should reveal differences which may explain the variation in virulence seen in the two species as a result of inactivation of Ace2 (Kamran et al. 2004; Kelly et al. 2004).

IV. Conclusion

The analysis of fungal transcriptional regulation has revealed many insights into the pathobiology of *C. albicans* and is beginning to shed light on disease causation by *C. glabrata*. This analysis is essential to understand the host–*Candida* relationship. This relationship can be meaningfully assessed also by using information available for other species, especially *S. cerevisiae*, and no doubt many attributes will be shared. However, the observation that *C. albicans* and *C. glabrata ace2* mutants display dramatically different virulence phenotypes suggests, not surprisingly, that species-specific processes are also at work in maintaining host–pathogen interactions.

Acknowledgements. I would like to thank all members of the Imperial College Molecular Mycology Group, in particular Elaine Bignell and Helen Findon, and Herb Arst for helpful discussions. This work was supported by grants from the BBSRC (BBS/B/10331) and the Wellcome Trust (072420).

References

Alarco AM, Raymond M (1999) The bZip transcription factor Cap1p is involved in multidrug resistance and oxidative stress response in *Candida albicans*. J Bacteriol 181:700–708

Aparicio OM, Billington BL, Gottschling DE (1991) Modifiers of position effect are shared between telomeric and silent mating-type loci in *S. cerevisiae*. Cell 66:1279–1287

Asmundsdottir LR, Erlendsdottir H, Gottfredsson M (2002) Increasing incidence of candidemia: results from a 20-year nationwide study in Iceland. J Clin Microbiol 40:3489–3492

Bagnat M, Simons K (2002) Cell surface polarization during yeast mating. Proc Natl Acad Sci USA 99:14183–14188

Beck-Sague C, Jarvis WR (1993) Secular trends in the epidemiology of nosocomial fungal infections in the United States, 1980–1990. National Nosocomial Infections Surveillance System. J Infect Dis 167:1247–1251

Benni ML, Neigeborn L (1997) Identification of a new class of negative regulators affecting sporulation-specific gene expression in yeast. Genetics 147:1351–1366

Bensen ES, Filler SG, Berman J (2002) A forkhead transcription factor is important for true hyphal as well as yeast morphogenesis in *Candida albicans*. Eukaryot Cell 1:787–798

Bignell E, Negrete-Urtasun S, Calcagno AM, Haynes K, Arst HN Jr, Rogers T (2005) The *Aspergillus* pH-responsive transcription factor PacC regulates virulence. Mol Microbiol 55(4):1072–1084

Biswas K, Rieger KJ, Morschhauser J (2003) Functional analysis of *CaRAP1*, encoding the Repressor/activator protein 1 of *Candida albicans*. Gene 307:151–158

Borneman AR, Hynes MJ, Andrianopoulos A (2001) An STE12 homolog from the asexual, dimorphic fungus *Penicillium marneffei* complements the defect in sexual development of an *Aspergillus nidulans steA* mutant. Genetics 157:1003–1014

Brand A, MacCallum DM, Brown AJ, Gow NA, Odds FC (2004) Ectopic expression of *URA3* can influence the virulence phenotypes and proteome of *Candida albicans* but can be overcome by targeted reintegration of *URA3* at the *RPS10* locus. Eukaryot Cell 3:900–909

Braun BR, Johnson AD (1997) Control of filament formation in *Candida albicans* by the transcriptional repressor *TUP1*. Science 277:105–109

Brockert PJ, Lachke SA, Srikantha T, Pujol C, Galask R, Soll DR (2003) Phenotypic switching and mating type switching of *Candida glabrata* at sites of colonization. Infect Immun 71:7109–7118

Calcagno A, Bignell E, Warn P, Jones MD, Denning DW, Mühlschlegel FA, Rogers TR, Haynes K (2003) *Candida glabrata STE12* is required for wild-type virulence and nitrogen starvation induced filamentation. Mol Microbiol 50:1309–1318

Calderone RA (2002) *Candida* and Candidiasis. ASM Press, Washington, DC

Cappellaro C, Mrsa V, Tanner W (1998) New potential cell wall glucanases of *Saccharomyces cerevisiae* and their involvement in mating. J Bacteriol 180:5030–5037

Caracuel Z, Roncero MI, Espeso EA, Gonzalez-Verdejo CI, Garcia-Maceira FI, Di Pietro A (2003) The pH signalling transcription factor PacC controls virulence in the plant pathogen *Fusarium oxysporum*. Mol Microbiol 48:765–779

Chang YC, Wickes BL, Miller GF, Penoyer LA, Kwon-Chung KJ (2000) *Cryptococcus neoformans STE12alpha* regulates virulence but is not essential for mating. J Exp Med 191:871–882

Chang YC, Penoyer LA, Kwon-Chung KJ (2001) The second *STE12* homologue of *Cryptococcus neoformans* is *MAT***a**-specific and plays an important role in virulence. Proc Natl Acad Sci USA 98:3258–3263

Chen YC, Chang SC, Sun CC, Yang LS, Hsieh WC, Luh KT (1997) Secular trends in the epidemiology of nosocomial fungal infections at a teaching hospital in Taiwan, 1981 to 1993. Infect Control Hosp Epidemiol 18:369–375

Colman-Lerner A, Chin TE, Brent R (2001) Yeast Cbk1 and Mob2 activate daughter-specific genetic programs to induce asymmetric cell fates. Cell 107:739–750

Cormack BP, Falkow S (1999) Efficient homologous and illegitimate recombination in the opportunistic yeast pathogen *Candida glabrata*. Genetics 151:979–987

Cormack BP, Ghori N, Falkow S (1999) An adhesin of the yeast pathogen *Candida glabrata* mediating adherence to human epithelial cells. Science 285:578–582

Davis D, Edwards JE, Mitchell AP, Ibrahim AS (2000) *Candida albicans* RIM101 pH response pathway is required for host-pathogen interactions. Infect Immun 68:5953–5959

Davis DA, Bruno VM, Loza L, Filler SG, Mitchell AP (2002) *Candida albicans* Mds3p, a conserved regulator of pH responses and virulence identified through insertional mutagenesis. Genetics 162:1573–1581

De Bernardis F, Muhlschlegel FA, Cassone A, Fonzi WA (1998) The pH of the host niche controls gene expression in and virulence of *Candida albicans*. Infect Immun 66:3317–3325

De Las Penas A, Pan SJ, Castano I, Alder J, Cregg R, Cormack BP (2003) Virulence-related surface glycoproteins in the yeast pathogen *Candida glabrata* are encoded in subtelomeric clusters and subject to *RAP1*- and *SIR*-dependent transcriptional silencing. Genes Dev 17:2245–2258

Drori N, Kramer-Haimovich H, Rollins J, Dinoor A, Okon Y, Pines O, Prusky D (2003) External pH and nitrogen source affect secretion of pectate lyase by *Colletotrichum gloeosporioides*. Appl Environ Microbiol 69:3258–3262

Eck R, Stoyan, T, Kunkel W (2001) The centromere-binding factor Cbf1p from *Candida albicans* complements the methionine auxotrophic phenotype of *Saccharomyces cerevisiae*. Yeast 18:1047–1052

Frieman MB, McCaffery JM, Cormack BP (2002) Modular domain structure in the *Candida glabrata* adhesin Epa1p, a beta1,6 glucan-cross-linked cell wall protein. Mol Microbiol 46:479–492

Gale CA, Bendel CM, McClellan M, Hauser M, Becker JM, Berman J, Hostetter MK (1998) Linkage of adhesion, filamentous growth, and virulence in *Candida albicans* to a single gene, *INT1*. Science 279:1355–1358

Gancedo JM (2001) Control of pseudohyphae formation in *Saccharomyces cerevisiae*. FEMS Microbiol Rev 25:107–123

Gudlaugsson O, Gillespie S, Lee K, Vande Berg J, Hu J, Messer S, Herwaldt L, Pfaller M, Diekema D (2003) Attributable mortality of nosocomial candidemia, revisited. Clin Infect Dis 37:1172–1177

Gustin MC, Albertyn J, Alexander M, Davenport K (1998) MAP kinase pathways in the yeast *Saccharomyces cerevisiae*. Microbiol Mol Biol Rev 62:1264–1300

Harbison CT, Gordon DB, Lee TI, Rinaldi NJ, Macisaac KD, Danford TW, Hannett NM, Tagne JB, Reynolds DB, Yoo J et al. (2004) Transcriptional regulatory code of a eukaryotic genome. Nature 431:99–104

Haw R, Yarragudi AD, Uemura H (2001) Isolation of a *Candida glabrata* homologue of *RAP1*, a regulator of transcription and telomere function in *Saccharomyces cerevisiae*. Yeast 18:1277–1284

Heiman MG, Walter P (2000) Prm1p, a pheromone-regulated multispanning membrane protein, facilitates plasma membrane fusion during yeast mating. J Cell Biol 151:719–730

Hensel M, Shea JE, Gleeson C, Jones MD, Dalton E, Holden DW (1995) Simultaneous identification of bacterial virulence genes by negative selection. Science 269:400–403

Hoyer LL (2001) The *ALS* gene family of *Candida albicans*. Trends Microbiol 9:176–180

Inglis DO, Johnson AD (2002) Ash1 protein, an asymmetrically localized transcriptional regulator, controls filamentous growth and virulence of *Candida albicans*. Mol Cell Biol 22:8669–8680

Ishii N, Yamamoto M, Yoshihara F, Arisawa M, Aoki Y (1997) Biochemical and genetic characterization of Rbf1p, a putative transcription factor of *Candida albicans*. Microbiology 143:429–435

Jarvis WR (1995) Epidemiology of nosocomial fungal infections, with emphasis on *Candida* species. Clin Infect Dis 20:1526–1530

Kadosh D, Johnson AD (2001) Rfg1, a protein related to the *Saccharomyces cerevisiae* hypoxic regulator Rox1, controls filamentous growth and virulence in *Candida albicans*. Mol Cell Biol 21:2496–2505

Kagi JH, Schaffer A (1988) Biochemistry of metallothionein. Biochemistry 27:8509–8515

Kaiser B, Munder T, Saluz HP, Kunkel W, Eck R (1999) Identification of a gene encoding the pyruvate decarboxylase gene regulator CaPdc2p from *Candida albicans*. Yeast 15:585–591

Kamran M, Calcagno AM, Findon H, Bignell E, Jones MD, Warn P, Hopkins P, Denning DW, Butler G, Rogers T et al. (2004) Inactivation of transcription factor gene *ACE2* in the fungal pathogen *Candida glabrata* results in hypervirulence. Eukaryot Cell 3:546–552

Kelly R, Kwon-Chung KJ (1992) A zinc finger protein from *Candida albicans* is involved in sucrose utilization. J Bacteriol 174:222–232

Kelly MT, MacCallum DM, Clancy SD, Odds FC, Brown AJ, Butler G (2004) The *Candida albicans CaACE2* gene affects morphogenesis, adherence and virulence. Mol Microbiol 53:969–983

Khalaf RA, Zitomer RS (2001) The DNA binding protein Rfg1 is a repressor of filamentation in *Candida albicans*. Genetics 157:1503–1512

Khoo B, Brophy B, Jackson SP (1994) Conserved functional domains of the RNA polymerase III general transcription factor BRF. Genes Dev 8:2879–2890

Koch KA, Thiele DJ (1999) Functional analysis of a homopolymeric (dA-dT) element that provides nucleosomal access to yeast and mammalian transcription factors. J Biol Chem 274:23752–23760

Kuranda MJ, Robbins PW (1991) Chitinase is required for cell separation during growth of *Saccharomyces cerevisiae*. J Biol Chem 266:19758–19767

Lachke SA, Srikantha T, Tsai LK, Daniels K, Soll DR (2000) Phenotypic switching in *Candida glabrata* involves phase-specific regulation of the metallothionein gene MT-II and the newly discovered hemolysin gene HLP. Infect Immun 68:884–895

Lane S, Birse C, Zhou S, Matson R, Liu H (2001) DNA array studies demonstrate convergent regulation of virulence factors by Cph1, Cph2, and Efg1 in *Candida albicans*. J Biol Chem 276:48988–48996

Lee TI, Rinaldi NJ, Robert F, Odom D, Bar-Joseph Z, Gerber GK, Hannett NM, Harbison CT, Thompson CM, Simon I et al. (2002) Transcriptional regulatory networks in *Saccharomyces cerevisiae*. Science 298:799–804

Limjindaporn T, Khalaf RA, Fonzi WA (2003) Nitrogen metabolism and virulence of *Candida albicans* require the GATA-type transcriptional activator encoded by *GAT1*. Mol Microbiol 50:993–1004

Liu H, Kohler J, Fink GR (1994) Suppression of hyphal formation in *Candida albicans* by mutation of a *STE12* homolog. Science 266:1723–1726

Lo HJ, Kohler JR, DiDomenico B, Loebenberg D, Cacciapuoti A, Fink GR (1997) Nonfilamentous *C. albicans* mutants are avirulent. Cell 90:939–949

Lorenz RT, Parks LW (1992) Cloning, sequencing, and disruption of the gene encoding sterol C-14 reductase in *Saccharomyces cerevisiae*. DNA Cell Biol 11:685–692

Manning BD, Barrett JG, Wallace JA, Granok H, Snyder M (1999) Differential regulation of the Kar3p kinesin-related protein by two associated proteins, Cik1p and Vik1p. J Cell Biol 144:1219–1233

McBride HJ, Yu Y, Stillman DJ (1999) Distinct regions of the Swi5 and Ace2 transcription factors are required for specific gene activation. J Biol Chem 274:21029–21036

McCaffrey G, Clay FJ, Kelsay K, Sprague GF Jr (1987) Identification and regulation of a gene required for cell fusion during mating of the yeast *Saccharomyces cerevisiae*. Mol Cell Biol 7:2680–2690

McCreath KJ, Specht CA, Robbins PW (1995) Molecular cloning and characterization of chitinase genes from *Candida albicans*. Proc Natl Acad Sci USA 92:2544–2548

Moreno I, Pedreno Y, Maicas S, Sentandreu R, Herrero E, Valentin E (2003) Characterization of a *Candida albicans* gene encoding a putative transcriptional factor required for cell wall integrity. FEMS Microbiol Lett 226:159–167

Muhlschlegel FA, Fonzi WA (1997) *PHR2* of *Candida albicans* encodes a functional homolog of the pH-regulated gene *PHR1* with an inverted pattern of pH-dependent expression. Mol Cell Biol 17:5960–5967

Murad AM, Leng P, Wishart JA, Straffon M, Schnell NF, Talibi D, Marechal D, d'Enfert C, Gaillardin C, Brown AJ (2001) *NRG1* represses yeast-hypha morphogenesis and hypha-specific gene expression in *Candida albicans*. EMBO J 20:4742–4752

Nantel A, Dignard D, Bachewich C, Harcus D, Marcil A, Bouin AP, Sensen CW, Hogues H, Van Het Hoog M, Gordon P et al. (2002) Transcription profiling of *Candida albicans* cells undergoing the yeast-to-hyphal transition. Mol Biol Cell 13:3452–3465

Navarro-Garcia F, Sanchez M, Nombela C, Pla J (2001) Virulence genes in the pathogenic yeast *Candida albicans*. FEMS Microbiol Rev 25:245–268

Nelson B, Kurischko C, Horecka J, Mody M, Nair P, Pratt L, Zougman A, McBroom LD, Hughes TR, Boone C et al. (2003) RAM: a conserved signaling network that regulates Ace2p transcriptional activity and polarized morphogenesis. Mol Biol Cell 14:3782–3803

Olaechea PM, Palomar M, Leon-Gil C, Alvarez-Lerma F, Jorda R, Nolla-Salas J, Leon-Regidor MA (2004) Economic impact of *Candida* colonization and *Candida* infection in the critically ill patient. Eur J Clin Microbiol Infect Dis 23:323–330

Penalva MA, Arst HN Jr (2004) Recent advances in the characterisation of ambient pH regulation of gene expression in filamentous fungi and yeasts. Ann Rev Microbiol 58:425–451

Pfaller MA, Diekema DJ (2002) Role of sentinel surveillance of candidemia: trends in species distribution and antifungal susceptibility. J Clin Microbiol 40:3551–3557

Pfaller MA, Diekema DJ, Jones RN, Sader HS, Fluit AC, Hollis RJ, Messer SA (2001) International surveillance of bloodstream infections due to *Candida* species: frequency of occurrence and in vitro susceptibilities to fluconazole, ravuconazole, and voriconazole of isolates collected from 1997 through 1999 in the SENTRY Antimicrobial Surveillance Program. J Clin Microbiol 39:3254–3259

Pfaller MA, Messer SA, Boyken L, Hollis RJ, Rice C, Tendolkar S, Diekema DJ (2004) In vitro activities of voriconazole, posaconazole, and fluconazole against 4169 clinical isolates of *Candida* spp. and *Cryptococcus neoformans* collected during 2001 and 2002 in the ARTEMIS global antifungal surveillance program. Diagn Microbiol Infect Dis 48:201–205

Pi H, Chien CT, Fields S (1997) Transcriptional activation upon pheromone stimulation mediated by a small domain of *Saccharomyces cerevisiae* Ste12p. Mol Cell Biol 17:6410–6418

Pittet D, Li N, Woolson RF, Wenzel RP (1997) Microbiological factors influencing the outcome of nosocomial bloodstream infections: a 6-year validated, population-based model. Clin Infect Dis 24:1068–1078

Ramon AM, Porta A, Fonzi WA (1999) Effect of environmental pH on morphological development of *Candida albicans* is mediated via the PacC-related transcription factor encoded by *PRR2*. J Bacteriol 181:7524–7530

Ren B, Robert F, Wyrick JJ, Aparicio O, Jennings EG, Simon I, Zeitlinger J, Schreiber J, Hannett N, Kanin E et al. (2000) Genome-wide location and function of DNA binding proteins. Science 290:2306–2309

Rine J, Herskowitz I (1987) Four genes responsible for a position effect on expression from HML and HMR in *Saccharomyces cerevisiae*. Genetics 116:9–22

Roberts RL, Mosch HU, Fink GR (1997) 14-3-3 proteins are essential for RAS/MAPK cascade signaling during pseudohyphal development in *S. cerevisiae*. Cell 89:1055–1065

Roberts CJ, Nelson B, Marton MJ, Stoughton R, Meyer MR, Bennett HA, He YD, Dai H, Walker WL, Hughes TR et al. (2000) Signaling and circuitry of multiple MAPK pathways revealed by a matrix of global gene expression profiles. Science 287:873–880

Rollins JA (2003) The *Sclerotinia sclerotiorum pac1* gene is required for sclerotial development and virulence. Mol Plant Microbe Interact 16:785–795

Saville SP, Lazzell AL, Monteagudo C, Lopez-Ribot JL (2003) Engineered control of cell morphology in vivo reveals distinct roles for yeast and filamentous forms of *Candida albicans* during infection. Eukaryot Cell 2:1053–1060

Schweizer A, Rupp S, Taylor BN, Rollinghoff M, Schroppel K (2000) The TEA/ATTS transcription factor CaTec1p regulates hyphal development and virulence in *Candida albicans*. Mol Microbiol 38:435–445

Shore D (1994) *RAP1*: a protean regulator in yeast. Trends Genet 10:408–412

Shore D (1997) Telomere length regulation: getting the measure of chromosome ends. Biol Chem 378:591–597

Shore D, Nasmyth K (1987) Purification and cloning of a DNA binding protein from yeast that binds to both silencer and activator elements. Cell 51:721–732

Simon I, Barnett J, Hannett N, Harbison CT, Rinaldi NJ, Volkert TL, Wyrick JJ, Zeitlinger J, Gifford DK, Jaakkola TS et al. (2001) Serial regulation of transcriptional regulators in the yeast cell cycle. Cell 106:697–708

Singh P, Ganesan K, Malathi K, Ghosh D, Datta A (1994) *ACPR*, a *STE12* homologue from *Candida albicans*, is a strong inducer of pseudohyphae in *Saccharomyces cerevisiae* haploids and diploids. Biochem Biophys Res Commun 205:1079–1085

Smith HE, Su SS, Neigeborn L, Driscoll SE, Mitchell AP (1990) Role of *IME1* expression in regulation of meiosis in *Saccharomyces cerevisiae*. Mol Cell Biol 10:6103–6113

Staab JF, Bradway SD, Fidel PL, Sundstrom P (1999) Adhesive and mammalian transglutaminase substrate properties of *Candida albicans* Hwp1. Science 283:1535–1538

Stead D, Findon H, Yin Z, Walker J, Selway L, Cash P, Dujon BA, Hennequin C, Brown AJP, Haynes K (2005) Proteomic changes associated with inactivation of the *Candida glabrata ACE2* virulence-moderating gene. Proteomics 5:1838–1848

Stoldt VR, Sonneborn A, Leuker CE, Ernst JF (1997) Efg1p, an essential regulator of morphogenesis of the human pathogen *Candida albicans*, is a member of a conserved class of bHLH proteins regulating morphogenetic processes in fungi. EMBO J 16:1982–1991

Su SS, Mitchell AP (1993) Molecular characterization of the yeast meiotic regulatory gene *RIM1*. Nucleic Acids Res 21:3789–3797

Talibi D, Raymond M (1999) Isolation of a putative *Candida albicans* transcriptional regulator involved in pleiotropic drug resistance by functional complementation of a *pdr1 pdr3* mutation in *Saccharomyces cerevisiae*. J Bacteriol 181:231–240

Thorvaldsen JL, Sewell AK, McCowen CL, Winge DR (1993) Regulation of metallothionein genes by the *ACE1* and *AMT1* transcription factors. J Biol Chem 268:12512–12518

Tortorano AM, Peman J, Bernhardt H, Klingspor L, Kibbler CC, Faure O, Biraghi E, Canton E, Zimmermann K, Seaton S et al. (2004) Epidemiology of candidaemia in Europe: results of 28-month European Confederation of Medical Mycology (ECMM) hospital-based surveillance study. Eur J Clin Microbiol Infect Dis 23:317–322

Tripathi G, Wiltshire C, Macaskill S, Tournu H, Budge S, Brown AJ (2002) Gcn4 co-ordinates morphogenetic and metabolic responses to amino acid starvation in *Candida albicans*. EMBO J 21:5448–5456

Vallim MA, Miller KY, Miller BL (2000) *Aspergillus* SteA (sterile12-like) is a homeodomain-C_2/H_2-Zn^{+2} finger transcription factor required for sexual reproduction. Mol Microbiol 36:290–301

Wang Z, Wilson WA, Fujino MA, Roach PJ (2001) The yeast cyclins Pc16p and Pc17p are involved in the control of glycogen storage by the cyclin-dependent protein kinase Pho85p. FEBS Lett 506:277–280

Wang Y, Shirogane T, Liu D, Harper JW, Elledge SJ (2003) Exit from exit: resetting the cell cycle through Amn1 inhibition of G protein signaling. Cell 112:697–709

Weigel D, Jurgens G, Kuttner F, Seifert E, Jackle H (1989) The homeotic gene fork head encodes a nuclear protein and is expressed in the terminal regions of the *Drosophila* embryo. Cell 57:645–658

Weiss EL, Kurischko C, Zhang C, Shokat K, Drubin DG, Luca FC (2002) The *Saccharomyces cerevisiae* Mob2p-Cbk1p kinase complex promotes polarized growth and acts with the mitotic exit network to facilitate daughter cell-specific localization of Ace2p transcription factor. J Cell Biol 158:885–900

Wey SB, Mori M, Pfaller MA, Woolson RF, Wenzel RP (1988) Hospital-acquired candidemia. The attributable mortality and excess length of stay. Arch Intern Med 148:2642–2645

Wheeler RT, Kupiec M, Magnelli P, Abeijon C, Fink GR (2003) A *Saccharomyces cerevisiae* mutant with increased virulence. Proc Natl Acad Sci USA 100:2766–2770

White TC, Agabian N (1995) *Candida albicans* secreted aspartyl proteinases: isoenzyme pattern is determined by cell type, and levels are determined by environmental factors. J Bacteriol 177:5215–5221

Whiteway M, Dignard D, Thomas DY (1992) Dominant negative selection of heterologous genes: isolation of *Candida albicans* genes that interfere with *Saccharomyces cerevisiae* mating factor-induced cell cycle arrest. Proc Natl Acad Sci USA 89:9410–9414

Wickes BL, Edman U, Edman JC (1997) The *Cryptococcus neoformans STE12alpha* gene: a putative *Saccharomyces cerevisiae STE12* homologue that is mating type specific. Mol Microbiol 26:951–960

Wisplinghoff H, Seifert H, Tallent SM, Bischoff T, Wenzel RP, Edmond MB (2003) Nosocomial bloodstream infections in pediatric patients in United States hospitals: epidemiology, clinical features and susceptibilities. Pediatr Infect Dis J 22:686–691

Wolfe KH, Shields DC (1997) Molecular evidence for an ancient duplication of the entire yeast genome. Nature 387:708–713

Yang X, Talibi D, Weber S, Poisson G, Raymond M (2001) Functional isolation of the *Candida albicans FCR3* gene encoding a bZip transcription factor homologous to *Saccharomyces cerevisiae* Yap3p. Yeast 18:1217–1225

Yeong FM, Lim HH, Surana U (2002) MEN, destruction and separation: mechanistic links between mitotic exit and cytokinesis in budding yeast. Bioessays 24:659–666

Young LY, Lorenz MC, Heitman J (2000) A *STE12* homolog is required for mating but dispensable for filamentation in *Candida lusitaniae*. Genetics 155:17–29

Zaragoza O, Rodriguez C, Gancedo C (2000) Isolation of the *MIG1* gene from *Candida albicans* and effects of its disruption on catabolite repression. J Bacteriol 182:320–326

Zhang X, Lester RL, Dickson RC (2004) Pil1p and Lsp1p negatively regulate the 3-phosphoinositide-dependent protein kinase-like kinase Pkh1p and downstream signaling pathways Pkc1p and Ypk1p. J Biol Chem 279:22030–22038

Zhou PB, Thiele DJ (1991) Isolation of a metal-activated transcription factor gene from *Candida glabrata* by complementation in *Saccharomyces cerevisiae*. Proc Natl Acad Sci USA 88:6112–6116

Zhou P, Thiele DJ (1993) Rapid transcriptional autoregulation of a yeast metalloregulatory transcription factor is essential for high-level copper detoxification. Genes Dev 7:1824–1835

Zhu Z, Thiele DJ (1996) A specialized nucleosome modulates transcription factor access to a *C. glabrata* metal responsive promoter. Cell 87:459–470

Zhu G, Spellman PT, Volpe T, Brown PO, Botstein D, Davis TN, Futcher B (2000) Two yeast forkhead genes regulate the cell cycle and pseudohyphal growth. Nature 406:90–94

12 Using Genomics to Study the Life Cycle of *Histoplasma capsulatum*

A. Sil[1]

CONTENTS

I. Introduction

Histoplasma capsulatum is a fungal pathogen on the brink of a molecular revolution. Years of tool-building (Goldman 1995; Magrini and Goldman 2001; Woods 2002) and ongoing genome-sequencing projects have set the stage for rich molecular experimentation with this organism. The goal of this chapter is to review the past, present, and future of functional genomics in *H. capsulatum*.

H. capsulatum, the etiologic agent of histoplasmosis, is a primary fungal pathogen that infects healthy as well as immunocompromised individuals. Approximately 500,000 infections are thought to occur every year in the US alone (Eissenberg and Goldman 1991; Marques et al. 2000; Woods 2003; Wheat and Kauffman 2003). Immunocompromised individuals tend to develop progressive, disseminated disease that can be fatal. Specifically, *H. capsulatum* has been documented to cause severe disease in patients with lymphoreticular neoplasms and those undergoing immunosuppressant chemotherapy (Kauffman et al. 1978; Bryan and DiSalvo 1979; Brown 1990; Samonis and Bafaloukos 1992; Bradsher 1996). *H. capsulatum* is endemic in the Ohio River Valley through the mid-western United States into Texas, and is a leading pathogen affecting AIDS patients in these endemic areas (Sternberg 1994).

A description of the *H. capsulatum* life cycle reveals that this organism responds to its environment (either the soil or a mammalian host) in a distinct manner (Fig. 12.1). *H. capsulatum* is a haploid dimorphic fungus that exists in a hyphal form in soil. This mycelial form of *H. capsulatum* is competent for mating if it encounters a strain of opposite mating type (see Sect. IV.D), and the resultant diploid undergoes meiosis to produce eight meiotic spores per ascus (Kwon-Chung 1973). Additionally, the mycelial form of the organism produces vegetative spores, or conidia. Conidia and hyphal fragments become airborne when the soil is disrupted, and are then inhaled by the host and taken up by macrophages and other phagocytic cells (Eissenberg and Goldman 1991; Bullock 1993; Woods 2003). Once inside the host, conversion of the mycelial form to a budding yeast form is triggered within hours. Yeast cells evade phagocytic killing and multiply within alveolar macrophages. Subsequently, yeast cells use phagocytic cells as vehicles to spread to multiple organs of the reticuloendothelial system such as the spleen, liver, lymph nodes, and bone marrow. In patients with disseminated disease, other organs such as the skin, heart,

[1] Department of Microbiology and Immunology, University of California, San Francisco, 513 Parnassus, P.O. Box 0414, San Francisco, CA 94143-0414, USA

The Mycota XIII
Fungal Genomics
Alistair J.P. Brown (Ed.)

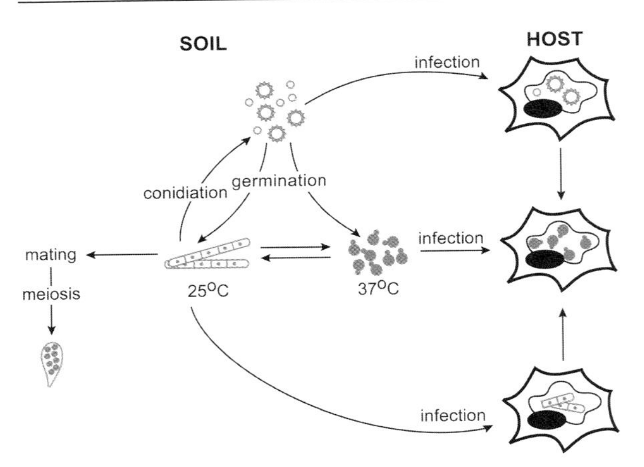

Fig. 12.1. Life cycle of *Histoplasma capsulatum*. The figure depicts different stages of the *H. capsulatum* life cycle in the soil and in the host. The mycelial form grows in the soil (or at 25 °C in the laboratory), and can mate and undergo meiosis to form ascospores. The mycelia can also generate vegetative conidiospores by a process known as conidiation. There are at least two types of conidiospores, macro- or microconidia. These cells can germinate and adopt either the mycelial form or the yeast form. The natural route of infection occurs when the host inhales conidia or hyphal fragments that aerosolize when the soil is disrupted. Once inside the host, these cells convert to the yeast form. The yeast form is infectious if introduced into an animal in the laboratory, but there is normally no host-to-host transmission of the yeast form. Schematic courtesy of Davina Hocking Murray

brain, adrenal glands, and gastrointestinal tract can be colonized. In the majority of hosts, the immune system curtails the infection by means that are not yet clearly defined, but that depend on the development of cell-mediated immunity and the corresponding activation of macrophages (Eissenberg and Goldman 1991; Newman 1999; Huffnagle and Deepe 2003).

One of the most interesting aspects of *H. capsulatum* biology is that it changes morphology and associated characteristics in response to its environment, as described above. All of the systemic dimorphic fungal pathogens, including *H. capsulatum*, *Coccidioides immitis*, *Blastomyces dermatidis*, and *Paracoccidioides brasiliensis*, use temperature as a key signal to determine morphology and cell specialization (Maresca and Kobayashi 1989). The transformation of *H. capsulatum* mycelial cells to yeast cells, or vice versa, can be recapitulated in culture simply by manipulating the growth temperature (Maresca and Kobayashi 1989; Maresca et al. 1994). When *H. capsulatum* cells are grown at 25 °C, they grow in the mycelial form. When these cells are shifted to 37 °C, they shift to the budding yeast form. Though this behaviour is characteristic of all the systemic dimorphic fungi, little is understood about how any of these organisms sense temperature and trigger the appropriate response. For *H. capsulatum*, the ability to grow in both forms is key to establishing and maintaining disease in the host. The mycelial form facilitates the initiation of infection because it becomes airborne and produces infectious conidiospores. Transformation to the yeast form is thought to be essential for the fungus to survive and proliferate in the host, since mycelial cells that are artificially blocked from shifting to the yeast form are unable to cause disease (Maresca et al. 1977; Medoff et al. 1986b). This type of dimorphic switch is a key virulence determinant for many pathogenic fungi (Borges-Walmsley and Walmsley 2000). Once inside the host cell, the *H. capsulatum* yeast cell subverts the normal response of macrophages to microbes by unknown means, as described in Sect. IV. A below. Thus, *H. capsulatum* can be thought of as a model system for studying temperature-regulated morphology as well as interactions between macrophages and intracellular eukaryotic pathogens.

With the advent of a genome sequence for *H. capsulatum*, we are at the dawn of the heyday of functional genomics in this organism. Especially because previous genetic studies have been limited, functional genomics is a powerful methodology for identifying candidate genes that are implicated in a particular process by virtue of regulated gene expression. Because it is still early days for genome-wide analyses in *H. capsulatum*, this chapter will cover (1) previous uses of functional genomics, (2) the status of the genome projects for this organism, (3) future uses of functional genomics for studies of interesting aspects of *H. capsulatum* biology, and (4) molecular tools that interface with genomic analyses.

II. Previous Use of Functional Genomics in *H. capsulatum*

A. Generation of a Shotgun Genomic Microarray

H. capsulatum is one of the few organisms for which functional genomic analysis preceded a genome sequence. Hence, this organism serves as a testimonial for fungal genomics in both sequenced and unsequenced organisms. Hwang et al. (2003) compared the gene expression profiles of yeast and

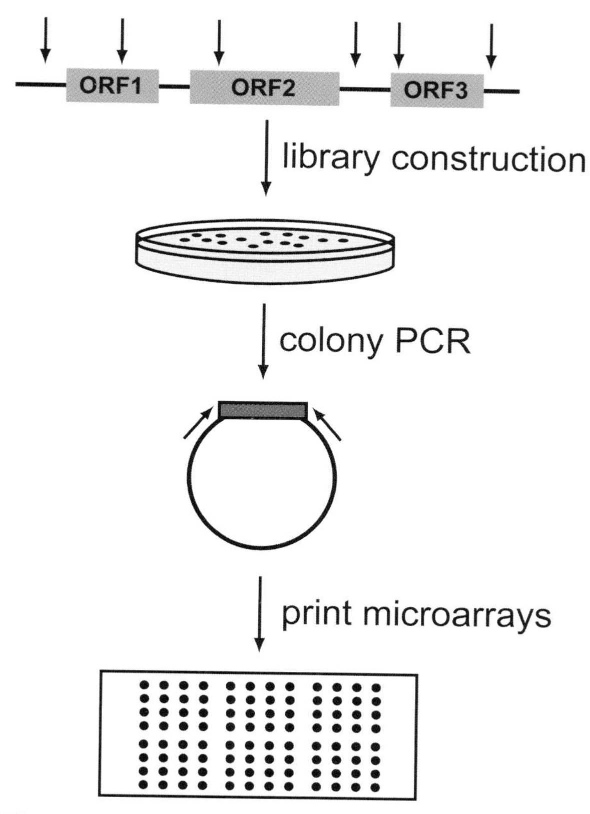

Fig. 12.2. Construction of a shotgun genomic microarray. *H. capsulatum* genomic DNA is shown at the *top* of the figure. This DNA was subjected to a partial restriction enzyme digestion, size selected, cloned into a vector, and transformed into *E. coli*. Approximately 10,000 independent colonies were inoculated into 96-well plates, and colony PCR was performed on each culture using common primers in the vector. The resultant PCR products were spotted on glass slides to generate shotgun genomic microarrays. Reprinted with permission from Hwang et al. (2003)

mycelial cells by building a shotgun genomic microarray (Fig. 12.2) at a time when only a handful of genes had been cloned from *H. capsulatum*. To construct the array, 1–2 kb genomic DNA fragments were generated by partial restriction enzyme digestion of genomic DNA from the virulent G217B strain. This pool of fragments was ligated into the *E. coli* vector pBLUESCRIPT to generate a library of genomic fragments. The genomic inserts from 9600 individual library clones were subjected to amplification by polymerase chain reaction (PCR) using a common pair of primers that flanked the site of genomic DNA insertion in the vector. Each PCR product was precipitated and then printed on glass slides to generate the first *H. capsulatum* microarray. Thus, each PCR product corresponded to a single "spot" on the microarray, and will be referred to hereafter as an array element. Based on the estimated genome size at that time, it was thought that the array covered approximately 1/3 of the genome. Hwang et al. (2003) isolated RNA from yeast and mycelial cells, generated differentially labeled cDNA, and subjected these labeled probes to a competitive hybridization on the microarray to identify a set of yeast-specific and mycelial-specific genes (discussed further below).

Hwang et al. (2003) chose to generate a microarray from a genomic library, rather than a cDNA library, for several reasons. Most importantly, a random genomic microarray provides better opportunity for equivalent representation of genes. Because cDNAs are generated from cells grown under a discrete number of conditions, genes that are not transcribed under these conditions will not be represented in the corresponding cDNA library. Additionally, unless one makes a normalized cDNA library, highly expressed genes will be over-represented in the cDNA library and, hence, on the microarray. In a random genomic library, most genes have an equivalent chance of being represented, with the exception of the minority of genes that fall in genomic repeat sequences, which will be over-represented.

The random genomic library approach (or "shotgun approach") is unlikely to work if (1) an organism has very large intron sequences that would interfere with the ability of a cDNA probe to bind to its cognate genomic fragment in an array experiment, and/or (2) the genome contains large intergenic regions that will clutter the microarray with non-coding sequence. Because the average intron size in *H. capsulatum* is small (in the order of 100 nucleotides), the shotgun approach was likely to work. To validate that the microarray was mainly comprised of coding sequence rather than intergenic sequence, Hwang et al. (2003) carried out a competitive hybridization on a trial random genomic microarray with genomic DNA, and with a cDNA probe generated from RNA from *H. capsulatum* cells growing in the yeast form. Genomic DNA will give a signal for all elements on the random genomic microarray, whereas cDNA will only give a signal for elements that correspond to coding sequence. They observed that 75% of the spots lit up with the cDNA probe, indicating that a minimum of 75% of the array elements contained coding sequence. Because the cDNAs represent genes expressed under a single growth condition, it is likely that even more than 75% of the array elements contained coding sequence.

B. Studying Differences Between Yeast and Mycelia

Hwang et al. (2003) used their shotgun genomic microarray to compare the transcriptional profile of yeast and mycelial cells. Since morphology is regulated by temperature, they reasoned that temperature triggers a signal transduction pathway resulting in either a yeast transcriptional program at 37 °C, or a mycelial transcriptional program at 25 °C. They were interested in identifying two main classes of genes: (1) master transcriptional regulators that govern morphology-related gene expression, and (2) yeast and mycelial-specific genes that might shed light on phenotypic differences between the two morphologic phases.

Previous work from a variety of laboratories had identified a number of phase-specific genes (Harris et al. 1989a, b; Keath et al. 1989; Keath and Abidi 1994; Di Lallo et al. 1994; Gargano et al. 1995; Patel et al. 1998; Tian and Shearer 2001, 2002; Johnson et al. 2002), but no large-scale analysis of differential gene expression had been attempted. Gene expression profiling is a powerful way to identify both regulatory and target genes, and thus was a good choice to probe differences between yeast and mycelial cells.

To perform this experiment, Hwang et al. (2003) grew cells at 25 and 37 °C, and isolated RNA from the resultant mycelial and yeast-form cells. Differentially labeled cDNAs were generated and subjected to a competitive hybridization on the shotgun genomic microarray. Approximately 500 clones from the microarray were expressed at least fivefold higher in one morphology compared to the other, and a small subset of these clones were validated as being differentially expressed by Northern blot analysis (Fig. 12.3). These clones were annotated as follows:

1. sequence information from each array clone was used to map its location on contigs from an ongoing genome-sequencing project (described below);
2. each contig was subjected to Basic Local Alignment Search Tool (BLAST) analysis against the non-redundant (nr) database of genes at the National Center for Bioinformatics to identify the location of putative genes based on homologs in other organisms; and
3. elements that overlapped with putative genes were annotated. Approximately half of the differentially expressed genes were annotated by this method.

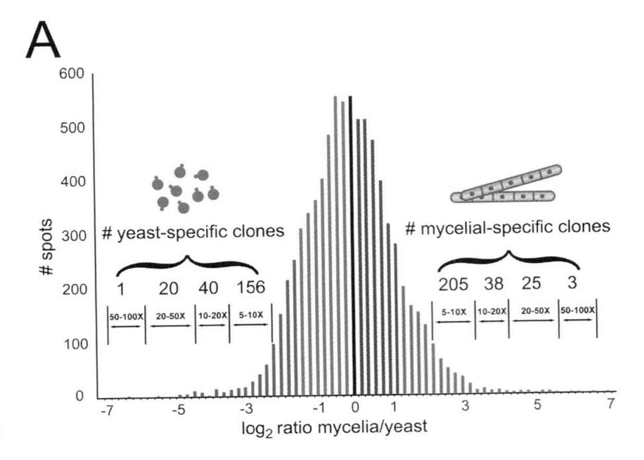

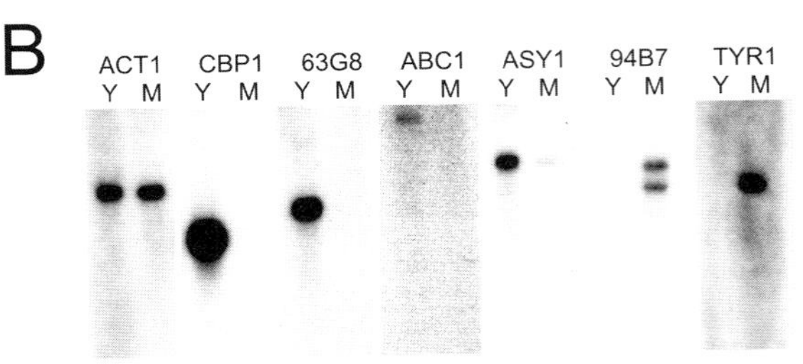

Fig. 12.3. A, B Identification of yeast- or mycelial-specific genes. **A** The histogram depicts the number of spots on the microarray vs. the $\log_2$ of the ratio of the mycelial signal to the yeast signal. Spots that are equivalently expressed have a $\log_2$ ratio of zero. The numbers of clones that show differential expression from five- to 100-fold in one morphologic form compared to the other are shown in the figure. **B** A Northern blot analysis of gene expression in yeast and mycelia. Each blot has yeast total RNA (*Y*) in the *left lane* and mycelial total RNA (*M*) in the *right lane*. *ACT1* (actin) is equivalently expressed between the two samples, *CBP1* (calcium-binding protein), 63G8 (unknown microarray clone), *ABC1*, and *ASY1* are yeast-specific. 94B7 (unknown microarray clone) and *TYR1* are mycelial-specific. Reprinted with permission from Hwang et al. (2003)

A variety of genes were identified as specifically expressed in one morphology or the other, and the identity of many of these genes generated interesting hypotheses about phenotypic differences between yeast and mycelia (Fig. 12.4). For example, only the mycelial form of *H. capsulatum* is able to produce vegetative spores, or conidia. Several orthologs of genes that regulate conidiation in *Aspergillus* species were identified as mycelial-specific in *H. capsulatum*. Hence, these orthologs are candidates for regulators of conidiation in *H. capsu-*

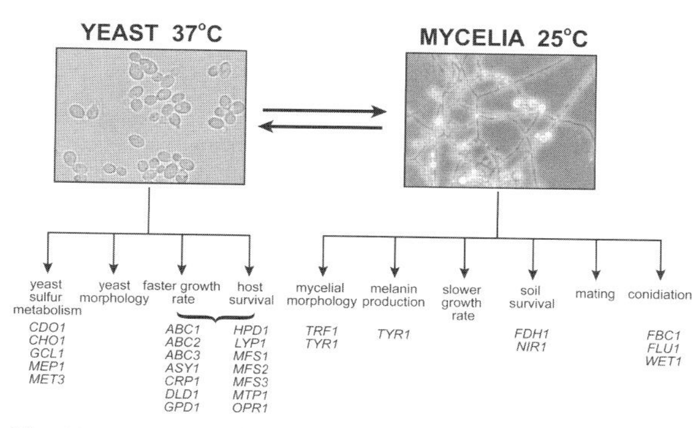

Fig. 12.4. Putative functions for differentially expressed genes identified by functional genomics. The figure shows *H. capsulatum* cells growing in either the yeast form at 37 °C or the mycelial form at 25 °C. A number of yeast-specific genes were annotated as potentially being involved in sulfur metabolism and growth rate/host survival based on the function of their ortholog in a different organism. Similarly, mycelial-specific genes included those implicated in polarized cell growth, melanin production, soil survival, and conidiation. Adapted with permission from Hwang et al. (2003). Cell images, Anita Sil

latum, which can be tested further by molecular genetic experiments. Since no molecular information was previously available about conidiation in *H. capsulatum*, these genes provide the first entrée into understanding how *H. capsulatum* might regulate spore production (see Sect. IV.B below).

Other mycelial-specific genes included orthologs of genes involved in polarized cell growth and melanin production, which under normal media conditions is confined to mycelia. Other yeast-specific genes included orthologs of genes involved in cell cycle regulation and nutrient metabolism. Since the yeast form of the organism is the pathogenic form in the host, all of the yeast-specific genes are candidates for genes that could be important for host survival. Although no conclusions can be made about the function of any gene based simply on sequence homology and gene expression data, the rapid identification of morphology-specific genes provided a number of valuable hypotheses that can be tested in future experiments. In a system where few genes had been identified prior to this work, these data highlight the value of genomics for opening up a molecular analysis of biological problems. As discussed below, the current construction of a whole-genome microarray for *H. capsulatum* will continue to enhance the value of genomics in dissecting *H. capsulatum* biology.

It should be noted, of course, that every genomics experiment has its caveats. Because temperature is a key regulator of morphology in *H. capsulatum*, the yeast phase sample used for microarray comparison was grown at 37 °C, whereas the mycelial phase sample was grown at 25 °C. Some genes identified as differentially expressed may be temperature-regulated, rather than morphology-regulated. Future studies with cells that are trapped in one morphologic phase independent of temperature would facilitate further categorization of these differentially expressed genes.

III. Current Genome-Sequencing Projects for *H. capsulatum*

A. Choice of Genomes to be Sequenced

Although the shotgun approach described above was useful for identifying interesting genes, the genome project for *H. capsulatum* is making a genomics approach much more straightforward. At the time of writing, two separate projects to sequence three different *H. capsulatum* strains are ongoing. To appreciate the choice of strains for sequencing, it is first necessary to understand the differences between these strains.

Molecular studies of *H. capsulatum* biology and pathogenesis have largely taken place in two distinct strains: G217B, a North American clinical isolate, and G186A, a clinical isolate from Panama. These strains were originally differentiated on the basis of polysaccharide composition of their cell walls (Reiss 1977; Davis et al. 1977; Reiss et al. 1977). G217B, designated "chemotype 1", lacks α-(1,3)-glucan in its cell wall, whereas G186A, designated "chemotype II", has copious amounts of α-(1,3)-glucan in its cell wall, much like the other systemic dimorphic fungal pathogens *Blastomyces dermatitidis* and *Paracoccidioides brasiliensis*. Interestingly, the G217B strain (which lacks α-(1,3)-glucan) is virulent, but variants of G186A that lack α-(1,3)-glucan are avirulent (Klimpel and Goldman 1987, 1988).

A second North American isolate, the Downs strain, has also been the subject of many research studies. The Downs strain, which was originally a clinical isolate from an octogenarian who was probably immunocompromised, is less thermotolerant than G217B or G186A (Medoff et al. 1986a; Keath et al. 1989). Furthermore, unlike G217B and G186A, the Downs strain is not virulent in standard animal models (Lambowitz et al. 1983). The literature indicates that the genomic structure of the three strains may be different. DNA renaturation kinetics suggested that the genome of the Downs strain is considerably larger than the genome of G186A (Carr and Shearer 1998). Additionally, though it was difficult to visualize chromosomes with high resolution, contour-clamped homogenous electric field gels and field-inversion gel electrophoresis suggested that the Downs strain and G217B may have chromosomes of different size and number (Steele et al. 1989). Restriction fragment length polymorphism analysis of genomic and mitochondrial DNA also confirms that Downs and G217B are distinct strains (Vincent et al. 1986; Spitzer et al. 1989). Ideally, because G217B, G186A, and the Downs strain have markedly different phenotypic characteristics, a comparison of the three strains at the sequence level will probably generate some testable hypotheses about how genetic variation results in different biological traits.

The most recent and extensive phylogenetic analyses of 137 different *H. capsulatum* isolates confirm that G217B, G186A, and Downs are members of different clades (distinct phylogenetic groups of organisms that include the most recent common ancestor of all of its members, and all of the common ancestor's descendants; Kasuga et al. 1999, 2003). In the most recent analysis, the authors used sequence comparison of four unlinked loci to identify 80 multilocus genotypes. Seventy-three of the 80 fell into eight clades: North American class 1 (NAm 1), North American class 2 (NAm 2), Latin American group A (LAm A), Latin American group B (LAm B), Australian, Dutch, Eurasian, and African. All clades except the Eurasian clade were genetically isolated, and thus can be defined as different phylogenetic species. The Eurasian clade, in contrast, was derived from Latin American group A. The seven isolates that did not belong to any of the clades were also distinct from each other, and were designated as lone lineages.

G217B and Downs are members of the NAm 2 and NAm 1 clades, respectively. G186A, which was originally isolated from Panama as stated above, did not fall into one of the eight clades, and instead is a member of a lone lineage. Interestingly, the isolates of *H. capsulatum* from Latin America, including G186A, showed the most phylogenetic diversity. These isolates (from Mexico, Guatemala, Surinam, Brazil, Argentina, Colombia, and Panama) mainly fell into two clades (LAm A and LAm B), but also included seven lone lineages. Since G186A appears to be only distantly related to G217B and Downs, and yet both G186A and G217B are virulent strains, it is likely that determining similarities and differences between the genomes will be quite informative.

The Fungal Genome Initiative at the Broad Institute (http://www.broad.mit.edu/annotation/fungi/fgi/) is sequencing WU24, a virulent NAm 1 strain that is likely to be much more closely related to the Downs strain than G217B and G186A. Although no information about the sequence was available at the time of this writing, genome information is predicted to be publicly available by early 2005. The Genome Sequencing Center (GSC) at Washington University in St. Louis was funded by the National Institute for Allergy and Infectious Disease to sequence both G217B and G186A (http://www.genome.wustl.edu/projects/hcapsulatum/), using a combination of whole-genome shotgun sequencing, fosmid end sequencing (and physical mapping; Magrini et al. 2004), and cDNA sequencing (the latter to contribute to gene-finding and annotation of the genome). Interestingly, the genomes of these two strains appear to be quite different, as predicted from the phylogenetic analysis described above. The genome size of G217B appears to be about 41 Megabases (Mb), whereas the genome size of G186A is approximately 30 Mb. A comparison of the two genomes is currently in progress. It appears that they have a strikingly different repeat content: 34% of the G217B genome is composed of repeat sequence (including transposon sequences), whereas only 7.5% of the G186A genome is made up of repeats. The exact nature of the repeat sequence is currently being analyzed. Final annotation of both genomes is now underway. It will be of great interest to determine whether there are genes that are unique to each of these strains.

Although the sequencing of G217B, G186A, and WU24 will make for interesting and informative comparisons, there are other isolates of *H. capsulatum* that will be fascinating subjects for future sequence analysis. For example, several of the clades described above included *Histoplasma* isolates that cause disease in horses (*H. capsulatum* var. *farciminosum*), but there is little understanding of what allows these isolates, but not others, to cause equine disease. Additionally, the African clade includes isolates of *H. capsulatum* var.

dubosii, which cause a disease known as African histoplasmosis. These infections are typified by lesions in the skin and bone, rather than the pulmonary infection caused by *H. capsulatum* var. *capsulatum* (Rippon 1988). Sequence analysis and functional studies of these isolates would shed some light on *Histoplasma* characteristics that influence the clinical manifestations of infection.

B. From Genome Sequence to Functional Genomics

A whole-genome oligonucleotide array will be built for the G217B and G186AR strains as part of the GSC genome-sequencing project, thereby easing the transition from genome sequence to functional genomics. This oligonucleotide array will represent each putative gene in these two strains as a unique 70-mer. As part of the genome project, this microarray will be used to determine the global gene expression profile of *H. capsulatum* cells grown in the yeast form and the mycelial form, as described above (Sect. II.B). Because this array will be a whole-genome array, rather than the previously used shotgun genomic array, this analysis will significantly extend the identification of genes that are differentially expressed between the two forms (or regulated by temperature). One goal of this analysis is to contribute to genome annotation. It will be possible to confirm that putative open reading frames truly represent genes because they encode an expressed transcript. In addition, because the mycelial form is specialized for growth in the soil and the yeast form is specialized for growth in the host, a truly global understanding of the gene expression profile of each of these forms will shed light on *H. capsulatum* biology and gene function.

These experiments will determine whether each gene is more highly expressed in yeast or mycelia, but this analysis is only the starting point. The ultimate goal is to determine the regulatory networks that are required to establish and maintain the yeast and mycelial forms. It may be possible to use pattern-finding algorithms such as MEME to identify putative regulatory elements that subsets of yeast-specific or mycelial-specific genes have in common (http://meme.sdsc.edu/meme/website/intro.html). By combining sequence homology with expression profile, it will be possible to pick out putative yeast-specific transcription factors, or mycelial-specific transcription factors, and study their function using molecular genetic approaches. Additionally, by determining the gene expression profile as cells transit from the yeast form to the mycelial form, and vice versa, it will be possible to use cluster analysis (Eisen et al. 1998) to identify classes of co-regulated genes (by virtue of having similar transcriptional profiles during the morphologic transition). Some of these classes of genes will be targets of common transcription factors. Furthermore, it may be possible to use known groups of co-regulated genes in other fungi to identify regulatory circuits in *H. capsulatum* (Gasch et al. 2004). Finally, by comparing the morphology-specific gene expression profiles in G217B and G186, it will be possible to determine similarities and differences in how these strains regulate their morphology. Ultimately, this type of comparative analysis will be extended to include WU24.

IV. Other Applications of Functional Genomics in *H. capsulatum*

Whole-genome microarrays, and other related technologies described below, will open up a number of areas of molecular exploration in *H. capsulatum*.

A. Macrophages

Macrophages are the main host cell for *H. capsulatum*, and functional genomics is an excellent tool to probe the interaction between *H. capsulatum* and the macrophage. To understand how functional genomics can be used to study the relationship between microbe and host cell, it is necessary to first describe the different steps of interaction between *H. capsulatum* and the macrophage.

H. capsulatum is phagocytosed by macrophages; once inside the macrophage, the microbe continues to divide (Fig. 12.5). Phagocytosis of *H. capsulatum* by human monocyte-derived macrophages is mediated by the CD18 family of adhesion-promoting glycoproteins, including CD11b/CD18 (otherwise known as CR3; Newman et al. 1990; Newman 1999). Recent work suggests that HSP60 on the cell surface of *H. capsulatum* mediates recognition by these receptors (Long et al. 2003). Once phagocytosis has occurred, the intracellular fate of *H. capsulatum* has been studied

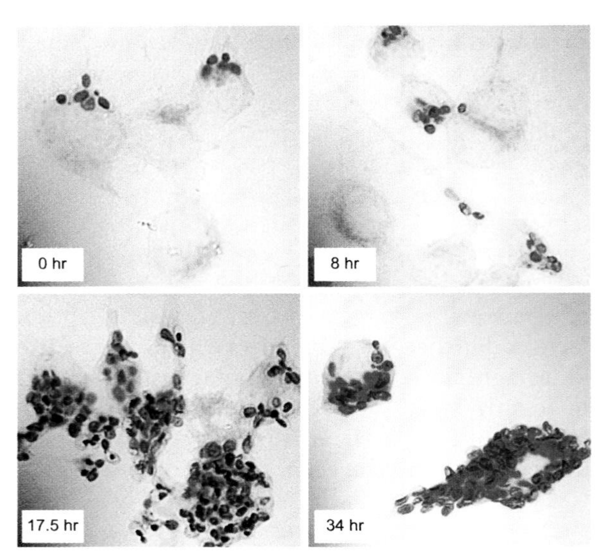

Fig. 12.5. Infection of RAW264.7 cells with *H. capsulatum*. Monolayers of macrophages were infected with *H. capsulatum*. At different time points, the monolayer was fixed and fungal cells were stained with periodic acid/Schiff base. The time point is indicated in the *lower left* of each quadrant. At later time points, the *Histoplasma* fills the macrophage. Cell images, Anita Sil, and courtesy of Dervla Isaac

in a variety of macrophages, including human monocyte-derived macrophages, human alveolar macrophages, murine peritoneal or alveolar macrophages, and murine macrophage-like cell lines such as RAW264.7 and P388D1 (Eissenberg and Goldman 1991; Newman 1999; Woods 2003; Fig. 12.5). In some but not all of these host cells, a respiratory burst is triggered after phagocytosis of *H. capsulatum*, but the viability of *H. capsulatum* seems to be unaffected by reactive oxygen intermediates.

Once internalized, the intracellular fate of *H. capsulatum* seems to differ depending on the type of macrophage studied. For the average nonpathogenic microbe, the nascent phagosome undergoes a series of events known as phagosome maturation, whereby the contents of the phagosome and phagosomal membrane are altered to promote destruction of the microbe (Duclos and Desjardins 2000; Vieira et al. 2002). Maturation ultimately results in the formation of a phagolysosome, which displays several degradative properties including acidic pH, the presence of acid-activated lysosomal hydrolases, anti-microbial peptides, and toxic oxidative compounds. In the case of infection of human macrophages and the murine RAW264.7 cell line, *H. capsulatum* resides in a phagosome that does not fuse with the lysosome, and displays reduced levels of the host vacuolar membrane ATPase/proton pump that is responsible for acidification of the phagosome during maturation (Newman et al. 1997; Strasser et al. 1999). Thus, *H. capsulatum* interferes with normal phagosome maturation in these cells. In other types of macrophage host cells such as the murine P388D1 cell line, *H. capsulatum* resides in the phagolysosome (Eissenberg et al. 1988). In either case, *H. capsulatum* prevents acidification of either the phagosome or the phagolysosome, instead maintaining the pH at approximately 6.0–6.5 (Eissenberg et al. 1993; Newman 1999; Strasser et al. 1999). Inhibition of acidification is thought to interfere with the function of lysosomal hydrolases. Furthermore, *H. capsulatum* is thought to maintain the pH at 6–6.5 to maximize acquisition of iron from host transferrin (Newman 1999). This hypothesis is supported by the fact that *H. capsulatum* cannot survive in macrophages treated with chloroquine, a weak base that raises the endocytic pH such that iron cannot be released by transferrin. This detrimental effect of chloroquine on intracellular growth of *H. capsulatum* is reversed by addition of iron compounds that are soluble at neutral or alkaline pH, but not by iron donors that require acidic pH to release iron (Newman et al. 1994). Thus, *H. capsulatum* is thought to survive and grow in macrophages by virtue of its ability to prevent acidification of the phagosome while maintaining an optimum pH for iron acquisition. Interestingly, *H. capsulatum* is known to neutralize the pH of both acidic and basic culture media, suggesting that the organism can sense and modify the ambient pH of its environment (Berliner 1973).

Functional genomics could be used in a number of different ways to probe the interaction between the host cell and *H. capsulatum*. One obvious experiment is to perform a time course of infection of macrophages with *H. capsulatum*. At each time point, it is possible to lyse host cells in a guanidinium thiocyanate buffer, pellet the *H. capsulatum* cells in the lysate by centrifugation, prepare RNA from these cells, and use microarray analysis to determine how the gene expression profile evolves with infection (M. Paige Nittler and Margareta Andersson, personal communication; Anita Sil, personal observation). These experiments are best done coupled with an analysis of the cell biology of infection, so that it is possible to correlate the timing of gene expression with trafficking through and replication within the host cell. The goal is to identify putative virulence factors by identifying genes whose expression changes during infection. For example, genes that are induced when *H. capsulatum* is present in early phagosomes might be important for blocking phagosome maturation. Because the fate of *H. capsulatum* seems to be somewhat different in different types of macrophages (as described above), gene expression profiling of *H. capsulatum* in these different types of host macrophages may provide a sensitive assay to determine whether the microbe is actually responding differently in each case.

This experimental setup also allows the purification of host cell RNA from the lysate supernatant after pelleting the *H. capsulatum*. Transcriptional profiling of host cells (using mouse or human microarrays) has been employed to understand how macrophages are manipulated by pathogens (Detweiler et al. 2001; Boldrick et al. 2002; Nau et al. 2002, 2003; McCaffrey et al. 2004). By monitoring changes in the expression profiles of both the host and *Histoplasma*, it will be possible to monitor the communication between host and pathogen.

Functional genomics can also be used to identify potential virulence factors by subjecting *H. capsulatum* to environmental conditions that mimic those experienced during infection. This type of strategy has been used successfully to identify virulence factors in *Salmonella*, by performing gene expression profiling after exposing the microbe to conditions that mimic the environment of the macrophage phagosome (Detweiler et al. 2003). In the case of *H. capsulatum*, it could be informative to determine the response to exogenous anti-microbial stimuli that mimic the anti-microbial activities of the macrophage, such as reactive oxygen and nitrogen species (Nittler et al. 2005). Additionally, stimuli such as low iron levels or low pH could trigger gene expression programs that are important for survival in host cells. These experiments may identify candidate genes whose roles in infection must be explored by further experimentation.

B. Understanding the Development and Germination of Conidia

The mycelial form of *Histoplasma* undergoes asexual sporulation to give rise to at least two types of conidia, macro- and microconidia, which are distinguished mainly on the basis of size (Pine 1960). Microconidia range in size from 2 to

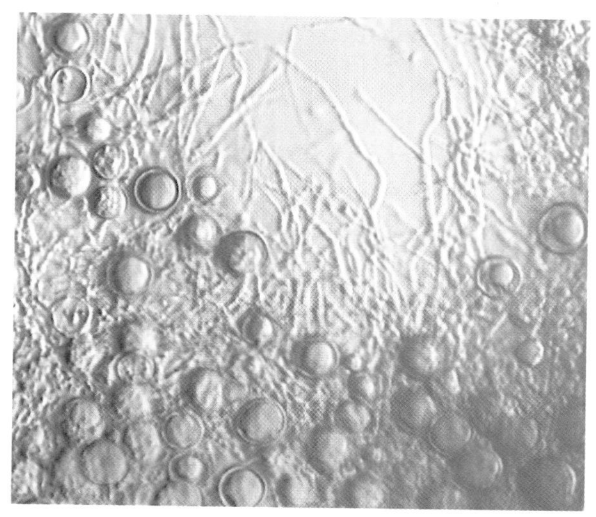

Fig. 12.6. G186AR macroconidia. The G186AR strain was grown under conidiating conditions. The image shows masses of tuberculate macroconidia. Image, Anita Sil, and courtesy of Diane Inglis

6 μm, whereas macroconidia (Fig. 12.6) have been reported to range in size from 8 to 14 μm or 10 to 25 μm, depending on the strain and growth conditions. It is thought that the microconidia are the appropriate size to lodge in the alveoli of the lungs. Therefore, microconidia are thought to be the most common infectious *Histoplasma* particle. Interestingly, even a single viable spore is thought to be capable of causing disease in the mouse (Ajello and Runyon 1953), and infection of mice with spores results in high mortality (Procknow et al. 1960). A number of studies published prior to 1973 discuss conditions that promote conidiation, as well as the morphology of the resultant spores (Pine 1960; Artis and Baum 1963; Goos 1964; Smith 1964; Neilsen and Evans 1964; Smith and Furcolow 1964; Anderson and Marcus 1968; Garrison and Lane 1973). Additionally, there have been cell biological studies monitoring the germination of micro- and macroconidia (Howard 1959; Procknow et al. 1960; Goos 1964; Garrison and Boyd 1977, 1978). However, nothing is known about molecules that are required for conidial development or germination.

Gene expression profiling would be an ideal method to identify genes that influence spore formation and germination in *H. capsulatum*. Conidiation has been studied extensively in the fungus *Aspergillus nidulans*, and several steps are known to be regulated at the level of transcription (Timberlake 1991). By determining the gene expression profile of conidiating mycelia, it will be possible to identify genes required for conidial development. Indeed, the comparison of yeast and conidiating mycelia using functional genomics as described above (Sect. II.B) revealed that several orthologs of *Aspergillus* conidiation genes are specifically expressed in *H. capsulatum* mycelia, as compared to yeast (Hwang et al. 2003). Extending this transcriptional analysis to the whole *H. capsulatum* genome will be very valuable. Gene expression profiling of pure populations of micro- and macroconidia (as compared to yeast and mycelia) will allow the identification of conidial-specific transcripts. These data will shed light on the molecular properties of conidia. Of particular interest is a greater molecular understanding of the pathogenicity of microconidia, which could be revealed by studying genes that are specifically expressed in these cells. Other functional genomics experiments should include determination of gene expression of conidia that are germinating at 25 °C, at 37 °C, and within host cells.

C. Clinical Isolates

A wide variety of *H. capsulatum* strains have been isolated as clinical samples from all over the world, as evidenced in the phylogenetic study described above (Kasuga et al. 2003). One goal of functional genomics will be to analyze molecular differences between these strains, and correlate these data to phenotypic differences. Individual *H. capsulatum* isolates will exhibit sequence differences, and so it will be optimal to employ a microarray that will tolerate mismatches between the sequence displayed on the array and the sequence of the given isolate. Thus, these studies may be better accomplished with a microarray that represents the entire ORF sequence, rather than a unique 70-mer oligonucleotide. It will be of interest to compare strains at both the DNA and the RNA level. In the former case, for example, a comparison of genomic DNA from different clinical isolates might be used to determine whether different isolates have the same or different complement of genes as the reference strain used to design the microarray. This type of analysis can be particularly valuable when combined with electrophoretic analysis of chromosomes to study whether particular chromosomal rearrangements are manifested in different isolates. For example, pulsed-field gel electrophoresis can be used to separate individual chromosomes. The DNA from these chromosomes can then be labeled and subjected to microarray analysis to determine whether a particular chromosome from one isolate contains the same complement of genes as the analogous chromosome from a different isolate. These experiments are of particular interest, since the precedent from bacterial pathogens suggests that genome rearrangement is found more frequently in clinical strains (Hughes 2000). Comparison of different isolates at the RNA level will also be of interest. Gene expression profiling of individual isolates grown under similar conditions (e.g., in the yeast form, the mycelial form, within macrophages, etc.) will reveal whether a given isolate expresses a similar set of genes as the reference strain on which the array is based. It would be interesting to extend this analysis to fresh soil isolates of *H. capsulatum* to see if these strains have a different expression pattern than that of fresh clinical isolates.

These types of experiments may contribute to an understanding of resistance to anti-fungal treatments. Previous studies have documented the evolution of drug resistance in a clinical setting (Wheat et al. 2001). Acquired Immunodeficiency Syndrome (AIDS) patients were treated with fluconazole for histoplasmosis in sequential clinical trials, and the authors were able to monitor the emergence of fluconazole-resistant strains. By comparing the gene expression profile of the resistant strains to that of the parent strain, it may be possible to understand some of the molecular changes that contribute to drug resistance.

Finally, this type of approach could assist in a molecular dissection of *H. capsulatum* virulence. For example, a comparison of the gene expression profile of isogenic avirulent and virulent strains could shed light on changes in transcript accumulation that influence virulence. An ideal opportunity to perform this type of experiment arises because extended laboratory passaging of *H. capsulatum* strain G217B results in reduction of virulence in the mouse model of histoplasmosis, but passaging through an animal results in restoration of virulence (George Deepe, personal communication). One could use microarray technology to compare the gene expression profile of yeast cells grown in culture from an avirulent and virulent pair to determine if the cells have significant differences in gene expression outside of the host. Similarly, since it is feasible to monitor the gene expression profile of microbes in a host animal (Talaat et al. 2004), it should be possible to compare the gene expression profile of the two strains once they are in the animal, to analyze differences in how each strain responds to the host.

D. Study of Mating Type

H. capsulatum has two mating types: – and +. Though the two mating types are thought to be equivalently represented in the soil, the – mating type is more prevalent in immunocompetent hosts (Kwon-Chung et al. 1974). Both mating types are equally represented in immunocompromised hosts, however, suggesting that each has equivalent access to the host (Kwon-Chung et al. 1984). Taken as a whole, these data suggest that the – mating type is more virulent than the + mating type in immunocompetent hosts. However, next to nothing is known about how these mating types differ, and the mating-type locus itself has not been identified. Additionally, since strains seem to lose the ability to mate through serial passaging in the laboratory, many of the laboratory strains (such as G217B and G186A) are of unknown mating type. However, + and – mating-type strains are available at the American Type Culture Collection; these strains are a wonderful starting point for genomic analysis of mating type.

By comparing the gene expression profile of + and – strains by microarray, it will now be possible to judge which genes are differentially expressed between the two mating types. (Again, as described in Sect. IV.C, these studies are better accomplished with a microarray that represents the entire ORF sequence, so that sequence mismatches between the mating strains and the reference strain will be better tolerated.) Mating-type-specific genes are possible candidates for genes in the mating-type locus, or for targets of the mating-type locus. One might expect to identify genes such as pheromones, pheromone receptors, transcription factors, and signaling components. If putative pheromone-encoding genes are identified, it could by very informative to synthesize pheromone in vitro, apply it to the opposite mating type, and determine the gene expression profile. Such experiments would make it possible to identify pheromone response genes (Bennett et al. 2003).

Additionally, since the – mating type seems to cause a higher frequency of clinical cases than the + mating type, any differentially expressed genes are candidates that might influence virulence in the host. It may be informative to examine the gene expression profile of each mating type in macrophages to see if – strains induce a different set of genes than + strains. This information could initiate an understanding of how mating type might influence virulence.

V. Technologies that Interface with Genomics

The point of functional genomics is to implicate genes in a particular process. Analogous to a large-scale genetic screen, one ends up with candidates that might directly or indirectly be involved in a biological function. It is extremely useful to use functional genomics to identify candidate genes, but of course it is essential to examine the function of these candidates by other means. A number of technical advances now make it possible to make the most of functional genomics data in *H. capsulatum.*

A. Gene Disruption

To evaluate the role of candidate genes, it is desirable to disrupt the gene and examine the resultant phenotype. Only a small number of genes have been disrupted in *H. capsulatum.* The first gene disrupted by homologous recombination was the *URA5* gene (Woods et al. 1998). A *URA5* disruption construct was generated by replacing an internal portion of the *URA5* gene with the hygromycin resistance gene. *H. capsulatum* cells were transformed with a linear fragment of DNA containing the *URA5* disruption, and rare homologous recombination at the genomic *URA5* locus was observed. More recently, the laboratory of William Goldman improved this disruption technology by placing the gene disruption construct on a telomeric vector (Sebghati et al. 2000). Because linear plasmids with telomere ends can be maintained extrachromosomally in *H. capsulatum*, this strategy allowed cells to be grown for multiple generations in the presence of the disruption construct. Essentially, the telomeres reduced ectopic integration of the disruption construct at non- homologous loci, and thereby increased the relative frequency of the homologous targeting event. This technology was used to show that the *H. capsulatum CBP1* gene is required for virulence (Sebghati et al. 2000). Despite this improvement in the gene disruption technology, however, it remains fairly arduous to disrupt genes in *H. capsulatum*, and only a handful of gene disruption strains have been generated.

B. RNA Interference

RNA interference (RNAi) technology is a relatively fast and powerful alternative to gene disruption in other organisms (Meister and Tuschl 2004). The expression of a double-stranded RNA causes specific gene silencing in a number of systems. The ability to "knock down" gene function with RNAi in *H. capsulatum* would provide a relatively quick way to screen for phenotypes for candidate genes identified in microarray experiments. The laboratory of William Goldman recently developed RNAi technology for *H. capsulatum* (Rappleye et al. 2004). In these experiments, RNA hairpins were expressed from a strong promoter in wild-type cells. These hairpins consisted of a large piece of coding sequence separated from its reverse complement by a spacer region, such that a stem-loop structure is formed by the expressed RNA. These constructs were introduced into *H. capsulatum* on telomeric vectors, which allowed them to be maintained extrachromosomally. The authors were able to reduce expression of both exogenous (green fluorescent protein) and endogenous genes in a specific and stable fashion. In other systems, it is clear that expression of short interfering RNAs (as compared to the longer sequences used by Rappleye et al. 2004) is quite effective at triggering silencing (McManus et al. 2002). These constructs are easier to generate than are the long hairpin constructs. If they are sufficient to trigger RNAi in *H. capsulatum*, they will make going from gene sequence to phenotype even more expeditious.

C. Insertional Mutagenesis

Until recently, it has been difficult to generate insertion mutants in *H. capsulatum*, since the introduction of DNA on non-telomeric vectors often results in complex integration events and multiple integrations per genome (Worsham and Goldman 1990). However, as described below, it has now been possible to generate simpler insertions in *H. capsulatum*. I will first describe how this is done, and then discuss the potential interface between this technology and functional genomics.

Sullivan et al. (2002) used *Agrobacterium tumefaciens*, a bacterial pathogen of plants, to generate insertions in *H. capsulatum* and the related systemic dimorphic fungus *Blastomyces dermatiditis* (Sullivan et al. 2002). *A. tumefaciens* can transfer DNA from a 200-kb-virulence plasmid (the Ti plasmid) to a variety of hosts (Zhu et al. 2000; Christie 2001; Sullivan et al. 2002). The transferred DNA, or T-DNA, integrates into the host genome in a random fashion, most often at a single site. To generate *Agrobacterium*-mediated insertions in *H. capsulatum*, Sullivan et al. (2002) used a specialized system where the functions of the Ti plasmid have been split into two binary vectors (allowing easy manipulation of these plasmids). One plasmid is a partial Ti plasmid that retains only the virulence genes necessary for transfer of T-DNA and integration of this DNA into the host genome, and the other plasmid contains the T-DNA and a selectable marker (such as the hygromycin resistance marker) in the region of DNA that is transferred to the host. The latter allows *H. capsulatum* recipients of the T-DNA to be selected as hygromycin-resistant strains. Sullivan et al. (2002) successfully used either T-DNA carrying this hygromycin resistance marker or T-DNA carrying a *URA5* marker (for transformation of *ura5*-mutant strains).

The exact nature of the T-DNA boundaries that are transferred to the host genome can vary from transformant to transformant, although the majority of insertions span the region between the left border and right border of the T-DNA (Zhu et al. 2000; Christie 2001; Sullivan et al. 2002). On occasion, the whole T-DNA plasmid can be transferred and integrated, or concatemers of T-DNA can be inserted in the host genome. Sullivan et al. (2002) examined 15 *H. capsulatum* insertions by Southern blot. They determined that 12 transformants showed single sites of integration, six had integrated concatemerized T-DNA, and seven had integrated the entire T-DNA plasmid. The sizes of bands on the Southern blot were consistent with different sites of integration in the genome, implying that integration of the T-DNA occurs at random sites in the genome. These results have since been expanded to an analysis of over 100 independent insertion strains (Van Nguyen, personal communication; Anita Sil, personal observation).

This insertion technology could interface with functional genomics in a couple of different ways. First, the introduction of gene disruption constructs via *A. tumefaciens* instead of telomeric vectors may increase the frequency of targeted gene knockouts, as has been shown for other fungi (Das 1998; Gouka et al. 1999). As mentioned above, facilitating gene disruption will speed up the analysis of candidate genes that are identified by expression analyses. Second, it is possible that *Agrobacterium*-mediated insertion could mimic transposon site hybridization (TraSH), which has been an extremely productive means of using genomics to identify avirulent insertion mutants (Sassetti and Rubin 2002, 2003; Sassetti et al. 2003). To perform TraSH analysis, a library of transposon-generated insertion mutants is pooled and subjected to growth under two conditions – for example, in vitro culture versus infection of an animal. The transposon contains an outward-facing RNA polymerase promoter (such as T7 polymerase) that will transcribe RNA that is complementary to the chromosomal DNA adjacent to the transposon insertion. By making genomic DNA from the pooled strains, performing an in vitro transcription reaction to make probes that represent the transposon insertion site, generating differentially labeled cDNA from each condition, and hybridizing the resultant probes to a microarray, one identifies insertion strains that are not present in the pool of interest (for example, mutant strains that fail to colonize an animal). The sites of these insertions will implicate genes that are important for virulence. The hope for *H. capsulatum* is that one could use the T-DNA in an analogous fashion to the TraSH transposon. To do so, one would introduce an RNA polymerase promoter adjacent to a boundary of the T-DNA. One difficulty is that a fraction of *Agrobacterium*-mediated insertions contain not only the T-DNA, but other plasmid sequences as well, which means that the RNA polymerase promoter may no longer flank *H. capsulatum* genomic sequence in the insertion. Thus, these insertion strains would not be detectable by generating T7 RNA probes. Nonetheless, because *Agrobacterium*-mediated insertion is an easy method for generating large numbers of insertion strains in *H. capsulatum*, a small number of undetectable strains may not interfere with adapting the technology for a TraSH-like purpose.

VI. Conclusions

As discussed above, it is a propitious time to study *H. capsulatum* biology. Previous experiments have validated the power of functional genomic studies in this organism: a comparison of gene expression profiles in two major stages of the *H. capsulatum* life cycle yielded a large number of differentially expressed genes of varied putative function. Currently, three diverse *H. capsulatum* isolates are being sequenced, providing ample fodder for comparative genomics. Finally, the rich biology of *H. capsulatum*, coupled with the current state of molecular genetic tools in this system, bode well for the future of functional genomic analyses in this fungal pathogen.

Acknowledgements. I would like to thank Joseph DeRisi and all members of the Sil laboratory for interesting discussions. I especially thank Davina Hocking Murray for tireless and invaluable assistance with all the figures.

References

Ajello L, Runyon LC (1953) Infection of mice with single spores of Histoplasma capsulatum. J Bacteriol 66:34–40

Anderson KL, Marcus S (1968) Sporulation characteristics of Histoplasma capsulatum. Mycopathol Mycol Appl 36:179–187

Artis D, Baum GL (1963) Tuberculate spore formation by thirty-two strains of Histoplasma capsulatum. Mycopathol Mycol Appl 21:29–35

Bennett RJ, Uhl MA, Miller MG, Johnson AD (2003) Identification and characterization of a Candida albicans mating pheromone. Mol Cell Biol 23:8189–8201

Berliner MD (1973) Histoplasma capsulatum: effects of pH on the yeast and mycelial phases in vitro. Sabouraudia 11:267–270

Boldrick JC, Alizadeh AA, Diehn M, Dudoit S, Liu CL, Belcher CE, Botstein D, Staudt LM, Brown PO, Relman DA (2002) Stereotyped and specific gene expression programs in human innate immune responses to bacteria. Proc Natl Acad Sci USA 99:972–977

Borges-Walmsley MI, Walmsley AR (2000) cAMP signalling in pathogenic fungi: control of dimorphic switching and pathogenicity. Trends Microbiol 8:133–141

Bradsher RW (1996) Histoplasmosis and blastomycosis. Clin Infect Dis 22 Suppl 2:102–111

Brown AE (1990) Overview of fungal infections in cancer patients. Semin Oncol 17:2–5

Bryan CS, DiSalvo AF (1979) Overwhelming opportunistic histoplasmosis. Sabouraudia 17:209–212

Bullock WE (1993) Interactions between human phagocytic cells and Histoplasma capsulatum. Arch Med Res 24:219–223

Carr J, Shearer G Jr (1998) Genome size, complexity, and ploidy of the pathogenic fungus Histoplasma capsulatum. J Bacteriol 180:6697–6703

Christie PJ (2001) Type IV secretion: intercellular transfer of macromolecules by systems ancestrally related to conjugation machines. Mol Microbiol 40:294–305

Das A (1998) DNA transfer from Agrobacterium to plant cells in crown gall tumor disease. Subcell Biochem 29:343–363

Davis TE Jr, Domer JE, Li YT (1977) Cell wall studies of Histoplasma capsulatum and Blastomyces dermatitidis using autologous and heterologous enzymes. Infect Immun 15:978–987

Detweiler CS, Cunanan DB, Falkow S (2001) Host microarray analysis reveals a role for the Salmonella response regulator phoP in human macrophage cell death. Proc Natl Acad Sci USA 98:5850–5855

Detweiler CS, Monack DM, Brodsky IE, Mathew H, Falkow S (2003) virK, somA and rcsC are important for systemic Salmonella enterica serovar Typhimurium infection and cationic peptide resistance. Mol Microbiol 48:385–400

Di Lallo G, Gargano S, Maresca B (1994) The Histoplasma capsulatum cdc2 gene is transcriptionally regulated during the morphologic transition. Gene 140:51–57

Duclos S, Desjardins M (2000) Subversion of a young phagosome: the survival strategies of intracellular pathogens. Cell Microbiol 2:365–377

Eisen MB, Spellman PT, Brown PO, Botstein D (1998) Cluster analysis and display of genome-wide expression patterns. Proc Natl Acad Sci USA 95:14863–14868

Eissenberg LG, Goldman WE (1991) Histoplasma variation and adaptive strategies for parasitism: new perspectives on histoplasmosis. Clin Microbiol Rev 4:411–421

Eissenberg LG, Schlesinger PH, Goldman WE (1988) Phagosome-lysosome fusion in P388D1 macrophages infected with Histoplasma capsulatum. J Leukoc Biol 43:483–491

Eissenberg LG, Goldman WE, Schlesinger PH (1993) Histoplasma capsulatum modulates the acidification of phagolysosomes. J Exp Med 177:1605–1611

Gargano S, Di Lallo G, Kobayashi GS, Maresca B (1995) A temperature-sensitive strain of Histoplasma capsulatum has an altered delta 9-fatty acid desaturase gene. Lipids 30:899–906

Garrison RG, Boyd KS (1977) The fine structure of microconidial germination and vegetative cells of Histoplasma capsulatum. Ann Microbiol (Paris) 128:135–149

Garrison RG, Boyd KS (1978) Electron microscopy of yeastlike cell development from the microconidium of Histoplasma capsulatum. J Bacteriol 133:345–353

Garrison RG, Lane JW (1973) Scanning-beam electron microscopy of the conidia of the brown and albino filamentous varieties of Histoplasma capsulatum. Mycopathol Mycol Appl 49:185–191

Gasch AP, Moses AM, Chiang DY, Fraser HB, Berardini M, Eisen MB (2004) Conservation and evolution of cis-regulatory systems in ascomycete fungi. PLoS Biol 2(e398):1–18

Goldman WE (1995) Molecular genetic technology transfer to pathogenic fungi. Arch Med Res 26:437–440

Goos RD (1964) Germination of the macroconidia of Histoplasma capsulatum. Mycologia 56:662–671

Gouka RJ, Gerk C, Hooykaas PJ, Bundock P, Musters W, Verrips CT, de Groot MJ (1999) Transformation of Aspergillus awamori by Agrobacterium tumefaciens-mediated homologous recombination. Nat Biotechnol 17:598–601

Harris GS, Keath EJ, Medoff J (1989a) Characterization of alpha and beta tubulin genes in the dimorphic fungus Histoplasma capsulatum. J Gen Microbiol 135(7):1817–1832

Harris GS, Keath EJ, Medoff J (1989b) Expression of alpha- and beta-tubulin genes during dimorphic-phase transitions of Histoplasma capsulatum. Mol Cell Biol 9:2042–2049

Howard DH (1959) Observations on tissue cultures of mouse peritoneal exudates inoculated with Histoplasma capsulatum. J Bacteriol 78:69–78

Huffnagle GB, Deepe GS (2003) Innate and adaptive determinants of host susceptibility to medically important fungi. Curr Opin Microbiol 6:344–350

Hughes D (2000) Evaluating genome dynamics: the constraints on rearrangements within bacterial genomes. Genome Biol 1 REVIEWS0006:1–8

Hwang L, Hocking-Murray D, Bahrami AK, Andersson M, Rine J, Sil A (2003) Identifying phase-specific genes in the fungal pathogen Histoplasma capsulatum using a genomic shotgun microarray. Mol Biol Cell 14:2314–2326

Johnson CH, Klotz MG, York JL, Kruft V, McEwen JE (2002) Redundancy, phylogeny and differential expression of Histoplasma capsulatum catalases. Microbiology 148:1129–1142

Kasuga T, Taylor JW, White TJ (1999) Phylogenetic relationships of varieties and geographical groups of the human pathogenic fungus Histoplasma capsulatum Darling. J Clin Microbiol 37:653–663

Kasuga T, White TJ, Koenig G, McEwen J, Restrepo A, Castaneda E, Da Silva Lacaz C, Heins-Vaccari EM, De Freitas RS, Zancope-Oliveira RM et al. (2003) Phylogeography of the fungal pathogen Histoplasma capsulatum. Mol Ecol 12:3383–3401

Kauffman CA, Israel KS, Smith JW, White AC, Schwarz J, Brooks GF (1978) Histoplasmosis in immunosuppressed patients. Am J Med 64:923–932

Keath EJ, Abidi FE (1994) Molecular cloning and sequence analysis of yps-3, a yeast-phase-specific gene in the dimorphic fungal pathogen Histoplasma capsulatum. Microbiology 140(4):759–767

Keath EJ, Painter AA, Kobayashi GS, Medoff G (1989) Variable expression of a yeast-phase-specific gene in Histoplasma capsulatum strains differing in thermotolerance and virulence. Infect Immun 57:1384–1390

Klimpel KR, Goldman WE (1987) Isolation and characterization of spontaneous avirulent variants of Histoplasma capsulatum. Infect Immun 55:528–533

Klimpel KR, Goldman WE (1988) Cell walls from avirulent variants of Histoplasma capsulatum lack alpha-(1,3)-glucan. Infect Immun 56:2997–3000

Kwon-Chung KJ (1973) Studies on Emmonsiella capsulata. I. Heterothallism and development of the ascocarp. Mycologia 65:109–121

Kwon-Chung KJ, Weeks RJ, Larsh HW (1974) Studies on Emmonsiella capsulata (Histoplasma capsulatum). II. Distribution of the two mating types in 13 endemic states of the United States. Am J Epidemiol 99:44–49

Kwon-Chung KJ, Bartlett MS, Wheat LJ (1984) Distribution of the two mating types among Histoplasma capsulatum isolates obtained from an urban histoplasmosis outbreak. Sabouraudia 22:155–157

Lambowitz AM, Kobayashi GS, Painter A, Medoff G (1983) Possible relationship of morphogenesis in pathogenic fungus, Histoplasma capsulatum, to heat shock response. Nature 303:806–808

Long KH, Gomez FJ, Morris RE, Newman SL (2003) Identification of heat shock protein 60 as the ligand on Histoplasma capsulatum that mediates binding to CD18 receptors on human macrophages. J Immunol 170:487–494

Magrini V, Goldman WE (2001) Molecular mycology: a genetic toolbox for Histoplasma capsulatum. Trends Microbiol 9:541–546

Magrini V, Warren WC, Wallis J, Goldman WE, Xu J, Mardis ER, McPherson JD (2004) Fosmid-based physical mapping of the Histoplasma capsulatum genome. Genome Res 14:1603–1609
Maresca B, Kobayashi GS (1989) Dimorphism in Histoplasma capsulatum: a model for the study of cell differentiation in pathogenic fungi. Microbiol Rev 53:186–209
Maresca B, Medoff G, Schlessinger D, Kobayashi GS (1977) Regulation of dimorphism in the pathogenic fungus Histoplasma capsulatum. Nature 266:447–448
Maresca B, Carratu L, Kobayashi GS (1994) Morphological transition in the human fungal pathogen Histoplasma capsulatum. Trends Microbiol 2:110–114
Marques SA, Robles AM, Tortorano AM, Tuculet MA, Negroni R, Mendes RP (2000) Mycoses associated with AIDS in the Third World. Med Mycol 38 Suppl 1:269–279
McCaffrey RL, Fawcett P, O'Riordan M, Lee KD, Havell EA, Brown PO, Portnoy DA (2004) A specific gene expression program triggered by Gram-positive bacteria in the cytosol. Proc Natl Acad Sci USA 101:11386–11391
McManus MT, Petersen CP, Haines BB, Chen J, Sharp PA (2002) Gene silencing using micro-RNA designed hairpins. RNA 8:842–850
Medoff G, Maresca B, Lambowitz AM, Kobayashi G, Painter A, Sacco M, Carratu L (1986a) Correlation between pathogenicity and temperature sensitivity in different strains of Histoplasma capsulatum. J Clin Invest 78:1638–1647
Medoff G, Sacco M, Maresca B, Schlessinger D, Painter A, Kobayashi GS, Carratu L (1986b) Irreversible block of the mycelial-to-yeast phase transition of Histoplasma capsulatum. Science 231:476–479
Meister G, Tuschl T (2004) Mechanisms of gene silencing by double-stranded RNA. Nature 431:343–349
Nau GJ, Richmond JF, Schlesinger A, Jennings EG, Lander ES, Young RA (2002) Human macrophage activation programs induced by bacterial pathogens. Proc Natl Acad Sci USA 99:1503–1508
Nau GJ, Schlesinger A, Richmond JF, Young RA (2003) Cumulative Toll-like receptor activation in human macrophages treated with whole bacteria. J Immunol 170:5203–5209
Neilsen GE, Evans RE (1964) A study of the sporulation of Histoplasma capsulatum. J Bacteriol 68:261–264
Newman SL (1999) Macrophages in host defense against Histoplasma capsulatum. Trends Microbiol 7:67–71
Newman SL, Bucher C, Rhodes J, Bullock WE (1990) Phagocytosis of Histoplasma capsulatum yeasts and microconidia by human cultured macrophages and alveolar macrophages. Cellular cytoskeleton requirement for attachment and ingestion. J Clin Invest 85:223–230
Newman SL, Gootee L, Brunner G, Deepe GS Jr (1994) Chloroquine induces human macrophage killing of Histoplasma capsulatum by limiting the availability of intracellular iron and is therapeutic in a murine model of histoplasmosis. J Clin Invest 93:1422–1429
Newman SL, Gootee L, Kidd C, Ciraolo GM, Morris R (1997) Activation of human macrophage fungistatic activity against Histoplasma capsulatum upon adherence to type 1 collagen matrices. J Immunol 158:1779–1786
Nittler MP, Hocking-Murray D, Foo CK, Sil A (2005) Identification of Histoplasma capsulatum transcripts. Induced in response to reactive nitrogen species. Mol Biol Cell (in press)
Patel JB, Batanghari JW, Goldman WE (1998) Probing the yeast phase-specific expression of the CBP1 gene in Histoplasma capsulatum. J Bacteriol 180:1786–1792
Pine L (ed) (1960) Morphological and physiological characteristics of Histoplasma capsulatum. Thomas, Springfield, IL Procknow JJ, Page MI, Loosli CG (1960) Early pathogenesis of experimental histoplasmosis. Arch Pathol 69:413–426
Rappleye CA, Engle JT, Goldman WE (2004) RNA interference in Histoplasma capsulatum demonstrates a role for alpha-(1,3)-glucan in virulence. Mol Microbiol 53:153–165
Reiss E (1977) Serial enzymatic hydrolysis of cell walls of two serotypes of yeast-form Histoplasma capsulatum with alpha(1 leads to 3)-glucanase, beta(1 leads to 3)-glucanase, pronase, and chitinase. Infect Immun 16:181–188
Reiss E, Miller SE, Kaplan W, Kaufman L (1977) Antigenic, chemical, and structural properties of cell walls of Histoplasma capsulatum yeast-form chemotypes 1 and 2 after serial enzymatic hydrolysis. Infect Immun 16:690–700
Rippon JW (1988) Medical mycology. W.B. Saunders, Philadelphia, PA Samonis G, Bafaloukos D (1992) Fungal infections in cancer patients: an escalating problem. In Vivo 6:183–193
Sassetti C, Rubin EJ (2002) Genomic analyses of microbial virulence. Curr Opin Microbiol 5:27–32
Sassetti CM, Rubin EJ (2003) Genetic requirements for mycobacterial survival during infection. Proc Natl Acad Sci USA 100:12989–12994
Sassetti CM, Boyd DH, Rubin EJ (2003) Genes required for mycobacterial growth defined by high density mutagenesis. Mol Microbiol 48:77–84
Sebghati TS, Engle JT, Goldman WE (2000) Intracellular parasitism by Histoplasma capsulatum: fungal virulence and calcium dependence. Science 290:1368–1372
Smith CD (1964) Ii. Evidence of the presence in yeast extract of substances which stimulate the growth of Histoplasma capsulatum and Blastomyces dermatitidis similarly to that found in starling manure extract. Mycopathol Mycol Appl 22:99–105
Smith CD, Furcolow ML (1964) The demonstration of growth stimulating substances for Histoplasma capsulatum and Blastomyces dermatitidis in infusions of starling (Sturnis vulgaris) manure. Mycopathol Mycol Appl 22:73–80
Spitzer ED, Lasker BA, Travis SJ, Kobayashi GS, Medoff G (1989) Use of mitochondrial and ribosomal DNA polymorphisms to classify clinical and soil isolates of Histoplasma capsulatum. Infect Immun 57:1409–1412
Steele PE, Carle GF, Kobayashi GS, Medoff G (1989) Electrophoretic analysis of Histoplasma capsulatum chromosomal DNA. Mol Cell Biol 9:983–987
Sternberg S (1994) The emerging fungal threat. Science 266:1632–1634
Strasser JE, Newman SL, Ciraolo GM, Morris RE, Howell ML, Dean GE (1999) Regulation of the macrophage vacuolar ATPase and phagosome-lysosome fusion by Histoplasma capsulatum. J Immunol 162:6148–6154
Sullivan TD, Rooney PJ, Klein BS (2002) Agrobacterium tumefaciens integrates transfer DNA into single

chromosomal sites of dimorphic fungi and yields homokaryotic progeny from multinucleate yeast. Eukaryot Cell 1:895–905

Talaat AM, Lyons R, Howard ST, Johnston SA (2004) The temporal expression profile of Mycobacterium tuberculosis infection in mice. Proc Natl Acad Sci USA 101:4602–4607

Tian X, Shearer G Jr (2001) Cloning and analysis of mold-specific genes in the dimorphic fungus Histoplasma capsulatum. Gene 275:107–114

Tian X, Shearer G Jr (2002) The mold-specific MS8 gene is required for normal hypha formation in the dimorphic pathogenic fungus Histoplasma capsulatum. Eukaryot Cell 1:249–256

Timberlake WE (1991) Temporal and spatial controls of Aspergillus development. Curr Opin Genet Dev 1:351–357

Vieira OV, Botelho RJ, Grinstein S (2002) Phagosome maturation: aging gracefully. Biochem J 366:689–704

Vincent RD, Goewert R, Goldman WE, Kobayashi GS, Lambowitz AM, Medoff G (1986) Classification of Histoplasma capsulatum isolates by restriction fragment polymorphisms. J Bacteriol 165:813–818

Wheat LJ, Kauffman CA (2003) Histoplasmosis. Infect Dis Clin N Am 17:1–19

Wheat LJ, Connolly P, Smedema M, Brizendine E, Hafner R (2001) Emergence of resistance to fluconazole as a cause of failure during treatment of histoplasmosis in patients with acquired immunodeficiency disease syndrome. Clin Infect Dis 33:1910–1913

Woods JP (2002) Histoplasma capsulatum molecular genetics, pathogenesis, and responsiveness to its environment. Fungal Genet Biol 35:81–97

Woods JP (2003) Knocking on the right door and making a comfortable home: Histoplasma capsulatum intracellular pathogenesis. Curr Opin Microbiol 6:327–331

Woods JP, Retallack DM, Heinecke EL, Goldman WE (1998) Rare homologous gene targeting in Histoplasma capsulatum: disruption of the URA5Hc gene by allelic replacement. J Bacteriol 180:5135–5143

Worsham PL, Goldman WE (1990) Development of a genetic transformation system for Histoplasma capsulatum: complementation of uracil auxotrophy. Mol Gen Genet 221:358–362

Zhu J, Oger PM, Schrammeijer B, Hooykaas PJ, Farrand SK, Winans SC (2000) The bases of crown gall tumorigenesis. J Bacteriol 182:3885–3895

13 *Cryptococcus neoformans* Pathogenicity

R.T. Nelson[1,2], J.K. Lodge[1]

CONTENTS

Abbreviations: BAC, bacterial artificial chromosome; CHEF, contour-clamped homogeneous electric field; CSF, cerebral spinal fluid; EST, expressed sequence tag; GalXM, galactoxylomannan; CRAG, cryptococcal capsular polysaccharide antigen; GXM, glucuronoxylomannan; HAART, highly active antiretroviral therapy; HR, homologous recombination; MP, mannoprotein; SAGE, serial analysis of gene expression; STM, signature tagged mutagenesis

I. Introduction

Cryptococcosis has become a major problem since the advent of AIDS (for review, see Casadevall and Perfect 1998). In the Western world, the impact of this systemic fungal disease has been mitigated by antifungal agents and therapies that reduce HIV and prevent the severe reduction of T-cells that is associated with late stages of AIDS and the onset of opportunistic infections. However, in other parts of the world, cryptococcosis is still a major problem and it has been estimated that 30% of HIV-infected individuals in sub-Saharan Africa will succumb to cryptococcal meningitis. Recent outbreaks of *Cryptococcus neoformans* among immunocompetent individuals highlight the need for vigilance and continued investigation. Treatments are not completely effective or are highly toxic, and so novel drug targets are needed. Genomics has accelerated this field of research, facilitating rapid identification of genes

[1] Edward A. Doisy Department of Biochemistry and Molecular Biology, Saint Louis University School of Medicine, 1402 S. Grand Blvd., St. Louis, MS 63104, USA
[2] Current address: G329 Agronomy Hall, Iowa State University, Ames, IA 50011, USA

The Mycota XIII
Fungal Genomics
Alistair J.P. Brown (Ed.)

important for virulence, and permitting identification of new fungal-specific genes that would be excellent candidates for antifungal therapies.

II. Human Disease

A. Patient Population

Disseminated cryptococcosis is a disease primarily associated with individuals whose cellular immunity has been compromised by viral infection, suppression due to tissue transplantation or antineoplastic chemotherapy. An estimated 6%–10% of AIDS patients acquire cryptococcosis during the course of their HIV disease (Eng et al. 1986; Currie et al. 1994). A more recent figure for the incidence of cryptococcosis in AIDS patients indicates that the annual incidence rate for HIV-infected persons ranges from 17 to 66 in every 1000 (Hajjeh et al. 1999). This same study concluded that the incidence rate for non-HIV-infected people in the same metropolitan areas was between 0.2 and 0.9 per 100,000. This estimate, if applied to the US population as a whole (about 250 million), indicates that as many as 2250 HIV-negative people will acquire the disease each year.

Fungal infection is a significant threat to patients undergoing tissue transplantation, with as many as 59% acquiring a fungal disease at some centers. However, infection with *Cryptococcus* is a relatively rare event, with a mean incidence of about 2.8% (Husain et al. 2001). Susceptibility to cryptococcosis does not appear to be influenced by the tissue being transplanted, with 2% of thoracic organ transplant patients (Grossi et al. 2000), 3% of liver transplant recipients (Rabkin et al. 2000), and 3.9% of renal transplant patients acquiring the disease (Bach et al. 1973). The susceptibility of transplantation patients to cryptococcosis can apparently be influenced by the type and duration of immune suppression, most infections occurring more than 2 months after transplantation or as a result of increased suppression due to rejection (Snydman 2001). Variation in the incidence of cryptococcosis associated with different transplantation groups may be a product of variation in environmental exposure to this pathogen during the post-transplantation period (Snydman 2001), and not a function of the type of tissue transplanted, as has been seen in outbreaks of histoplasmosis in renal transplant patients (Wheat et al. 1983).

Immune suppression due to anti-neoplastic therapy is also a risk factor for the disease, with a projected incidence rate of approximately 18 of every 100,000 cancer patients becoming infected (Kontoyiannis et al. 2001). In a retrospective study of HIV-negative patients with malignancy at the M.D. Anderson Cancer Center, 65% of patients with cryptococcosis also had a hematologic malignancy, indicating the close association of the disease with the status of the host's cellular immunity (Kontoyiannis et al. 2001). Long-term heavy corticosteroid use is also a risk factor for the disease (Cunha 2001a).

B. Disease Presentation

In the HIV-compromised host, patients with cryptococcosis generally present with symptoms of meningitis such as fever, headache, and malaise with or without a stiff neck (reviewed in Casadevall and Perfect 1998). Nausea or altered mentation may or may not be present. Meningioencephalitis caused by *Cryptococcus* presents as an indolent infection with insidious onset, fever, and often a headache. It is the most frequent presenting syndrome of AIDS patients, representing approximately 60%–85% of all cryptococcosis cases. Neurologic symptoms are present in approximately 50% of patients, with cranial nerve palsies being the most frequent manifestation (Moosa and Coovadia 1997; Cunha 2001a). However, computed tomographic analyses of these patients are normal in approximately 60% of cases (Mitchell et al. 1995). The most common abnormalities visualized by computed tomographic analysis are single mass lesions (22%) and cerebral atrophy (19%; Graybill et al. 2000). Once formed, these lesions may persist many years after apparent disease eradication (Hospenthal and Bennett 2000). The persistence of these lesions will undoubtedly complicate the diagnosis of a possible relapse, and may thus be of limited use as indication of therapeutic response.

Definitive diagnosis of cryptococcal meningitis is made by isolating organisms from the CSF. Therefore, lumbar puncture of these patients is critical to the proper diagnosis and appraisal of their infection. A number of CSF parameters are of particular diagnostic and prognostic value, namely, opening pressure, mycological culture, cryptococcal antigen titer, and India ink staining (Fig. 13.1).

Opening pressure of the CFS has been shown to be a prognostic indictor of cryptococcal meningi-

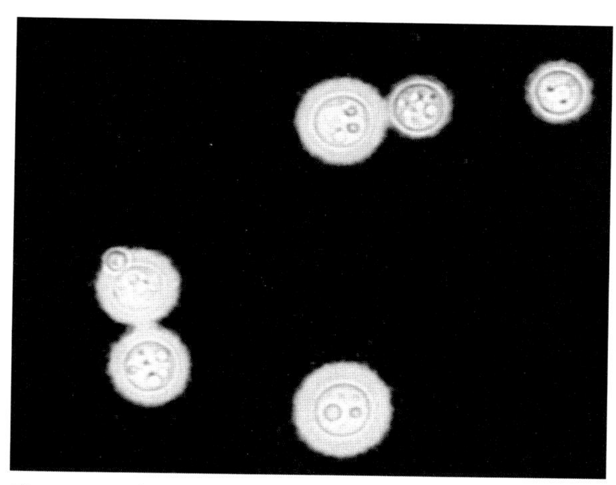

Fig. 13.1. India ink stain of *Cryptococcus neoformans* that illustrates the large polysaccharide capsule

tis, with a high opening pressure (>250 mm H_2O) an indication of a poor outcome (Graybill et al. 2000). Management of these patients may require repeated lumbar punctures to reduce intercranial pressure (Graybill et al. 2000; Saag et al. 2000). High intercranial pressure may also be treated by shunting excess CSF from the brain. In a study of four patients that had intracranial hypertension that was uncontrollable by pharmacologic therapy alone, Liliang et al. (2002) demonstrated that the installation of a ventriculoperitoneal shunt could reduce intercranial pressure in these patients, and presumably increase their chances of survival.

Cerebral spinal fluid parameters may be normal or near normal due to an absence of an inflammatory response in patients with advanced HIV disease (Graybill et al. 2000), and are thus of lesser diagnostic value. However, CRAG titers in infected CSF can also be very high, ranging up to 1:8000 or higher (Eng et al. 1986; Graybill et al. 2000). Because of this, testing CSF for the presence of CRAG in these patients is reliable, with a sensitivity of approximately 90%–100% in some studies (Chuck and Sande 1989; Tanner et al. 1994). Monitoring CRAG titers in the CSF as an indication of treatment response may be of some predictive value. In a study of 73 cryptococcal meningitis patents, 83% of patients who ultimately responded to treatment had a reduction in CRAG titer (Powderly et al. 1994). However, only 57% of those who did not respond to treatment had no change or an increase in their CRAG titer. Therefore, CSF CRAG titer reduction is a better indicator of treatment success than an increase in antigen titer is of treatment failure.

Cerebrospinal CRAG titers are also subject to false positive reactions. HIV-positive patients have been identified as having CSF CRAG titers without evidence of the growth of the organism from serum or urine, and without evidence of serum CRAG titers (Manfredi et al. 1996). This may be a result of the penetration of polysaccharide antigen in the brain tissue of infected patients (Lee and Casadevall 1996). These deposits could be the source of antigen detected in the CSF long after apparent clinical cure has occurred.

India ink staining of the CSF from these patients is also a fairly reliable diagnostic test, with 76%–95% of meningitis patients having a positive test (Graybill et al. 2000) when the test was performed on the sedimented cell pellet (Saito et al. 1999). One of the limitations of this test is that it would not readily allow one to differentiate between live and dead cells, and thus is of limited value in determining treatment progress.

These tests are good indicators of the presence of *Cryptococcus* in the brain. However, they are of lesser value as indicators of the amount of the fungus in the brain. In addition, only CSF culture provides indications of the metabolic state of the fungus, i.e., whether fungal cells are alive or dead. Thus, a test that is more sensitive to the burden of live or quiescent cryptococcal cells in the brain is needed in order to properly assess the effectiveness of anti-cryptococcal treatment in these patients.

Cryptococcal pneumonia is the next most frequent manifestation of cryptococcosis in AIDS patients. It occurs as a primary infection in approximately 4% of cases, and is associated with general symptoms of pneumonia such as fever, cough, pleuritic chest pain and/or dyspnea (Wasser and Talavera 1987; White and Armstrong 1994; Cunha 2001b). However, approximately 50% of all cryptococcosis patients have lung involvement. Lung infection is associated with the presence of subacute or chronic infiltrates on chest X-ray examination (Cunha 2001b; Sax 2001). The findings can include multifocal nodular or perihilar infiltrates with consolidation. Lung function, measured as pO_2, is generally normal, which may explain why as many as 25% of patients with pulmonary involvement may not be symptomatic (Pappas et al. 2001). The definitive diagnosis of cryptococcal pneumonia requires the isolation of the organism from bronchoalveolar lavage or pleural fluid (White and Armstrong 1994). This test is fairly sensitive, with a positive culture obtained in approximately 85% of cases (Cameron et al. 1991).

In patients with disseminated disease, measurable serum cryptococcal antigen (CRAG) is present 60% or more of the time. Graybill et al. (2000) found CRAG titers in the patients examined could be very high, ranging from 1:1037 to 1:16,384. Testing of serum from patients with disseminated disease for CRAG has a similar sensitivity as that of testing CSF (Tanner et al. 1994). To reduce the interpretation of false positive reactions, a titer of less than1:8 has been proposed as a useful cutoff point (White and Armstrong 1994).

Serum CRAG titers can also be misleading. In animal models, circulating polysaccharide and purified glucuronoxylomannan are both quickly removed from the circulation by macrophages and concentrated in the tissues of the reticuloendothelial system (Kappe and Muller 1991; Lendvai et al. 1998; Grinsell et al. 2001). If this were the case with human patients, then there is a potential source of antigen present in the body possibly long after the elimination of the organism. If this store of antigen were transiently released by the death of the macrophage, then this could explain the identification of patients with antigenemia yet without culture-proven disease. Conversely, a newly established infection may go undiagnosed by serum testing because the reticuloendothelial system can efficiently remove circulating antigens and yeast cells. This notion is bolstered by the observation that *Cryptococcus* cells may survive in macrophages that engulf them (Feldmesser et al. 2000). Eventually, the macrophage fills with polysaccharide and dies, liberating the organism (Tucker and Casadevall 2002). Thus, a test that does not rely on antigenic molecules of the capsule and yet is very sensitive to the presence of the organism is obviously needed.

DNA- or RNA-based detection methods have been demonstrated to be both sensitive and accurate tests for the presence of a number of pathogens. The use of DNA-based detection methods has been demonstrated using medically important yeasts including *Cryptococcus* (Rappelli et al. 1998; Aoki et al. 1999; Posteraro et al. 2000; Lindsley et al. 2001). DNA-based detection methods, while sensitive and specific, currently only provide evidence of the presence of the organism. They do not provide information regarding the metabolic state of the organism. Polymerase chain reaction amplification of the cDNA of genes associated with the vegetative or pathogenic growth of the organism would provide an indication both of the presence of the organism as well as its metabolic state.

Cryptococcosis in children is relatively rare, even in HIV-infected patients (Gonzalez et al. 1996). On the basis of a retrospective study of 1478 pediatric AIDS cases, Abadi et al. (1999) inferred that the 10-year point prevalence of cryptococcosis in these patients was 1.4% (Abadi et al. 1999), compared to an estimated 6%–10% rate for adults. This disparity may be influenced by exposure to this pathogen. In a study of urban children (Goldman et al. 2001), less than 50% of children under 2 years had serum reactive to cryptococcal proteins. This figure increased to 70% in children 5 or more years old. In another study of HIV-negative adults from the same urban area, 100% ($n = 26$) had serum reactive to cryptococcal antigens, indicating that most adults in this urban setting have been immunologically exposed to *C. neoformans* (Chen et al. 1999). The symptoms associated with infection appear to be similar to those seen with adults, with headache and fever being the most frequent (Abadi et al. 1999).

Cryptococcal infection in HIV-negative adults is similar to that seen in HIV-positive patients, with the exception of a slightly reduced frequency of CNS infection (51%) and elevated frequency of pulmonary infection (36%; Pappas et al. 2001). The symptoms of both pulmonary and CNS infection also appear to be similar to that seen in HIV-positive patients, with the exception of a higher frequency of nausea (72.3%) and altered mental status (52%) associated with CNS infection in HIV-negative compared to HIV-positive patients.

C. Therapy

Practice guidelines for the management of cryptococcal disease have been issued periodically. For a complete treatment of the current recommendations, the reader should see Saag et al. (2000). The recommendations for management of cryptococcosis vary depending on the immune status of the patient, and the site and extent of disease. Treatment options include amphotericin B, a combination of amphotericin B and flucytocine, fluconazole or itraconazole. For patients with renal disease, a lipid formulation of amphotericin B can be substituted in the induction regimen. Cryptococcal meningitis in any patient is potentially life threatening and should be treated aggressively.

In the immune-competent host, asymptomatic patients with pulmonary disease may be closely monitored for disease progression or treated with fluconazole for 3–6 months. Patients with symptomatic mild pulmonary disease should be treated with fluconazole orally for 6–12 months. Meningeoenchephalits of HIV-negative patients can be treated with a combination of amphotericin and flucytocine, possibly followed by oral fluconazole for at least 8–10 weeks. To assess whether the chosen therapy has achieved sterilization of the CSF, lumbar puncture should be performed. If CSF sterility has not been achieved, additional therapies for 6–12 months could be instituted. If oral fluconazole is not tolerated, itraconazole can be substituted.

The treatment regimen for HIV-positive individuals with CNS involvement is different than that of HIV-negative patients, in that treatment durations are much longer owing to the profound and lifelong immune suppression experienced by these patients. Even with effective, highly active antiretroviral therapy (HAART), lifelong maintenance therapy is recommended. However, in a study of six patients who responded to HAART of more than 12-month duration with an increase in total CD4+ count (>150 cells/μL), Aberg et al. (2002) demonstrated that anti-cryptococcal treatment could be discontinued without relapse. If this observation is supported by more studies, then the recommendation to continue anti-cryptococcal treatment for the life of the patient may be revised in the future.

Due to the severity of the disease, it is recommended that HIV-compromised individuals with pneumonia and CD4 lymphocyte counts of less than 200 cells/ml always be tested for disseminated cryptococcosis. For HIV-positive patients with asymptomatic culture-proven pulmonary disease or those with mild to moderate symptoms, treatment of fluconazole for life is recommended. Severe or progressive disease should be treated by the use of amphotericin B, followed by fluconazole or itraconazole therapy. These patients should also be evaluated for extra-neural disease, which is present in up to 50% of cases (Kovacs et al. 1985).

Maintenance therapy in HIV-positive patients currently consists of a lifelong course of oral fluconazole. Amphotericin B can be substituted for patients who cannot tolerate fluconazole or have had multiple relapses while on azole maintenance therapy. Maintenance therapy is necessary because of a 20%–60% relapse rate following initial anti-cryptococcal therapy (Bozzette et al. 1991; Powderly et al. 1992; Powderly 1996).

In HIV-negative patients (cancer and transplant) the relapse rate is variable. In an early study, the relapse or treatment failure rate approached 100% (Kaplan et al. 1977). In this study of 46 cryptococcosis patients, all patients died within 600 days of diagnosis. This may be due to the rather advanced disseminated disease these patients had at the time of diagnosis. In more contemporary studies, the relapse rate appears to approach 0% (White et al. 1992; Kontoyiannis et al. 2001), probably due to earlier diagnosis and better maintenance therapy.

III. Models for Studying Pathogenesis

A. Animal Models

The pathogenesis of *Cryptococcus* has been studied using various animal models of pulmonary, disseminated and CNS disease. Disseminated and pulmonary models of disease have been established in mice and rats (Casadevall and Perfect 1998). The disseminated mouse model is based on the injection of live cells into the circulation via the lateral tail vein. This leads to the hemotogenous dissemination of the cells to all parts of the body.

A mouse model of pulmonary infection has been established using two different inoculation routes. The first involves the surgical resection of the trachea of a mouse. The inoculum is delivered to the lungs in a 30-GA needle inserted into the trachea between adjacent cartilage rings.

A second inoculation route is through inhalation. This is thought to more closely mimic the natural course of infection. In this procedure, the mice are anesthetized and the animals suspended by their superior incisors on a string. An inoculum is then delivered to each animal by applying a cryptococcal cell suspension to one nare of the animal. The inoculum flows through the nasal cavity down the back of the animal's throat and is aspirated into the lungs. This type of inoculation is a very efficient: greater than 90% of the inoculum can be recovered from the lung 2 hours after inoculation, and the procedure is less technically demanding than are the surgical or tail vein techniques.

A cerebral model of cryptococcal infection has also been established in corticosteroid-treated rabbits (Perfect et al. 1980). The rabbits are immunosuppressed by methyl-prednisone treatments. The rabbits are then infected by inter-cisternal inoculation. The type and extent of infection in each model of cryptococcosis is subject to variation related to strain differences between *Cryptococcus* strains and the strain of animals used in the experiment.

B. Alternative Models

Non-mammalian models of pathogenesis have been developed for *C. neoformans* and include amoeba, slime mold, worms and flies. Although these models will probably be useful, none will completely mimic human disease, and it is unlikely that they will replace mammalian models. Because *C. neoformans* is commonly found in soil and does not have an obligate life-cycle stage that requires passage through a mammalian host, it has been puzzling why environmental isolates generally retain the capacity to cause disease. It has been hypothesized that many *C. neoformans* virulence mechanisms have been maintained by constant selective pressure from various soil organisms (Steenbergen et al. 2001). The interactions with amoeba and slime mold are very similar to the interactions of *C. neoformans* with macrophages (Steenbergen et al. 2001, 2003). *C. neoformans* has the ability to survive phagocytosis by *Acanthamboeba castillanii* and *Dictyostelium discoideum*, replicate in the phagocytic vacuole, and cause killing of these phagocytic hosts. The presence of the polysaccharide capsule on *C. neoformans* was shown to inhibit phagocytosis and be important for fungal survival in the presence of the hosts.

Other alternative models include *Caenorhabditis elegans* (Mylonakis et al. 2002) and *Drosophila melanogaster* (Apidianakis et al. 2004). Both of these organisms are killed when they ingest *C. neoformans*, whereas related fungal species do not have this effect. Although some similarities with mammalian models systems are evident, there are some clear distinctions, for example, the capsule does not appear to be important for virulence in either *C. elegans* or *D. melanogaster*.

IV. Biology of *Cryptococcus neoformans*

A. Taxonomy

Cryptococcus neoformans is a member of the Basidiomycetes. Its telomorph state was recognized in 1975 and assigned to the genus *Filobasidiella* by Kwon-Chung (1975). *Cryptococcus neoformans* is believed to grow in a yeast form in the environment, similar to the growth form in the tissues. It has a bipolar mating system with α(Mat α) and a (Mat a) mating type cells. This fungus is saprobic and has a strong association with pigeon droppings and some plants. In part because of its association with the excreta of the cosmopolitan pigeon and other birds, *C. neoformans* has a worldwide distribution.

Based on the antigens of the polysaccharide capsule, *Cryptococcus neoformans* has been separated into four serotypes (A, B, C, D; Wilson et al. 1968). There are three varieties of *C. neoformans* currently recognized – var. *neoformans*, var. *grubii* (Franzot et al. 1999) and var. *gattii*. Variety *grubii* is composed of *C. neoformans* of serotype A whereas variety *neoformans* is composed of *C. neoformans* of serotype D. Variety *gattii* is composed of *C. neoformans* serotypes B and C, and has been found associated with eucalyptus trees. Varieties *grubii* and *neoformans* (serotypes A and D) are responsible for the majority of human infections in the United States and Europe whereas variety *gattii* is more commonly seen in infections of immunocompetent individuals from Australia, Southeast Asia and other tropical areas of the world. Infection of immunocompromised individuals is predominantly due to variety *grubii*, regardless of geographic origin. A recent outbreak of var. *gattii* on Vancouver Island has been puzzling (Stephen et al. 2002). At least 66 individuals have been diagnosed with *Cryptococcus* since 1999, a tenfold increase in the normal rate, and many of these individuals have been otherwise healthy. The organism responsible has been identified as a serotype B strain, and rather than being found in conjunction with eucalyptus trees, it has been found in the soil or associated with other trees. This expansion of habitat by this primary pathogen and the associated increase of exposure of humans makes it imperative to understand the similarities and differences between the varieties.

The clinical presentation and outcome differ between infection with var. *gattii* and var. *grubii*. Infection with var. *gattii* tends to occur in immune a competent host, which separates it from infection with var. *neoformans*. The clinical presentation of infections with variety *gattii* in these patients tends to differ somewhat from that of variety *grubii*. Individuals infected with var. *gattii* have a higher frequency of abnormal CT findings (78%), with more multiple ring enhancing lesions (40%), focal CNS involvement (29%), pulmonary infection (65%), papilledema (50%), decreased mentation

(31%) and seizure (38%) than those infected with var. *grubii* (Mitchell et al. 1995; Speed and Dunt 1995). Conversely, they have a lower frequency of urinary tract infection (≤ 8%) and cryptococcemia (≤ 8%). The outcome from infection with var. *gattii* is significantly poorer than that for infections with var. *grubii*. Patients with var. *gattii* infections require more days of treatment, have more neurological sequelae, require more surgery, and suffer a higher frequency of relapse than those infected with var. *grubii*.

B. Life Cycle

This fungus is normally haploid and reproduce asexually by budding or haploid fruiting. Haploid fruiting was first discovered in Mat α cells that were subjected to severe nitrogen and moisture limitation (Wickes et al. 1996), but it can also occur in Mat a cells (Tscharke et al. 2003). In response to this stress, cells form monokaryotic hyphae with pseudo-clamp connections. These hyphae then produce basidiospores on which haploid basidiospores form. Some of these basidiospores are less than 3 μm and are thus capable of deep alveolar penetration. Under similar conditions, *C. neoformans* is capable of sexual reproduction. Under nitrogen starvation, if hyphae from opposite mating types encounter each other, the hyphae fuse and dikaryotic hyphae are produced. Sexual development continues and haploid basidiospores are produced.

C. Genetics

The mating type locus in *C. neoformans* is found on a chromosomal segment of about 100 kb (Karos et al. 2000; Lengeler et al. 2002), which is much larger than previously thought. It contains at least 20 genes, both specific for mating as well as of other functions. In this region, homologs of the mating-related genes *STE20*, *STE11*, *STE12*, MF1 (pheromone precursor) and *CPR1* (pheromone receptor) are located. The MF1 and *CPR1* homologs are not mating type-specific whereas *STE20*, *STE11*, and *STE12* have mating type-specific alleles for Mat and Mat a cells (Lengeler et al. 2000a, 2002; Clarke et al. 2001). The *STE12* homolog is important for haploid fruiting but does not appear to significantly affect mating or virulence of var. *grubii* (serotype A) strains using the rabbit model of CNS disease and the mouse model of disseminated disease (Yue et al. 1999). However, *STE12* does appear to affect the virulence phenotype in var. *neoformans* (serotype D) strains (Chang et al. 2000). The only alpha mating type-specific gene is *SXI1*, which has been identified as a protein that is essential for cell identity and sexual development, but not for virulence (Hull et al. 2002, 2004).

In a recent study of 358 clinical isolates from the United States, no Mat a cells of var. *grubii* were isolated (Yan et al. 2002). These data support the notion that Mat a cells of var. *grubii* must be exceedingly rare (Kwon-Chung and Bennett 1978), but it is not extinct. In a survey of clinical isolates from Tanzania, a single isolate of *C. neoformans* var. *grubii* was identified as being Mat a (Lengeler et al. 2000b), indicating that this mating type, while rare, still exists in the environment. Other serotype A Mat a strains were identified from the environment (Viviani et al. 2001) and from a clinical isolate (Viviani et al. 2003). Additional Mat a strains were identified from Botswana and shown to have engaged in sexual recombination in the environment (Litvintseva et al. 2003).

Recent work has taken the rare Mat a strains and crossed them to a hyper-responsive serotype A Mat α strain that was particularly prolific at mating (Neilson et al. 2003). The resulting strains were crossed with the commonly used clinical isolate H99, and then backcrossed ten times to produce a congenic mating pair for var. *grubii*. This mating pair is proving highly useful for many genetic applications, including crossing strains with different mutations and for linking virulence traits with mutations.

D. Molecular Genetics

C. neoformans is normally haploid, which facilitates certain reverse genetic manipulations such as gene deletion and replacement experiments. Over the last 15 years, tools have been developed, such as the selectable markers, transformation systems, reporter genes and regulatable promoters, which make genetic manipulation possible.

A number of several selectable markers have been developed for use with *C. neoformans*. These include hygromycin B resistance conferred by the hygromycin B phosphotransferase gene (*hpt*) from *Escherichia coli* (Cox et al. 1996), G418 resistance conferred by the kanamycin gene (*kan*) from the transposable element Tn5 (Hua et al. 2000), resistance to the antibiotic phleomycin from the bleomycin resistance gene (*ble*) also from the transposable element Tn5 (Hua et al. 2000), and resistance to the aminoglycoside nourseothricin conferred by the nourseothricin acetyltransferase gene (*nat1*) from *Streptomyces noursei* (McDade and Cox 2001). In addition to these dominant selectable markers, the auxotrophic markers of phosphoribosylaminoimidazole carboxylase activity conferred by the *ADE2* gene isolated from var. *grubii* (Toffaletti et al. 1993) and var. *neoformans* (Sudarshan et al. 1999), and orotidine monophosphate pyrophosphorylase activity conferred by the *URA5* gene isolated from var. *neoformans* (Edman and Kwon-Chung 1990) have been developed. In order to use these auxotrophic markers, appropriate recipient strains have been created.

To drive the expression of these genes as well as the expression of other genes, the promoter and terminator region of a number of several genes have been isolated and incorporated into numerous vectors. These include the promoter and terminator of the inducible *C. neoformans* var. *neoformans GAL7* gene (Wickes and Edman 1995), the constitutive promoter of the *C. neoformans* var. *grubii* actin gene (Cox et al. 1995), the terminator region of the N-myristoyltransferase (*NMT1*) gene of var. *grubii* (Hua et al. 2000) or the *TRP1* gene (McDade and Cox 2001), the constitutive promoter of the glyceraldehyde-3-phosphate dehydrogenase gene from var. *neoformans* (Varma and Kwon-Chung 1999), and the inducible promoter from the copper-regulated *CTR4* gene (Ory et al. 2004).

In addition to these selectable markers, green fluorescent protein (GFP) technology has also been developed for use with *C. neoformans*. A yeast-optimized green fluorescent protein (yGFP) was fused to the *GAL7* and the MFα promoter to demonstrate differential expression in the rabbit model of CNS infection (Del Poeta et al. 1999a).

Along with the development of vectors for gene expression and disruption, technologies for the transfer of the genes to recipient cells are required. Three transformation procedures have been developed to transfer recombinant DNA into the genome of *C. neoformans*. The first is the use of electroporation to transform yeast cells (Edman and Kwon-Chung 1990). The efficiency of this procedure varies depending on the locus interrupted, the marker used, the vector construction, and differences in the recipient strain. Published transformation efficiencies vary between 0 and >10,000 per μg of transforming DNA, resulting in both integrated and episomal forms of the transforming DNA. The relative frequency of each form also varied among the transformation procedures.

The second transformation procedure for *C. neoformans* is biolistic transformation (Toffaletti et al. 1993). In this procedure, transforming DNA is coated onto 0.6-μm gold beads. These beads are then shot out of a device using high-pressure helium gas. The beads are accelerated by the gas and strike recipient cells on agar plates, transferring the coated beads to the nucleus of the cell and liberating the attached DNA. The efficiency of this form of transformation is subject to the same sources of variation as that of the electroporation procedure. Published transformation efficiencies also vary from 0 to >1000 transformants per μg of transforming DNA due to those factors.

Recently, *Agrobacterium tumafaciens*-mediated transformation has been demonstrated for *C. neoformans* (Idnurm et al. 2004). Transforming DNA is inserted between the flanking T-DNA ends in the *A. tumafaciens* plasmid. The bacteria are co-cultured with *Cryptococcus*, and then selection is applied to eliminate the bacteria and identify fungal transformants. The advantages of this system is that it is relatively inexpensive to set up, it can be used to make random insertions, it does not seem to generate unstable extrachromosomal transformants, tandem repeats of the inserted DNA do not seem to occur, and because the ends of the inserted DNA are defined, cloning the flanking sequences is easier. The disadvantages are that constructing the *Agrobacterium* plasmid is still relatively time-consuming, multiple insertion events in a single cell can still occur, resulting phenotypes are not always linked to the inserted DNA, and it has not been demonstrated to integrate by homologous recombination.

V. Genome Project

A. Description of the Genome

The genome project for *C. neoformans* began in 1999 (Heitman et al. 1999), and by mid-2004, five *Cryptococcus* genomes have been sequenced. As preliminary studies for the genome sequencing, several features of the genome were determined to help anticipate timeframe and potential pitfalls.

To estimate the size of the genome, *C. neoformans* chromosomes were separated by pulsed field electrophoresis using contour-clamped homogeneous electric field (CHEF) gels (Perfect et al. 1989; Wickes et al. 1994). Examination of 20 strains by pulsed field electrophoresis suggested that there are between nine and 15 chromosomes, depending on the strain (Wickes et al. 1994). Most bands on the CHEF gels were counted as single chromosomes, but bands with more intense ethidium bromide staining were counted as two or three chromosomes, depending on the intensity of the staining. Within varieties, the number of chromosomes was not constant. Five var. *grubii* strains varied between 10 and 13 chromosomes, five var. *neoformans* strains had 12–15 chromosomes, and ten var. *gattii* strains had 9–14 chromosomes. Careful size analysis of chromosome bands from five strains representing all serotypes suggested that the genome is between 21 and 25 megabases (Wickes et al. 1994), although the sizes of the larger chromosomes had to be estimated because the largest available marker was smaller than the larger *C. neoformans* chromosomes.

Analysis of 24 genomic clones of *C. neoformans* genes deposited in Genbank suggested that the A+T content was close to 50%, indicating that there should be little difficulty with sequencing. Comparison of genomic clones to cDNA clones suggested that typical *C. neoformans* genes contained an average of six introns per gene, and that the introns were generally short, with an average size of 50–60 nt. This relatively high number of introns suggested that a concerted effort be made to delineate the *C. neoformans* intron/exon boundaries, and to develop gene prediction software specifically for *C. neoformans*.

B. Strain Choices

Five strains are currently the subject of genome projects: var. *neoformans*, strains JEC21 and B3501, var. *grubii*, strain H99, and var. *gattii*, strains WM276 and R265. JEC21 is a MAT α strain that has a congenic mating partner, and B3501 shares 50% of the JEC21 genome, but is significantly more thermotolerant and virulent. From var. *grubii*, the strain H99 was the unanimous choice and is a clinical isolate that has been used extensively for reverse genetics and virulence assays. It is highly virulent in all animal models, particularly the immunocompromised rabbit model (Perfect et al. 1980). WM276 was selected as the type strain for var. *gattii* since it was used extensively in the laboratory and in virulence assays. Finally, R265 is the var. *gattii* strain that is causing the outbreak on Vancouver Island and being sequenced at lower coverage.

C. Physical Mapping

The genomes of JEC21, H99, B3501, and WM276 have been physically mapped using fingerprinting of bacterial artificial chromosome (BAC) libraries at the University of British Columbia (Schein et al. 2002; James Kronstad, personal communication). BAC libraries were constructed for each strain. The inserts for the JEC21 library averaged 108 kb, and the inserts for the H99 library averaged 107 kb. The BACs were restricted with HindIII and the fragment sizes carefully analyzed. Overlapping BACs were identified by having overlapping HindIII fragments. The BACs have been assembled into 20 contigs for each strain. The contigs are long enough to be chromosomes, and they cover 15.79 Mb of JEC21 genomic DNA and 15.55 Mb of H99 genomic DNA. These sizes are smaller than the original predicted sizes of the genomes based on migration of chromosomes on CHEF gels. One likely explanation for some of the discrepancy is that there are a few gaps in the assembly, since there are more contigs than chromosome bands on the CHEF gels. These gaps may represent centromers or other repetitive elements that are difficult to clone. The physical maps now provide an excellent and useful framework for the assembly and finishing of JEC21 and H99.

D. Meiotic Map

A genetic linkage map for the MAT α strain B3501 and its sibling MAT a strain B3502 was constructed (Marra et al. 2004). Sixty-three polymorphic microsatellite loci, 228 restriction fragment length polymorphisms (RFLPs), two indel markers, and six mating type-specific genes were identified and used to create the linkage map. The 301 genetic markers were placed into 20 linkage groups that were subsequently mapped to the 14 chromosomes in both B3501 and B3502 on CHEF gels. Six chromosomes contained two linkage groups. Relating the linkage groups to the separated chromosomes revealed two reciprocal translocations.

E. Shotgun Sequencing

Shotgun sequencing has been completed for the five *Cryptococcus neoformans* strains, and the data are available for BLAST analysis and downloading from the different genome center websites (see Table 13.1). The coverage is greater than 10× for each of the strains, except WM272 that has 6× coverage. The genome of var. *neoformans* strain JEC21 genome is considered the "type" strain for *C. neoformans*, and has been finished by The Institute for Genome Research (TIGR; Loftus et al. 2005). TIGR made JEC21 libraries using small and medium inserts in plasmid vectors, so that paired reads could be obtained. Paired reads are assembled by sequencing each end of an inserted genomic fragment. The advantage of paired reads is that if there is a gap in the assembled sequence, some of the paired reads may fall into two different contigs and thus can be used to link the contigs together and provide a template for closing the gap. Contigs from the assembled genome sequence have been mapped to the physical BAC map (Schein et al. 2002) and to the meiotic map (Marra et al. 2004) described above.

The genome of the other var. *neoformans* strain, B3501, has been completed by Stanford University (Loftus et al. 2005). The data for this project have been generated by sequencing a combination of plasmid clones and M13 clones. M13 is a single-stranded vector, so only one side of the insert can be sequenced. The clones are rapidly prepared, but lack the advantage of the paired end reads. Because the virulence of JEC21 and B3501 are very different, with JEC21 being weakly pathogenic and B3501 being much more virulent, comparison of the genomes of these two closely related strains could provide some clues as to genes' important for virulence. Sequencing and annotation of both strains revealed that there are only a few strain-specific genes.

The var. *grubii* H99 genome was cloned by random shotgun cloning into plasmids. This has been a joint effort by Duke University and the Broad Institute, whereas sequencing var. *gattii* strain WM276 has been performed at the University of British Columbia. Completion of these projects will provide valuable comparative information about strain differences. Because the varieties have different epidemiologies, it may be possible to dissect parameters that are responsible for the different host and environmental ranges by comparison of the genomes and by functional characterization of gene expression.

F. EST Sequencing

C. neoformans genes have an average of about six introns each. The identification of intron splice signals is an important prelude to efficient and accurate gene finding. To this end, an expressed sequence tag (EST) sequencing project has been completed at the University of Oklahoma (Kupfer et al. 2004), and the data from that project can be viewed and downloaded from their website (see Table 13.1). cDNA libraries have been constructed for H99 and for B3501. Both ends of several thousand clones have been sequenced. Analyses of the cDNA sequences have resulted in the assembly of complete coding sequences. Comparison of the cDNA sequences with the genomic sequence

Table. 13.1. Websites of *C. neoformans* genome databases

Website	Genome center	Strain	Task
www.tigr.org/tdb/e2k1/cna1/	The Institute for Genome Research (TIGR)	JEC21, var. *neoformans*	Sequencing, assembly, annotation
www.sequence.stanford.edu/group/C.neoformans/	Stanford Genome and Technology Center	B3501, var. *neoformans*	Sequencing, assembly, annotation
cgt.genetics.duke.edu/data/index.html	Duke University Center for Genome Technology	H99, var. *grubii*	Sequencing, assembly, annotation
www.broad.mit.edu/annotation/fungi/cryptococcus_neoformans/	Broad Institute	H99, var. *grubii*	Sequencing, assembly, annotation
rcweb.bcgsc.bc.ca/cgi-bin/cryptococcus/cn.pl	University of British Columbia	H99, var. *grubii*; B3501, var. *neoformans*; JEC21, var. *neoformans*; WM276, var. *gattii*	Physical mapping
www.bcgsc.ca/about/news/crypto_public		WM276, var. *gattii*	Sequencing, assembly, annotation
www.genome.ou.edu/cneo.html	University of Oklahoma	H99, var. *grubii*; B3501, var. *neoformans*	EST sequencing

has pinpointed the location of introns within the genomic sequence, and analysis of these introns have identified likely splice site signals, including a 5′ splice junction of GU(A/G)(A/C)G(U/C), a 3′ splice junction of (C/U)AG, and a branch point signal of (C/U)U(G/A)A(C/U).

Importantly, 45,000 clones from a normalized cDNA library from JEC21 were sequenced (Loftus et al. 2005). The RNA was isolated from *C. neoformans* that was grown under eight different conditions including different mating types, nutrients, stresses, pHs and temperatures. Sequencing of a large number of cDNAs provided a substantial framework for annotation of the genome. EST coverage is estimated to be about 80% of the *C. neoformans* gene space. Sequencing this large number of ESTs also provided the interesting observation that *C. neoformans* has a significant level of alternative slicing and antisense RNA. Over 270 genes expressed differentially spliced RNAs, using the stringent criteria that three independent clones were found to contain the alternatively spliced form. Spliced, antisense transcripts were detected for over 50 genes. The function of alternative splicing or the potential for regulation of gene expression by these two mechanisms is yet to be determined.

G. Annotation

Genomic and cDNA sequences from the genome projects have been subjected to BLAST analysis using the non-redundant databases at NCBI, and the results are available on their respective websites. The data on gene splicing signals derived from the EST projects have been used to train gene-finding software, including TwinScan, specifically for *C. neoformans* (Tenney et al. 2004). TwinScan utilizes ab initio signals like splice junctions, but incorporates comparative genomic data from closely related species. The availability of multiple, but diverged strains of *Cryptococcus* has enhanced the gene-finding capabilities. Exons and splice junctions are more likely to be conserved between related species than are introns and other intergenic regions. Incorporating the comparison of the H99 genome to the JEC21 gene predictions improved the gene predictions by 10%. The TwinScan predictions were empirically tested by cloning and sequencing predicted transcripts. RT-PCR was used to amplify transcripts that were predicted by TwinScan that differed from manually annotated genes in GenBank. Sequencing showed that when the manually annotated genes followed the GA/UG or GC/UG splice junctions, the manual annotations were correct, but that when they did not, TwinScan identified the correct junction, suggesting that the splice junction sequence should be an important consideration of manual annotation (Tenney et al. 2004). Tests of over 100 genes predicted by TwinScan suggested that it could correctly identify introns about 90% of the time. This direct testing of gene prediction is valuable for several reasons: it is complementary to EST datasets, it facilitates assessment of the quality of the annotation, and it permits refinement of the gene prediction programs.

Another valuable approach to gene prediction has been to use multiple gene-finder programs and to compile the results (Loftus et al. 2005). By compiling the results of multiple gene-finding programs with the EST sequencing data, a relatively accurate set of predicted proteins has been generated. It is estimated that there are about 6500 protein-coding genes in the JEC21 genome, 80% of which are well supported by EST data. The number of genes is likely to change as gene prediction software is improved, and as the presence or absence of genes are experimentally determined.

VI. Analysis of Genes that Affect Pathogenicity of *C. neoformans* and the Prospects for Genomic Impact

Many genes that influence virulence have been identified in *Cryptococcus* using a variety of techniques, including predictions based on other pathogenic organisms and the analysis of mutants with obvious in vitro phenotypes. These factors include the ability to produce a polysaccharide capsule, melanin, urease, and phospholipase. Signal transduction pathways that regulate the production of these virulence factors have been studied extensively. The ability to produce melanin, urease and phospholipase are single-gene traits whereas capsule production requires a multi-step biosynthetic pathway requiring many genes in its biosynthesis. The fungi must also be able to survive in the phagolysosome and resist the reactive oxygen, nitrogen, and chlorinating species that are produced as microbicidal agents. The availability of genomic sequences is already having a major impact on facilitating the analysis of the contribution of many more genes to virulence. Having a complete set of predicted proteins facilitates rapid identification of paralogs of known virulence genes, and homologs of virulence genes from other organisms. Comparison of gene sets of fungal pathogens to non-pathogens may allow us to identify pathogen-specific gene sets. Transcriptome or proteomic analysis of virulence mutants will probably lead to the identification of other genes that impact the expression or actively of the virulence traits. The availability of mutants, either by insertional mutagenesis or by targeted deletion, will enable us to screen for genes that affect the expression or activity of these virulence factors.

A. Capsule Biosynthesis

Capsule production is a major virulence determinant of *Cryptococcus* (for review, see Bose et al. 2003). The ability of *C. neoformans* isolates to produce a thick polysaccharide capsule is unique among the pathogenic fungi. Production of the capsule is regulated by environmental cues such as CO_2 concentration, iron availability, and the presence of asparagine and glucose in the culture medium (Granger et al. 1985; Vartivarian et al. 1993). This capsule is composed of the polysaccharides glucuronoxylomannan (GXM), galactoxylomannan (GalXM) and mannoprotein (MP; Cherniak et al. 1980, 1988, 1992; Bhattacharjee et al. 1984). The GXM is composed of mannose, xylose and glucuronic acid residues, and the minor polysaccharide antigens GalXM and MP. The majority of the soluble cryptococcal polysaccharide consist of GXM and a lesser amount of GalXM and MP antigens (Cherniak and Sundstrom 1994). Besides being an impediment to phagocytosis (Kozel and Gotschlich 1982; Richardson et al. 1993) and subsequent vacuolar killing (Goldman et al. 2000; Tucker and Casadevall 2002), the capsule of *Cryptococcus* has immune-modulating properties as well.

Capsular polysaccharide has been demonstrated to affect the immune system in a number of different ways. Soluble polysaccharide interferes with the cell-mediated immunity by disturbing the chemotactic and proliferative response of phagocytic cells (Mody and Syme 1993; Fujihara et al. 1997; Syme et al. 1999; Coenjaerts et al. 2001). It also reduces the antigen-presenting capacity of these cells (Retini et al. 1998). Injections of purified polysaccharide can also reduce the immune response to subsequent cryptococcal antigen immunization (Kozel et al. 1977). This tolerance is associated with T-cell-dependent and -independent pathways (Sundstrom and Cherniak 1993).

Polysaccharide can also modulate the innate immune system. Whole cells or purified polysaccharide induce the production of the complement factor C3 in mouse peritoneal cells (Blackstock and Murphy 1997) as well as C3 binding in both mouse and human serum (Wilson and Kozel 1992; Pfrommer et al. 1993). Factor C3 is converted to C3b by contact with the capsule (Pfrommer et al. 1993). The active C3b fragment is then quickly converted to the inactive iC3b by contact with the capsule. Complement factor C5 and C5a are also involved in the reaction with whole cells or polysaccharide of *Cryptococcus* (Vecchiarelli et al. 1998). In addition, capsular polysaccharides may be capable of interfering with the C5a receptor on neutrophils, which also reduces the cell-mediated immune response (Monari et al. 2002).

Mannoprotein is a relatively minor constituent of the capsule. However, it is capable of producing immune-modulating effects on its own (Pietrella et al. 2001). Mannoprotein induces a strong protective response when injected into mice, leading to the production of the pro-inflammatory cytokines IL-12 and INF-γγ by macrophages. Recently, two mannoprotein antigens have been isolated and identified. Both were deacetylases that were capable of producing a protective response in animal models when used as an immunogen (Levitz et al. 2001; Biondo et al. 2002).

There are at least eight genes known to affect capsule biosynthesis. *CAP59*, *CAP64*, *CAP60* and *CAP10* were discovered on the basis of their ability to complement capsule-deficient mutants (Chang and Kwon-Chung 1994, 1998, 1999; Chang et al. 1996). The function of these genes is unknown, but recent work suggests that *cap59* mutants cannot transport the capsular material to the cell surface (Garcia-Rivera et al. 2004). By predicting the biochemical reactions that would be required for capsule synthesis, Doering and colleagues have identified another gene important for capsule synthesis, a UDP-xylose synthase encoded by *UXS1* (Bar-Peled et al. 2001). *uxs1Δ* strains are completely avirulent (Moyrand et al. 2002). Perfect and colleagues (Wills et al. 2001) cloned and characterized a phosphomannose isomerase gene (*MAN1*) that affects capsule biosynthesis and virulence, since mannose is a major constituent of the *C. neoformans* capsule. Janbon and colleagues have identified mutants that make an altered capsule, and have identified the corresponding genes by complementation. One of these, *CAS1*, is required for GXM O-acetylation (Janbon et al. 2001; Kozel et al. 2003), and the deletion of this gene results in a more virulent phenotype. This is the first example of the capsule structure playing a role in the pathobiology of disease progression. Using predictions of biochemical pathways important for capsule synthesis and the identification of potential paralogs of genes known to affect capsule synthesis, genomic analysis reveals at least 30 additional genes that could be important for capsule biosynthesis. Two gene families related to *CAP64* and *CAP10* are found in the *C. neoformans* genome, and these can now be tested for their impact on capsule biosynthesis.

It was shown recently that the capsule requires alpha-1,3-glucan in the cell wall in order to bind to the cell surface of *C. neoformans* (Reese and Doering 2003). Glucanase treatments that included alpha glucanases reduced binding of capsular material. Availability of the genome sequence made it possible to rapidly define the single gene responsible for alpha-glucan synthesis, and to show by RNAi that reduction of expression of the alpha-glucan synthase reduced the binding of capsule to *C. neoformans*.

B. Melanin Biosynthesis

Cryptococcus neoformans is the only member of the genus *Cryptococcus* that produces significant quantities of melanin, a brown to black pigment, when grown on medium containing diphenolic compounds (for review, see Zhu and Williamson 2004). Other species of *Cryptococcus* are apparently capable of producing lesser amounts of the pigment (Ikeda et al. 2002). This trait has been used to differentiate *Cryptococcus neoformans* from other *Cryptococcus* species as well as other yeasts of medical importance such as *Candida albicans* (Korth and Pulverer 1971; Shaw and Kapica 1972). Melanin is produced by *Cryptococcus* in the environment, is assumed to protect the cells against oxidative stress and UV light damage (Wang and Casadevall 1994; Nosanchuk et al. 1999a), and may help protect against antifungal agents (van Duin et al. 2002).

The ability to produce melanin has been demonstrated to be a virulence factor in experimental models of cryptococcosis (Kwon-Chung et al. 1982; Rhodes et al. 1982; Kwon-Chung and Rhodes 1986). Melanin is produced by the laccase gene (*CNLAC1*) of *Cryptococcus* (Williamson 1994). This enzyme is found in the cell wall of the fungus as well as the plasma membrane (Polacheck et al. 1982; Zhu et al. 2001). Specific inactivation of this gene has been demonstrated to confer a significantly less virulent phenotype to the recipient strain (Salas et al. 1996).

Melanin production has been demonstrated to protect *C. neoformans* from oxidative stresses (Jacobson and Emory 1991; Jacobson and Tinnell 1993). This property has been suggested to protect *Cryptococcus* from oxidative attack by phagocytic cells (Liu et al. 1999a). In addition, melanin may also protect the yeast from the effect of microbiocidal proteins (Doering et al. 1999), and may interfere with the host's cell mediated immune reactions (Huffnagle et al. 1995). The phenoloxidase activity of the laccase enzyme is responsible for converting neurochemical compounds such as the catecholamines epinephrine, norepinephrine and dopamine to melanin (Shaw and Kapica 1972; Polacheck et al. 1982). This activity has led some to postulate that the presence of melanogenic substrates in the brain contributes to the cerebral tropism of *Cryptococcus*. However, although *Cryptococcus* can produce melanin in the brain (Nosanchuk et al. 1999b, 2000; Rosas et al. 2000), there is evidence for the production of other oxidative compounds being produced (Liu et al. 1999b).

Expression of the laccase enzyme is negatively regulated by temperature and glucose concentration. Cnlac1 production is repressed by the presence of glucose in the culture medium (Polacheck et al. 1982). The derepression of laccase activity in medium deficient in glucose was sensitive to cycloheximide, indicating that protein synthesis is involved in the increase in *Cnlac1* activity, and thus glucose does not interfere with the activity of the enzyme. Similarly, laccase activity is reduced in cells incubated at 37 and 40 °C (Jacobson and Emory 1991), but the catalytic activity of the enzyme is not significantly affected at these temperatures. Genomic analysis of *C. neoformans* has revealed a second laccase gene that contributes to melanin production (Missall et al. 2005a; Pukkila-Worley et al. 2005).

C. Phospholipase

The ability to produce extracellular phospholipase has been identified as a virulence factor in both bacteria and fungi. Extracellular phospholipase activity has been identified in *Cryptococcus* strains collected from clinical sources (Chen et al. 1997a, b). The amount of phospholipase activity produced by individual isolates has been correlated with their ability to cause disseminated disease. In addition, a phospholipase gene from *Cryptococcus* (*PLB1*) has been cloned and its activity shown to be necessary for the full pathogenic potential of the organism (Cox et al. 2001). Animals injected with *plb1* deletion strains survived longer than those injected with wild-type strains. In the rabbit model of CNS disease, animals injected inter-cisternally with the mutant strains had significantly smaller fungal burden in the CSF than did those of control animals. *C. neoformans* produces eicosanoids that may act as signaling molecules between fungal cells, or between the host and the fungus (Noverr et al. 2001). Phospholipase has been implicated in prostaglandin production, *plb1* mutants exhibit a defect in eicosanoid production (Noverr et al. 2003), and it has been hypothesized that secreted phospholipases may help cleave the substrates needed for prostaglandin production from the host membranes.

D. Urease

Urease production has been shown to be involved in the pathogenesis of *Cryptococcus*. A *Cryptococcus* urease gene (*URE1*) has been cloned and sequenced (Cox et al. 2000). The deletion of this gene produced mutants that were significantly less virulent than were the wild-type or replacement strains when infected by the lateral tail vein or the inhalation route in mice. In contrast, rabbits infected inter-cisternally with mutant or wild-type strains did not have significantly different fungal burdens in their CSF. Thus, the *ure1* phenotype may be subject to species- or organ-type variation. Recent studies in mice have shown that *ure1* strains grow well in the brain when they are inoculated directly, but do not disseminate to the brain from other inoculation sites (Olszewski et al. 2004), suggesting a role for urease in dissemination.

E. Signal Transduction Pathways

Signal transduction pathways control expression of virulence factors including capsule biosynthesis and melanin production (for review, see Lengeler et al. 2000a). Extensive work has shown that the cyclic AMP- dependent protein kinase signaling pathway is a major pathway that controls both mating and the production of virulence factors in var. *grubii* (D'Souza et al. 2001; Pukkila-Worley and Alspaugh 2004; Hicks et al. 2004). A G-protein coupled receptor detects the signal (Alspaugh et al. 1997). An adenylyl cyclase is not essential for vegetative growth in *C. neoformans*, but has been shown to regulate mating, production of melanin and capsule, and virulence in a mouse model (Alspaugh et al. 2002). A catalytic subunit of the cAMP-dependent protein kinase (*PKA1*) was shown to be essential for production of virulence factors, virulence and mating, and the deletion of the regulatory subunit of the protein kinase (*PKR1*) was shown to have constitutive hyper-expression of capsule and melanin production, and was hyper-virulent in mouse models of cryptococcal infection (D'Souza et al. 2001). Interestingly, there is a striking difference in this pathway between the var. *grubii* strain, H99 and the var. *neoformans* strain JEC21 (Hicks et al. 2004). Deletion of the *PKA1* gene in JEC21 did not result in the reduction of melanin and capsule. Searches of the genome databases revealed a second PKA paralog (*PKA2*). Deletion of *PKA2* in H99 did not affect virulence factor production, but deletion of *PKA2* in JEC21 reduced expression of both capsule biosynthesis and melanin. Interestingly, deletion of *PKA2* in JEC21 did not affect virulence in a mouse model, nor did deletion of *PKR1* result in hyper-virulence.

The calcium/calmodulin signal transduction pathway also impacts virulence. Strains deleted for calcineurin are avirulent and cannot grow at higher temperatures (reviewed in Lengeler et al. 2000a; Kraus and Heitman 2003). The protein kinase C (PKC) signal transduction pathway has been extensively studied in other fungi for its role in cell wall biogenesis. Deletion of the MAP kinase MPK1 reduces cell integrity and virulence in *C. neoformans* (Kraus et al. 2003). The deletion of homologs of other genes in the PKC pathway including *MKK2*, *ROM2*, *SIT4*, and *LRG1* results in sensitivity to cell wall inhibitors and reduced virulence in a mouse model (Gerik and Lodge, unpublished data).

F. Mating Type

Virulence and mating type have been linked in var. *neoformans*, but not in var. *grubii*. The vast majority of clinical isolates are Mat a, suggesting that there was a significant difference in virulence between the two mating types. The congenic mating pair from var. *neoformans*, JEC21 and JEC20, were tested for virulence in a mouse model, and it was shown that JEC21 was more virulent in a mouse model (Kwon-Chung et al. 1992). This result has been repeated in other laboratories. However, congenic pairs of var. *grubii* strain K99a and K99α had identical virulence in mouse models and rabbit models of pathogenesis (Neilson et al. 2003). Genomic analyses may help to elucidate the difference in virulence between the serotype D, var. *neoformans* mating type and the serotype A, var. *grubii* strains.

G. Taxonomy

The epidemiology and environmental niche of the *C. neoformans* and *C. gattii* are significantly different. *C. gattii* is largely found in subtropical parts of the world, is found in conjunction with eucalyptus trees, and infects immunocompetent hosts. Var. *grubii* and var. *neoformans* are found worldwide, are often associated with soil contaminated with bird droppings, and infect immunocompromised hosts. A recent study examined var. *grubii* clinical isolates that were obtained from apparently healthy individuals (D'Souza et al. 2004). The authors showed that the isolates were not hybrids of *C. gattii* and var. *grubii*, nor did they observe *PKR1* mutations that would be consistent with hypervirulence. However, perturbations in cAMP signaling, which is known to regulate expression of virulence traits, were present in these strains.

Var. *grubii* and var. *neoformans* are also distinct from each other. In general, the var. *grubii* strains are more virulent in animal models and are more prevalent as clinical isolates. Var. *neoformans* strains require higher inoculums in order to cause the same level of disease as that of a var. *grubii* strain, and var. *neoformans* is virulent in the rabbit model of meningitis. There is an ongoing effort to compare the genomes from the three subspecies in order to identify candidate genes that might play a major role in these phenotypic differences. Genome comparison of two var. *neoformans* strains with very different virulence properties, JEC21 and B3501, is complete. There are only a few genes that are unique to each strain, suggesting that the differences in virulence are probably due either to single nucleotide polymorphisms between the strains, or to differences in gene expression (Loftus et al. 2005).

H. Survival in a Phagocytic Environment and Resistance to Oxidative and Nitrosative Stress

An additional virulence-related trait of *C. neoformans* is the ability of cells to survive in phagocytic cells and invade endothelial and epithelial cells. *C. neoformans* has been demonstrated to survive in the phagolysosome of human and rat macrophages (Levitz et al 1999; Goldman et al. 2000). Persistence in this compartment appears to rely on the fusion of lysosomes with the phagosome containing the engulfed fungal cell. This fusion leads to an acidification of the phagolysosome, which is necessary for the continued survival of the organism (Levitz et al. 1997). *C. neoformans* grows well in medium with ambient pHs as low as 3.0 (Lodge, unpublished data). *C. neoformans* cells persist inside phagolysosomes where they grow and produce extracellular polysaccharide that is contained in additional vacuoles (Feldmesser et al. 2000, 2001). Eventually, the host macrophage is weakened and dies, liberating the resident fungal cells. *C. neoformans* is also capable of adhering to and invading endothelial and epithelial cells (Merkel and Cunningham 1992; Goldman et al. 1994, 2000; Ibrahim et al. 1995). Once bound, the cells are internalized by phagocytosis. The phagosome then is transported through the cell and presumably exits from the cell. The incorporation of the fungus into the epithelial cell is associated with cell injury and death. This injury appears to be associated in part with the active production of some factor, and is increased by the presence of capsular constituents. Thus, *C. neoformans* has the capacity to evade the immune system by surviving in phagocytic cells. Once liberated, these cells can then escape the confining granulomas and invade adjacent epithelial cells. Because *C. neoformans* cells are also capable of traversing endothelial cells, *C. neoformans* can invade small blood vessels as well as escape from them. These phenotypes undoubtedly play pivotal roles in Cryptococcal pathogenesis.

One aspect of survival inside a phagocytic cell that is amenable to genomic approaches, and has been characterized to a certain extent, is the ability to resist oxidative and nitrosative stresses. Macrophages and neutrophils produce numerous reactive oxygen, reactive nitrogen and reactive chlorinating species in order to kill invading microorganisms. Since *C. neoformans* can survive inside this environment, it must have mechanisms to resist the reactive molecular species (reviewed in Missall et al. 2004a). Based on work in other systems, one can predict the kinds of proteins that may be important for resistance to various reactive species. Analysis of the *C. neoformans* genomic sequences shows that there are two superoxide dismutases to protect from superoxide. To reduce peroxides, there are four catalases, several classes of peroxidases including thiol peroxidases, glutathione peroxidases, glutaredoxins and cytochrome c peroxidases. To protect against reactive nitrogen species, *C. neoformans* has a flavohemoglobin denitrosylase and an S-nitrosoglutathione (GSNO) reductase.

C. neoformans has a Cu, Zn superoxide dismutase and a Mn superoxide dismutase. The Cu,Zn superoxide dismutase gene (*SOD1*) has been extensively characterized. It was cloned, and *sod1* deletions were made in both var. *grubii* (Cox et al. 2003) and *C. gattii* (Narasipura et al. 2003). *sod1Δ* mutants are less resistant to superoxide in vitro, less virulent in a mouse model, and less able to survive in macrophage-like cell lines or human neutrophils.

As described below, a thiol peroxidase was shown by proteomic analysis to be up-regulated at high temperature, and has been shown to be important for resistance to oxidative and nitrosative stress, survival in macrophages, and virulence in a mouse model (Missall et al. 2004b). This thiol peroxidase is predicted to require thioredoxin for recycling. The entire thioredoxin pathway, including Tsa1, two thioredoxins and thioredoxin reductase, and the glutathione pathway, including three glutaredoxins and two glutathione peroxidases, are being analyzed for their contributions to resistance to oxidative and nitrosative stresses. Both pathways play a role in resistance to oxidative and nitrosative stress, and they both influence the survival of *C. neoformans* in a macrophage-like cell line (Missall and Lodge 2005b; Missall et al. 2005b).

Gene deletion of the *C. neoformans* flavohemoglobin denitrosylase gene (*FHB1*) and the GSNO reductase (*GNO1*) demonstrated that the *FHB1* gene was important for resistance to NO (de Jesus-Berrios et al. 2003). Deletion of the *FHB1* gene also reduced survival in macrophages and virulence in a mouse model, and this reduction was dependent on host NO production. A wild-type mouse strain was less susceptible to the *fhb1Δ* strain than to wild-type *C. neoformans*, but an iNOS mouse was equally susceptible to the *fhb1Δ* and wild-type fungus. In contrast, the *GNO1* gene was not required for in vitro NO resistance or for virulence (de Jesus-Berrios et al. 2003).

Taken together, many virulence-related genes have been identified in *C. neoformans* by more traditional methods, and genomic information is accelerating and enhancing these studies. The ability to proliferate in the lung, parasitize phagocytic cells, escape from the lung, cross the blood–brain barrier, and proliferate in the brain most probably requires more cryptococcal genes that just those that have already been described. Genome-wide analysis such as transcriptome or proteomic studies of expression during macrophage engulfment or during pathogenesis will probably reveal genes that could be important for *C. neoformans* survival in vivo.

VII. Post-Genomic Approaches for Identification of Virulence Genes and Their Use in *C. neoformans*

A. Functional Genomic Techniques

Post-genomic approaches to analyzing biological function, regulatory networks and processes often include techniques that permit global analysis of gene expression either at the RNA level or at the protein level. The analysis of protein or mRNA expression during the modulation of external conditions or during particular developmental states can provide useful clues about what genes might be important for particular functions. Another useful approach is the systematic generation of gene knockouts to analyze the contribution of many genes to a particular phenotype.

1. Transcriptome Analysis

a) Microarrays

Microarray analysis is a powerful technique that allows one to simultaneously analyze comparative RNA expression levels from thousands of genes. Essentially, DNA corresponding to the different genes is spotted on a solid support, often a glass slide. The DNA can be PCR products from genomic clones, cDNA clones or oligonucleotides. RNA is isolated from organisms or tissues grown under particular conditions. The RNA is labeled with fluorescent dyes, and mixed with a differentially labeled control RNA and hybridized to the DNA on the slide. The ratio of the experimental RNA to the control RNA is measured by comparing the different fluorescent signals. Microarrays have been used in many other pathogenic organisms to delineate their developmental program and their response to environmental stimuli such as macrophage engulfment, iron limitation, antibiotics, or stress (reviewed in Lorenz 2002; see, for example, Whiteway and Nantel, Chap. 8, this volume).

The first report of microarray analysis was recently published. Kraus et al. (2004) used a >6000 element microarray consisting mostly of random genomic clones to probe the expression profiles of the strain H99 at 25 and 37 °C. They reported that the genes that were up-regulated at 37 °C included those important for cell wall biogenesis (chitin synthases and a WSC domain protein), genes involved in resistance to oxidative stress (catalases, superoxide dismutase, and other oxidases), genes in trehalose biosynthesis, a voltage-gated chloride channel important for melanin biosynthesis (*CLC1*), and the transcription factor *MGA2*. The deletion of *MGA2* resulted in slower growth and a slight temperature sensitivity at 37 °C. Microarray analysis of the *mga2Δ* mutant suggested that this transcription factor is important for the regulation of lipid biosynthesis.

Several laboratories have initiated microarray studies using PCR products based on EST sequences (J. Murphy, personal communication). A spotted oligonucleotide microarray is being developed, and it should be available to the entire academic community by 2005. The first arrays will carry 70-mer oligonucleotides representing genes in both serotype A and serotype D strains (C. Hull, T. Doering and J. Lodge, personal communication).

b) SAGE

*S*erial *a*nalysis of *g*ene *e*xpression (SAGE) is another technique that is used to quantify changes in mRNA expression. This technique is diagrammed in Fig. 13.2. RNA is isolated from cells grown under different conditions and cDNA is made from the RNA. SAGE tags from the cDNAs are generated and ligated together. The tags from many genes are ligated into one large fragment, which is then cloned into a plasmid vector and sequenced. The tags can be uniquely linked to a specific gene, and the number of times that a tag is found is directly proportional to the expression level of that gene. SAGE has the advantage that the libraries can be made before all genes in the genome are annotated, but the construction of the libraries is labor-intensive, and therefore the analysis of multiple replicates, either biological or technical, is rare.

Two studies have focused on the SAGE analysis of *C. neoformans*. In the first, the transcriptome of *C. neoformans* grown at 25 °C was compared to that at 37 °C in two strains: the var. *grubii* clinical isolate H99 and the var. *neoformans* strain B3501

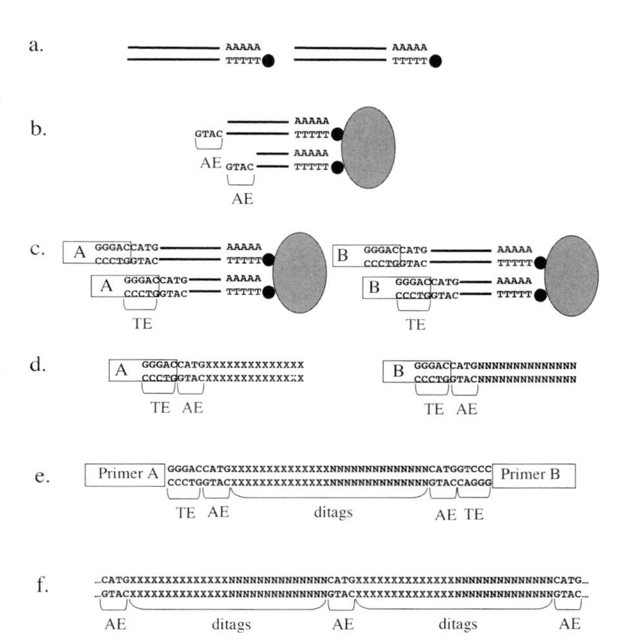

Fig. 13.2.a–f Serial analysis of gene expression (SAGE) library construction. **a** mRNA from cells or tissue is converted to cDNA using biotinylated primers (*dots*). **b** DNA copy of mRNA is cut with the anchoring enzyme (*AE*, in this example, NlaIII) to make cohesive ends, and are purified by binding to streptavidin coated magnetic beads. **c** The sample is split in two, and PCR primers A and B are ligated to the cDNA. They contain a recognition site for a type IIs restriction enzyme (tagging enzyme, *TE*, in this example, BsmF1) and *AE* cohesive ends. The tagging enzyme cleaves 10 and 14 bp from the recognition site. **d** These primers are ligated to the cDNA and then cut with the tagging enzyme that cleaves some distance away (9–14 bp) from the recognition site, releasing the tagged end from the magnetic bead. The tagged ends are purified using magnetic beads, and blunt-ended. **e** The blunt tags are ligated together to form ditags. Ditags are amplified by PCR using primers A and B. The tags are then released by digestion with anchoring enzyme and primers removed with avidenated magnetic beads. **f** Ditags are ligated into concatamers and cloned into a sequencing vector. These clones are sequenced, the number of times each tag is sequenced recorded indicating the abundance of the mRNA in the original RNA pool

(Steen et al. 2002). For H99, libraries containing over 30,000 tags were used in the analysis, and the 50 most abundant transcripts at 25 and 37 °C were identified, but the analysis of H99 was difficult due to the lack of available genome data at that time. Libraries from B3501 were also generated. B3501 does not grow as well at 37 °C, and so the library from the higher temperature was smaller. There were 15,363 tags from the 37 °C library and 65,399 tags from the 25 °C library. The analysis of genes that changed significantly showed that at 25 °C, genes for histones, sterol and lipid metabolism, and transporters were upregulated. Changes in the expression of lipid metabolism genes might be responsible for altered membrane fluidity at different temperatures. At 37 °C, genes for heat shock proteins, translational machinery, mitochondrial proteins, and stress response proteins were up-regulated.

The second SAGE analysis focused on *C. neoformans* gene expression in the rabbit model of meningitis (Steen et al. 2003). *C. neoformans* cells were isolated from the cerebral spinal fluid of rabbits. A SAGE library was generated and it was compared to the in vitro libraries from the previous study. Surprisingly, the in vivo library shared more in common with the 25 °C in vitro library with respect to the classes of genes that were represented, sharing 1885 tags with the 25 °C library and only 632 with the 37 °C library. Classes of genes that were abundant in vivo included those that encode proteins involved in protein biosynthesis and catabolism, stress responses, respiration, signal transduction and transport. A few genes of particular interest were abundant in vivo, including mannitol-phosphate dehydrogenase, myoinositol 1-phosphate synthase, and MP98, a secreted mannoprotein that is a chitin deacetylase.

A third study examined expression differences between *C. neoformans* in the presence and absence of iron (Lian et al. 2005). In this study, genes that encoded putative components for iron transport and homeostasis, including the *FTR1* (iron permease) gene, had higher transcript levels in the low-iron condition. In addition, lack of iron increased expression for putative extracellular mannoproteins including *CIG1*, and a glycosylphosphatidylinositol transamidase, *GPI8*. Functional analyses of mutants in these genes showed that the *ftr1Δ* mutant and *cig1Δ* mutant had impaired growth in low-iron medium, and altered capsule regulation. The *GPI8* gene appeared to be essential. Lastly, iron-replete conditions led to elevated transcripts for genes for iron storage, nitrogen metabolism, glycolysis, mitochondrial function, lipid metabolism and calmodulin-calcineurin signaling.

c) *Other Methods of Global Gene Expression Profiling*

Methods to identify genes that were alternatively regulated under specific conditions, such as differential display or subtractive hybridization, have been used extensively to discover genes that could be important for virulence. These methods do not require prior knowledge of the genome sequence, but genome sequence helps to identify the entire gene once a fragment is isolated. Once microarrays are readily available, these methods are likely to become less popular. Differential display uses nonspecific PCR primers to amplify sequences from RNA isolated from cells grown under different conditions. The appearance of unique bands in one of the samples suggests that the RNA template corresponding to that band is up-regulated in that condition.

Using differential display, Rude et al. (2002) were able to show that the gene encoding isocitrate lyase (*ICL1*), a component of the glyoxylate pathway, was up-regulated in strain H99 during experimental cryptococcal meningitis in the rabbit model. Further characterization of this gene in *C. neoformans* showed that it was not required for virulence, in contrast to results from other pathogenic organisms.

Del Poeta and colleagues (Luberto et al. 2001) had tested the role of inositol-phosphoryl ceramide synthase (*IPC1*) in pathogenesis, and found that reduced *IPC1* expression resulted in a strain that grew more slowly in macrophages and was less virulent. Differential display of the *IPC1* down-regulated strain was performed to identify targets that might be responsible for this phenotype. This resulted in the identification of *APP1*, a novel regulator of phagocytosis (Luberto et al. 2003).

Lastly, a comparison of *C. neoformans* grown at 37 versus 25 °C using a subtractive cDNA library showed that alternative oxidase was up-regulated at 37 °C. The deletion of the *AOX1* gene resulted in a strain that was sensitive to oxidative stress and was less virulent in the mouse model than wild type (Akhter et al. 2003).

2. Gene Expression Modulation

With the advent of genome data, expression of specific genes can be modulated systematically. Genes can be "knocked out" in several ways. Genes can be targeted for disruption by homologous recombination, their expression can be modulated by RNAi or antisense RNA, or an inducible promoter can be used to control their transcription.

a) *Manipulation of mRNA Levels*

Technology for gene expression modulation has also been developed for *Cryptococcus*. Cryptococcal gene expression can be post-transcriptionally modulated by two methods: antisense repression and RNA interference. RNA antisense technology relies on the ability of RNA that is complementary to a target gene's mRNA to bind to that message and render it incompetent for translation and target it for degradation. The expression of the calcineurin A gene (*CNA1*) from var. *neoformans* and the laccase (*LAC1*) gene from var. *grubii* have been repressed using antisense RNA to each message (Gorlach et al. 2002). In this system, cDNA for each gene was cloned in an antisense orientation in a vector containing the inducible promoter *GAL7*. This construct was used to transform the corresponding variety of *C. neoformans*. When the transformed cells were grown on medium containing galactose, mRNA levels of both genes were significantly reduced, producing the expected phenotypes of reduced melanin production (*LAC1*) and a growth defect at 37 °C (*CNA1*).

Gene expression can also be modified by RNA interference (RNAi). RNAi relies on an organism's ability to degrade double-stranded RNA (dsRNA). In RNAi competent organisms, dsRNA is specifically degraded into short 21–25 bp fragments. Any message that is homologous to these fragments is targeted for degradation (reviewed in Hutvagner and Zamore 2002). RNAi has been used to modulate the expression of the *CAP59* and *ADE2* genes in variety *neoformans* (Liu et al. 2002). Inverted repeats of portions of the *CAP59* or *ADE2* genes were cloned into a vector separated from each other by an unrelated sequence. Expression of this sequence was driven by the promoter of the constitutively expressed cryptococcal actin (*ACT*) gene. Cells with the appropriate mutant phenotypes were recovered after transformation with the RNAi constructs. Analysis of these cells indicated that mRNA expression for each targeted gene was significantly reduced compared to wild type. The ability to specifically and temporally modulate gene expression in *Cryptococcus* will allow us to identify genes that are required for the viability of the organism as well as to characterize the effects of individual genes on the development of this organism. Isolation of a tightly regulated promoter from the copper-regulated CTR4 gene (Ory et al. 2004) allows one to control RNAi expression (Missall and Lodge, unpublished data). An effort to generate libraries expressing RNAi for *C. neoformans* genes is underway (T. Doering, personal communication).

A third method of analyzing the contribution of specific genes to virulence or vegetative growth is to modulate their expression by placing an inducible promoter in front of the protein-coding sequence. Use of an inducible promoter requires that the start site for translation of the protein is known, and that the locations of genes adjacent to the gene of interest are known so that the normal regulation of neighboring genes is unaffected. The gene encoding topoisomerase was placed under the control of the GAL7 promoter, and although the expression was reduced, the cells were still viable (Del Poeta et al. 1999b). However, with a more tightly regulated promoter, the thioredoxin reductase gene has been shown to be essential in *C. neoformans* (Missall and Lodge 2005a).

b) *Large-Scale Gene Deletion*

In comparison with another common fungal pathogen, *Candida albicans*, which has two alleles of each gene, *C. neoformans* is haploid and this facilitates targeted gene disruptions because there is only one allele of each gene. In order to perform reliable gene deletion experiments, the disrupting DNA must be specifically targeted to the appropriate gene. Targeted gene disruption can be performed at high frequencies in the yeast *Saccharomyces cerevisiae* by homologous recombination (HR). However, the efficiency of homologous recombination in *C. neoformans* is significantly less than that seen in *S. cerevisiae*.

The reported frequency of homologous recombination in *C. neoformans* varies considerably due to some of the same factors as those listed above for variation in transformation efficiencies, i.e., the recipient strain, the origin of the transforming DNA, the transforming vector, and the transformation procedure. Published HR frequencies vary from 0.008 to 50% (Salas et al. 1996; Fox et al. 2001), but because of the various methodologies used in published gene replacement experiments, direct comparison of the HR frequencies may lead to incorrect inferences. It is clear that higher rates of homologous recombination are obtained using biolistic transformation. HR rates in var. *neoformans* using electroporation vary from 0.008 to 0.1%, but rates using biolistic transformation in var. *neoformans* vary from 1 to 87% (Davidson et al. 2000, 2002).

The effects of varying the parameters of a homologous recombination experiment are largely unknown. The minimum flanking sequence requirement for efficient homologous recombination in serotype A has been investigated (Nelson et al. 2003). A systematic homologous recombination experiment using the selectable marker for hygromycin B resistance was performed. In these experiments two independent loci, *LAC1* and *CAP59*, were chosen for disruption. The *CAP59* gene is involved with the production of the polysaccharide capsule and *cap59* mutants have a rough and dry colonial phenotype, whereas *lac1* mutants are unable to produce melanin when grown on L-dopa medium. Five constructs of each disruption vector were created with 400, 300, 200, 100, and 50 bp of flanking sequence on each side of the selectable marker. In addition, two asymmetric disruption vectors were created. Constructs with 50 bp or less of flanking sequence had an HR frequency of less than 1% of the total transformants whereas constructs with flanking sequences longer than 200 bp had HR frequencies greater than 8%. If unstable transformants were removed from the total transformants, the HR frequency of these constructs varied between 10 and 75%. Taken together, these data indicate that *C. neoformans* has a relatively efficient HR system that can be relied upon for the efficient creation of targeted gene disruptions.

A system has been devised for efficiently generating constructs for use in targeted gene disruption using homologous recombination (Davidson et al. 2002). In this system, homologous sequences from *C. neoformans* var. *neoformans* and var. *grubii* were PCR amplified from genomic DNA using primers with short (35–42 bp) overlaps to a selectable marker cassette at their 5′ ends. These PCR products were then "linked" together with the selectable marker cassette separating the two flanking sequences in a third PCR reaction. This product was then used to disrupt the genomic copy of the target gene. Because this method is not dependent on the cloning of the target locus to create the knockout vector, it is providing a starting point for the high-throughput generation of gene deletions.

As demonstrated in *S. cerevisiae*, systematic gene deletions have been a valuable tool for analysis of gene function (Winzeler et al. 1999; Giaever et al. 2002). Gene deletion experiments are most informative when the boundaries of the protein-coding sequence of the gene in question are known, so that the entire protein can be eliminated without deleting genetic information of the surrounding genes. Deletions also require the development of efficient protocols for transformation, homologous recombination, and screening. In a haploid organism, gene deletions cannot be used to analyze the function of genes essential for vegetative growth. However, a stable diploid strain has been made for var. *neoformans* (Sia et al. 2000), and this has proved useful for the analysis of essential genes.

A large set of gene deletions is being generated by the Madhani, Janbon, Heitman, Lodge and other laboratories, with the long-term goal of deleting every nonessential gene in the serotype A strain H99. Each of the deletions is being tagged with a unique DNA sequence so that pools of mutants can be analyzed in vivo, similar to signature tagged mutagenesis strategies (Hensel et al. 1995; Nelson et al. 2001). In vitro and in vivo analysis of these gene deletions have already revealed that the *PKC1* pathway is important for cell integrity and for virulence, and that the deletion of a homolog of *SSD1* makes *C. neoformans* more virulent (Gerik and Lodge, unpublished data).

c) Genome-Wide Insertional Mutagenesis

In addition to targeted gene disruption, random insertional mutagenesis strategies have been used to disrupt genes in *C. neoformans*. A major advantage of insertional mutagenesis is that insertions in essential genes can be obtained if they alter expression or only partially abrogate function. Also, the construction of libraries of insertion mutants is relatively fast. One disadvantage can be the difficulty in cloning the flanking sequence around the inserted DNA. *C. neoformans* can insert multiple copies of the transforming DNA in tandem, making cloning the flanking sequences difficult and time-consuming, although use of linear vectors has been more successful.

A form of random mutagenesis has been used to identify var. *grubii* mutants with altered virulence in a mouse model of disseminated disease (Nelson et al. 2001). In this study, biolistic transformation circular plasmids containing signature tag sequences (Hensel et al. 1995) were used to produce random insertion mutants. Signature tagged mutagenesis (STM; Fig. 13.3) uses a series of uniquely tagged plasmid vectors to mutagenize a target genome, producing libraries of mutants. The strength of STM is the ability to pool mutants, each with a distinct tag, and analyze them simultaneously for any observable phenotype. Comparisons of a pool grown in vitro to the same pool following growth in the mouse will identify mutants that were eliminated in vivo. The power of STM is the ability to screen many individual mutants in a single mouse, thus increasing the efficiency of identifying rare mutants. In this study, several avirulent mutants and one hyper-virulent mutant were identified. However, the method of transformation produced tandem arrays of plasmid, making identification of the flanking sequences difficult.

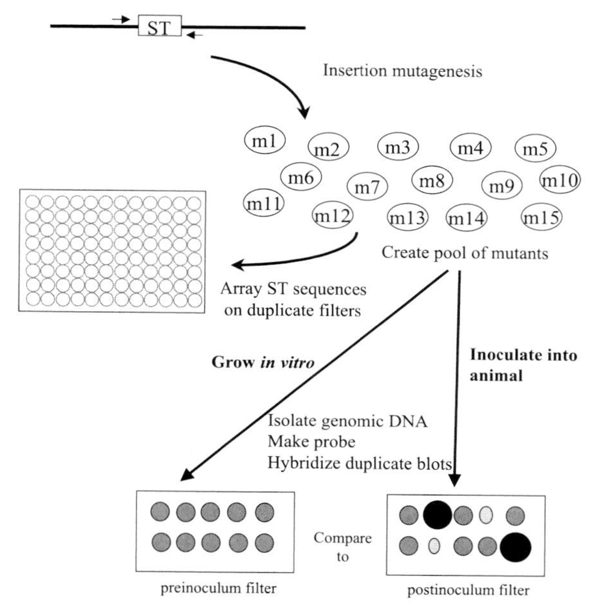

Fig. 13.3. Signature tagged mutagenesis and analysis of pooled mutants. A unique signature tag (*ST*) oligonucleotide, (NK)20, is synthesized and incorporated into a plasmid vector. Each "tagged" vector is used to transform cells producing pools of uniquely tagged mutants. Single mutants from each pool (96 total) are assembled into an input pool. Tags present in the input pool are PCR amplified, labeled and used to probe a filter containing dot blots of each tag, forming the preinoculum filter. The pooled organisms are injected into a mouse and the infection allowed to proceed. Organisms are recovered from tissue and the tags are amplified using common primers, labeled and used to probe a duplicate filter (postinoculum filter). Tags missing from the pool are identified by a loss of signal on the blot. Mutants that over proliferate can also be identified by having a much stronger signal compared to the input blot. Mutants that did not have a change in their virulence have a similar signal to that of the input blot

Random insertion mutagenesis was also used to isolate genes associated with the production of laccase in var. *grubii* (Erickson et al. 2001; Zhu and Williamson 2003). In this study, a component of the vesicular protein pump, *Vph1p*, and a CLC-type chloride channel were identified. The inactivation of the *VPH1* gene by insertional mutagenesis resulted in a reduced virulence phenotype characterized by reduced production of the major virulence factors capsule and laccase as well as urease.

Newer methods for transformation and the availability of genome sequence are improving the recovery of inserted sequences and the rapidity with which the insertion sites are identified. Mutagenesis using *A. tumafaciens*-mediated transformation has been highly successful for generating and recovering insertional mutants (Idnurm et al. 2004). There are major advantages of *A. tumafaciens*-mediated transformation.

1. All transformants are stable and have DNA integrated into the genome, rather than being present as unstable, extrachromosomal copies of the transformation vector.
2. The ends of the insertion are defined by T-DNA borders.
3. Single copies of the transforming DNA are integrated into the genome.

These advantages reduce screening and allow for easy cloning of the insertion. The limited data available at this time suggest that insertion of the T-DNA may not be completely random, as there may be a slight sequence bias toward insertion into promoter regions. Also, the phenotype was only linked to the inserted marker in two of three mutants tested.

A major improvement in the analysis of random insertions is the recent availability of mating pairs for serotype A. This has permitted rapid analysis to determine whether the marker is linked to the phenotype, enabling one to confirm that the phenotype is caused by the insertion. A comparison of a limited set of mutants generated by biolistic transformation of plasmid DNA and *A. tumafaciens*-mediated transformation suggested that the rate of generating unlinked phenotypes was similar (Idnurm et al. 2004).

Screening 590 biolistically transformed mutants for defects in melanin biosynthesis, sensitivity to NO and temperature sensitivity produced four mutants with defects in melanin formation, two that were unable to grow at 37 °C, and two mutants that were hypersensitive to NO (Idnurm et al. 2004). Of these eight, five had phenotypes that were linked to the selectable marker. The mutants with melanin defects had an insertion into a predicted gene of unknown function or into a mitogen activated kinase gene (*MPK2*). The NO-hypersensitive mutants had insertions into the *FHB1* gene, which encodes a flavohemoglobin shown to be important for resistance to NO stress. The temperature-sensitive mutant had an insertion between a predicted protein and a helicase. Screening 576 mutants generated by *Agrobacterium*-mediated transformation resulted in the isolation of three melanin-defective strains. Two of these mutants had phenotypes that co-segregated with the selectable marker. One mutant had an insertion in the promoter region of the laccase gene, and the other in a voltage-gated chloride channel.

B. Proteomics

Direct analysis of protein expression levels is complementary to the analysis of mRNA expression levels. For studies of *C. neoformans* virulence, the discovery of proteins that are regulated during pathogenesis will be important. Many genes are regulated by transcriptional control, but others are regulated by translational control, protein degradation or by regulation of activities by post-translational modifications. These would not be detected by transcriptome analysis. It is clear that there is not always a correlation between mRNA abundance and protein abundance (Gygi et al. 1999). However, the use of proteomics to determine changes in protein expression in vivo can be problematic due to contamination by host proteins.

Techniques such as two-dimensional (2D) gel electrophoresis or liquid chromatography that can separate and quantitate proteins can be used to determine which proteins are regulated under specific conditions. There are advantages and disadvantages to each system. 2D gels are relatively simple to run and have been the mainstay of proteomic analysis, but the types of proteins that are analyzed can be limited on 2D gels. Proteins with high or low molecular weights, or proteins with pIs of less than 3 or greater than 10 can be difficult to resolve. Different solubilization procedures have helped with separation of membrane or cell wall proteins. The detection of proteins by staining is based on abundance, and so typically, only the more abundant cellular proteins are analyzed. Newer methods hold promise for analyzing differential expression of proteins. The isotope coded affinity tag (ICAT) method covalently attaches tags to specific amino acids (e.g., cysteines). Two different tags are used, so that two different protein lysates can be labeled. The proteins are cleaved into peptides by a protease, and the labeled peptides purified. The two samples are mixed together and separated by liquid chromatography. Peptides that are present in a different ratio than most of the other peptides are chosen for sequencing.

Peptide mass fingerprinting or internal sequencing can be used to identify the protein. For identification of *C. neoformans* proteins, both of these methods require a high-quality database of predicted proteins. Peptide mass fingerprinting uses the molecular mass of tryptic peptides from the protein in question and matches them to a database of proteins. With highly conserved proteins, a heterologous database could provide a good match, but with divergent or unique proteins, other databases would not be useful. A database generated from translating all of the open reading frames in the *C. neoformans* genome was also reasonably useful, but due to the large number of introns, it failed to match many proteins that were present in the new annotated database (Pusateri and Lodge, unpublished data).

Proteomics has been used extensively in other fungal systems to identify cell wall proteins, to identify response to drugs, and to identify response to stresses. To date, only one study has been published using proteomics in *C. neoformans*, but with the creation of a well-annotated set of predicted proteins, proteomic studies are likely to become more common.

A proteomic comparison of *C. neoformans* strain H99 grown at 25 vs. 37 °C using 2D gel analysis revealed that two proteins were highly upregulated at 37 °C (Missall et al. 2004b; Fig. 13.4). Both of these proteins were identified as thiol peroxidases, proteins that have been shown to be important for resistance to oxidative stress in other systems. The deletion of both genes, as well as a third thiol peroxidase that was identified through a genomic search, demonstrated that only one of these, Tsa1, had a role in protection against exogenous oxidative stresses and in virulence. The *tsa1Δ* strain also was sensitive to nitrosative stress. In addition, 2D gels of the lysates from the deletion strains demonstrated conclusively that

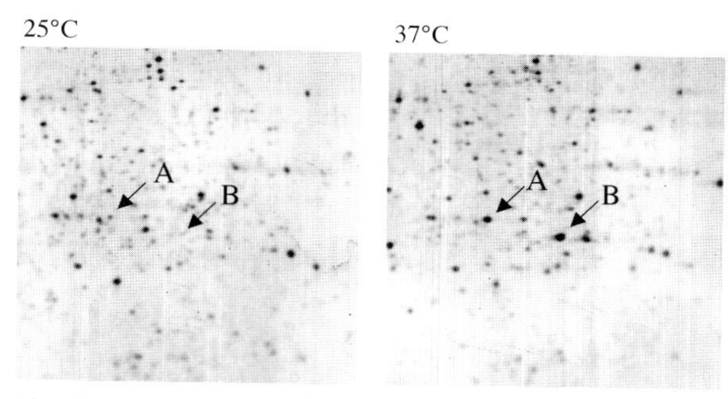

Fig. 13.4. Comparison of lysates from *C. neoformans* grown at 25 vs. 37 °C. The two thiol peroxidases that are upregulated at higher temperature are indicated by *arrows*

the correct identification of the protein spots had been made – the spots that had been identified as Tsa1 and Tsa3 were absent in the lysates from the appropriate deletion strain. Proteomic analyses of the response to oxidative and nitrosative stress is ongoing (T. Missall and J. Lodge, unpublished data).

VIII. Conclusions and Future Impact

The *Cryptococcus neoformans* genome project has already had a major impact on research on the virulence of this important fungal pathogen, but we have only seen the tip of the iceberg. We have fairly complete sequence datasets on five different genomes, and an outstanding set of annotations based on multiple gene finders and a high level of EST coverage. Investigators have begun taking advantage of this information by comparing the JEC21 genome with the B3501 genome. Tools for molecular analysis are in place including transformation, multiple positive selectable markers, homologous recombination, regulatable promoters, and isogenic mating pairs. Studies of global gene expression analysis have begun, a systematic set of gene deletions has been initiated, comprehensive sets of random insertion mutants are being constructed, and databases are ready for analysis of proteomic data. Within a year or two, there will be comparisons of the different serotypes that have different epidemiology and environmental niches. Microarrays will be available for the analysis of growth in an animal or of responses to specific stimuli or genotypes. Once specific genes are identified by microarray or proteomic analysis, well-defined mutants will be available to test their contributions to the process under study. Because of the tools and genomic information that are available, *C. neoformans* is and will continue to be a good model for fungal pathogenesis. Although some features of *C. neoformans* are unique, many will be shared with other fungi, and so many of the lessons learned while examining the mechanisms of *C. neoformans* pathogenesis will be applicable to other, less tractable systems.

Acknowledgements. The genome project of *Cryptococcus neoformans* has been a community effort, with many genome centers contributing to the work, including the Institute for Genome Research, the Stanford Genome and Technology Center, Duke University Center for Genome Technology, the Broad Institute, the University of British Columbia, and University of Oklahoma. Many thanks to Joe Heitman, John Perfect, Juneann Murphy, Jim Kronstad, June Kwon-Chung, Fred Dietrich, Richard Hyman, Brendan Loftus, and James Galagan for their major contributions to the genome project. Research in J.K. Lodge's laboratory is supported by NIH grants R01 AI50184 and R01 AI051209.

References

Abadi J, Nachman S, Kressel AB, Pirofski L (1999) Cryptococcosis in children with AIDS. Clin Infect Dis 28:309–313

Aberg JA, Price RW, Heeren DM, Bredt B (2002) A pilot study of the discontinuation of antifungal therapy for disseminated cryptococcal disease in patients with acquired immunodeficiency syndrome, following immunologic response to antiretroviral therapy. J Infect Dis 185:1179–1182

Akhter S, McDade HC, Gorlach JM, Heinrich G, Cox GM, Perfect JR (2003) Role of alternative oxidase gene in pathogenesis of Cryptococcus neoformans. Infect Immun 71:5794–5802

Alspaugh JA, Perfect JR, Heitman J (1997) Cryptococcus neoformans mating and virulence are regulated by the G-protein alpha subunit GPA1 and cAMP. Genes Dev 11:3206–3217

Alspaugh JA, Pukkila-Worley R, Harashima T, Cavallo LM, Funnell D, Cox GM, Perfect JR, Kronstad JW, Heitman J (2002) Adenylyl cyclase functions downstream of the Galpha protein Gpa1 and controls mating and pathogenicity of Cryptococcus neoformans. Eukaryot Cell 1:75–84

Aoki FH, Imai T, Tanaka R, Mikami Y, Taguchi H, Nishimura NF, Nishimura K, Miyaji M, Schreiber AZ, Branchini ML (1999) New PCR primer pairs specific for Cryptococcus neoformans serotype A or B prepared on the basis of random amplified polymorphic DNA fingerprint pattern analyses. J Clin Microbiol 37:315–320

Apidianakis Y, Rahme LG, Heitman J, Ausubel FM, Calderwood SB, Mylonakis E (2004) Challenge of Drosophila melanogaster with Cryptococcus neoformans and role of the innate immune response. Eukaryot Cell 3:413–419

Bach MC, Sahyoun A, Adler JL, Schlesinger RM, Breman J, Madras P, P'eng F, Monaco AP (1973) High incidence of fungus infections in renal transplantation patients treated with antilymphocyte and conventional immunosuppression. Transplant Proc 5:549–553
Bar-Peled M, Griffith CL, Doering TL (2001) Functional cloning and characterization of a UDP-glucuronic acid decarboxylase: the pathogenic fungus Cryptococcus neoformans elucidates UDP-xylose synthesis. Proc Natl Acad Sci USA 98:12003–12008
Bhattacharjee AK, Bennett JE, Glaudemans CP (1984) Capsular polysaccharides of Cryptococcus neoformans. Rev Infect Dis 6:619–624
Biondo C, Beninati C, Delfino D, Oggioni M, Mancuso G, Midiri A, Bombaci M, Tomaselli G, Teti G (2002) Identification and cloning of a cryptococcal deacetylase that produces protective immune responses. Infect Immun 70:2383–2391
Blackstock R, Murphy JW (1997) Secretion of the C3 component of complement by peritoneal cells cultured with encapsulated Cryptococcus neoformans. Infect Immun 65:4114–4121
Bose I, Reese AJ, Ory JJ, Janbon G, Doering TL (2003) A yeast under cover: the capsule of Cryptococcus neoformans. Eukaryot Cell 2:655–663
Bozzette S, Larsen RA, Chiu J, Leal MA, Jacobsen J, Rothman P, Robinson P, Gilbert G, McCutchan JA, Tilles JG et al. (1991) A placebo-controlled trial of maintenance therapy with fluconazole after treatment of cryptococcal meningitis in the acquired immunodeficiency syndrome. N Engl J Med 324:580–584
Cameron ML, Bartlett JA, Gallis HA, Waskin HA (1991) Manifestations of pulmonary cryptococcosis in patients with acquired immunodeficiency syndrome. Rev Infect Dis 13:64–67
Casadevall A, Perfect JR (1998) Cryptococcus neoformans. ASM Press, Washington, DC
Chang YC, Kwon-Chung KJ (1994) Complementation of a capsule-deficient mutation of Cryptococcus neoformans restores its virulence. Mol Cell Biol 14:4912–4919
Chang YC, Kwon-Chung KJ (1998) Isolation of the third capsule-associated gene, CAP60, required for virulence in Cryptococcus neoformans. Infect Immun 66:2230–2236
Chang YC, Kwon-Chung KJ (1999) Isolation, characterization, and localization of a capsule-associated gene, CAP10, of Cryptococcus neoformans. J Bacteriol 181:5636–5643
Chang YC, Penoyer LA, Kwon-Chung KJ (1996) The second capsule gene of Cryptococcus neoformans, CAP64, is essential for virulence. Infect Immun 64:1977–1983
Chang YC, Wickes BL, Miller GF, Penoyer LA, Kwon-Chung KF (2000) Cryptococcus neoformans STE12alpha regulates virulence but is not essential for mating. J Exp Med 191:871–882
Chen SC, Muller M, Zhou JZ, Wright LC, Sorrell TC (1997a) Phospholipase activity in Cryptococcus neoformans: a new virulence factor? J Infect Dis 175:414–420
Chen SC, Wright LC, Santangelo RT, Muller M, Moran VR, Kuchel PW, Sorrell TC (1997b) Identification of extracellular phospholipase B, lysophospholipase, and acyltransferase produced by Cryptococcus neoformans. Infect Immun 65:405–411
Chen LC, Goldman DL, Doering TL, Pirofski A, Casadevall A (1999) Antibody response to Cryptococcus neoformans proteins in rodents and humans. Infect Immun 67:2218–2224
Cherniak R, Sundstrom JB (1994) Polysaccharide antigens of the capsule of Cryptococcus neoformans. Infect Immun 62:1507–1512
Cherniak R, Reiss E, Slodki ME, Plattner RD, Blumer SO (1980) Structure and antigenic activity of the capsular polysaccharide of Cryptococcus neoformans serotype A. Mol Immunol 17:1025–1032
Cherniak R, Jones RG, Reiss E (1988) Structure determination of Cryptococcus neoformans serotype A-variant glucuronoxylomannan by 13C-n.m.r. spectroscopy. Carbohydr Res 172:113–138
Cherniak R, Morris LC, Turner SH (1992) Glucuronoxylomannan of Cryptococcus neoformans serotype D: structural analysis by gas-liquid chromatography-mass spectrometry and by 13C-nuclear magnetic resonance spectroscopy. Carbohydr Res 223:263–269
Chuck SL, Sande MA (1989) Infections with Cryptococcus neoformans in the acquired immunodeficiency syndrome. N Engl J Med 321:794–799
Clarke DL, Woodlee GL, McClelland CM, Seymour TS, Wickes BL (2001) The Cryptococcus neoformans STE11alpha gene is similar to other fungal mitogen-activated protein kinase kinase kinase (MAPKKK) genes but is mating type specific. Mol Microbiol 40:200–213
Coenjaerts FE, Walenkamp AM, Mwinzi PN, Scharringa J, Dekker HA, Van Strijp AG, Cherniak R, HoepelmanAI (2001) Potent inhibition of neutrophil migration by cryptococcal mannoprotein-4-induced desensitization. J Immunol 167:3988–3995
Cox GM, Rude TH, Dykstra CC, Perfect JR (1995) The actin gene from Cryptococcus neoformans: structure and phylogenetic analysis. J Med Vet Mycol 33:261–266
Cox GM, Toffaletti DL, Perfect JR (1996) Dominant selection system for use in Cryptococcus neoformans. J Med Vet Mycol 34:385–391
Cox GM, Mukherjee J, Cole GT, Casadevall A, Perfect JR (2000) Urease as a virulence factor in experimental cryptococcosis. Infect Immun 68:443–448
Cox GM, McDade HC, Chen SC, Tucker SC, Gottfredsson M, Wright LC, Sorrell TC, Leidich SD, Casadevall A, Ghannoum MA et al. (2001) Extracellular phospholipase activity is a virulence factor for Cryptococcus neoformans. Mol Microbiol 39:166–175
Cox GM, Harrison TS, McDade HC, Taborda CP, Heinrich G, Casadevall A, Perfect JR (2003) Superoxide dismutase influences the virulence of Cryptococcus neoformans by affecting growth within macrophages. Infect Immun 71:173–180
Cunha BA (2001a) Central nervous system infections in the compromised host: a diagnostic approach. Infect Dis Clin N Am 15:567–590
Cunha BA (2001b) Pneumonias in the compromised host. Infect Dis Clin N Am 15:591–612
Currie BP, Freundlich LF, Casadevall A (1994) Restriction fragment length polymorphism analysis of Cryptococcus neoformans isolates from environmental (pigeon excreta) and clinical sources in New York City. J Clin Microbiol 32:1188–1192

Davidson RC, Moore TD, Odom AR, Heitman J (2000) Characterization of the MFalpha pheromone of the human fungal pathogen Cryptococcus neoformans. Mol Microbiol 38:1017–1026

Davidson RC, Blankenship JR, Kraus PR, de Jesus Berrios M, Hull CM, D'Souza C, Wang P, Heitman J (2002) A PCR-based strategy to generate integrative targeting alleles with large regions of homology. Microbiology 148:2607–2615

de Jesus-Berrios M, Liu L, Nussbaum JC, Cox GM, Stamler JS, Heitman J (2003) Enzymes that counteract nitrosative stress promote fungal virulence. Curr Biol 13:1963–1968

Del Poeta M, Toffaletti DL, Rude TH, Sparks SD, Heitman J, Perfect JR (1999a) Cryptococcus neoformans differential gene expression detected in vitro and in vivo with green fluorescent protein. Infect Immun 67:1812–1820

Del Poeta M, Toffaletti DL, Rude TH, Dykstra CC, Heitman J, Perfect JR (1999b) Topoisomerase I is essential in Cryptococcus neoformans: role in pathobiology and as an antifungal target. Genetics 152:167–178

Doering TL, Nosanchuk JD, Roberts WK, Casadevall A (1999) Melanin as a potential cryptococcal defense against microbicidal proteins. Med Mycol 37:175–181

D'Souza CA, Alspaugh JA, Yue C, Harashima T, Cox GM, Perfect JR, Heitman J (2001) Cyclic AMP-dependent protein kinase controls virulence of the fungal pathogen Cryptococcus neoformans. Mol Cell Biol 21:3179–3191

D'Souza CA, Hagen F, Boekhout T, Cox GM, Heitman J (2004) Investigation of the basis of virulence in serotype A strains of Cryptococcus neoformans from apparently immunocompetent individuals. Curr Genet 46:92–102

Edman JC, Kwon-Chung KJ (1990) Isolation of the URA5 gene from Cryptococcus neoformans var. neoformans and its use as a selective marker for transformation. Mol Cell Biol 10:4538–4544

Eng RH, Bishburg E, Smith SM, Kapila R (1986) Cryptococcal infections in patients with acquired immune deficiency syndrome. Am J Med 81:19–23

Erickson T, Liu L, Gueyikian A, Zhu X, Gibbons J, Williamson PR (2001) Multiple virulence factors of Cryptococcus neoformans are dependent on VPH1. Mol Microbiol 42:1121–1131

Feldmesser M, Kress Y, Novikoff P, Casadevall A (2000) Cryptococcus neoformans is a facultative intracellular pathogen in murine pulmonary infection. Infect Immun 68:4225–4237

Feldmesser M, Tucker S, Casadevall A (2001) Intracellular parasitism of macrophages by Cryptococcus neoformans. Trends Microbiol 9:273–278

Fox DS, Cruz MC, Sia RA, Ke H, Cox GM, Cardenas ME, Heitman J (2001) Calcineurin regulatory subunit is essential for virulence and mediates interactions with FKBP12-FK506 in Cryptococcus neoformans. Mol Microbiol 39:835–849

Franzot SP, Salkin IF, Casadevall A (1999) Cryptococcus neoformans var. grubii: separate varietal status for Cryptococcus neoformans serotype A isolates. J Clin Microbiol 37:838–840

Fujihara H, Kagaya K, Fukazawa Y (1997) Anti-chemotactic activity of capsular polysaccharide of Cryptococcus neoformans in vitro. Microbiol Immunol 41:657–664

Garcia-Rivera J, Chang YC, Kwon-Chung KJ, Casadevall A (2004) Cryptococcus neoformans CAP59 (or Cap59p) is involved in the extracellular trafficking of capsular glucuronoxylomannan. Eukaryot Cell 3:385–392

Giaever G, Chu AM, Ni L, Connelly C, Riles L et al. (2002) Functional profiling of the Saccharomyces cerevisiae genome. Nature 418:387–391

Goldman D, Lee SC, Casadevall A (1994) Pathogenesis of pulmonary Cryptococcus neoformans infection in the rat. Infect Immun 62:4755–4761

Goldman DL, Lee SC, Mednick AJ, Montella L, Casadevall A (2000) Persistent Cryptococcus neoformans pulmonary infection in the rat is associated with intracellular parasitism, decreased inducible nitric oxide synthase expression, and altered antibody responsiveness to cryptococcal polysaccharide. Infect Immun 68:832–838

Goldman DL, Khine H, Abadi J, Lindenberg DJ, Pirofski L, Niang R, Casadevall A (2001) Serologic evidence for Cryptococcus neoformans infection in early childhood. Pediatrics 107:E66

Gonzalez CE, Shetty D, Lewis LL, Mueller BU, Pizzo PA, Walsh TJ (1996) Cryptococcosis in human immunodeficiency virus-infected children. Pediatr Infect Dis J 15:796–800

Gorlach JM, McDade HC, Perfect JR, Cox GM (2002) Antisense repression in Cryptococcus neoformans as a laboratory tool and potential antifungal strategy. Microbiology 148:213–219

Granger DL, Perfect JR, Durack DT (1985) Virulence of Cryptococcus neoformans. Regulation of capsule synthesis by carbon dioxide. J Clin Invest 76:508–516

Graybill JR, Sobel J, Saag MS, van der Horst C, Powderly WG, Cloud G, Riaser L, Hamill R, Dismukes WE (2000) Diagnosis and management of increased intracranial pressure in patients with AIDS and cryptococcal meningitis. Clin Infect Dis 30:47–54

Grossi P, Farina C, Fiocchi R, Dalla Gasperina D (2000) Prevalence and outcome of invasive fungal infections in 1963 thoracic organ transplant recipients: a multicenter retrospective study. Italian Study Group of Fungal Infections in Thoracic Organ Transplant Recipients. Transplantation 70:112–116

Gygi SP, Rochon Y, Franza BR, Aebersold R (1999) Correlation between protein and mRNA abundance in yeast. Mol Cell Biol 19:1720–1730

Hajjeh RA, Conn LA, Stephens DS, Baughman W, Hamill R, Graviss E, Pappas PG, Thomas C, Reingold A, Rothrock G et al. (1999) Cryptococcosis: population-based multistate active surveillance and risk factors in human immunodeficiency virus-infected persons. Cryptococcal Active Surveillance Group. J Infect Dis 179:449–454

Heitman J, Casadevall A, Lodge JK, Perfect JR (1999) The Cryptococcus neoformans genome sequencing project. Mycopathologia 148:1–7

Hensel M, Shea JE, Gleeson C, Jones MD, Dalton E, Holden DW (1995) Simultaneous identification of bacterial virulence genes by negative selection. Science 269:400–403

Hicks JK, D'Souza CA, Cox GM, Heitman J (2004) Cyclic AMP-dependent protein kinase catalytic subunits have divergent roles in virulence factor production in two varieties of the fungal pathogen Cryptococcus neoformans. Eukaryot Cell 3:14–26

Hospenthal DR, Bennett JE (2000) Persistence of cryptococcomas on neuroimaging. Clin Infect Dis 31:1303–1306

Hua J, Meyer JD, Lodge JK (2000) Development of positive selectable markers for the fungal pathogen Cryptococcus neoformans. Clin Diagn Lab Immunol 7:125–128

Huffnagle GB, Chen GH, Curtis JL, McDonald RA, Strieter RM, Toews GB (1995) Down-regulation of the afferent phase of T cell-mediated pulmonary inflammation and immunity by a high melanin-producing strain of Cryptococcus neoformans. J Immunol 155:3507–3516

Hull CM, Davidson RC, Heitman J (2002) Cell identity and sexual development in Cryptococcus neoformans are controlled by the mating-type-specific homeodomain protein Sxi1alpha. Genes Dev 16:3046–3060

Hull CM, Cox GM, Heitman J (2004) The alpha-specific cell identity factor Sxi1alpha is not required for virulence of Cryptococcus neoformans. Infect Immun 72:3643–3645

Husain S, Wagener MM, Singh N (2001) Cryptococcus neoformans infection in organ transplant recipients: variables influencing clinical characteristics and outcome. Emerg Infect Dis 7:375–381

Hutvagner G, Zamore PD (2002) RNAi: nature abhors a double-strand. Curr Opin Genet Dev 12:225–232

Ibrahim AS, Filler SG, Alcouloumre MS, Kozel TR, Edwards JE, Ghannoum MA (1995) Adherence to and damage of endothelial cells by Cryptococcus neoformans in vitro: role of the capsule. Infect Immun 63:4368–4374

Idnurm A, Reedy JL, Nussbaum JC, Heitman J (2004) Cryptococcus neoformans virulence gene discovery through insertional mutagenesis. Eukaryot Cell 3:420–429

Ikeda R, Sugita T, Jacobson ES, Shinoda T (2002) Laccase and melanization in clinically important Cryptococcus species other than Cryptococcus neoformans. J Clin Microbiol 40:1214–1218

Jacobson ES, Emory HS (1991) Catecholamine uptake, melanization, and oxygen toxicity in Cryptococcus neoformans. J Bacteriol 173:401–403

Jacobson ES, Tinnell SB (1993) Antioxidant function of fungal melanin. J Bacteriol 175:7102–7104

Janbon G, Himmelreich U, Moyrand F, Improvisi L, Dromer F (2001) Cas1p is a membrane protein necessary for the O-acetylation of the Cryptococcus neoformans capsular polysaccharide. Mol Microbiol 42:453–467

Kaplan MH, Rosen PP, Armstrong D (1977) Cryptococcosis in a cancer hospital: clinical and pathological correlates in forty-six patients. Cancer 39:2265–2274

Kappe R, Muller J (1991) Rapid clearance of Candida albicans mannan antigens by liver and spleen in contrast to prolonged circulation of Cryptococcus neoformans antigens. J Clin Microbiol 29:1665–1669

Karos M, Chang YC, McClelland CM, Clarke DL, Fu J, Wickes BL, Kwon-Chung KJ (2000) Mapping of the Cryptococcus neoformans MATalpha locus: presence of mating type-specific mitogen-activated protein kinase cascade homologs. J Bacteriol 182:6222–6227

Kontoyiannis DP, Peitsch WK, Reddy BT, Whimbey EE, Han XY, Bodey GP, Rolston KV (2001) Cryptococcosis in patients with cancer. Clin Infect Dis 32:E145–150

Korth H, Pulverer G (1971) Pigment formation for differentiating Cryptococcus neoformans from Candida albicans. Appl Microbiol 21:541–542

Kovacs JA, Kovacs AA, Polis M, Wright WC, Gill VJ, Tuazon CU, Gelmann EP, Lane HC, Longfield R, Overturf G et al. (1985) Cryptococcosis in the acquired immunodeficiency syndrome. Ann Intern Med 103:533–538

Kozel TR, Gotschlich EC (1982) The capsule of Cryptococcus neoformans passively inhibits phagocytosis of the yeast by macrophages. J Immunol 129:1675–1680

Kozel TR, Gulley WF, Cazin J Jr (1977) Immune response to Cryptococcus neoformans soluble polysaccharide: immunological unresponsiveness. Infect Immun 18:701–707

Kozel TR, Levitz SM, Dromer F, Gates MA, Thorkildson P, Janbon G (2003) Antigenic and biological characteristics of mutant strains of Cryptococcus neoformans lacking capsular O acetylation or xylosyl side chains. Infect Immun 71:2868–2875

Kraus PR, Heitman J (2003) Coping with stress: calmodulin and calcineurin in model and pathogenic fungi. Biochem Biophys Res Commun 311:1151–1157

Kraus PR, Fox DS, Cox GM, Heitman J (2003) The Cryptococcus neoformans MAP kinase Mpk1 regulates cell integrity in response to antifungal drugs and loss of calcineurin function. Mol Microbiol 48:1377–1387

Kraus PR, Boily MJ, Giles SS, Stajich JE, Allen A, Cox GM, Dietrich FS, Perfect JR, Heitman J (2004) Identification of Cryptococcus neoformans temperature-regulated genes with a genomic-DNA microarray. Eukaryot Cell 3:1249–1260

Kupfer DM, Drabenstot SD, Buchanan KL, Lai H, Zhu H, Dyer DW, Roe BA, Murphy JW (2004) Introns and splicing elements of five diverse fungal organisms. Eukaryot Cell 3:1088–1100

Kwon-Chung KJ (1975) A new genus, filobasidiella, the perfect state of Cryptococcus neoformans. Mycologia 67:1197–1200

Kwon-Chung KJ, Bennett JE (1978) Distribution of alpha and alpha mating types of Cryptococcus neoformans among natural and clinical isolates. Am J Epidemiol 108:337–340

Kwon-Chung KJ, Rhodes JC (1986) Encapsulation and melanin formation as indicators of virulence in Cryptococcus neoformans. Infect Immun 51:218–223

Kwon-Chung KJ, Polacheck I, Popkin TJ (1982) Melanin-lacking mutants of Cryptococcus neoformans and their virulence for mice. J Bacteriol 150:1414–1421

Kwon-Chung KJ, Edman JC, Wickes BL (1992) Genetic association of mating types and virulence in Cryptococcus neoformans. Infect Immun 60:602–605

Lee SC, Casadevall A (1996) Polysaccharide antigen in brain tissue of AIDS patients with cryptococcal meningitis. Clin Infect Dis 23:194–195

Lendvai N, Casadevall A, Liang Z, Goldman DL, Mukherjee J, Zuckier L (1998) Effect of immune mechanisms on the pharmacokinetics and organ distribution of cryptococcal polysaccharide. J Infect Dis 177:1647–1659

Lengeler KB, Davidson RC, D'Souza C, Harashima T, Shen WC, Wang P, Pan X, Waugh M, Heitman J (2000a) Signal transduction cascades regulating fungal development and virulence. Microbiol Mol Biol Rev 64:746–785

Lengeler KB, Wang P, Cox GM, Perfect JR, Heitman J (2000b) Identification of the MATa mating-type locus of Cryptococcus neoformans reveals a serotype A MATa strain thought to have been extinct. Proc Natl Acad Sci USA 97:14455–14460

Lengeler KB, Fox DS, Fraser JA, Allen A, Forrester K, Dietrich FS, Heitman J (2002) Mating-type locus of Cryptococcus neoformans: a step in the evolution of sex chromosomes. Eukaryot Cell 1:704–718

Levitz SM, Harrison TS, Tabuni A, Liu X (1997) Chloroquine induces human mononuclear phagocytes to inhibit and kill Cryptococcus neoformans by a mechanism independent of iron deprivation. J Clin Invest 100:1640–1646

Levitz SM, Nong SH, Seetoo KF, Harrison TS, Speizer RA, Simons ER (1999) Cryptococcus neoformans resides in an acidic phagolysosome of human macrophages. Infect Immun 67:885–890

Levitz SM, Nong S, Mansour MK, Huang C, Specht CA (2001) Molecular characterization of a mannoprotein with homology to chitin deacetylases that stimulates T cell responses to Cryptococcus neoformans. Proc Natl Acad Sci USA 98:10422–10427

Lian T, Simmer MI, D'Souza CA, Steen BR, Zuyderduyn SD, Jones SJ, Marra MA, Kronstad JW (2005) Iron-regulated transcription and capsule formation in the fungal pathogen Cryptococcus neoformans. Mol Microbiol 55:1452–1472

Liliang PC, Liang CL, Chang WN, Lu K, Lu CH (2002) Use of ventriculoperitoneal shunts to treat uncontrollable intracranial hypertension in patients who have cryptococcal meningitis without hydrocephalus. Clin Infect Dis 34:e64–e68

Lindsley MD, Hurst SF, Iqbal NJ, Morrison CJ (2001) Rapid identification of dimorphic and yeast-like fungal pathogens using specific DNA probes. J Clin Microbiol 39:3505–3511

Litvintseva AP, Marra RA, Nielsen K, Heitman J, Vilgalys R, Mitchell TG (2003) Evidence of sexual recombination among Cryptococcus neoformans serotype A isolates in sub-Saharan Africa. Eukaryot Cell 2:1162–1168

Liu L, Tewari RP, Williamson PR (1999a) Laccase protects Cryptococcus neoformans from antifungal activity of alveolar macrophages. Infect Immun 67:6034–6039

Liu L, Wakamatsu K, Ito S, Williamson PR (1999b) Catecholamine oxidative products, but not melanin, are produced by Cryptococcus neoformans during neuropathogenesis in mice. Infect Immun 67:108–112

Liu H, Cottrell TR, Pierini LM, Goldman WE, Doering TL (2002) RNA interference in the pathogenic fungus Cryptococcus neoformans. Genetics 160:463–470

Loftus B, Eula Fung E, Roncaglia P, Rowley D, Amedeo P, Bruno D, Vamathevan J, Miranda M, Anderson I, Fraser JA et al. (2005) The genome and transcriptome of Cryptococcus neoformans, a basidiomycetous fungal pathogen of humans. Science 307:1321–1324

Lorenz MC (2002) Genomic approaches to fungal pathogenicity. Curr Opin Microbiol 5:372–378

Luberto C, Toffaletti DL, Wills EA, Tucker SC, Casadevall A, Perfect JR, Hannun YA, Del Poeta M (2001) Roles for inositol-phosphoryl ceramide synthase 1 (IPC1) in pathogenesis of *C. neoformans*. Genes Dev 15:201–202

Luberto C, Martinez-Marino B, Taraskiewicz D, Bolanos B, Chitano P, Toffaletti DL, Cox GM, Perfect JR, Hannun YA, Balish E et al. (2003) Identification of App1 as a regulator of phagocytosis and virulence of Cryptococcus neoformans. J Clin Invest 112:1080–1094

Manfredi R, Moroni M, Mazzoni A, Nanetti A, Donati M, Mastroianni A, Coronado OV, Chiodo F (1996) Isolated detection of crytpcococcal polysaccharide antigen in cerebrospinal fluid samples from patients with AIDS. Clin Infect Dis 23:849–850

Marra RE, Huang JC, Fung E, Nielsen K, Heitman J, Vilgalys R, Mitchell TG (2004) A genetic linkage map of Cryptococcus neoformans variety neoformans serotype D (Filobasidiella neoformans). Genetics 167:619–631

McDade HC, Cox GM (2001) A new dominant selectable marker for use in Cryptococcus neoformans. Med Mycol 39:151–154

Merkel GJ, Cunningham RK (1992) The interaction of Cryptococcus neoformans with primary rat lung cell cultures. J Med Vet Mycol 30:115–121

Missall TA, Lodge JK (2005a) Thioredoxin reductase is essential for viability in the fungal pathogen, Cryptococcus neoformans. Eukaryot Cell 4:487–489

Missall TA, Lodge JK (2005b) Function of the thioredoxin proteins in Cryptococcus neoformans during stress or virulence and regulation by putative transcriptional modulators. Mol Microbiol 57:847–858

Missall TA, Lodge JK, McEwen JE (2004a) Mechanisms of resistance to oxidative and nitrosative stress: implications for fungal survival in mammalian hosts. Eukaryot Cell 3:835–846

Missall TA, Pusateri ME, Lodge JK (2004b) Thiol peroxidase is critical for virulence and resistance to nitric oxide and peroxide in the fungal pathogen, Cryptococcus neoformans. Mol Microbiol 51:1447–1458

Missall TA, Moran JM, Corbett JA, Lodge JK (2005a) Distinct stress responses of two functional laccases in Cryptococcus neoformans is revealed in the absence of the thiol-specific antioxidant, Tsa1. Eukaryot Cell 4:202–208

Missall TA, Cherry-Harris JF, Lodge JK (2005b) Two glutathione peroxidases in the fungal pathogen, Cryptococcus neoformans, are expressed in the presence of specific substrates. Microbiology 151:2573–2591

Mitchell DH, Sorrell TC, Allworth AM, Heath CH, McGregor AR, Papanaoum K, Richards MJ, Gottlieb T (1995) Cryptococcal disease of the CNS in immunocompetent hosts: influence of cryptococcal variety on clinical manifestations and outcome. Clin Infect Dis 20:611–616

Mody CH, Syme RM (1993) Effect of polysaccharide capsule and methods of preparation on human lymphocyte proliferation in response to Cryptococcus neoformans. Infect Immun 61:464–469

Monari C, Kozel TR, Bistoni F, Vecchiarelli A (2002) Modulation of C5aR expression on human neutrophils by encapsulated and acapsular Cryptococcus neoformans. Infect Immun 70:3363–3370

Moosa MY, Coovadia YM (1997) Cryptococcal meningitis in Durban, South Africa: a comparison of clinical features, laboratory findings and outcome for human immunodeficiency virus (HIV)-positive and HIV-negative patients. Clin Infect Dis 24:131–134

Moyrand F, Klaproth B, Himmelreich U, Dromer F, Janbon G (2002) Isolation and characterization of capsule structure mutant strains of Cryptococcus neoformans. Mol Microbiol. 2045:837–849

Mylonakis E, Ausubel FM, Perfect JR, Heitman J, Calderwood SB (2002) Killing of Caenorhabditis elegans by Cryptococcus neoformans as a model of yeast pathogenesis. Proc Natl Acad Sci USA 99:15675–15680

Narasipura SD, Ault JG, Behr MJ, Chaturvedi V, Chaturvedi S (2003) Characterization of Cu,Zn superoxide dismutase (SOD1) gene knock-out mutant of Cryptococcus neoformans var. gattii: role in biology and virulence. Mol Microbiol 47:1681–1694

Neilson K, Cox GM, Wang P, Toffaletti DL, Perfect JR, Heitman J (2003) Sexual cycle of Cryptococcus neoformans var. grubii and virulence of congenic a and alpha isolates. Infect Immun 71:4831–4841

Nelson RT, Hua J, Pryor B, Lodge JK (2001) Identification of virulence mutants of the fungal pathogen Cryptococcus neoformans using signature-tagged mutagenesis. Genetics 157:935–947

Nelson RT, Pryor BA, Lodge JK (2003) Sequence length required for homologous recombination in Cryptococcus neoformans. Fungal Genet Biol 38:1–9

Nosanchuk JD, Rudolph J, Rosas AL, Casadevall A (1999a) Evidence that Cryptococcus neoformans is melanized in pigeon excreta: implications for pathogenesis. Infect Immun 67:5477–5479

Nosanchuk JD, Valadon P, Feldmesser M, Casadevall A (1999b) Melanization of Cryptococcus neoformans in murine infection. Mol Cell Biol 19:745–750

Nosanchuk JD, Rosas AL, Lee SC, Casadevall A (2000) Melanisation of Cryptococcus neoformans in human brain tissue. Lancet 355:2049–2050

Noverr MC, Phare SM, Toews GB, Coffey MJ, Huffnagle GB (2001) Pathogenic yeasts Cryptococcus neoformans and Candida albicans produce immunomodulatory prostaglandins. Infect Immun 69:2957–2963

Noverr MC, Cox GM, Perfect JR, Huffnagle GB (2003) Role of PLB1 in pulmonary inflammation and cryptococcal eicosanoid production. Infect Immun 71:1538–1547

Olszewski MA, Noverr MC, Chen GH, Toews GB, Cox GM, Perfect JR, Huffnagle GB (2004) Urease expression by Cryptococcus neoformans promotes microvascular sequestration, thereby enhancing central nervous system invasion. Am J Pathol 164:1761–1771

Ory JJ, Griffith CL, Doering TL (2004) An efficiently regulated promoter system for Cryptococcus neoformans utilizing the CTR4 promoter. Yeast 21:919–926

Pappas PG, Perfect JR, Cloud GA, Larsen RA, Pankey GA, Lancaster DJ, Henderson H, Kauffman CA, Haas DW, Saccente M et al. (2001) Cryptococcosis in human immunodeficiency virus-negative patients in the era of effective azole therapy. Clin Infect Dis 33:690–699

Perfect JR, Lang SD, Durack DT (1980) Chronic cryptococcal meningitis: a new experimental model in rabbits. Am J Pathol 101:177–194

Perfect JR, Magee BB, Magee PT (1989) Separation of chromosomes of Cryptococcus neoformans by pulsed field gel electrophoresis. Infect Immun 57:2624–2627

Pfrommer GS, Dickens SM, Wilson MA, Young BJ, Kozel TR (1993) Accelerated decay of C3b to iC3b when C3b is bound to the Cryptococcus neoformans capsule. Infect Immun 61:4360–4366

Pietrella D, Cherniak R, Strappini C, Perito S, Mosci P, Bistoni F, Vecchiarelli A (2001) Role of mannoprotein in induction and regulation of immunity to Cryptococcus neoformans. Infect Immun 69:2808–2814

Polacheck I, Hearing VJ, Kwon-Chung KJ (1982) Biochemical studies of phenoloxidase and utilization of catecholamines in Cryptococcus neoformans. J Bacteriol 150:1212–1220

Posteraro B, Sanguinetti M, Masucci L, Romano L, Morace G, Fadda G (2000) Reverse cross blot hybridization assay for rapid detection of PCR-amplified DNA from Candida species, Cryptococcus neoformans, and Saccharomyces cerevisiae in clinical samples. J Clin Microbiol 38:1609–1614

Powderly WG (1996) Recent advances in the management of cryptococcal meningitis in patients with AIDS. Clin Infect Dis 22:S119–123

Powderly WG, Saag MS, Cloud GA, Robinson P, Meyer RD, Jacobson JM, Graybill JR, Sugar AM, McAuliffe VJ, Follansbee SE et al. (1992) A controlled trial of fluconazole or amphotericin B to prevent relapse of cryptococcal meningitis in patients with the acquired immunodeficiency syndrome. The NIAID AIDS Clinical Trials Group and Mycoses Study Group. N Engl J Med 326:793–798

Powderly WG, Cloud G, Dismukes WE, Saag MS (1994) Measurement of cryptococcal antigen in serum and cerebralspinal fluid: value in the management of AIDS-associated cryptococcal meningitis. Clin Infect Dis 18:789–792

Pukkila-Worley R, Alspaugh JA (2004) Cyclic AMP signaling in Cryptococcus neoformans. FEMS Yeast Res 4:361–367

Pukkila-Worley R, Gerrald QD, Kraus PR, Boily MJ, Davis MJ, Giles SS, Cox GM, Heitman J, Alspaugh JA (2005) Transcriptional network of multiple capsule and melanin genes governed by the Cryptococcus neoformans cyclic AMP cascade. Eukaryot Cell 4:190–201

Rabkin JM, Oroloff SL, Corless CL, Benner KG, Flora KD, Rosen HR, Olyaei AJ (2000) Association of fungal infection and increased mortality in liver transplant recipients. Am J Surg 179:426–430

Rappelli P, Are R, Casu G, Fiori PL, Cappuccinelli P, Aceti A (1998) Development of a nested PCR for detection of Cryptococcus neoformans in cerebrospinal fluid. J Clin Microbiol 36:3438–3440

Reese AJ, Doering TL (2003) Cell wall alpha-1,3-glucan is required to anchor the Cryptococcus neoformans capsule. Mol Microbiol 50:1401–1409

Retini C, Vecchiarelli A, Monari C, Bistoni F, Kozel TR (1998) Encapsulation of Cryptococcus neoformans with glucuronoxylomannan inhibits the antigen-presenting capacity of monocytes. Infect Immun 66:664–669

Rhodes JC, Polacheck I, Kwon-Chung KJ (1982) Phenoloxidase activity and virulence in isogenic strains of Cryptococcus neoformans. Infect Immun 36:1175–1184

Richardson MD, White LJ, McKay IC, Shankland GS (1993) Differential binding of acapsulate and encapsulated strains of Cryptococcus neoformans to human neutrophils. J Med Vet Mycol 31:189–199

Rosas AL, Nosanchuk JD, Feldmesser M, Cox GM, McDade HC, Casadevall A (2000) Synthesis of polymerized melanin by Cryptococcus neoformans in infected rodents. Infect Immun 68:2845–2853

Rude TH, Toffaletti DL, Cox GM, Perfect JR (2002) Relationship of the glyoxylate pathway to the pathogenesis of Cryptococcus neoformans. Infect Immun 70:5684–5694
Saag MS, Graybill RJ, Larsen RA, Pappas PG, Perfect JR, Powderly WG, Sobel JD, Dismukes WE (2000) Practice guidelines for the management of cryptococcal disease. Infectious Diseases Society of America. Clin Infect Dis 30:710–718
Saito Y, Osabe S, Kuno H, Kaji M, Oizumi K (1999) Rapid diagnosis of cryptococcal meningitis by microscopic examination of centrifuged cerebrospinal fluid sediment. J Neurol Sci 164:72–75
Salas SD, Bennett JE, Kwon-Chung KJ, Perfect JR, Williamson PR (1996) Effect of the laccase gene CNLAC1, on virulence of Cryptococcus neoformans. J Exp Med 184:377–386
Sax PE (2001) Opportunistic infections in HIV disease: down but not out. Infect Dis Clin N Am 15:433–455
Schein JE, Tangen KL, Chiu R, Shin H, Lengeler KB, MacDonald WK, Bosdet I, Heitman J, Jones SJ, Marra MA et al. (2002) Physical maps for genome analysis of serotype A and D strains of the fungal pathogen Cryptococcus neoformans. Genome Res 12:1445–1453
Shaw CE, Kapica L (1972) Production of diagnostic pigment by phenoloxidase activity of Cryptococcus neoformans. Appl Microbiol 24:824–830
Sia RA, Lengeler KB, Heitman J (2000) Diploid strains of the pathogenic basidiomycete Cryptococcus neoformans are thermally dimorphic. Fungal Genet Biol 29:153–163
Snydman DR (2001) Epidemiology of infections after solid-organ transplantation. Clin Infect Dis 33 suppl 1:S5–S8
Speed B, Dunt D (1995) Clinical and host differences between infections with the two varieties of Cryptococcus neoformans. Clin Infect Dis 21:28–34
Steen BR, Lian T, Zuyderduyn S, MacDonald WK, Marra M, Jones SJ, Kronstad JW (2002) Temperature-regulated transcription in the pathogenic fungus Cryptococcus neoformans. Genome Res 12:1386–1400
Steen BR, Zuyderduyn S, Toffaletti DL, Marra M, Jones SJ, Perfect JR, Kronstad JW (2003) Cryptococcus neoformans gene expression during experimental cryptococcal meningitis. Eukaryot Cell 2:1336–1349
Steenbergen JN, Shuman HA, Casadevall A (2001) Cryptococcus neoformans interactions with amoebae suggest an explanation for its virulence and intracellular pathogenic strategy in macrophages. Proc Natl Acad Sci USA 98:15245–15250
Steenbergen JN, Nosanchuk JD, Malliaris SD, Casadevall A (2003) Cryptococcus neoformans virulence is enhanced after growth in the genetically malleable host Dictyostelium discoideum. Infect Immun 71:4862–4872
Stephen C, Lester S, Black W, Fyfe M, Raverty S (2002) Multispecies outbreak of cryptococcosis on southern Vancouver Island, British Columbia. Can Vet J 43:792–794
Sudarshan SR, Davidson RC, Heitman J, Alspaugh JA (1999) Molecular analysis of the Cryptococcus neoformans ADE2 gene, a selectable marker for transformation and gene disruption. Fungal Genet Biol 27:36–48
Sundstrom JB, Cherniak R (1993) T-cell-dependent and T-cell-independent mechanisms of tolerance to glucuronoxylomannan of Cryptococcus neoformans serotype A. Infect Immun 61:1340–1345
Syme RM, Bruno TF, Kozel TR, Mody CH (1999) The capsule of Cryptococcus neoformans reduces T-lymphocyte proliferation by reducing phagocytosis, which can be restored with anticapsular antibody. Infect Immun 67:4620–4627
Tanner DC, Weinstein MP, Fedorciw B, Joho KL, Thorpe JJ, Reller L (1994) Comparison of commercial kits for detection of cryptococcal antigen. J Clin Microbiol 32:1680–1684
Tenney AR, Brown RH, Vaske C, Lodge JK, Doering TL, Brent MR (2004) Gene prediction and verification in a compact genome with numerous small introns. Genome Res 14:2330–2335
Toffaletti DL, Rude TH, Johnston SA, Durack DT, Perfect JR (1993) Gene transfer in Cryptococcus neoformans by use of biolistic delivery of DNA. J Bacteriol 175:1405–1411
Tscharke RL, Lazera M, Chang YC, Wickes BL, Kwon-Chung KJ (2003) Haploid fruiting in Cryptococcus neoformans is not mating type alpha-specific. Fungal Genet Biol 39:230–237
Tucker SC, Casadevall A (2002) Replication of Cryptococcus neoformans in macrophages is accompanied by phagosomal permeabilization and accumulation of vesicles containing polysaccharide in the cytoplasm. Proc Natl Acad Sci USA 99:3165–3170
van Duin D, Casadevall A, Nosanchuk JD (2002) Melanization of Cryptococcus neoformans and Histoplasma capsulatum reduces their susceptibilities to amphotericin B and caspofungin. Antimicrob Agents Chemother 46:3394–3400
Varma A, Kwon-Chung KJ (1999) Characterization of the glyceraldehyde-3-phosphate dehydrogenase gene [correction of glyceraldehyde-3-phosphate gene] and the use of its promoter for heterologous expression in Cryptococcus neoformans, a human pathogen. Gene 232:155–163
Vartivarian SE, Anaissie E, Cowart RE, Sprigg HA, Tingler MJ, Jacobson ES (1993) Regulation of cryptococcal capsular polysaccharide by iron. J Infect Dis 167:186–190
Vecchiarelli, A, Retini C, Casadevall A, Monari C, Pietrella D, Kozel TR (1998) Involvement of C3a and C5a in interleukin-8 secretion by human polymorphonuclear cells in response to capsular material of Cryptococcus neoformans. Infect Immun 66:4324–4330
Viviani MA, Esposto MC, Cogliati M, Montagna MT, Wickes BL (2001) Isolation of a Cryptococcus neoformans serotype A MATa strain from the Italian environment. Med Mycol 39:383–386
Viviani MA, Nikolova R, Esposto MC, Prinz G, Cogliati M (2003) First European case of serotype A MATa Cryptococcus neoformans infection. Emerg Infect Dis 9:1179–1180
Wang Y, Casadevall A (1994) Decreased susceptibility of melanized Cryptococcus neoformans to UV light. Appl Environ Microbiol 60:3864–3866
Wasser L, Talavera W (1987) Pulmonary cryptococcosis in AIDS. Chest 92:692–695

Wheat LJ, Smith EJ, Sathapatayavongs B, Batteiger B, Filo RS, Leapman SB, French MV (1983) Histoplasmosis in renal allograft recipients: two large urban outbreaks. Arch Intern Med 143:703–707
White MH, Armstrong D (1994) Cryptococcosis. Infect Dis Clin N Am 8:383–398
White A, Cirrincione C, Blevins A, Armstrong D (1992) Cryptococcal meningitis: outcome in patients with AIDS and patients with neoplastic disease. J Infect Dis 165:960–963
Wickes BL, Edman JC (1995) The Cryptococcus neoformans GAL7 gene and its use as an inducible promoter. Mol Microbiol 16:1099–1109
Wickes BL, Moore TD, Kwon-Chung KJ (1994) Comparison of the electrophoretic karyotypes and chromosomal location of ten genes in the two varieties of Cryptococcus neoformans. Microbiology 140:543–550
Wickes BL, Mayorga ME, Edman U, Edman JC (1996) Dimorphism and haploid fruiting in Cryptococcus neoformans: association with the alpha-mating type. Proc Natl Acad Sci USA 93:7327–7331
Williamson PR (1994) Biochemical and molecular characterization of the diphenol oxidase of Cryptococcus neoformans: identification as a laccase. J Bacteriol 176:656–664
Wills EA, Roberts IS, Del Poeta M, Rivera J, Casadevall A, Cox GM, Perfect JR (2001) Identification and characterization of the Cryptococcus neoformans phosphomannose isomerase-encoding gene, MAN1, and its impact on pathogenicity. Mol Microbiol 40:610–620
Wilson MA, Kozel TR (1992) Contribution of antibody in normal human serum to early deposition of C3 onto encapsulated and nonencapsulated Cryptococcus neoformans. Infect Immun 60:754–761
Wilson DE, Bennett JE, Bailey JW (1968) Serologic grouping of Cryptococcus neoformans. Proc Soc Exp Biol Med 127:820–823
Winzeler EA, Shoemaker DD, Astromoff A, Liang H, Anderson K, Andre B, Bangham R, Benito R, Boeke JD, Bussey H et al. (1999) Functional characterization of the *S. cerevisiae* genome by gene deletion and parallel analysis. Science 285:901–906
Yan Z, Li X, Xu J (2002) Geographic distribution of mating type alleles of Cryptococcus neoformans in four areas of the United States. J Clin Microbiol 40:965–972
Yue C, Cavallo LM, Alspaugh JA, Wang P, Cox GM, Perfect JR, Heitman J (1999) The STE12alpha homolog is required for haploid filamentation but largely dispensable for mating and virulence in Cryptococcus neoformans. Genetics 153:1601–1615
Zhu X, Williamson PR (2003) A CLC-type chloride channel gene is required for laccase activity and virulence in Cryptococcus neoformans. Mol Microbiol 50:1271–1281
Zhu X, Williamson PR (2004) Role of laccase in the biology and virulence of Cryptococcus neoformans. FEMS Yeast Res 5:1–10
Zhu X, Gibbons J, Garcia-Rivera J, Casadevall A, Williamson PR (2001) Laccase of Cryptococcus neoformans is a cell wall-associated virulence factor. Infect Immun 69:5589–5596

Biosystematic Index

Subject Index